INVERTEBRATE ZOOLOGY

Edited by D. T. Anderson
Emeritus Professor of Biology
in the University of Sydney

Melbourne
Oxford University Press
Oxford   Auckland   New York

OXFORD
UNIVERSITY PRESS
253 Normanby Road, South Melbourne, Australia

Oxford University Press is a department of the University of Oxford.
It furthers the University's objective of excellence in research, scholarship,
and education by publishing worldwide in

Oxford New York

Athens Auckland Bangkok Bogotá Buenos Aires Calcutta Cape Town
Chennai Dar es Salaam Delhi Florence Hong Kong Istanbul Karachi
Kuala Lumpur Madrid Melbourne Mexico City Mumbai Nairobi Paris
Port Moresby São Paulo Singapore Taipei Tokyo Toronto Warsaw

with associated companies in Berlin Ibadan

OXFORD is a trade mark of Oxford University Press
in the UK and in certain other countries

© D.T. Anderson 1998
First published 1998
Reprinted 1999
Authors retain copyright for the contributions to this volume

This book is copyright. Apart from any fair dealing for the purposes of
private study, research, criticism or review as permitted under the
Copyright Act, no part may be reproduced, stored in a retrieval system,
or transmitted, in any form or by any means, electronic, mechanical,
photocopying, recording or otherwise without prior written permission.
Enquiries to be made to Oxford University Press.

Copying for educational purposes
Where copies of part or the whole of the book are made under Part VB of
the Copyright Act, the law requires that prescribed procedures be followed.
For information, contact the Copyright Agency Limited.

National Library of Australia
Cataloguing-in-Publication data:

Invertebrate zoology.

       Bibliography.
       Includes index.
       ISBN 0 19 553941 9 (pbk)
       1. Invertebrates. I. Anderson, D.T. (Donald Thomas), 1931–.

592

Edited by David Meagher
Text design by Derrick I Stone Design
Cover design by Steve Randles
Typeset by Derrick I Stone Design
Printed through Bookpac Production Services, Singapore

INVERTEBRATE ZOOLOGY

Preface

Invertebrate biodiversity has been a focus of scientific attention for more than 200 years. Like other sciences, the biology of invertebrate animals can be approached at two levels: the level of the specialist, delving into the abstruse minutiae of complex animal groups; and the level of the general student, seeking to grasp the broad scope of the subject.

Following the acceptance of the concept of evolution in the mid 19th century, the broad view of invertebrate zoology became based on the analysis of the phylogenetic evolution and interrelationships of invertebrate phyla through comparative morphology and embryology. This approach provided a logical entry into invertebrate zoology for any student willing to devote sufficient time and effort to the task. It was generally accepted until about 1970 that the appropriate amount of time was a year of full-time study. Thus endowed, the student was considered to be well equipped to enter many biological fields in which a knowledge of invertebrates was deemed useful — marine biology, entomology, parasitology and ecology, to name a few, as well as the more academic pursuits of zoology departments and natural history museums.

Almost three decades on, the prerequisite has not changed. It is still not possible to be knowledgeable and effective in many of the current pursuits of animal biology and related fields without a basic knowledge of invertebrates; yet three factors have militated against such achievement in recent years. The first is a more crowded curriculum. Dramatic new biological advances in genetics, molecular biology and analytical ecology have had to be accommodated, usually at the expense of more traditional ongoing aspects of zoology. The second factor, which is a direct outcome of the first, is that the application of the newer molecular techniques to invertebrate biology has revealed uncertainties about the scientific rigour of the traditional conspectus. Old theories of animal phylogeny and relationships have, in the main, been discredited, and new proposals based on testable evidence are emerging to take their place. The third factor is that, as specialists in every animal group have continued their work, there has been an explosion of new knowledge and a major revision of the systematics and biology of every phylum.

The essence of these dramatic developments has now to be conveyed and comprehended in a course of study usually occupying, not a full year, but one semester among other demanding studies. This book has been designed to take account of these changes. The multi-author approach acknowledges that specialist expertise is required to identify what is now important in the biology of each animal group. Individual authors have been responsible for the content and illustration of their chapter(s), working within general guidelines agreed at the outset among the contributors. Foremost among these was an agreement to emphasise the dynamic aspects of modern invertebrate biology and to reduce the traditional emphasis on comparative morphology and embryology. Since the book has been written mainly by authors from Australia and New Zealand, some attention is paid to examples drawn from the fauna of this part of the world, but the general themes of the book are those of international invertebrate biology. We have also endeavoured to convey recent advances in the classification of invertebrate groups, stemming from the application of cladistic analysis. The major classification of every phylum has been changed, and continues to be changed, as a consequence of this revolution in taxonomic methodology.

Because the chapters emphasise the dynamics of each group rather than comparative morphology, the subject of phylogenetic relationships among the invertebrate phyla is addressed in a culminating chapter, which incorporates the newer evidence of molecular biology. We hope that this approach will make the major aspects of contemporary invertebrate biology accessible and stimulating to students.

D.T. Anderson
Wamberal, 1997

Contents

Preface *v*
Acknowledgements *ix*

1   The invertebrate phyla *1*
    D.T. Anderson

2   The Porifera *10*
    P.R. Bergquist

3   The Cnidaria and Ctenophora *28*
    R.T. Hinde

4   The Platyhelminthes, Nemertea, Entoprocta and Gnathostomulida *58*
    J.C. Walker and D.T. Anderson

5   The Aschelminths *86*
    W.L. Nicholas

6   The Sipuncula and Priapula *116*
    D.T. Anderson

7   The Mollusca *122*
    J.M. Healy

8   The Annelida *174*
    G. Rouse

9   The Onychophora and Tardigrada *204*
    N.N. Tait

10 Introduction to arthropods *222*
    D.T. Anderson

11 The Hexapoda *228*
    D.F. Hales

12 The Myriapoda *269*
    N.N. Tait

13 The Crustacea   *286*
    P. Greenaway

14 The Chelicerata   *319*
    D.T. Anderson

15 The lophophorates: Phoronida, Brachiopoda and Ectoprocta   *343*
    P.J. Doherty

16 The Echinodermata   *366*
    M. Byrne

17 The invertebrate Chordata, Hemichordata and Chaetognatha   *396*
    L. Stocker

18 Metazoan phylogeny   *416*
    R.A. Raff

Bibliography   *439*
Glossary   *445*
Index   *453*
About the authors   *466*

Acknowledgements

Contributions towards the preparation of this book have been made by many people. The editor wishes to thank:

Joanne T. Anderson for unstinting support and assistance in all aspects of editorial work, and for contributing to the preparation of the figures of chapters 1, 4, 6, 10 and 14.

The contributing authors for their courtesy and cooperation at all times.

The staff of Oxford University Press, who have provided so much expertise to the production of the book; especially the Publisher — Academic Division for Australia and New Zealand, Jill Lane, who initiated the concept and has been the guiding hand throughout; David Meagher for scrupulous attention to detail in editing the text; Steve Randles for meticulous preparation of the final published figures; Derrick Stone for the design of the book; Andrea Allan for editorial assistance; and Maggie Way and Rebecca Coates for secretarial support.

Dr T.R. New of Latrobe University and Dr R.G. Wear of The Victoria University of Wellington provided cogent criticism of an early draft of the book and made many valuable suggestions for its improvement.

Dr Noel Tait of Macquarie University read a draft of the glossary, identified a number of errors and provided several definitions.

Thanks are also due to all those who have contributed to the development of individual chapters: Vivian Ward, University of Auckland, for illustrations, and Iain MacDonald, University of Auckland, for photographic assistance (chapter 2); Barry Little for preparation of illustrations; Eric Dorfman, University of Sydney, for line drawings; Malcom Ricketts, University of Sydney, for photographic assistance; Meredith Peach for fig. 3.3a, and Anya Salih for fig. 3.5 (chapter 3); Penny Berents, Australian Museum, Sydney, and Dr Lester Cannon, Queensland Museum, Brisbane, for loans of specimen; Ross Boadle, Westmead Hospital, Sydney, for electron micrographs (chapter 4); C. Turnbull, Macquarie University, for preparation of figs 9.2, 9.3, 9.9 and 12.1b; B. Thorn, Macquarie University, for figs 12.1a, 12.6a, 12.7a, b; S. Doyle and A. Gregor, Macquarie University, for the SEMs of fig. 9.4; S. Claxton for the SEMs of fig. 9.8; J. Norman and V. Brown, Macquarie University, for photographic assistance; Dr M. Harvey and G. Milledge for the use of figs 12.2a, e, 12.4, 12.5a, b (chapters 9 and 12); Barbara Duckworth, Macquarie University, for line drawings; Lucy Sorrentino, Macquarie University, for fig. 11.8a; Vianney Brown, Macquarie University, for photographic assistance (chapter 11); Nikki McDonald, University of New South Wales, for assistance in the preparation of figures and for drawing figs 13.1b, 13.11g and 13.15e (chapter 13); Professor J.E. Morton, who encouraged the author's early work on brachiopods and drew fig. 15.7 (chapter 15); A. Cerra, A. Feldman and R. Smith, University of Sydney, for technical assistance (chapter 16).

Many of the illustrations for the book were newly prepared, based either on original material or on sources in the zoological literature. All of these sources are acknowledged in the legends to the figures. A number of illustrations have been reproduced directly from previously published works, permission for which is gratefully acknowledged as follows:

Academic Press Ltd: Fig 8.9a, from Mill, P.J., *Physiology of Annelids*, Academic Press, London, 1978; fig. 11.18, from Ashburner, M. (ed.), *The Genetics and Biology of Drosophila*, Volume 2c, 1978; fig. 13.13b, c, from Atwood, H.L. and D.C. Sandeman, *The Biology of Crustacea*. Volume 3, Neurobiology: Structure and Function, 1982.

The American Physiological Society: Fig. 11.11a, from Phillips, J., *American Journal of Physiology*, 241, R241–R257, 1981.

Prof. D.T. Anderson: Figs 9.7, 12.10 from Anderson, D.T., *Embryology and Phylogeny in Annelids and Arthropods*, Pergamon, 1973.

A.A. Balkema Ltd: Figs 12.2b, c, from Lawrence, R.F., *The Centipedes and Millipedes of South Africa. A Guide*, 1983.

The Biological Bulletin: Fig. 16.2d, from Young, C.M., and R.H.Emson, *Biological Bulletin*, 188, 88–97, 1995; fig. 16.19, from Byrne, M., and M.F. Barker, *Biological Bulletin*, 180, 332–345, 1991; fig. 16.20, from Byrne, M., and A. Cerra, *Biological Bulletin*, 191, 17–26, 1996.

Blackwell Science Ltd: Fig. 11.2b, from Neville, A.C. (ed.), *Insect Ultrastructure*. Symposia of the Royal Entomological Society of London, No. 5, 1970; fig. 11.7, from Mordue, W., G.J. Goldsworthy, J. Brady and W.M. Blaney, *Insect Physiology*, 1980; figs 13.3a, 13.9b, c, from Dales, R.P. (ed.), *Practical Invertebrate Zoology*, 1981.

The Canadian Journal of Zoology: Fig. 16.2c, 16.9a, from Byrne, M., and A.R. Fontaine, *Canadian Journal of Zoology*, 59, 11–18, 1981.

Chapman and Hall, London: Fig. 11.17, from Gullan, P.J. and P.S. Cranston, *The Insects: an Outline of Entomology*, 1994.

The Crustacean Society: Fig. 13.2a, from Yager, J., *Journal of Crustacean Biology*, 14, 752-762, 1994; fig. 13.14c, from Wehrtman, I.S., L. Albornoz, D. Veliz and L.M. Pardo, *Journal of Crustacean Biology*, 16, 730–747, 1996.

Elsevier Science Ltd: Fig. 11.11b, from Kerkut, G.A. and L.I. Gilbert (eds), *Comprehensive Insect Physiology, Biochemistry and Pharmacology*, Volume 4, Pergamon, 1985.

Evolution: Fig. 16.1a, from Byrne, M., and M.J. Anderson, *Evolution*, 48, 564–576, 1994.

The Food and Agricultural Organisation of the United Nations: Fig. 13.7a, b, from Holthuis, L.B., *FAO Fisheries Synopses*, 125, 1–261.

Prof. H. Greven: Fig. 9.9c, from Greven, H., *Die Bärtierchen* A. Ziemsen Verlag Wittenberg Lutherstadt, 1980.

Kendall/Hunt Publishing Co.: Figs 17.1c, d, 17.5d, from Smith, D.L., *A guide to Marine Coastal Plankton and Marine Invertebrate Larvae*, 1977.

The Linnean Society of New South Wales: Fig. 17.3c, from Anderson, D.T. et al., *Proceedings of the Linnean Society of New South Wales*, 100, 205–217, 1976.

Macmillan Press Ltd: Fig. 17.1a, from Bullough, W.S., *Practical Invertebrate Anatomy*, 1970.

The McGraw-Hill Companies: Fig. 17.3b, from Berrill, N. and G. Karp, *Development*, 1976.

Melbourne University Press: Figs 11.3, 11.5, 11.6, 11.9, from *The Insects of Australia*, CSIRO, 1973 and 1991.
Dr. B. Mesibov: Fig. 12.2d, from Mesibov, B., *A Guide to Tasmanian Centipedes*, 1986.
NRC Research Press: Fig. 13.3b, from Gnewuch, W.T. and Crocker, R.A., *Canadian Journal of Zoology*, 51, 1011–1020, 1973.
The National Institute of Water and Atmospheric Research, Wellington: Fig. 17.1e, g, from Millar, R.H., Ascidiacea, New Zealand Oceanographic Institute Memoir 85, 1982.
Natural History: Fig. 12.6c, from Eisner, T. and H.E. Eisner, *Natural History*, 74(3), 30–37, 1965.
Ophelia: Figs 16.16a,b, 16.17a, from Byrne, M., *Ophelia*, 24, 75–90, 1985.
Oxford University Press: Figs 12.2a, e, 12.4, 12.5a, b, from Harvey, M.S. and A.L. Yen, *Worms to Wasps. An Illustrated Guide to Australia's Terrestrial Invertebrates*, with illustrations by G. Milledge, 1989; fig. 12.3, from Hopkin, S.P. and H.J. Read, *The Biology of Millipedes*, 1992; fig. 13.16, from Meglitsch, P.A. and F.W. Schram, *Invertebrate Zoology*, 3rd edition, 1991.
Prentice Hall Ltd: Figs 8.3d, 17.2a, 17.5a, c, from Sherman, I.W. and V.G. Sherman, *The Invertebrates: Function and Form. A Laboratory Guide*, Macmillan, 1972.
Nelson H. Prentiss, Administrator of the Estate of Thomas S. Prentiss: Fig. 13.2c, from Caldwell, R.L. and H. Dingle, *Scientific American*, 234, 81–89, 1976.
Princeton University Press: Fig. 13.6d, from Crane, J., *Fiddler Crabs of the World*, 1975.
The Queensland Museum Board of Trustees: Fig. 17.1f, from Kott, P., *Memoirs of the Queensland Museum*, 23, 1–440, 1985; figs 17.2b, 17.3a, from Kott, P., *Memoirs of the Queensland Museum*, 29, 1–298, 1990; fig. 17.3d, from Kott, P., *Memoirs of the Queensland Museum*, 32, 375–620, 1992.
The Royal Society of London: Fig. 8.9b, from Nilsson, D.E. *Philosophical Transactions of the Royal Society*, B346, 195–212; fig. 9.5, from Anderson, D.T. and S.M. Manton *Philosophical Transactions, of the Royal Society*, B264, 161–189, 1972.
Dr W.D Russell-Hunter: Figs 8.7c, 12.5d, from Russell-Hunter, W.D., *A Life of Invertebrates*, Macmillan, 1979.
Saunders College Publishing: Figs 8.3c, 8.5b, 8.6a, 8.11f, from Barnes, R.D. *Invertebrate Zoology*, 3rd edition, 1974; figs 13.4a and 13.9a from Boolootian, R.A. and Heyneman, D. *An Illustrated Laboratory Text in Zoology*, updated fourth edition.
Sinauer Associates, Inc.: Fig. 3.3b, from Brusca, R.C., and G. Brusca, *Invertebrates*, with illustrations by Nancy. J. Haver, 1990.
SIR Publishing: Fig. 13.8a, from Knox, G.A., *Journal of the Royal Society of New Zealand*, 7, 425–432, 1977.
The Smithsonian Institution Press: Fig. 13.7d, from Martin, J.W and L.G. Abele, *Smithsonian Contributions to Zoology*, 453, 1-46.
Springer Verlag GmbH & Co. KG, Heidelberg: Fig. 8.1b, from Bartolomaeus, T., Zoomorphology 115, 161–177, 1995; fig. 16.5a, from Stauber, M., and K. Markel, *Zoomorphology*, 108, 137–148, 1988; figs 16.9b, c, from Byrne, M., and A.R. Fontaine, *Zoomorphology*, 102, 177-187, 1983; fig. 16.14a, from Byrne, M., *Marine Biology*, 125, 551–567, 1996.

The University of Chicago Press: Figs 18.4, 18.5, 18.6, 18.9, from Raff, R.A., The Shape of Life, 1996.
The University of New South Wales Press: Figs 13.5a, 16.7b, 16.13, 16.15, from Anderson, D.T., Atlas of Invertebrate Anatomy, 1996.
Wiley-Liss, Inc., a subsidiary of John Wiley & Sons, Inc.: Figs 8.5c, d, 9.2e, from Harrison, F.W. and M. Rice (eds) Microscopic Anatomy of Invertebrates, volume 12, Onychophora, Chilopoda and Lesser Protostomata, 1993; fig. 13.12f, from Harrison, F.W and A.G. Humes (eds), Microscopic Anatomy of Invertebrates. Volume 10, Decapod Crustacea., 1992; figs 16.8, 16.14b, from Harrison, F.W. and F.-S. Chia (eds), Microscopic Anatomy of Invertebrates, volume 14, Echinoderms, 1996.

Chapter 1

The invertebrate phyla

D.T. Anderson

Introduction 2

Radially symmetrical invertebrates 3

Bilaterally symmetrical invertebrates 4
   Acoelomates 4
   Body cavities (pseudocoel, coelom and haemocoel) 4

Pseudocoelomates 5
Coelomates 5
Haemocoelic coelomates 8
Embryonic development 8
   Spiral cleavage development 8
   Deuterostome development 9
   Phylogenetic consequences 9

Introduction

Animal diversity is one of the great manifestations of life on Earth. More than a million species of multicellular animals have been described in the global fauna. Several more millions are likely to be discovered. Comparable numbers of species have existed in the past and become extinct, leaving only a remnant indication of their presence in the fossil record. Almost all of these animals are invertebrates. The diversity of vertebrate life, for all that it includes the most impressive kinds of animals in terms of size, organisation and capabilities, is limited to less than 50 000 species living and some equivalent number extinct. Vertebrate diversity is based on modifications of a single body plan. Invertebrate diversity includes about 32 basic body plans, each characterising a distinctive animal group or phylum, each with its own range of diversity. These body plans exhibit a number of organisational levels ranging from simple to complex:

1. Non-symmetrical, with a cellular construction (Porifera)
2. Radially symmetrical, with a tissue construction (Cnidaria, Ctenophora)
3. Bilaterally symmetrical, with an organ construction:
   - 3a Acoelomate: unsegmented (Platyhelminthes, Nemertea, Entoprocta, Gnathostomulida) (Whether, other than the platyhelminths, these phyla are truly acoelomate is a matter of debate.)
   - 3b Pseudocoelomate: unsegmented (Nematoda, Nematomorpha, Gastrotricha, Kinorhyncha, Rotifera, Loricifera, Acanthocephala, Priapula)
   - 3c Coelomate: unsegmented (Sipuncula), trimeric (Echinodermata, Chordata, Hemichordata, Chaetognatha, Phoronida, Brachiopoda, Ectoprocta) or metameric (Annelida, Echiura, Pogonophora, Vestimentifera)
   - 3d Haemocoelic: unsegmented (Mollusca) or metameric (Onychophora, Tardigrada, Hexapoda, Myriapoda, Crustacea, Chelicerata).

All of these animals are definable as metazoans. They are multicellular animals with a significant degree of cellular differentiation. Their epithelial cells are joined by septate or tight junctions, a fact of particular importance in providing continuous epithelia at body surfaces. They have collagen as a component of an intercellular matrix and share an intercellular communication mechanism based on acetylcholine and cholinesterase.

One marine phylum, the Porifera or sponges, retains a simple level of metazoan cellular integration. Sponges (Fig. 1.1a) are loosely constructed, porose animals through which water is driven by the action of numerous flagellated cells. All sponges are attached to a substratum, except in the motile larval stage, and grow in a vegetative form, without obvious symmetry. Nevertheless, the Porifera are a diverse phylum of animals, of great importance in marine benthic communities (see Chapter 2).

The invertebrate phyla other than sponges share several additional features in their organisation. Cells are grouped together as tissues, which are arranged in layers. The surface epithelial layer is supported by an underlying basement membrane. Gap junctions are present between the cells, facilitating intercellular communication. There is a greater range of cellular differentiation, including myofibrils and a connected network of neurons, permitting neuromuscular activity. The body is symmetrical and encloses an enteron, opening to the exterior by a closable mouth

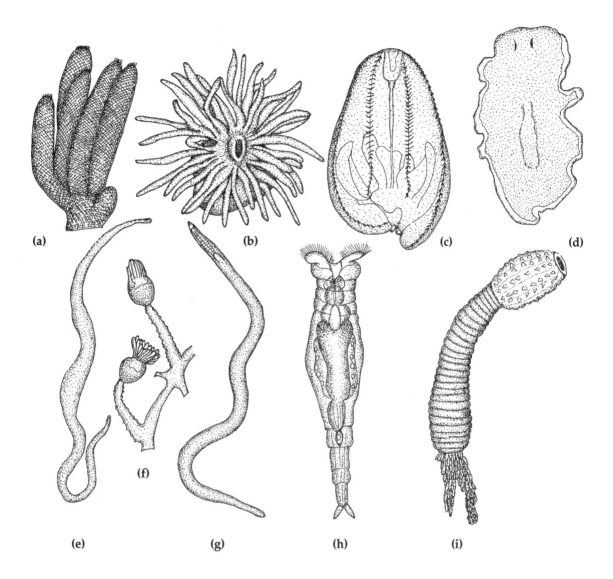

Figure 1.1
Body form in representative non-coelomate phyla.
(a) Porifera. (b) Cnidaria.
(c) Ctenophora.
(d) Platyhelminthes.
(e) Nemertea. (f) Entoprocta.
(g) Nematoda. (h) Rotifera.
(i) Priapula.

Radially symmetrical invertebrates

Two marine phyla, the Cnidaria and Ctenophora, manifest this organisation in a relatively simple way. Both have a layered wall structure, a single opening to the enteron, and a low level of neuro-muscular differentiation. Flagellated cells occur on various body surfaces. The Cnidaria (hydroids, medusae, anemones (Fig. 1.1b), corals, etc.) are radially symmetrical, attached or floating animals which depend on specialised cellular effectors called cnidae for food capture. Like sponges, the cnidarians are a diverse component of the marine benthos, but the phylum also includes many planktonic species. The Ctenophora (Fig 1.1c) are biradially symmetrical, gelatinous animals of the plankton. Ctenophores use flagella in propulsion and rely for food capture on cellular effectors called colloblasts. Cnidarians and ctenophores are discussed in more detail in Chapter 3.

Bilaterally symmetrical invertebrates

The other invertebrate phyla have a more active motility, associated with a bilaterally symmetrical body, some degrees of cephalisation, a more elaborate musculature, and a concentration of part of the neuronal network as a brain and nerve cords. Monociliated or multiciliated cells (or both) at the body surface and elsewhere may assist locomotion, generate water currents, move food, and act in other ways. Internally, the tissues are grouped together as organs performing specific functions.

Acoelomates

The morphologically simplest bilateral metazoans also have a single internal cavity, the enteron, opening by a mouth. The regions between the organs are filled with an undifferentiated tissue (the mesenchyme). These animals are called acoelomate — strictly, lacking a coelom; in practice, lacking any form of permanent body cavity between the body wall and the enteron.

The acoelomate phylum Platyhelminthes (flatworms) displays this level of organisation in its clearest form. Free-living flatworms (Fig. 1.1d) are slow-moving, marine or freshwater worms of carnivorous habit; many species in the phylum are parasites on or in the bodies of larger animals (see Chapter 4).

All other bilaterally symmetrical metazoans have a tubular gut with an anus, as well as a mouth, and a through-put of food. Three of these phyla — Nemertea, Entoprocta and Gnathostomulida — have traditionally been said to retain the acoelomate level of organisation, although current investigations are raising questions about this interpretation.

The Nemertea are cylindrical worms (Fig. 1.1e), almost all of marine habit. Their major specialisation is a dorsal, eversible proboscis, situated in a cavity above the gut called the rhynchocoel. Nemerteans also have a network of blood channels in the mesenchyme, but have no pumping vessel or heart. Some researchers have proposed that the rhynchocoel and blood spaces of nemerteans should be interpreted as coelomic cavities (see below and Chapters 4 and 18). Centralisation of the nervous system is evident in the well-developed brain and paired lateral nerve cords.

The Entoprocta (see Chapter 4) are marine acoelomates of a quite different kind (Fig. 1.1f). The basic body is an upright, sessile zooid with a circlet of ciliated tentacles around the apex. The gut is U-shaped, and both the mouth and anus are situated within the tentacle ring. Entoprocts grow in multiple zooid form.

The Gnathostomulida (see Chapter 4) are minute, vermiform acoelomates, which live in the spaces within marine sands. Although these animals are small, they have a pharynx armed with jaws.

These four phyla — Platyhelminthes, Nemertea, Entoprocta and Gnathostomulida — show major differences in adult form and mode of life, but share a mode of embryonic development which begins with a sequence of cell divisions known as spiral cleavage (Fig. 4.1). The divisions take place in such a way that the resulting cells become arranged in a spiral pattern in the wall of the blastula. We shall see later that several other phyla of invertebrates have spiral cleavage development.

Body cavities (pseudocoel, coelom and haemocoel)

The body plans of the majority of bilaterally symmetrical invertebrates include an internal cavity between the body wall and the gut wall.

Functionally, a 'body cavity' in this position confers several advantages. The body wall and gut wall can act independently, increasing the scope for muscular activities by both; the fluid in the cavity can act as a deformable hydrostatic skeleton during these activities; and other organ systems can be further elaborated within the fluid-filled space. Three kinds of body cavity have been distinguished in the bilateral metazoans: pseudocoel, coelom and haemocoel.

Pseudocoelomates
Several phyla of worm-like invertebrates have a pseudocoel, or false coelom, an internal space that is not bounded by a cellular epithelium and is not associated with a circulatory system. The major pseudocoelomate phylum is the Nematoda (Fig. 1.1g), consisting of cylindrical worms with a well-developed external cuticle. Small nematodes are very numerous in sands, muds and soils. The phylum also includes many parasites of plants and animals, some of which are quite large (see Chapter 5). Other pseudocoelomate phyla, the Nematomorpha, Gastrotricha, Kinorhyncha, Loricifera and Rotifera (Fig. 1.1h) (see Chapter 5) and the Priapula (Fig. 1.1i) (see Chapter 6), are mainly microscopic animals of marine or fresh water sediments, although the Nematomorpha and some Priapula are quite large. One phylum of pseudocoelomates, the Acanthocephala, is wholly parasitic (see Chapter 5).

Coelomates
The remaining invertebrate phyla, including all those with a complex body structure, have a coelomic body cavity. This is a fluid-filled space developed within mesoderm and lined by a coelomic epithelium. In many coelomates there is a series of channels or blood spaces outside the coelomic epithelium, forming a circulatory system, often with contractile walls to the larger vessels which act as pumps. The coelomate phyla are of two basic organisational types. In the first type the coelom forms an extensive perivisceral cavity, and the blood system (if present) is a closed system of narrow channels; in the second type the coelom is greatly reduced, and the blood system is expanded to form a perivisceral haemocoel, retaining a pumping circulation. Coelomate phyla can therefore be referred to as having either a coelomic or a haemocoelic body plan.

Only one coelomic coelomate phylum has an unsegmented body. This is the phylum Sipuncula (Fig. 1.2a), a group of sedentary marine worms with an anterior end that can be introverted into the hinder part of the body, and a crown of small, ciliated tentacles at the anterior apex (see Chapter 6). The gut is U-shaped, with an anus located at the side of the body.

A significant number of coelomic coelomate phyla have a trimeric construction. In these animals, the body plan is based on three subdivisions (the protosome, mesosome and metasome) within which the coeolom is developed as three pairs of compartments (the protocoel, mesocoel and metacoel). Certain trimeric phyla are also deuterostomes; that is, animals in which the anus forms as a primary opening in the early embryo and the mouth develops later by secondary ingrowth. The two major deuterostome phyla are the Echinodermata (see Chapter 16), which display secondary radial symmetry, and the Chordata (see Chapter 17), which have a notochord and dorsal nerve cord. In these animals the trimeric construction is often obvious during development but becomes obscured in the adult

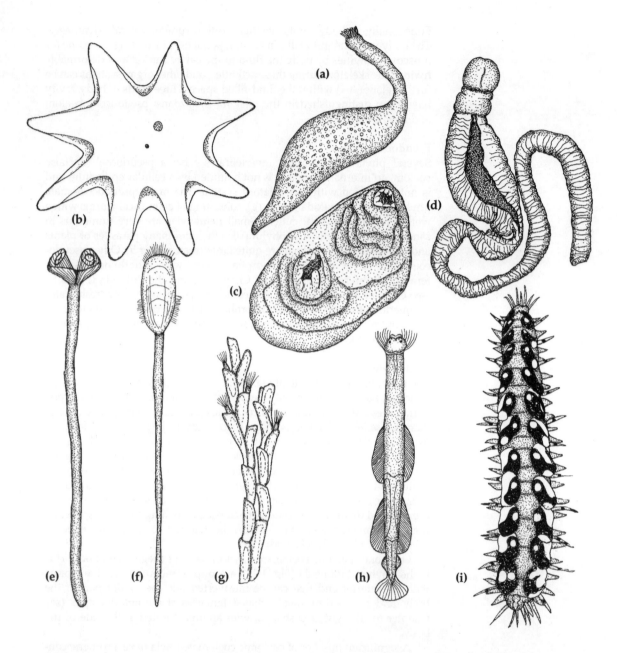

Figure 1.2
Body form in the coelomic coelomate phyla. (a) Sipuncula. (b) Echinodermata. (c) Chordata. (d) Hemichordata. (e) Phoronida. (f) Brachiopoda. (g) Ectoprocta. (h) Chaetognatha. (i) Annelida.

(Fig. 1.2b, c). The Hemichordata (Fig. 1.2d), a small deuterostome phylum of burrowing and sessile marine animals, has a clearly tripartite structure even in the adult (see Chapter 17).

There are also a number of other trimeric phyla whose relationships are more controversial (see Chapter 18). The Chaetognatha, also tripartite as adults (Fig. 1.2h), are small, fish-like, planktonic marine animals. Three phyla, the Phoronida (Fig. 1.2e), Brachiopoda (Fig. 1.2f) and Ectoprocta (Fig. 1.2g), are sedentary or sessile marine animals in which the mesosome is developed into an elaborate crown of ciliated tentacles, the lophophore, around the mouth (see Chapter 15). Phoronids are worm-like tube

THE INVERTEBRATE PHYLA

dwellers. Brachiopods have a bivalved, calcareous shell protecting the body and lophophore. Ectoprocts grow in a branching pattern of small zooids, each with a retractable lophophore. The ectoprocts have a superficial resemblance to the acoelomate entoprocts, but the anus in ectoprocts opens outside the lophophore, as in other lophophorates.

One major coelomic coelomate phylum, the Annelida (polychaetes, earthworms and leeches), has the body and coelom divided into a series of repeated segmental units, a condition known as metameric segmentation (Fig. 1.2i). Annelids are basically motile worms, although burrowing and tube-dwelling are common modes of life among them. Three groups of

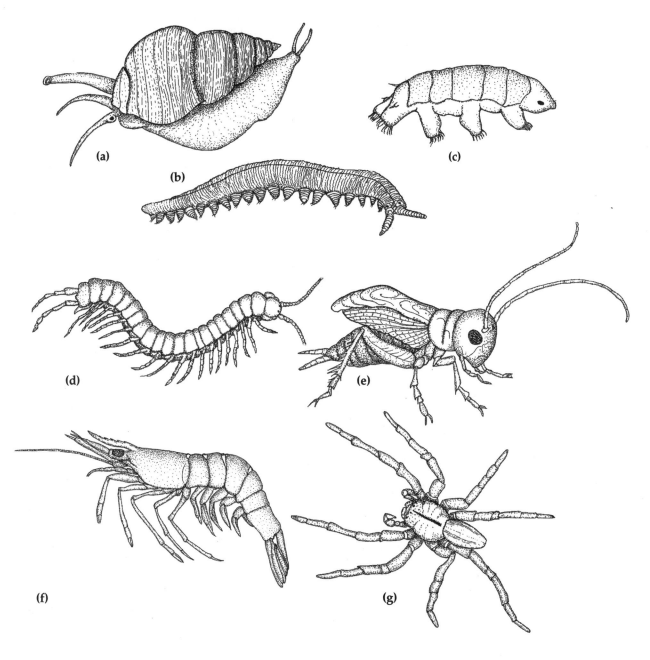

Figure 1.3
Body form in the haemocoelic coelomate phyla. **(a)** Mollusca. **(b)** Onychophora. **(c)** Tardigrada. **(d)** Arthropoda (subphylum Myriapoda). **(e)** Arthropoda (subphylum Hexapoda). **(f)** Arthropoda (subphylum Crustacea). **(g)** Arthropoda (subphylum Chelicerata).

more specialised segmented marine worms, the Echiura, Pogonophora and Vestimentifera, are closely related to annelids (see chapter 8). Annelids, of course, are not the only metamerically segmented coelomates. Metameric segmentation is a feature of the cephalochordates and vertebrates of the phylum Chordata, and of several phyla about to be mentioned.

Haemocoelic coelomates

The final organisational category of coelomates is the haemocoelic phyla, which include all of the most complex invertebrate animals. One haemocoelic phylum, the Mollusca, is the only group of complex animals to retain an unsegmented body plan. Molluscs (chitons, snails (Fig. 1.3a), slugs, bivalves, cephalopods, etc.) are characterised by an external, calcareous, dorsal shell and are basically slow-moving, although the molluscan subgroup Cephalopoda (squid, octopus, etc.) rivals the vertebrates in size and complexity of bodily organisation. Molluscs are discussed in detail in Chapter 7.

The other haemocoelic phyla comprise animals with a metamerically segmented body plan that includes a chitinous external cuticle and paired, haemocoel-filled limbs on the segments. Members of the phylum Onychophora (Fig. 1.3b) have a soft, flexible construction and stumpy, uniramous limbs known as lobopods. Modern onychophorans (see Chapter 9) are terrestrial animals of damp habitats, but the phylum is known to have had a marine lobopod ancestry. The phylum Tardigrada (Fig. 1.3c) consists of short-bodied microscopic animals which also have lobopods (see Chapter 9).

The majority of metamerically segmented, haemocoelic animals are characterised by a jointed, chitinous exoskeleton supporting the body and limbs, and are placed in the phylum Arthropoda. The exoskeleton is a basis for a vast range of form and function, making the arthropods the most diverse and ubiquitous of all types of invertebrate (see Chapter 10). Living arthropods fall into four distinct groups: the Hexapoda (Fig. 1.3e) (insects, etc.; see Chapter 11), the Myriapoda (Fig. 1.3d) (centipedes, millipedes, etc.; see Chapter 12), the Crustacea (prawns (Fig. 1.3f), shrimps, crabs, barnacles, etc.; see Chapter 13) and the Chelicerata (horseshoe crabs, scorpions, spiders (Fig. 1.3g), etc.; see Chapter 14). The relationships between these groups have been the subject of much discussion in recent years. Current opinion favours their classification as subphyla in the phylum Arthropoda, which is the arrangement we have followed in this book, but there are also arguments for the recognition of each group as a separate phylum. The different views on this question are reviewed in Chapter 18.

Embryonic development

Two distinct modes of embryonic development are recognisable among the bilateral phyla: spiral cleavage and deuterostome.

Spiral cleavage development

In addition to beginning with spiral cleavage, this pattern includes the development of mesoderm as a pair of ventrolateral mesodermal bands in the embryo. If coelomic cavities develop, they originate as splits within the mesodermal bands. Spiral cleavage development occurs, sometimes in a highly modified form, in the Platyhelminthes, Entoprocta, Gnathostomulida, Nemertea, Sipuncula, Annelida, Echiura, Pogonophora and Mollusca. Traces of the same pattern of development have also been detected in the embryos of the Onychophora, Hexapoda, Myriapoda and Crustacea, although not in the Chelicerata. Spiral cleavage development is also protostome, with the mouth forming before the anus (if present), but there are other protostome phyla which do not show spiral cleavage, as we shall see in later chapters.

Deuterostome development
The distinctive features of deuterostome development are the formation of the anus before the mouth and the development of mesoderm as hollow, dorsolateral, coelomic pouches, which usually become arrayed in a trimeric pattern (see Fig. 17.4). Deuterostome development occurs in the Echinodermata, Chordata, Hemichordata and Chaetognatha. Traces of this developmental pattern are also seen in the Phoronida, Brachiopoda and Ectoprocta, but their significance is still in dispute.

Phylogenetic consequences
The shared expression of distinctive pathways of embryonic development offers a first indication that about two-thirds of the invertebrate phyla appear to fall into two superphyla: the protostomes and deuterostomes. Each group encompasses a wide range of body plans. Other phyla do not appear to fit developmentally into either category. We shall have more to say about the complex problem of assessing the evolutionary or phylogenetic relationships between the invertebrate phyla in Chapter 18.

Chapter 2

The Porifera

P. R. Bergquist

PHYLUM PORIFERA

Introduction 11

Sponge organisation 11
   Body structure and canal
      organisation 12
   Cellular activities 14
   Ingestion, digestion, excretion
      14
   Integration 16
   Secretion 16
   Response to injury and invasion
      18
   Aggregation and totipotency
      18
   Reproduction 19

Sponge ecology 24
   Ecological processes 24
   Bioerosion 24
   Trophic dynamics 24
   Interaction with substrata 25
   Biotic interactions 25

Classification of the phylum
   Porifera 26
   Relationships within the
      Porifera 27

Phylum Porifera

Sessile metazoans without tissues, organs or symmetry; having epithelia without basement membranes; a canal system lined by choanocytes which generate water currents and collect food; and a supporting tissue (mesohyl) containing motile cells and skeletal elements (mineral spicules, spongin fibres). Mostly marine, with motile larvae.

Introduction

Sponges are sedentary, filter-feeding metazoans in which a single layer of flagellated cells, the choanocytes, pumps a unidirectional water current through the body. There are sponges that seem to be exceptions to this definition, but these are simply examples of how flexible sponges can be in response to extreme environments.

Sponges are isolated from the environment by perforated epithelia, one cell deep (see Fig. 2.3). Both the internal flagellated epithelium (the choanoderm) and the squamous epithelium (the pinacoderm) lack the basement membrane seen in other metazoan epithelia. The cells of these bounding layers are highly mobile and capable of many functions. Between the two epithelial layers is the mesohyl, a region which varies greatly in composition and extent. The mesohyl always contains some mobile phagocytic and secretory cells and has a collagenous matrix. Most frequently, mineral skeletal elements (spicules) supplement the collagen. Unique collagenous fibres can augment the skeleton. The mesohyl corresponds to the connective tissue of other metazoans.

Sponges are diverse. More than 9000 species are presently known and at least as many remain to be described. Sponges can grow to great size, some being among the largest living invertebrates. They occupy the full range of marine habitats, from the shallowest sea to the deep ocean, and some occur in fresh water. In marine benthic environments they occupy most niches. They excavate within calcareous substrata, attach to hard surfaces, bind soft sediments to provide attachment, or use skeletal elements as an anchor.

Sponge organisation

One way to view sponge organisation is to ask: What developments were necessary to become multicellular and then to achieve integration, in the absence of tissues and organs? The answer requires an introduction to sponge cells, their functions, capabilities and cellular organisation.

Although sponges are metazoans, their cells retain a high degree of independence, but the entire cell mass functions as a whole in pumping a sufficient volume of water through the body to effect all necessary exchanges. The sponge is thus capable of integrative cellular behaviour directed towards ensuring effective feeding, reproduction and dispersal.

The multicellular state (Box 2.1) requires that cells are mobile; that like and unlike cells can be recognised on contact with other cells; and that, following recognition of like, cells can adhere to each other. This adhesion may be transitory, permitting exchange of molecular signals, or more permanent when cells cooperate in the secretion of skeletal elements.

Cellular motility, recognition and adhesion require the existence and operation of complex intracellular and extracellular, surface-located molecular arrays and precise genetic controls of their expression and activity. These cellular processes were established in sponges and have been retained and refined in more complex metazoans. Sponges, representing a simple multicellular state, have provided important clues about the evolution of immune systems, integrative systems, adhesive specialisations and connective tissues.

> **Box 2.1**
> **What is required to become a multicellular organism?**
>
> - cell recognition systems;
> - cellular adhesion systems;
> - cellular integration systems;
> - elaboration of a matrix or connective tissue;
> - skeletal systems to permit increase in size;
> - feeding, reproductive and dispersal mechanism.

Whenever we describe the arrangement of individual cells that forms a functional sponge, the description applies only at one point in time. All cells are mobile, and most can change their structure and their function.

The epithelial pinacoderm and the flagellated choanoderm are more stable in terms of their location and cellular differentiation than the mobile mesohyl cells; but choanocytes leave the choanoderm to form sperm, and pinacocytes can phagocytose or become collagen-secreting cells. Of the many cell types in the mesohyl, only those with an obvious secretory role can be regarded as irreversibly differentiated.

Four attributes can be used to define the functions of sponge cells:
1. Their position. Location indicates function even though it may not be constant over time.
2. Their structure. Sponge cells are small; intracellular details can be seen only by electron microscopy.
3. The chemical characterisation of their cellular inclusions. Functions have been identified from inclusions for contractile cells (myocytes) and some secretory cells such as spherulous cells and granular rhabdiferous cells.
4. Their behaviour. This is best observed in the development of larvae, during cell aggregation, or in the response to invasion by foreign tissue.

Body structure and canal organisation

In order to understand basic sponge organisation it is best to start with the simple example found only in a few calcareous sponges, such as *Leucosolenia*, and described as an asconoid state. The flagellated cells, or choanocytes, each of which has a microvillous collar surrounding the base of an apical flagellum (Fig. 2.1), drive the water current. They are arranged as a single-layered, continuous epithelium lining the internal surface of the tubular body (Fig. 2.2a). A simple epithelium, the pinacoderm, lines all external surfaces.

Individual cells of this layer, called porocytes, contain simple channels through which water passes from the exterior to the choanocyte-lined cavity or spongocoel. Porocytes are single cells extending through the thin

Figure 2.1
Choanocytes with apical flagellum and collar tentacles extending from the cell coat.

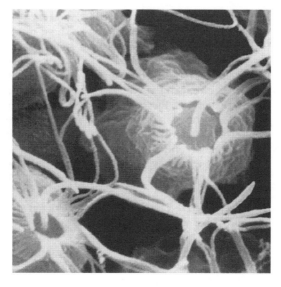

  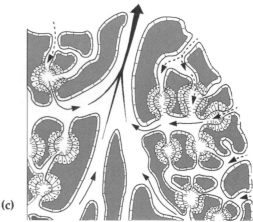

Figure 2.2
Organisation of the sponge body. (a) Half-section of a simple asconoid. (b) Half-section of a syconoid with slightly thickened mesohyl and folded choanoderm and pinacoderm. (c) Sector of a leuconoid, with divided choanoderm and extensive mesohyl, traversed by a complex system of canals. Arrows indicate the flow of water through the sponge.

intervening mesohyl which contains the skeleton, mobile cells and matrix. Water leaves the spongocoel by way of an apical opening, the osculum.

Almost all sponges have a more complex organisation, which is achieved by two processes: the folding of the pinacoderm and choanoderm (Fig. 2.2b), and the increase in volume of the mesohyl (Fig. 2.2c). These processes lead to the development of a system of canals which traverse the mesohyl to conduct water to the now subdivided choanoderm (incurrent canals) and to permit water flow from choanocyte chambers to the osculum (excurrent canals). Incurrent canals open to choanocyte chambers by one or several small openings, prosopyles. Each chamber is drained by a single larger opening, an apopyle, opening into an excurrent canal. The pinacodermal cells which cover the external surface are termed exopinacocytes. Those lining the canals are endopinacocytes and those attaching the sponge to the substratum are basopinacocytes. There are structural differences between the three cell types.

Some Calcarea (calcareous sponges) have a body in which folding has progressed only sufficiently to form large sac-like choanocyte chambers

(Fig. 2.2b). These are served by short incurrent canals and open directly to the spongocoel. Such sponges exhibit moderate mesohyl thickening and are termed syconoid. All other sponges have a greatly amplified choanoderm which is partitioned into small spherical or oval choanocyte chambers dispersed throughout an extensive mesohyl (Fig. 2.2c). Canal length and complexity are increased, giving a so-called leuconoid condition. The inner part of the body in which the chambers are present is termed the choanosome to differentiate it from a region immediately below the exopinacoderm, which lacks chambers and is often marked by wide exhalant canals and dense mesohyl. This region is the ectosome. Some groups of sponges have a very elaborate ectosome, referred to as a cortex.

The increase in size and strength of the body underlying the diversification of Porifera depended upon the increase in volume and complexity of the mesohyl. This facilitated the regional specialisation of the skeleton and provided for a greatly increased surface area occupied by feeding cells, or choanocytes.

Cellular activities

Body functions in the more complex Metazoa are carried out by tissues or organs. In sponges, individual cells acting singly or in small groups carry out all activities. Apart from feeding, which involves the choanoderm and pinacoderm, all cellular activities (Box 2.2) take place in the mesohyl.

Box 2.2
Cell types in sponges

Epithelial
    Choanocyte — flagellated
    Pinacocyte — squamous
        Exopinococyte — external
        Endopinacocyte — lining canals
        Basopinacocyte — basal, attaching
    Porocyte — pore-lining

In mesohyl
    Archaeocyte — totipotent; digestive, reproductive
    Myocyte — conducting; contractile
    Secretory cells
        Collencyte — collagen
        Lophocyte — collagen
        Sclerocyte — mineral spicules
        Spongocyte — spongin fibres
        Spherulous — bioactive and other glycoprotein metabolites
    Gray — as immunocytes
    Rhabdiferous — polysaccharides

Ingestion, digestion, excretion

Sponges are unselective particle feeders. The various apertures, in conjunction with the choanocyte collar tentacles and intertentacular mucous net, constitute a set of sieves of diminishing mesh size in the path of the water current (Fig. 2.3). A large volume of water is pumped through a sponge:

THE PORIFERA

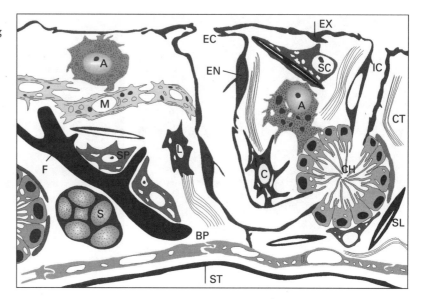

Figure 2.3
Stylised section of an encrusting leuconoid sponge, illustrating the major cell types and their relationships.
A — archaeocyte,
BP — basopinacocytes,
C — collencyte,
CH — choanocyte chamber,
CT — collagen tract,
EC — excurrent canal,
EN — endopinacocyte,
EX — exopinacocyte,
F — fibre,
IC — incurrent canal,
L — lophocyte,
M — myocyte,
SC — sclerocyte,
S — secretory cell,
SL — spicule in fibre,
SP — spongocyte,
ST — substratum.

experimental results have yielded estimates of 27 litres per day, cycling the net volume in 7.6 seconds. The flow through the sponge is laminar, without significant mixing; movement of the water within the choanocyte chambers can be modulated by specialised central cells.

Choanocyte entrapment accounts for 80% of particulate organic carbon intake. Particles of bacterial size (0.1–1.5 μm) are ingested with high efficiency. In addition to choanocyte uptake, most other cells at the surface are capable of direct phagocytosis of particles of around 50 μm. Absorption of dissolved or colloidal nutrients directly from sea water is also known to occur.

Digestion takes place predominantly in large nucleolate mesohyl cells, the archaeocytes (Fig. 2.3). These migrate to the base of a choanocyte, where vacuoles containing food are located 30 minutes after uptake. Waste products are discharged through canal linings. Diffusion ensures an oxygen supply to all cells, as the folding of the lining epithelia is sufficient to bring all cells close to the canal system.

The recent discovery of carnivory in a deep-water sponge, *Cladorhiza*, which has survived in caves trapped by sea-level changes, provides an excellent example of the flexibility of sponge cellular systems. The environment in which this tiny organism lives is low in suspended nutrients. The sponge has no choanocytes or canals. Nutrition depends on the ability of surface pinacocytes to move and phagocytose directly (Figs 2.4, 2.5). Long, filamentous extensions of the body contact small crustaceans, entangle them with grapnel-like spicules, and hold them while the cells flow over and finally internalise the prey.

The extreme environment of hydrothermal vents provides another variation involving the same genus. The sponge, in addition to carnivory, has developed a symbiotic association with methanotrophic bacteria. Intracellular digestion of the bacteria, with carnivory, allows the sponge to attain a much larger size.

These examples illustrate the potential for diversity in sponges provided by cell mobility, combined with multipotency, a low level of differentiation and cell independence.

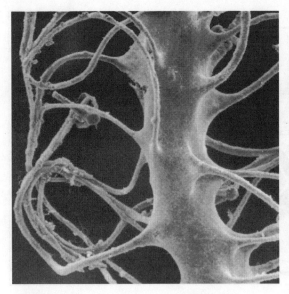
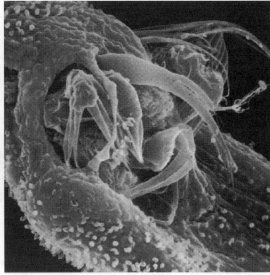

Figure 2.4
Carnivorous sponge, *Cladorhiza*, with tentacle-like protrusions and superficial chelate microscleres which entangle small Crustacea (courtesy Dr J. Vacelet).

Figure 2.5
Internalisation of trapped prey by flow of pinacocytes in *Cladorhiza*. (courtesy Dr J. Vacelet).

Integration

A sponge individual consists of all tissue enclosed within a continuous pinacoderm (Figs 2.6, 2.7). Such individuals are capable of functioning in an integrated fashion. The coordinated closure of oscula and incurrent pores (ostia), arrest of water currents, coordinated flagellar action in swimming larvae, contraction of mesohyl and pinacoderm, and complex, endogenously controlled patterns of pumping activity, all argue for the presence of a system that generates and conducts signals.

A conduction system controlling simultaneous arrest of flagellar activity has been demonstrated in the glass sponges (Hexactinellida). The spread of the response at 2 mm per second is considered too fast to be mediated by chemical diffusion, but is much slower than a typical neuronal response. The concept proposed to explain this phenomenon is that of a network of myocyte cells (Fig. 2.8) which links the external epithelia receiving the stimulus with the internal effectors, the choanocytes. This network is more established in thick-bodied leuconoid sponges. In thin-walled sponges, myocytes link pinacocytes and choanocytes directly. The myocytes also have a proven integrative role in oscular contraction. Myocytes arrayed around the oscula have numerous button-cell junctions and intracellular microtubules associated with acetycholinesterase activity.

Secretion

The secretion of a range of substances is important in the life of sponges. The most obvious is the secretion of the mineral, fibrous, and collagenous skeleton (Fig. 2.3). The secretion of the mesohyl matrix requires polysaccharide material to be deposited to form, in combination with collagen, the ground substance of this connective tissue layer. In addition, sponges secrete copious quantities of mucus into the exhalant current. This binds metabolites that are released by particular secretory cells. A large number of the secretions released by sponges have potent biological activity.

The secretion of polysaccharide is achieved by granular rhabdiferous cells. These cells have been observed to discharge their contents into the mesohyl, and histochemical tests reveal their contents to be polysaccharide. In two cases it has been demonstrated by X-ray microanalysis that bioactive

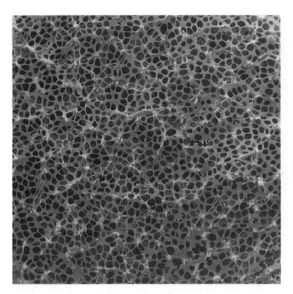
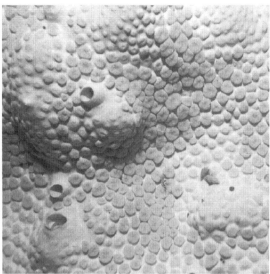

Figure 2.6
Surface of a fibrous sponge with a superficial fibre network, supporting a pinacoderm with numerous inhalant pores and ostia.

Figure 2.7
Surface of *Cliona*, a boring sponge. Contractile oscules and inhalant pore sieves are evident.

metabolites are synthesised in large anucleolate secretory cells, termed spherulous cells. These have been observed to release their contents into exhalant canals, along which they occur in large numbers.

In all sponges the mesohyl matrix is built upon a framework of fibrillar collagen identical to that found throughout the Metazoa. Cells known as collencytes (Fig. 2.3) secrete the bulk of this material, assisted in some cases by lophocytes — large mobile cells which trail a band of collagen in their wake. As well as the universal fibrillar collagen, sponges elaborate a number of other collagenous skeletal elements, each with distinctive ultrastructural characteristics, but referred to collectively as spongin. Fibrous spongin is secreted by special cells called spongocytes (Fig. 2.3), which act in groups to elaborate fibres or to surround mineral elements with fibre support. In some groups, spongin fibres are present to the exclusion of mineral elements and form elaborate strong skeletons (Fig. 2.9), well known as bath sponges. In freshwater sponges the overwintering reduction bodies, or gemmules, have a thick spongin coat.

The mineral skeleton of sponges is made up of either hydrated silica or calcium carbonate. Most frequently the two are not found together, but some sclerosponges combine massive calcareous skeletons with superficial siliceous elements. The most common mineral skeletons are composed of discrete spicules which, in the siliceous sponges, have diverse shapes and sizes (Figs 2.10–13). Spicules are secreted by sclerocytes (Fig. 2.3), either singly (in the case of small spicules or microsclere spicules) or in groups of up to four (in the case of large spicules or megascleres). The pattern of secretion is determined by an axial proteinaceous filament around which mineral elements are laid down. The number of filaments establish the number of axes or rays of the finished spicule. The main function of a megasclere skeleton is structural, maintaining the overall form of the sponge. Microscleres line and reinforce surfaces and canals.

A complex terminology has been developed to describe spicules. Terms which designate the number of axes end in –axon (e.g. triaxon, with three axes). Those designating the number of rays end in –actine (e.g. hexactine, with six rays). Further terms specify shape and ornamentation.

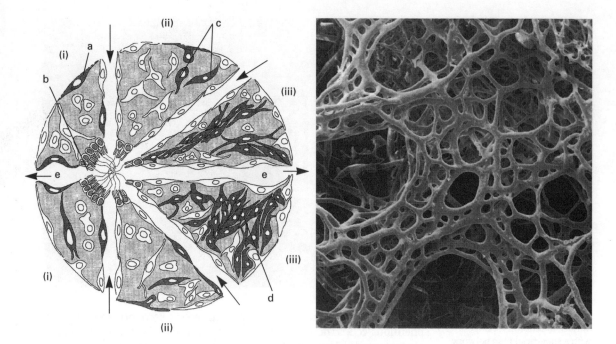

Figure 2.8
The relationship between myocytes, pinacocytes and choanocytes in sponges with thin, moderately developed, and thick mesohyl. (i) Thin-walled sponge. (ii) Moderately dense sponge. (iii) Thick-walled sponge. As mesohyl thickness increases, direct contact between surface cells and choanocytes cannot be effected and networks of myocytes in contact are evident around canals and oscules. The networks are replaced by pronounced myocyte tracts in heavily collagenous, thick-walled forms. In all cases pinacocytes can contribute to the contractile network.
a — pinacocyte,
b — choanocyte chamber,
c — loose myocyte network,
d — myocyte tract,
e — incurrent canal,
f — excurrent canal.
(After Pavans de Ceccatty 1974.)

Figure 2.9
A complex skeleton made up entirely of anastomosing spongin fibres.

Response to injury and invasion

It has been noted that sponges provide insights about how more complex processes found in other organisms have evolved. There is evidence that a histocompatibility system with many attributes of the vertebrate system operates in sponges. Experiments have demonstrated the rejection of grafts from other species and the acceptance of same-species grafts. Histological observations have also revealed that there is a substantial cellular flow to grafts or abutting boundaries, indicating cellular involvement in the response. In two species of marine sponges, heavily granular cells termed gray cells are the 'immunocytes'. In these species foreign material is quickly recognised and a signal is produced that suppresses movement of self cells in the zone of contact. This may reduce the entry of invading cells. If contact is prolonged, the immune system is activated and gray cells are primed to be responsive to an attractant released at the zone of allogeneic (foreign) contact. Cytotoxic responses are initiated and all cells in the contact zone are killed. The activated state of the immune system can last for three weeks. Archaeocytes may play a part by differentiating, once tissue is challenged, to augment the population of gray cells.

Aggregation and totipotency

Sponges were the first organisms in which the ability of dissociated cells to aggregate and reconstitute into functional organisms was discovered. Mixtures of cell suspensions of two species were observed to reconstitute into the separate species. The process involved cell–cell recognition, which is a basic attribute for building a multicellular body and retaining its integrity. The molecular mechanisms that guide the process are now well known and involve complex surface-located proteoglycans, surface-active diffusible elements, and many associated enzymes.

These molecular arrays are thought to be present on most sponge cells, although those that are terminally differentiated (such as the larval ciliated

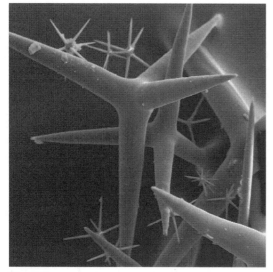

Figure 2.10

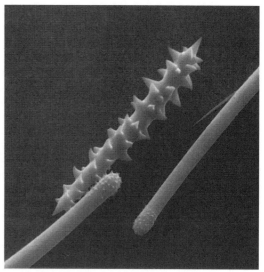

Figure 2.11

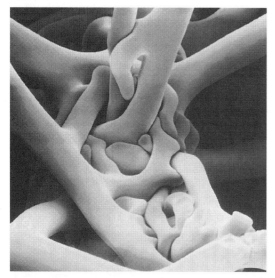

Figure 2.12

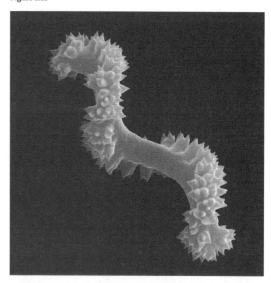

Figure 2.13

Figure 2.10
Tetractinal megascleres and asterose microscleres.

Figure 2.11
Acanthose megascleres.

Figure 2.12
Ankylosed megascleres, desmas.

Figure 2.13
Spiraster microsclere.

cells and secretory cells) may not continue to express them. During the process of aggregation, cell types can be separated in order to determine which cells are active. Archaeocytes and collencytes are essential to aggregation. All other cell types can be derived from them. The greatest contribution is made by the archaeocytes, which are referred to as totipotent, or capable of future differentiation in any direction. The essential elements of the sponge system have persisted in higher Metazoa, and cell aggregation studies have become a routine tool in biomedical research.

Reproduction

All sponges are capable of sexual and asexual reproduction. The diversity of sexual reproductive sequences in sponges is truly bewildering, and the following descriptions indicate only general trends.

Sexual reproductive sequences

Stages in sexual reproduction of sponges are elusive, as the lack of any specific location for the process in the body makes observation difficult. Also, the fact that cells change their function and migrate considerable distances during fertilisation, egg maturation and early embryo development makes precise sequences hard to describe. The following account is based mainly upon the largest group, the Demospongiae, mentioning the Calcarea only where they are of special interest.

Sperm are formed from choanocytes, and eggs from choanocytes or archaeocytes. Sperm are contained in spermatic cysts, which are choanocyte chambers transformed by spermatogenesis. Eggs are distributed throughout the mesohyl. Some sponges are oviparous (e.g. *Polymastia*, Fig. 2.18a). Following gamete release, fertilisation and development proceed externally. Other sponges are viviparous, with fertilisation and development both occurring in the mesohyl. In oviparous sponges, gamete release can be a sudden mass event. The way in which individuals down-current of each other spawn in rapid sequence suggests that a chemical trigger operates after an initial event, but none has yet been confirmed. In the case of *Xestospongia* from the Great Barrier Reef, gamete release always occurs close to the new moon in October. In viviparous forms, larval release is gradual over weeks or months, sometimes continuing at reduced levels throughout the year.

The sperm, once released, enter other sponges in the inhalant current. They are trapped by choanocytes which then lose their flagella and collars and, as carrier cells, migrate through the mesohyl to an adjacent egg. The carrier cell fuses with the egg, and the sperm is transferred.

In all cases cleavage of the embryo is total, but the details of the process are known for only a few species in which the blastomeres lack the usual granular material which tends to obscure cell boundaries. The growth of the oocytes (and later the embryo) is supported by nurse cells. These are formed by archaeocytes, which become densely granular, and migrate toward the embryo, where they are incorporated to provide material for growth. Cell division and differentiation proceed until the larva has a

Figure 2.14
Parenchymella larva. Ciliary waves and the posterior ciliary tuft are shown.

Figure 2.15
Blastula larva in cross-section. The flagella are external, and undifferentiated columnar cells surround the blastula cavity.

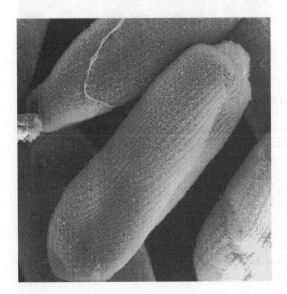
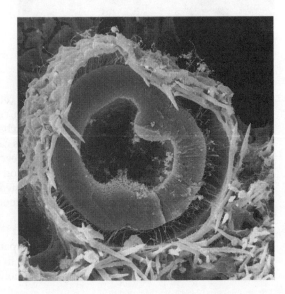

THE PORIFERA

superficial layer of small, columnar ciliated cells around a mass of cells in which most major adult types are present, in some cases including spicules and organised choanocyte chambers. A larva of this type is called a parenchymella (Fig. 2.14). The other common larval type in Demospongiae and Calcarea is a blastula (Fig. 2.15), in which a cavity is present at some stage prior to settlement. The parenchymella larva, when ready for release in the exhalant stream, is a mini-adult in terms of tissue differentiation. Blastulae have only minimal tissue differentiation.

The larvae either swim freely, or rotate near the substratum for two to forty-eight hours. Despite their simple structure, sponge larvae are capable of responses to light and gravity. Careful studies of behaviour during free life have provided evidence that larval behaviour can dictate settlement location and thus adult distribution. Settlement is preceded by a short period of creeping, after which the larva adheres by the anterior pole and cells flow from the central mass and spread on the attachment surface. Metamorphosis to a functional adult is rapid.

There are a number of peculiarities in cellular behaviour during the formation and metamorphosis of sponge larvae which preclude any close comparisons with gastrulation and germ-layer formation in other Metazoa. The fate of the larval ciliated layer at settlement has not always been clear. Some investigators have argued that these cells lose their flagella and migrate to the interior, where they become choanocytes. This process, termed 'reversal of layers', was so different from those followed in other Metazoa that it was the basis for the establishment of the subkingdom Parazoa to receive the Porifera. However, recent work which followed the settlement sequence using electron microscopy has shown that the larval ciliated cells are terminally differentiated locomotory structures that are either shed or phagocytosed. Reversal of layers does not take place. The taxon Parazoa is therefore based upon a false premise.

Another peculiar developmental sequence occurs in some Calcarea (e.g. *Sycon*, Fig. 2.16), where the larvae are blastulae with numerous small micromeres and eight large macromeres (Fig. 2.16a). By the time a mass of 64-cells has formed, the micromere derivatives have flagella which are

Figure 2.16
Development of the blastula larva of *Sycon* (Calcarea) depicted as sections through a spherical embryo. (a) Early blastula lying adjacent to a choanocyte chamber, from which it ingests cellular material (stomoblastula stage).
(b) Blastula with unflagellated anterior macromeres, and micromeres with inwardly directed flagella. (c) Blastula with 'cellules en croix' (cc) and with blastula cavity open at a gap in the macromere region, through which micromeres are beginning to evert. (d) Eversion of micromeres, bringing their flagella to the exterior.
(e) Eversion almost completed, with macromeres about to meet and fuse at the posterior pole.
(f) Vertical section through the amphiblastula. The 'cellules en croix' are tetraradially disposed. (After Tuzet 1973.)

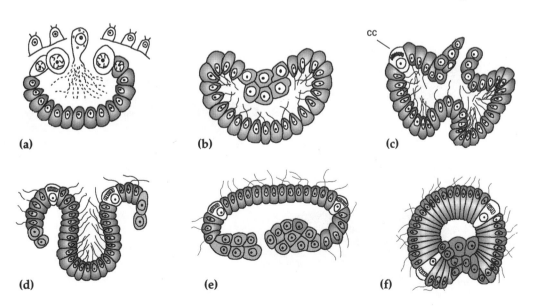

directed inward into the blastular cavity (Fig. 2.16b). The macromeres divide more slowly and remain unflagellated. A gap appears in the centre of the macromere region and adjacent choanocytes are ingested. The gap then closes, only to reopen and permit the micromeres to evert (Fig. 2.16c), bringing the flagella to the exterior. This process (Figs. 2.16d, e) places the macromeres at the posterior pole where, on settlement, they will produce all cells except choanocytes. Larvae of this type, which are called amphiblastulae (Fig. 2.16f), have a brief free-swimming life in the plankton. At settlement and metamorphosis the flagellated cells are overgrown by macromeres and inverted once more to form the choanocyte layer.

The developmental sequence of these calcareous sponges has no parallel in other animals. Other sponges produce blastulae called coeloblastulae. These are of several types, but show little cell differentiation. Some pass cells inwards from the flagellated layer to form a loose tissue in the blastular cavity.

Asexual reproduction

The asexual reproductive processes of sponges, in addition to serving a reproductive function, provide mechanisms for dispersal, for maintaining and extending attachment space, and for surviving extreme conditions. The example most frequently cited is gemmule formation in freshwater sponges, where tissue which has regressed to a mass of archaeocytes charged with vitelline material (thesocytes) is encapsulated within a thick spongin coat in which microsclere spicules are embedded. Gemmules serve as over-wintering devices. When environmental conditions permit, hatching proceeds and archaeocytes flow through the micropyle, a thinner region of the spongin coat, on to the substratum. All adult cells are derived from archaeocytes as a functional sponge is reconstituted. This is an excellent demonstration of totipotency. Gemmules of similar but less structured form occur in a number of marine sponges.

The production of surface or internal buds is another means of asexual reproduction. A variety of mechanisms are known. In *Tethya*, a common globe sponge, cells flow along extruded spicules, aggregate at the ends, round up and detach. In a related genus, *Aaptos*, and in *Cinachyra* (Fig 2.17),

Figure 2.17
Asexual reproduction in *Cinachyra*. (a) Adult sponge attached to a firm substratum. (b) The sponge tissue flows as a thin plate over the adjacent substratum, and small mounds of tissue arise at intervals from the plate. (c) Tissue plate regresses, leaving small, isolated sponges growing around the parent sponge.

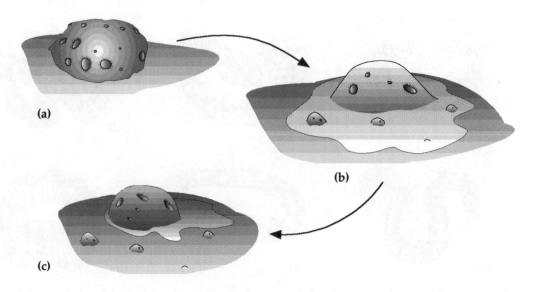

a thin plate of tissue flows out from the sponge base and, at intervals, new bodies rise from the plate. Eventually the plate regresses, isolating the individuals. These spreading basal mats can extend over large areas. Any species that shows an aggregated distribution and inhabits an environment with mobile sediment is likely to reproduce asexually in this way.

In *Tethya* and *Cinachyra*, internal buds are also produced. These are highly structured at the time of release, having the tissue and skeletal organisation of a small adult. The buds break through canal walls to exit via oscula on to the sponge surface, where they glide to the base and attach adjacent to the parent.

Many species produce groups of small buds lying in surface depressions just below the surface of the sponge. These are released by surface breakdown. In the case of *Polymastia* (Fig. 2.18b) the trigger for release is wave action. The wave surge moves the mobile sediment and clears attachment space for the buds. Such structures are best referred to as propagules, as they have no internal organisation and no connection with the parent. Despite this simplicity, some intricate and precise behavioural sequences have been revealed during the attachment period. The sponge propagules select fragments of sediment to which they attach themselves with great specificity with respect to both size and type. The fragments are used as anchors to provide orientation while the tissue establishes an oscular tube which can elevate above encroaching sediment. Propagules may also fuse, extend and undergo further asexual proliferation by breaking into a series of beads before attachment takes place.

Figure 2.18
Reproduction in *Polymastia*.
A. Sexual reproduction. **(i)** Egg and sperm from different individuals fuse. **(ii)** Cleavage. **(iii)** Creeping blastula larva. **(iv)** Settlement.
B. Asexual reproduction.
(a) Fragment detached from a sponge surface papilla.
(b, c) Free propagules. **(d)** Direct settlement of propagule, with beginning of oscular tube formation. **(e)** Fusion of propagules. **(f)** Extension of propagule. **(g)** Beading along length of propagule, and migration of detached elements.
(h) Migrating propagule.
(i) Settled propagule with oscular tube.

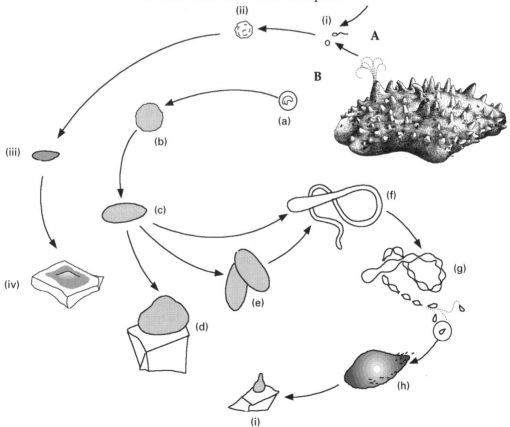

The great regenerative and differentiation potential of sponge cells ensures that tissue fragmentation, and more organised propagule formation, provide all species with asexual dispersal opportunities. Careful field observation will certainly reveal behaviours closely attuned to particular habitats. Sponges are not as simple as they have been portrayed to be.

Sponge ecology

While sponges have colonised all aquatic habitats, certain groups such as the Calcarea are limited to depths of less than 100 metres by physical factors which control the secretion of the calcium carbonate skeleton. Others, such as the Hexactinellida, mainly inhabit deep water. Demospongiae can be found from the upper intertidal regions to the greatest depths in the sea and in fresh and brackish water. Calcarea require a hard substratum for attachment. Hexactinellida can colonise firm and soft sediments. Demospongiae utilise any surface. The ecological dominance of the Demospongiae simply reflects their diversity in form, structure, reproductive processes and physiology.

One way to approach adult sponge ecology is to look at sponges as contributors to broad ecological processes in marine habitats and then to identify attributes that have made them successful competitors in so many microhabitats.

Ecological processes

Two major ecological processes in which sponges are involved are the bioerosion of calcareous substrates, mainly coral reefs, and the trophic dynamics of benthic communities.

Bioerosion

Bioerosion of coral reefs by sponges, mainly *Cliona*, has been calculated to remove 6–7 kg of material from 1 m^2 of calcareous substratum in 100 days. This is approximately equivalent to the removal of a layer between 0.1 mm and 1 mm deep per year, depending upon the density of the sponge population. Over geological time, boring by sponges has produced a marked subtidal notch on some reefs. The mechanism by which the complex galleries are bored is cellular etching. Archaeocytes transform to etching cells, which use a combination of enzymatic and mechanical processes to excavate small calcareous chips, which are then expelled. These multifaceted chips can be identified in sediments, so that estimates of past sponge-mediated erosion can be made with some accuracy. Large sponges also have an important role as interim binders of detached reef-frame material, a process which increases the rate of carbonate accretion.

Trophic dynamics

In certain coral reef environments, large sponges are second only to corals in biomass and their contribution to trophic exchanges is important. An interesting feature of the biology of a majority of these species is that they exist in symbiotic association with Cyanobacteria and Eubacteria, which are present in high densities in the mesohyl. Studies carried out on the Great Barrier Reef have shown that the sponges become net primary producers, contributing three times more oxygen to the sea water than they consume in respiration. Such sponges are often foliaceous, a growth form which exposes the maximum surface to light. The number of phototrophic sponges increases with distance from land, where waters become less productive. On shallow New Caledonian reefs there is a huge biomass of a boring sponge which extends beyond the galleries in the coral to form a thin continuous mat. This species has a cyanobacterial association and almost certainly makes a strong photosynthetic contribution to the trophic balance of the reefs. Interestingly, in the higher productivity waters on Caribbean reefs, sponges receive little nutrition from photosynthetic symbionts.

Interaction with substrata

In the Antarctic, sponges are the dominant component of the macrobenthos. In the still waters under the ice shelf the huge numbers of siliceous spicules that they extrude form mobile mats up to 2 m deep. This deposit renders the substratum unsuitable for the settlement of their echinoderm predators. Sponges contribute in this way to structuring their own microenvironment.

Biotic interactions

Many invertebrates produce secondary metabolites of diverse and novel types, but sponges elaborate by far the greatest number. In many cases the metabolites exhibit bioactivity in pharmacological, antiviral and antimicrobial tests and have a multitude of biotechnological applications. The production of natural antifouling agents is an example. While the utility of these compounds in medicine and technology is obvious, the role that they have in the life of the sponges is less clear. It is commonly stated that chemical defence against predation and overgrowth is their role, but the situation is far more complex. The great chronological age of the major groups of sponges — more than 550 million years — has provided a vast time for 'experimentation' with different chemical solutions to essential steps in evolution. For example, cell membranes in all Metazoa have the sterol cholesterol as the major lipid constituent. In addition to cholesterol, sponges produce a great range of novel sterols and terpenes, the latter sometimes to the exclusion of sterols. These molecules are membrane constituents, relics of past experiments in building a multifunctional cell membrane. Coincidentally, many of these terpenoid molecules are extremely toxic to other organisms. While they exist to build a membrane, perhaps they persist because their bioactivity can be deployed in other ways.

Interference competition: allelopathy

One way in which toxic compounds are very successfully deployed is in interference competition, which is best observed in species-diverse, benthic encrusting communities. In these communities, sponges, ectoprocts (Chapter 15) and ascidians (Chapter 17) are the major groups competing for attachment space. Careful study of boundary interactions taking place as species spread and abut on neighbours reveals that sponges are almost always successful in preventing overgrowth by other species. They use allelopathy, deploying toxic chemical agents as deterrents, thus ensuring that attachment space is maintained. This capability, in conjunction with great versatility in asexual reproduction, makes sponges efficient spatial competitors.

Predation

Predation of sponges by grazing fish is common in tropical and temperate seas. It is often stated that the presence of toxic compounds deters predation by making the sponge unpalatable. But recent studies using a sponge-grazing fish as an assay organism show that the toxicity of a sponge is not an indicator of palatability: fragments found to be unpalatable were as often non-toxic as toxic, and some palatable extracts were toxic. It cannot therefore be inferred that toxicity has a role in deterring predation. Predation is rarely fatal to a sponge. Fish predation generates fragments of tissue which can reattach and grow, and so could be assisting the spread of sponge species.

Classification of the phylum Porifera

Sponge classification will always be difficult because of the absence of polarity, symmetry, tissues and organs, but it has been made more difficult by an overwhelming emphasis upon the shape and ornamentation of the mineral skeletal elements, the spicules. These structures can be described precisely, but they are only one of many features which can be utilised to describe sponges and to build a classification. Recent broadly based classifications incorporate biology, histology and chemistry and take account of the discoveries of organisms thought to be long extinct.

There is now consensus over the classes of Porifera, of which three are recognised, and near-consensus over the subclasses. The classification of orders of sponges is still being debated.

The discovery of hexacts living in fiords at diveable depth has facilitated the study of their biology. The typical sponge pinacoderm is replaced in hexactinellids by a trabecular syncytial layer. The choanoderm is also syncytial and the collar structures are separated from the cell bodies. Conducting trabecular networks are present and weak action potentials have been detected. The histology of hexactinellids differs markedly from that of other Porifera. The best-known genus, *Euplectella*, has an ornate, structured apical sieve plate made up of fused megascleres. Such a patterned, organised structure certainly required homeobox genes to guide its development.

CLASS HEXACTINELLIDA

Exclusively marine sponges; siliceous mineral skeleton which always includes some spicules with hexactinal (six-rayed) structure; megascleres and microscleres always present (e.g. *Euplectella*)

CLASS CALCAREA

Exclusively marine sponges; mineral skeleton composed of calcium carbonate; spicules not differentiated into megascleres and microscleres; megascleres are monaxon, triaxon or tetraxon; calcareous skeleton can be massive, with or without associated spicules or plates; asconoid, syconoid and leuconoid types of organisation occur within the class. (e.g. *Sycon, Clathrina*)

Subclass Calcaronea

Spicules mainly diactines and/or sagittal triactines and tetractines; non-spicular calcareous skeleton can occur in addition to the spicules; during development, first spicules to be produced are diactines; choanocytes apinucleate; basal system of flagellum adjacent to the nucleus; first stage in embryogenesis a blastula with internally directed flagella; blastula everts, giving rise to an amphiblastula larva with flagellate and non-flagellate poles.

Subclass Calcinea
Regular triradiates and a basal system of quadriradiates; during development, triradiates are the first spicules to be secreted; choanocytes basinucleate, with spherical nuclei; basal body of flagellum not adjacent to the nucleus; coeloblastula larvae.

CLASS DEMOSPONGIAE
Predominantly marine sponges (three families living in fresh water); skeleton siliceous (with discrete megasclere and microsclere spicules), fibrous (spicules replaced by spongin fibres or occur in conjunction with them), calcareous (massive or basal calcareous structures occur with siliceous spicules and spongin fibre), or absent (matrix collagen provides the only support); megascleres are monaxon, tetraxon or triaxon; microscleres are of diverse types (e.g. *Chondrilla, Haliclona, Ircinia, Tethya*)

Subclass Homoscleromorpha
Megascleres (when present) triactines with diactine and tetractine modifications; all spicules very small, with no regional organisation; some genera lack spicules; embryos are incubated; larvae blastulae.

Subclass Tetractinomorpha
Megasclere spicules tetraxonid and monaxonid, together or separately, organised usually in a recognisable pattern which is either radial or axial; microscleres most frequently asterose, but sigmas and raphides also occur; typical reproductive pattern oviparous; larvae usually parenchymellae but blastula larvae occur in some genera, e.g. *Cliona* and *Polymastia*.

Subclass Ceratinomorpha
Megascleres diverse but always monaxonid; microscleres generally sigmoid or chelate, never asterose; spongin a universal component of the skeleton; skeletal organisation reticulate or dendritic; typical reproductive pattern is viviparous; parenchymella larvae; in two orders reproduction is oviparous, but larval type is unknown.

Relationships within the Porifera
Without a consideration of fossil forms it is difficult to comment on relationships within the phylum. The important events in sponge evolution took place in the Pre-Cambrian, more than 550 million years ago. By the early to mid Cambrian (550–525 mya), representatives of all recent classes were present, and the modern fauna was established by the late Mesozoic (100 mya). Patterns of numerical dominance and preferred habitats have changed over time, but the major groups have remained. Each class has thus a long separate history. Relationships within the classes are well reflected in the subclass arrangement. More detailed discussions on relationships within the Porifera can be found in the works listed in the Bibliography.

Chapter 3

The Cnidaria and Ctenophora

R.T. Hinde

Introduction 29
   Diploblastic structure 29
   Radial and biradial symmetry 29
   Differences between the Cnidaria and Ctenophora 30
   Cnidaria and Ctenophora as metazoans 30

PHYLUM CNIDARIA

Introduction 31
Polyp and medusa 31
Diversity 32
   Adding size and complexity to the diploblastic body plan 33
   Cnidocytes 35
   Locomotion 37
   Feeding 39
   Respiratory exchange 40
   Sense organs 41
Evolutionary history 41
Physiological processes 42
   Digestion 42
   Nutrition 43
   Circulation and excretion 43
   Control systems 43
   Regeneration 46

Reproduction and development 46
   Asexual reproduction 46
   Alternation of sexual and asexual reproduction 46
   Reproductive system 48
   Gametes and fertilisation 48
   Embryonic development 50
   Larval development 50
   Metamorphosis 50
Coral reefs 52
   Structure of coral reefs 52
   Roles of cnidarians in reef formation 53
   Importance of symbiosis in reef growth and productivity 53
Classification of the phylum Cnidaria 53

PHYLUM CTENOPHORA

Introduction 54
Morphology 55
Diversity 55
   Colloblasts 55
   Locomotion 55
   Feeding 56
   Bioluminescence 56
   Sense organs 57

Physiological processes 57
   Digestion 57
   Respiration, circulation and excretion 57
   Nervous system 57
Reproduction and development 57
Classification of the phylum Ctenophora 57

Introduction

The Cnidaria and Ctenophora were formerly placed together in a single phylum (Coelenterata) because of a number of similarities between them. These include being radially or biradially symmetrical, having only one major opening to the digestive cavity, having a simple net-like nervous system and having only two primary cell layers with a secreted layer of material between them. But recent research has shown that there are some important differences between these groups, leading to their being separated into two phyla, although they are still referred to collectively as 'coelenterates'. The Cnidaria will be the main topic of this chapter, as they are much more diverse and widespread than the Ctenophora, and are of major importance in the construction of coral reefs.

Diploblastic structure

Coelenterates have epithelia composed of cells which are tightly bound together by intercellular junctions and supported by a basement membrane containing collagen. Embryonic Cnidaria and Ctenophora have only two layers of cells — the outer ectoderm and the inner endoderm. In post-embryonic stages the body retains two epithelia: the outer epidermis, derived from the ectoderm; and the inner gastrodermis, derived from the endoderm. The epithelia also contain some non-epithelial cells (e.g. nerve cells and either cnidocytes or colloblasts) derived from the two embryonic layers. These cells may lie above or below the epithelial cells, so that there may be more than one layer of cells in places. The animals discussed in later chapters all have three layers of cells in their embryos: an outer ectoderm, an inner endoderm, and an intervening mesoderm (see page 59). Because they have only two primary cell layers, the Cnidaria and Ctenophora are said to be diploblastic, in contrast to the three-layered or triploblastic animals.

The two epithelia secrete a gelatinous layer, the mesogloea, which lies between them, holds them together and provides support for the body. In larger species the mesogloea may be thick and gelatinous or fibrous, and may contain cells which migrate into it from the primary cell layers. Some of these cells may secrete skeletal material which provides additional support.

The lack of a mesoderm and the sheet-like nature of the epithelial layers limits the development of complex organs in cnidarians and ctenophorans. While both phyla have fairly well-differentiated tissues, they tend to have fewer cell types than in triploblastic animals. Most species lack organs altogether, although large medusae have simple sense organs. Because of this, these two phyla are often said to be at the 'tissue grade' of organisation.

Radial and biradial symmetry

Cnidaria are primarily radially symmetrical, having body parts arranged symmetrically about a central, longitudinal axis, so that all longitudinal sections which pass through this axis are similar to each other (Fig. 3.1a). In fact, in most Cnidaria there is a limited number of radii (the perradii) about which the structure of the body is truly symmetrical. Some Cnidaria are biradially symmetrical, but this is a secondary condition. The Ctenophora are biradially symmetrical from early in their embryonic development. In the biradial forms, most structures are radially arranged but some are either paired or symmetrical about a particular longitudinal axis. Because of this, the body is symmetrical about two diameters only (Fig. 3.1b). Triploblastic animals are bilaterally symmetrical (Fig. 3.1c; see Chapter 4).

Radial symmetry allows equal opportunities for food gathering and defence in all directions around the body, but does not favour the differentiation of a specialised head (cephalisation) or of a favoured direction for movement.

Differences between the Cnidaria and Ctenophora

In the Cnidaria, individual cells have a single flagellum, whereas the cells of Ctenophora have large numbers of cilia. Unique stinging cells called cnidocytes are characteristic of Cnidaria but are not found in Ctenophora, which have unique cells of another type, the adhesive colloblasts. The Ctenophora swim by means of arrays of fused cilia arranged in eight rows, while the Cnidaria move by means of muscle contraction. Cnidaria lack true muscle cells. Their muscle fibres are always extensions of an epithelial cell. Ctenophora have true muscles. Unlike Cnidaria, the Ctenophora have gonoducts and gonopores by which gametes leave the body. There are also fundamental differences in development between the two groups.

Cnidaria and Ctenophora as metazoans

Sponges, as we have seen in Chapter 2, are metazoans lacking true symmetry and true tissues. The Cnidaria and Ctenophora are diploblastic, radially symmetrical metazoans with tissues, but have no organs (or, rarely, very simple organs). The remaining metazoans are triploblastic, primarily bilaterally symmetrical animals with tissues grouped into organs. The Cnidaria and Ctenophora are the most primitive extant phyla to have epithelia with basement membranes, a nervous system, an enteron and a mouth. Cnidae are unique to the Cnidaria and colloblasts to the Ctenophora, and each phylum has many other specialised features. Although it has been argued in the past that the coelenterates may have originated independently of other multicellular animals, recent gene sequence data refute this model (see Chapter 18). The contrasting features of cnidarians and ctenophores suggest that they diverged early in metazoan evolution.

Figure 3.1
Animal symmetry. **(a)** Radial symmetry. The body has a longitudinal axis, and all structures are arranged symmetrically about this axis. A longitudinal section along any diameter produces two similar halves. **(b)** Biradial symmetry. The body has a longitudinal axis and one diameter (the sagittal axis) that differs from the others, so that only two longitudinal planes of section produce similar halves. These are the sagittal axis (A–B) and the transverse axis (C–D).
(c) Bilateral symmetry. The dorsal and ventral surfaces of the animal are different. Only one longitudinal plane of section, the sagittal axis (A-B), produces similar (left and right) halves.

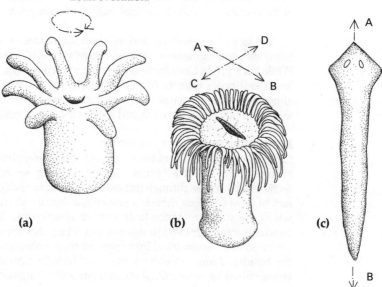

Phylum Cnidaria

Diploblastic metazoans with radial symmetry; tissues, but rarely organs; two basic body forms (polyp and medusa); ectodermal and gastrodermal epithelia, including epitheliomuscular cells, separated by mesogloea; single body cavity, mouth but no anus; nervous system a nerve net; no respiratory, circulatory or excretory system; cnidae, secreted by cnidocytes; sexual and asexual reproduction, planula larva.

Introduction

The Phylum Cnidaria includes many familiar animals, such as jellyfish, bluebottles, sea anemones and corals. Its members are diploblastic, primarily radially symmetrical animals with tentacles, a single body cavity with only one opening, and specialised stinging cells. Cnidarians have a pool of undifferentiated interstitial cells that enables them to regenerate after injury. The body cavity functions in digestion, absorption, gas exchange, excretion and circulation and, when the mouth is closed, as a hydrostatic skeleton. It is known as the gastrovascular cavity (but also called the enteron or the coelenteron). The mouth is used to take in water and food and to release wastes. The stinging cells (cnidocytes) are the main means of prey capture and defence. The name of the phylum, which is derived from the Greek word for nettle ('knide'), reflects this unique characteristic.

All Cnidaria are aquatic. Most of the approximately 9000 species are marine, but there are freshwater species. Cnidaria may be planktonic or benthic. There are two major body forms: the polyp (e.g. sea anemone; Fig. 3.2a,b); and the medusa (e.g. jellyfish; Fig. 3.2c). In general, polyps are benthic and medusae are planktonic. Many cnidarian species have a planktonic medusa and a benthic polyp stage in their life cycles. The typical cnidarian larva, the planula (Fig. 3.2d), may be solid or hollow.

Polyp and medusa

Both these forms consist of the two basic cell layers and the mesogloea, folded to form a cylindrical (polyp) or bowl-shaped (medusa) structure, with a mouth at the apex of the cylinder or in the centre of the flatter side of the bowl (Fig. 3.2). The mouth is ringed by tentacles which are used in feeding and defence and sometimes for locomotion. The epidermis covers all the outer surfaces, while the mouth opens into a gastrovascular cavity lined by the gastrodermis.

Cnidarian orientation is described in terms of the oral–aboral axis (Fig. 3.2). Polyps are normally attached to the substratum by the end opposite the mouth (the aboral end). This may be knob-like for attachment in sand or mud, an adhesive basal disc for attachment to hard surfaces, or a series of root-like branches called stolons. In the class Anthozoa, the oral end is flattened to form an oral disc around the mouth, but in other polyps the mouth opens at the tip of a conical extension called the hypostome. In anthozoan polyps, a sleeve-like, muscular pharynx extends into the gastrovascular cavity as an inverted extension of the column wall, and the true mouth is at the inner end of the pharynx (Fig. 3.2b). The wall of the

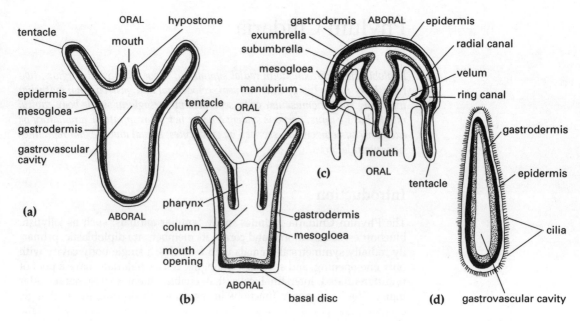

Figure 3.2
Cnidarian body forms.
(a) Hydrozoan polyp, with a terminal mouth on the hypostome. (b) Anthozoan polyp, with the body surface invaginated to form a pharynx, with the true mouth at the inner end. (c) Medusa, with shortened oral–aboral axis and thick mesogloea in the bell.
(d) Planula larva, showing epithelia and thin layer of mesogloea. (After Bayer and Owre, 1968.)

gastrovascular cavity may be folded to form mesenteries which increase its surface area. In small polyps the mesogloea is thin, but in larger ones it is thickened, supporting the thin layers of cells. Polyps extend the body column and tentacles into the water for feeding, but can retract them when disturbed.

In medusae the bell or umbrella is formed of thick, gelatinous mesogloea which makes it buoyant. The convex aboral surface is known as the exumbrella and the flat oral surface as the subumbrella (Fig. 3.2c). The subumbrella surface may be partially closed by a shelf-like muscular ingrowth, forming a velum or, in a different form, a velarium. The tentacles are on the rim of the bell. The mouth lies at the end of a tubular manubrium, which is lined with gastrodermis and may be extended to form tentacle-like oral arms. The mouth opens into a sac-like gastric cavity which may be divided into four gastric pouches. From this central cavity, radial canals extend to the edge of the bell, where they join a peripheral ring canal. Medusae normally swim or float with the mouth downwards and the tentacles and oral arms trailing in the water (Fig. 3.2c).

In cnidarian species which have both polyp and medusa stages in the life cycle, the medusae reproduce sexually and the polyps asexually. This is generally thought to be the primitive condition in Cnidaria. In many groups of Cnidaria, one form or the other has been lost and the remaining form reproduces sexually, whether it is a polyp or a medusa. Both forms may also reproduce asexually by fission or by budding.

Diversity

Cnidarian diversity is related to two major factors, the utilisation of both benthic and pelagic habitats, and the achievement of large body size in spite of the apparent limitations of the diploblastic body plan.

There are four classes in the phylum. In the class Hydrozoa (e.g. the tiny freshwater polyp *Hydra*, the colonial hydroid *Obelia*, the hydrozoan coral *Millepora* and the bluebottle *Physalia*) both polyps and medusae have relatively simple internal structures, without mesenteries. Many species grow as polymorphic colonies (Fig. 3.6). The class Scyphozoa includes the common jellyfish (e.g. *Catostylus* and the moon jelly *Aurelia*) and the class Cubozoa contains the box jellyfish (e.g. the sea wasp, *Chironex fleckeri*). In both these classes the medusa is dominant in the life cycle. The body cavity is divided by mesenteries and, if the polyps are colonial, the colonies are never polymorphic (Fig. 3.7). Both polyps (benthic) and medusae (pelagic) occur in the life cycles of most species of Hydrozoa, Scyphozoa and Cubozoa.

Members of the class Anthozoa (e.g. hard corals, sea anemones, gorgonians and soft corals) are benthic and have polyps only in the life cycle. All of them have mesenteries which increase the internal surface area. Many anthozoan species are colonial, and colonies may be monomorphic or polymorphic (Fig. 3.8).

Adding size and complexity to the diploblastic body plan

Cnidaria do not have specialised organs for respiration, excretion or digestion. These functions take place across the epidermis and gastrodermis. But in larger animals there is less surface per unit volume of cells for these processes, and the volume of the gastrovascular cavity increases faster than the surface area of its gastrodermal lining. Because of the absolute dependence of Cnidaria on diffusion, it might be expected that they would all be very small; yet there are jellyfish with bells more than 1 m in diameter, and polyps up to 1 m long. The size of the body is increased by thickening of the mesogloea and sometimes by having a skeleton, but not by increasing the thickness of the tissue layers. A bigger increase is required in the surface area of the cell layers, particularly the gastrodermis, than would follow as a direct result of the increase in the radius of the body. An increase in size also requires greater support for the thin epithelia at the body surfaces.

Skeletal systems and body size

In all large cnidarians the body consists mainly of mesogloea, which may be tough and fibrous or watery and gelatinous with a tougher outer membrane. As well as providing bulk, the mesogloea acts as a deformable skeleton. This is best seen in medusae, in which muscle contractions cause the bell to narrow, expelling water and propelling the medusa forward. The elasticity of the mesogloea in the bell restores its shape when the muscles relax. Since the mesogloea of medusae contains about 95% water, much of it in pockets within an organic matrix, this elasticity is largely hydrostatic; however, there are also elastic fibres in the mesogloea. When the mesogloea is bulky, it usually contains living cells which migrate into it from the two primary cell layers. Their functions include transport of nutrients and wastes, phagocytosis of invading organisms, and secretion of the mesogloeal matrix. In some cnidarians these cells also secrete skeletal elements which remain in the mesogloea, adding to its bulk and its strength.

All cnidarians use the fluid in the gastrovascular cavity as part of the hydrostatic skeleton and to add bulk; when the gastrovascular cavity is full, the body is inflated to its maximum size. In most polyps, the water in the gastrovascular cavity provides more bulk than the mesogloea, and with the mouth closed the gastrovascular cavity is the main element in the hydrostatic skeleton. This method of adding to size allows an animal to move rapidly if it is disturbed; the mouth is opened and muscles are contracted to expel the gastrovascular fluid, causing the body to collapse.

In Cnidaria there may be no skeleton (apart from the hydrostatic skeletal system); or there may be an internal skeleton secreted by the gastrodermis, as in gorgonians (Fig. 3.8d), or the mesogloea, as in soft corals (Fig. 3.8e); or there may be an external skeleton secreted by the epidermis, as in hydroids and hard corals. External skeletons may be secreted by both the upper and lower epidermis, enclosing the body completely except for pores through which polyps can extend, as in hydrocorals (Fig. 3.6e), or by the lower epidermis only, so that the tissue lies on top of the skeleton, as in scleractinian corals (Fig. 3.8c). Skeletons may be organic or inorganic. Organic skeletons may consist of chitin, as in hydroids (Fig. 3.6a), or of tanned protein, as in gorgonians (Fig. 3.8d). Inorganic skeletons are always calcareous, consisting of calcium carbonate laid down as spicules (e.g. soft corals) or large, solid masses (e.g. hard corals). Cnidarian skeletons provide support and protection and lift the animal higher into the water column.

Increasing the surface area to volume ratio

In both polyps and medusae the body wall may be folded, dividing the gastrovascular cavity into a series of inter-connected spaces. In medusae this division results in an orderly array of canals, lined with gastrodermis and separated by thickened mesogloea; the larger the medusa, the larger the number of canals (Fig. 3.7a). Being tubular, the canals have a large surface area to volume ratio, allowing food to come into contact with the digestive cells of the gastrodermis more frequently than in a single large chamber. In polyps, the body wall may be folded into a series of thin partitions, each containing an extension of the mesogloea and covered with gastrodermis (Fig. 3.8a). These structures are called mesenteries or septa; mesentery is preferable, as the term 'septa' is also used to describe the mineralised supporting blades in the skeletons of hard corals (Fig. 3.8c). Mesenteries extend from the column wall towards the centre of the gastrovascular cavity but do not meet in the middle, allowing the fluid in the cavity to circulate freely. In the class Anthozoa, the pharynx also adds to the area of the gastrodermis.

The primary or complete mesenteries are the first to develop and are the largest; they fuse with the wall of the pharynx, but have free edges below it. There may be openings in the upper parts of these mesenteries which improve water flow in the top of the gastrovascular cavity. The other, incomplete, mesenteries develop in pairs, between the pairs of primary mesenteries, and never fuse with the pharynx. The next widest mesenteries are the secondary mesenteries, and so on. Anthozoan mesenteries usually taper towards the bottom of the column (Fig. 3.8a). Surface area may be further increased by the development of extensions of the gastrodermis; for example, the gastric filaments of jellyfish (Fig. 3.7a, g) and the mesenterial filaments and acontia of Anthozoa (Fig. 3.8a, b). Mesenterial filaments are cord-like thickenings of the edges of the mesenteries which are involved in digestion. Usually the filaments have three lobes: a central lobe consisting of gland cells and cnidocytes, and two outer lobes of ciliated cells. The filaments may extend beyond the base of the mesentery as long, thread-like acontia. These filaments are also important in digestion, and can be extruded through the mouth or through openings in the body wall (cinclides) for prey capture and defence.

Increasing size by colony formation

Many cnidarian species form colonies in which a number of small individuals cooperate, so that they function much like a large animal. Colonies may consist entirely of polyps or of a mixture of polyps and medusae. All members share a common gastrovascular cavity and thus can share food. Because the individual members of the colony (called zooids) are small and the interconnecting parts of the gastrovascular cavity are flattened or tubular, the colony as a whole maintains a favourable surface area to volume ratio. Benthic colonies may spread out over the substratum, or grow up into the water column. There are also planktonic cnidarian colonies, such as the bluebottle, *Physalia*.

In monomorphic colonies (e.g. reef-building corals) the individuals are all identical, and all can feed, defend themselves and reproduce. *Physalia*, in contrast, is a polymorphic colony with several types of individual, each with a particular function. A modified medusa acts as a float, and specialised polyps are involved in catching prey (dactylozooids), digestion (gastrozooids) and reproduction (gonozooids) (Fig. 3.6d). Many benthic hydroid colonies are also polymorphic (Fig. 3.6a). Generally, dactylozooids are reduced polyps with one or more long tentacles but no mouth. Gastrozooids may have reduced tentacles, or none at all, but have a mouth and a large gastrovascular cavity. Gonozooids lack a mouth and are the site of development of medusae or of gonads. Polymorphic colonies are the most complex body form in cnidarians, and the various individuals are often considered to be analogous to organs. Simpler colonies, such as corals, may continue to add individuals, and thus to grow larger, indefinitely.

Cnidocytes
Structure and function

Cnidocytes (also called cnidoblasts or nematocytes) are the cells which secrete and house the cnidae; each cnidocyte contains a single cnida, consisting of a sac containing a coiled, usually hollow, eversible thread. These structures are unique to the Cnidaria. The cnida is secreted by the Golgi body of the cnidocyte and is the most complex secreted structure known. The sac wall and thread are composed of collagen-like protein. The wall of the thread is an extension of the sac wall, so that the structure is analogous to a glove with a finger inverted, except that the thread is often many times longer than the capsule and is coiled around itself (Fig. 3.3a, b). The cnida is filled with a fluid which contains toxins or sticky secretions (or both). The capsule often has a hinged operculum or three flaps covering the area where the thread will emerge. The cnidocyte may also bear a sensory cnidocil, which is a modified flagellum, often surrounded by microvilli. When the cnidocil is stimulated by food or predators, the cnida fires and the thread everts, penetrating or wrapping around the prey (or predator). Cnidae which wrap around prey are generally long and often bristly, while those which penetrate have pointed tips and are usually barbed (Fig. 3.3c). The hollow thread transmits the fluid from the inside of the capsule. Sticky secretions help secure the prey and toxins immobilise or kill it. Cnidae can be used only once. Prey is brought to the mouth and the cnidae are shed before it is ingested, releasing the prey from the tentacle. Attacking predators are also killed or irritated by cnidae and are eventually released. The cnidocytes differentiate from interstitial cells which migrate during or after differentiation to replace those which have been used.

Cnidae are found throughout the epidermis, but are especially common on the tentacles; they are also found in the gastrodermis in some groups of Cnidaria. On tentacles, and often on the rest of the body, the cnidae are

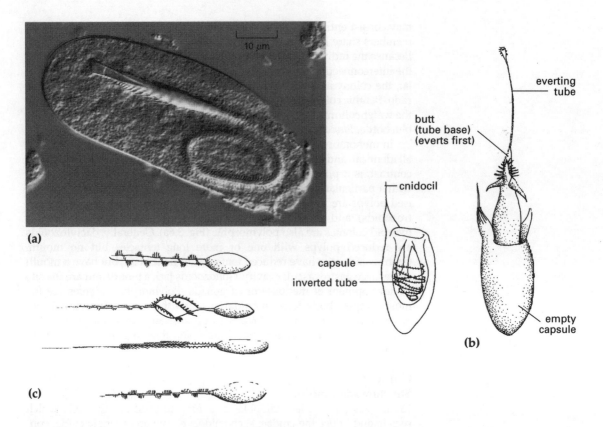

Figure 3.3
(a) Micrograph of an undischarged cnida, a macrobasic p-mastigophore from the coral *Goniopora tenuidens* (scale bar = 10 μm).
(b) The firing of a cnida. The butt of the tube everts first (after Brusca and Brusca 1990).
(c) Cnidae of various types (after Hyman 1940).

grouped in batteries which may contain a range of types of cnidae. In *Hydra*, each battery consists of an epithelial cell with pockets on its surface housing 10 to 20 cnidocytes, a nerve cell, and in some cases a sensory cell. The cells are all connected by intercellular junctions and synapse with the nerve cell. In addition, extensions of the nerve cell make contact with the sensory cells of other batteries. It is likely that stimulating one battery changes the threshold stimulus needed to fire the surrounding batteries. In the box jellyfish, *Chironex fleckeri*, cnidocytes are surrounded by seven to nine accessory cells. These cells and the cnidocytes are connected by radial fibres which can lift the cnida so that its cnidocil is exposed (preparing the cnida for firing), or lower it so the cnidocil is withdrawn into the cnidocyte.

There are three main categories of cnidae: nematocysts, spirocysts and ptychocysts. Nematocysts have been classified into about 27 types, and most species have several types. The pattern may vary with the stage in the life cycle, but is constant for any one species at a given stage. Since different taxa of Cnidaria have different types of cnidae, they form a useful taxonomic character (Fig. 3.3c). Nematocysts may bear spines and can usually inject a toxin into the prey. Spirocysts and ptychocysts are found only in the class Anthozoa. Spirocysts lack spines but are covered with a close, regular array of smaller, interconnected tubules, which stick to the prey without penetrating it or delivering venom. Ptychocysts are found in only one order of Anthozoa, the burrowing Ceriantharia. During burrowing, a tube is built up from the intertwined, everted threads of the ptychocysts.

The firing of cnidae is a complex process. In most cases both a chemical and a mechanical stimulus are required; this prevents cnidae from discharging when prey are out of reach or when the cnidocytes brush against inert objects. However, a very powerful mechanical or chemical stimulus delivered alone can cause firing. For example, cnidae will often fire if the surface of the tentacle is abraded. The animal's nervous system can modify the threshold needed for discharge, and may be able to cause firing without an external stimulus. Effective chemical stimuli include organic molecules, such as mucus and certain amino acids.

Stimulation of the cnidocil causes the operculum to open, and the invaginated thread then everts; in *Hydra*, the mean speed of movement of the base of the thread is 2 m/s. The mechanism of firing is still not understood, and at least four models have been proposed to account for it:

1. The osmotic model suggests that the permeability of the capsule changes when the cnidocyte is stimulated, allowing water to enter and increasing the internal pressure, leading to eversion of the thread.
2. Another model proposes that the contraction of the network of fibres surrounding the cnida capsule may cause firing.
3. It has been suggested for the sea wasp, *Chironex fleckeri*, that rapid polymerisation of the capsule contents may cause them to swell, leading to eversion.
4. It has also been proposed that the thread of the cnida is under tension, and that a sudden release of this tension drives the rapid eversion of the thread.

Given the diversity of cnidae, it is possible that different types function differently; their complexity and small size have so far defied definitive tests of these models.

Functions of cnidae

Cnidae are involved in feeding, defence, locomotion and attachment. They are used to capture and subdue prey before it is ingested. Cnidae in the gastrodermis help to hold and kill prey that have been ingested alive. There are three classes of toxins — neurotoxins, cytolysins (cellulolytic peptides, which cause pain) and proteinases, which inhibit blood clotting — each with many variants. Species which feed on fish have toxins that are particularly effective against vertebrates; these include the bluebottle *Physalia*, which can cause a very painful sting in humans, and the sea wasp *Chironex fleckeri*, which has caused human deaths. Many other Cnidaria can deliver an irritating or painful sting. The toxins which kill prey can also kill small predators and deter larger ones, so that cnidae are the main means of defence.

Some benthic polyps can move by using the cnidae in their tentacles to grip the substratum. When cnidarian larvae settle from the plankton, they may initially use cnidae to attach themselves.

Locomotion

While many benthic cnidarians, particularly colonial ones, are sessile and live permanently attached to the substratum, others can move slowly. Many polyps can move by 'crawling' with the basal disc, by which they are normally attached. Small polyps, such as *Hydra* and some anemones, can move by 'stepping' or 'looping'. The tentacles attach themselves to the substratum with the help of cnidae, and the base of the polyp is then brought close to the tentacles ready for another step, or flipped over the tentacles. Some polyps burrow in sand or mud.

Swimming is a more common method of locomotion in cnidarians. Among primarily benthic groups, certain species of anemones (e.g. *Phlyctenactis tuberculosa*) can swim either by 'rowing' with their tentacles or by strong bending movements of the column. Among the pelagic Cnidaria, all solitary medusae and all colonies which contain medusae with bells can swim. Because they are not strong swimmers, pelagic cnidarians cannot overcome ocean currents and are essentially planktonic. They use their swimming ability to catch prey and to move within the water mass carrying them. Medusae swim by contracting the bell so as to expel water from the subumbrella surface, thus propelling themselves upward or forward. The muscular contractions involved are highly rhythmical and are controlled by aggregations of nerves in the bell, which act as pacemakers for the muscles. The bell typically contracts 20 to 100 times per minute. In some medusae, the muscular, shelf-like ingrowth of tissue forming the velum (Hydrozoa) or velarium (Cubozoa) increases the rate of water flow during contractions (Figs 3.2c, 3.7g). The tilt of the bell determines the direction of travel. Statocysts at intervals around the margin of the bell enable the medusa to sense and control the degree of tilt. Some pelagic, colonial cnidarians (e.g. *Physalia* and *Velella*) lack swimming bells but secrete chitinous floats which project above the surface of the water and act as sails. In *Physalia* the float is filled with a mixture of nitrogen and carbon monoxide secreted by the gastrodermis.

All these forms of movement depend on muscular contractions; the mesogloea and the fluid in the gastrovascular cavity act as the antagonists to the muscle fibres. When the gastrovascular cavity is the major source of hydrostatic skeletal function, movement is possible only when the mouth is shut, as hydrostatic pressure is lost when the mouth opens. Although many cnidarians have mineralised skeletons, these never play a role in locomotion. Perhaps this is because they are usually massive in comparison with the animals themselves, and thus are found only in sessile species.

The muscles of Cnidaria are formed from sheets of myofibrils, which are extensions of the bases of epitheliomuscular cells and contain contractile fibres; the muscles are, therefore, extensions of either the gastrodermis or the epidermis (Fig. 3.4b). Primitively, the epitheliomuscular cells remain part of the general epithelium and the myofibrils are anchored to the surface of the mesogloea, so that their contractions change its shape. When the muscles relax, the elasticity of the mesogloea and the pressure in the gastrovascular fluid cause the animal to return to its resting shape. In some cases the epitheliomuscular cells sink below the rest of the epithelium, forming a second layer on the surface of the mesogloea. Some cnidarians have mesogloeal muscles that are formed by the migration of the epitheliomuscular cells into the mesogloea, so that the muscles lose contact with the epithelium. Many Cnidaria have large blocks of muscle, formed by extensive folding of the mesogloea and the associated epitheliomuscular cells. There are lateral and longitudinal intercellular junctions between myofibrils, as well as synapses between myofibrils and neurons. Hydrozoan polyps probably have smooth muscle fibres only, but hydrozoan medusae and the other classes of Cnidaria are the simplest animals to have both striated and smooth muscle fibres.

The arrangement of muscle fibres is very simple in small hydrozoan polyps, such as *Hydra*. The epidermal myofibrils all run longitudinally along the mesogloea and the gastrodermal myofibrils run around the column. Contracting the longitudinal fibres shortens the column, while con-

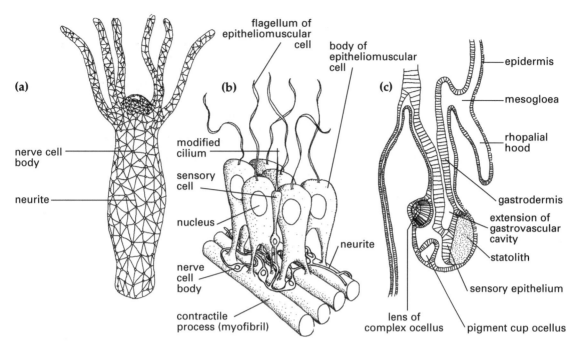

Figure 3.4
(a) Nerve net of *Hydra*.
(b) Epitheliomuscular cells of *Hydra*, showing innervation (modified from various sources).
(c) Rhopalium of cubozoan, showing statocyst and ocelli (after Bayer and Owre 1968).

tracting the circular fibres causes it to become longer and thinner. More complex movements are possible through simultaneous contractions of parts of each of these muscle fields. Tentacles always have longitudinal fibres but may lack circular fibres. In larger and more complex polyps there are radial muscles in the hypostome or the oral disc. These are used in feeding and for movements of the upper part of the column. Anthozoan polyps have rather diffuse, circular epidermal musculature and well-developed longitudinal gastrodermal muscles, particularly in the mesenteries. The retractor muscles lie longitudinally on one side of some or all mesenteries and are used to shorten the column. They can contract strongly enough to flatten the polyp against the substratum or to retract it into the skeleton for protection. Medusae have sheets (Hydrozoa) or bundles (Scyphozoa and Cubozoa) of muscles associated with the subumbrellar epidermis, but there is little or no musculature in the exumbrellar epidermis. Medusae use both circular and radial muscles in swimming and depend on the elasticity of the mesogloea as antagonist. The velum of hydrozoan medusae contains circular muscles and the cubozoan velarium has radial muscles; in both cases these structures contract at the same time as the bell, helping expel water through the narrowed aperture they themselves provide. The net effect is to increase the rate at which water is expelled from the bell. Longitudinal muscles predominate in the manubrium, oral arms and tentacles.

Feeding

Cnidaria are often said to be exclusively carnivorous feeders, and many are. However, some are suspension feeders and many species have intracellular symbiotic algae which provide them with organic nutrients. There is also evidence that many Cnidaria absorb and utilise dissolved organic matter. In most cases these sources of nutrition are combined with predation, but some Cnidaria (e.g. soft corals such as *Heteroxenia* and *Leptogorgia*) apparently depend almost entirely on their symbiotic algae and on the uptake of dissolved organic matter from the water.

Predatory Cnidaria use the tentacles to catch prey. The cnidae entangle or stick to the prey, preventing its escape, and those with toxins may stun or kill it. Planktonic colonies often have very long trailing dactylozooids which contract strongly to bring captured fish and zooplankton to the mouth. In *Physalia*, individual dactylozooids may be up to 20 m long. In medusae the true tentacles around the margin of the bell are often quite short, but the manubrium is usually extended to form tentacle-like 'oral arms' which may be involved in prey capture and transport of food to the mouth (Fig. 3.7a). These vary in form but may be long and frilled, and sometimes branched, giving them a large surface area. Food captured on the subumbrella or marginal tentacles is picked up by the oral arms, which may then be wiped across the manubrium, transferring the food to the mouth. Small food particles may be transported along the arms to the mouth by currents created by flagella. As in polyps, suspension feeding may be combined with predation, or may even be the only source of particulate food. Medusae may trap prey or suspended organic matter by swimming upwards, then spreading their oral arms and sinking through the water column.

Predatory polyps extend their tentacles upwards and outwards into the water column, leaving them expanded throughout the feeding period (which is often determined by the time of day or tide). When the tentacles have captured food they bend towards the mouth, which may be raised so that the food can be ingested. If there is a pharynx, it is muscular and helps hold and swallow the prey. Chemicals released by injured prey stimulate exploratory tentacle movements and the opening of the mouth, as well as providing the chemical stimulus which is necessary if cnidae are to fire on contact with prey. Amino acids, reduced glutathione and small peptides are the most important of these feeding stimulants, and amino acids also induce movements of the tentacles and manubrium in Cubozoa. In hydroid polyps (e.g. *Hydra*) a continuous sheet of epithelium is formed whenever the mouth closes. During feeding, muscular contractions stretch the epithelium of the hypostome until the intercellular junctions break. The mouth takes at least a minute to form.

Polyps may supplement their animal prey by ingesting particles that fall on the tentacles, or may be specifically adapted for suspension feeding. The commonest adaptations are increased numbers of tentacles and the production of mucus nets, which entangle organic particles and are then withdrawn and ingested, carrying the trapped particles with them.

Respiratory exchange

Cnidarians have no specialised respiratory organs. Gas exchange takes place across all surfaces, and the metabolic rate is potentially limited by the surface area to volume ratio (see page 34). However, mass-specific oxygen consumption is lower than in triploblastic animals of similar mass, due to the large contribution of the metabolically inactive mesogloea to body mass in Cnidaria.

The water in the gastrovascular cavity must be expelled at intervals in order to get rid of carbon dioxide which has diffused into it and to bring in oxygenated water. When the water is expelled, any food that has not yet been phagocytosed is also likely to be expelled. This is a major disadvantage of having a single body cavity with a single opening. Some Anthozoa have small ciliated grooves (siphonoglyphs) on one or both sides of the pharynx, so that ciliary currents through the siphonoglyphs can pump water in and out of the gastrovascular cavity without the mouth being opened. This improves gas exchange after feeding, and also allows the animal to control the pressure of the gastrovascular fluid without ejecting undigested food.

Many cnidarians have intracellular symbiotic algae, which produce oxygen during photosynthesis. This may lead to a build-up of oxygen in the cells or the gastrovascular fluid to potentially toxic concentrations. These species have particularly high levels of enzymes which prevent oxygen toxicity by destroying oxygen free radicals (e.g. superoxide dismutase, catalase, peroxidase) or of biochemical antioxidants (e.g. uric acid).

Sense organs

Simple and sessile cnidarians have no sense organs, but they do have sensory cells in both epithelia which perceive light, chemical stimuli or mechanical stimuli. These receptors are often structurally similar to those of vertebrates. They have one or more modified cilia forming a hair-like projection which protrudes into the water. The sensory cells synapse with nerve cells, allowing the animal to make a generalised response to stimuli rather than simply responding at the site of the stimulus.

Medusae and complex motile colonies (Siphonophora and Chondrophora) have more complex sense organs: the statocysts detect the degree of tilt of the body and the ocelli are light receptors. Statocysts vary in structure, but each essentially consists of a calcareous granule, the statolith, and one or more receptor cells with sensory cilia (Fig. 3.4c). As the body tilts, some of the statoliths come into contact with the sensory cilia, which relay impulses to the nerve net. The animal can then right itself by increasing the strength of the contractions of the bell on the low side. Ocelli consist of patches of light-sensitive cells, associated with pigmented cells which allow light to reach the receptors from a limited range of directions. This in turn allows the animal to orient itself to light. Cnidarian ocelli range from patches of photoreceptors alternating with pigment cells, to complex structures in which the light receptors have a cup-shaped shield of pigmented cells behind them and are covered by a lens formed from cytoplasmic extensions from neighbouring cells, as in Cubozoa (Fig. 3.4c). These lenses intensify the light falling on the receptors, but probably do not form images. Because statocysts and ocelli consist of more than one type of tissue (nervous and epithelial), they are regarded as simple organs.

In Scyphozoa and Cubozoa the ocelli and statocysts are located in the rhopalia, which are tubular extensions distributed regularly around the rim of the bell. In hydrozoan medusae, the statocysts and ocelli are on the exumbrella and are associated with the tentacular bulbs at the bases of the tentacles.

Chemoreception is also important to cnidarians. The presence of potential prey in the water causes them to start searching behaviour.

Evolutionary history

The oldest undisputed cnidarian fossils come from the fossil fauna of the Ediacara region of South Australia and date from the late Pre-Cambrian. These include hydrozoan and scyphozoan medusae, hydrozoan polyps with chitinous skeletons, and sea-pens (Pennatulacea). It seems likely that the cnidarian classes and the anthozoan subclasses Octocorallia and Hexacorallia diverged during the late Pre-Cambrian. The anemones first appear in the middle Cambrian, gorgonians and early corals in the early Ordovician, and hydrozoan corals in the late Cretaceous. There are several extinct coral-like orders, of which the Rugosa (Ordovician to late Permian) and Tabulata (Ordovician to Permian) were the most important. Together with other animals, these gave rise to early limestone reefs similar in many ways to modern coral reefs. Because coral skeletons often show diurnal, monthly and annual banding, coral fossils are used in research on changes in the Earth's rate of rotation and in palaeoclimatological research.

Physiological processes

Digestion

Digestion starts in the gastrovascular cavity, but once the food is reduced to particles small enough to be taken up by the digestive cells of the gastrodermis, digestion is completed intracellularly. Cnidaria, unlike sponges, can ingest large food items, but the extracellular phase of digestion is relatively short. This is important because, with only one opening to the gastrovascular cavity, most Cnidaria cannot feed again, remove wastes, or take in oxygenated water while food is being digested in the gastrovascular cavity.

In general, glandular cells in the gastrodermis around the mouth secrete a mixture of enzymes and mucus; the mucus is thought to protect the tissues from abrasion by the prey. For example, in Hydrozoa, the glandular cells secrete mucus, acid and alkaline phosphatases, proteases and peptidases. Cells deeper in the gastrovascular cavity also release these secretions, together with lipases and esterases. Some species produce chitinases which break down the skeletons of crustaceans. The particles and soluble molecules produced by extracellular digestion are taken into the digestive cells by phagocytosis or pinocytosis. There have been no detailed studies of intracellular digestion, but the process appears to be similar to digestion in protists and sponges. Soluble nutrients diffuse from the digestive cells to the rest of the body. In cnidarians with thick mesogloea, motile cells in the mesogloea absorb nutrients and carry them to the rest of the mesogloea and to the epidermis. Epidermal cells are often covered with microvilli and may take up dissolved organic compounds from the water, thus decreasing their dependence on the diffusion of nutrients from the gastrodermis. Indigestible wastes are returned to the gastrovascular cavity by exocytosis and eventually removed via the mouth. In Hydrozoa and some Anthozoa, wastes are consolidated into a bolus and ejected at intervals through the open mouth. In other anthozoans, and in Scyphozoa and Cubozoa, tracts of ciliated cells move wastes along defined paths that are quite separate from those used for undigested food, so that waste can be egested continuously without interrupting extracellular digestion.

In a gastrovascular cavity, enzymes may be diluted by the gastrovascular fluid and particles may not come into contact with digestive cells. These problems are lessened when the gastrovascular cavity is made up principally of narrow tubular spaces, or when its surface area is increased by mesenteries. In Anthozoa, the free ends of the mesenterial filaments wrap around the prey and secrete mucus as well as the enzymes. This ensures that high concentrations of digestive enzymes are maintained around the prey.

Nutrition

Many Cnidaria, including all the reef-building scleractinian corals, contain large numbers of symbiotic dinoflagellate algae called zooxanthellae. These symbionts are found inside the cells of their hosts (Fig. 3.5c), where they photosynthesise normally but transfer certain products of photosynthesis to the host. In some corals the host may receive more than 98% of the organic compounds produced by the algae, and this is often more than enough organic carbon to meet the respiratory needs of the coral. The algae release a mixture of glycerol, glucose and organic acids to the host, along with small amounts of non-essential amino acids. Although they may satisfy the host's energy needs, the algae clearly do not provide all the nutrients the animal requires, and symbiotic cnidarians must still feed in order to obtain enough organic nitrogen, essential amino acids and other nutrients.

Circulation and excretion

Cnidarians, as we have seen, lack specialised systems for circulation and excretion. Nutrients and oxygen absorbed from the water by the epidermis and from the gastrovascular fluid by the gastrodermis are distributed around the body by diffusion, by the circulation of the gastrovascular fluid, and by the movements of motile cells in the mesogloea. The gastrovascular cavity extends into all parts of the body, except where the tentacles are solid. In colonies, the gastrovascular cavities of the various individuals are linked. In most cases these links are tubular: for example, in hydroid colonies the linking stalks are cylindrical. In soft corals and gorgonians there are tubular extensions of the gastrovascular cavities of the polyps (called solenia), throughout the thickened mesogloea. Individual polyps of hard corals are connected by a layer called the coenenchyme, which includes an extension of the gastrovascular cavity. In siphonophoran and chondrophoran colonies there are convoluted, tubular, endodermal canals linking the individual zooids.

Ammonia is the main nitrogenous waste produced by cnidarians; being very soluble, it is able to diffuse into the water or the gastrovascular fluid. Uric acid, the main waste product of the metabolism of nucleic acids, is barely soluble in water and is stored in the cells of some cnidarians. In cnidarians with symbiotic algae, the algae take up ammonia from the host and use it to synthesise amino acids. Symbiotic corals and anemones excrete less ammonia than those without symbionts, and may even take up ammonia from sea water. There have been few studies of osmoregulation in cnidarians; many species can tolerate some variation in salinity, but they do not appear to regulate the osmotic potential of the gastrovascular fluid.

Control systems

Nervous system

Cnidarians are the simplest animals to have a nervous system composed of neurons, but their neurons differ from those of most other animals in several ways. Cnidarian neurons have two or more unmyelinated extensions called neurites, rather than a single, myelinated axon. The neurons can conduct impulses in more than one direction. While unipolar neurons occur in Cnidaria as in other animals, most of their neurons are bipolar (with two neurites) or multipolar (with more than two neurites). Most cnidarian neurons are through-conducting; that is, they conduct impulses

Figure 3.5 (opposite)
(a, b) Regeneration in the scleractinian coral *Plesiastrea versipora*. (a) Skeleton of coral after much of the tissue has been brushed off. Disrupted tissue remains in the calices (T) and acontia (A). (b) Freshly collected specimen (F) of *P. versipora*, showing partially retracted polyps (M, mouth; TE, tentacle) and coenenchyme (CE); and regenerated coral (R), about six weeks after it had been brushed, with fully regenerated polyps and coenenchyme. **(c)** Part of a tentacle of *P. versipora* showing epidermis (E) with cnidae (C) and gastrodermis (G) with zooxanthellae (Z). **(d)** Planula larva of *P. versipora*.

without any attenuation of the signal. In most other animal neurons the impulse becomes attenuated as it passes along the axon. The predominance of bipolar and multipolar neurons in cnidarians mean that impulses spread in all directions from their origin; and because impulses are not attenuated, generalised responses can result from local stimuli. Conduction tends to be slower than in most other animals, but giant, fast-conducting neurons do occur, particularly in medusae. The neurons lie among the bases of the epitheliomuscular cells and form synapses with sensory cells, muscle fibres and other effectors, particularly cnidocytes and battery cells.

The simplest known nervous systems are those of polyps, particularly hydrozoan polyps (e.g. *Hydra*, Fig. 3.4a). In these animals the nervous system consists of two networks of neurons, one in the epidermis and one in the gastrodermis. These networks are not connected across the mesogloea. There is little specialisation within the network and no equivalent of a brain. However, even in very simple polyps such as *Hydra*, the neural net is denser in the hypostome and tentacles than in the column. In larger polyps there is a ring-like aggregation of neurons associated with the base of the tentacles. There is at least one non-polarised net, and usually two or more; for example, anemones have one fast and two slow-conducting nets. There are usually both gastrodermal and epidermal nets. Conduction through the epithelia is also important for coordination in anthozoan polyps.

Medusae have more elaborate nervous systems, associated with the control of swimming. In hydrozoan medusae there is a generalised, diffuse nerve net in the epidermis of the tentacles, manubrium and subumbrella, and there are two marginal nerve rings associated with the sense organs of the tentacular bulbs. Scyphozoan medusae have two subumbrellar epidermal nerve nets. There are no nerve rings, but the two nets are interconnected via aggregations of nerve cells, or ganglia, associated with the rhopalia. One of the nerve nets is diffuse and widespread, and receives impulses from scattered sensory cells and the sense organs, relaying them to the marginal ganglia. The second network conducts impulses from the marginal ganglia to the swimming muscles. This separation of sensory and motor nerve cells is found in all animals except the simpler Cnidaria. The marginal ganglia also act as pacemakers for the swimming muscles. At any one time, one of the ganglia initiates regular, rhythmical action potentials which spread through the motor network. Each stimulates the bell to contract once. The output of the pacemaker, and thus the rate of contractions, can be altered by environmental cues. Cubozoa have a similar system with a subumbrellar net, marginal ganglia associated with the rhopalia, and pacemakers in the rhopalial ganglia.

Endocrine system
There are no endocrine glands in Cnidaria. However, there are peptides and some unidentified low molecular weight compounds which are important in the regulation of growth. For example, *Hydra* produces a 'head-activating peptide' which interacts with cells during growth and regeneration, ensuring the production of a hypostome and tentacles. A 'proportion-altering factor' has been isolated from the tissues of several colonial hydroid species. This also stimulates the formation of the oral end of the polyps during normal metamorphosis and growth, and during regeneration.

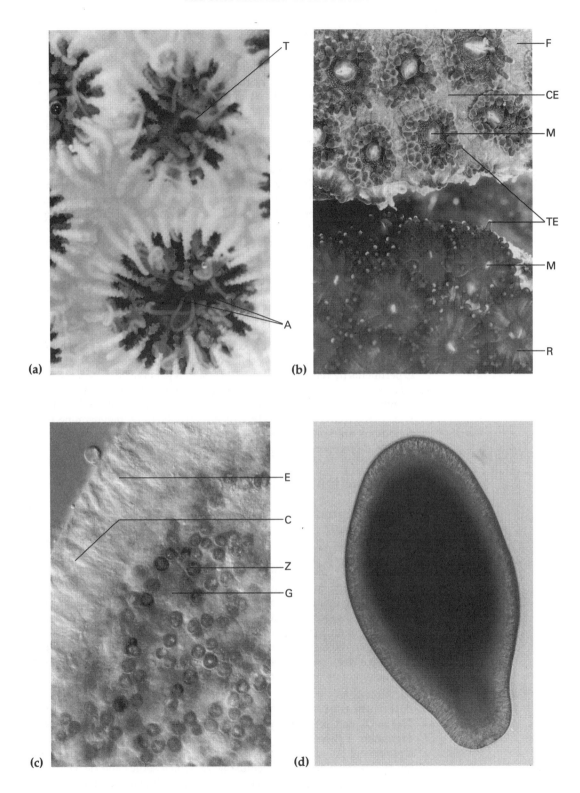

Regeneration

The ability of Cnidaria to regenerate has been known since Trembley observed the regeneration of both halves of a bisected *Hydra* in 1740. Subsequent studies of *Hydra* helped establish the role of polarity and induction in development. Wherever a hydra is cut, the end nearest the original hypostome will develop a new hypostome, while the proximal end of any fragment will develop a basal disc; thus the original polarity of the animal is maintained during regeneration. If a hypostome, or a fragment of a hypostome, is grafted anywhere on to a whole hydra, the hypostome tissue will induce the formation of a new hydranth (a bud) at the site of the graft. This is controlled by the head-activating peptide, which is produced in the hypostome. During regeneration, interstitial cells divide, migrate and differentiate to replace all types of cell except epitheliomuscular and digestive cells, which are replaced by the division of existing ones.

All types of Cnidaria share the ability to regenerate, often from small pieces or even from a suspension of dissociated cells. This allows them to recover from injury and to reproduce asexually. Some scleractinian corals can regenerate after extensive injuries. In studies of symbiosis in the coral *Plesiastrea versipora*, the tissue was routinely removed from the skeleton by brushing, leaving what appeared to be a bare skeleton. However, when some of these skeletons were put back into an aquarium, the tissue grew and reorganised itself, forming normal-looking colonies after two to six weeks (Fig. 3.5a, b). It is possible to 'recycle' the coral colonies used in experiments by returning them to the sea to regenerate and then collecting them again. Other corals recover from grazing and abrasion in the field. Many Cnidaria can form resistant stages (e.g. scyphozoan podocysts), in which they round up and reduce their metabolic rates, regenerating when conditions are suitable. Some mechanisms of asexual reproduction also depend on regeneration (see the following section).

Reproduction and development

Asexual reproduction

All Cnidaria can reproduce asexually by longitudinal or transverse fission, by budding, or by fragmentation, followed by regeneration. In budding, the column wall, oral disc or manubrium of either a polyp or a medusa grows out and develops tentacles and a mouth (Fig. 3.7c). When it is big enough to feed independently, the bud separates from the parent and becomes free-living. Colonies often form by budding, without the new 'individuals' separating from the parent. Anemones may reproduce by pedal laceration, in which a small part of the basal disc is autotomised (separated from the rest of the body by strong contractions) and then reorganised into a small polyp. Another form of asexual reproduction involves the production of planulae by parthenogenesis, as in some soft corals.

Alternation of sexual and asexual reproduction

The basic, and presumably primitive, life cycle of Cnidaria involves an alternation between a sexually reproducing medusa and an asexually reproducing polyp (Figs 3.6a–c, 3.7a–e). Medusae produce eggs or sperm; the zygote develops into a planula larva, which settles and develops into the polyp stage. The polyps may be solitary or colonial and may reproduce themselves asexually. Eventually the polyp phase will produce medusae

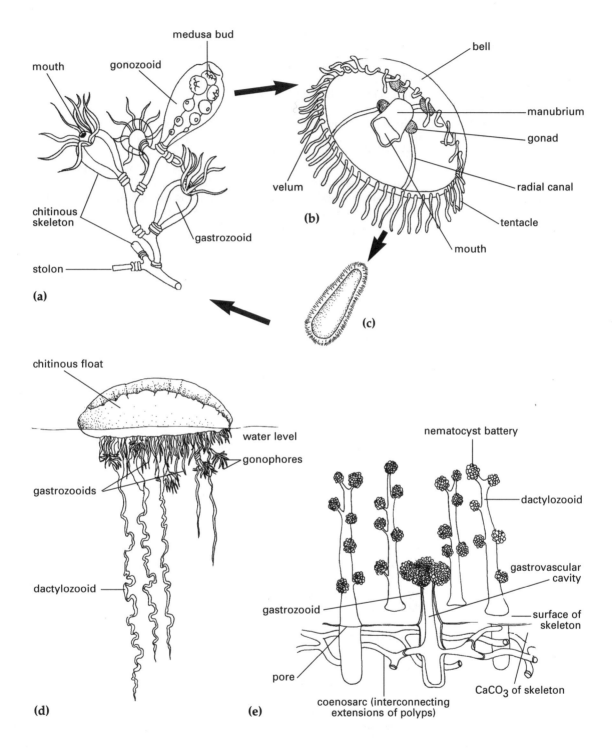

Figure 3.6
Hydrozoa. (a–c) Life-cycle of *Obelia*. (a) Polymorphic colony. (b) Free-swimming medusa. (c) Planula larva. (d) *Physalia*, a siphonophore. (e) *Millepora*, a hydrozoan coral.

Figure 3.7 (opposite)
Scyphozoa and Cubozoa.
(a–e) Life cycle of the scyphozoan *Aurelia*: (a) Adult medusa. (b) Planula.
(c) Scyphistoma. (d) Strobila.
(e) Ephyra. **(f)** Structure of a scyphozoan polyp, showing hypostome and four mesenteries projecting into the gastrovascular cavity
(after Pearse *et al.* 1987).
(g) Cubomedusan
(after various sources).

asexually, completing the cycle. Where there is a free-swimming medusa, it is the dispersal phase of the species and the planula is short-lived. Where there is no medusa, sexual reproduction occurs in polyps and the planula is the sole dispersal phase. Many species lack stages in this cycle, and the time spent in each stage is very variable.

Hydrozoa have a wide range of life cycles. In some (e.g. *Obelia*, Fig. 3.6b) the medusae are free-swimming, but in others they remain attached to the polypoid phase. In this case they may be obviously medusoid (gonophores) or be reduced to simple sacs (sporosacs) which no longer resemble medusae. Some Hydrozoa lack polyp stages entirely (e.g. Trachymedusae), and some have no medusae (e.g. *Hydra*).

In Cubozoa and Scyphozoa the medusa is the dominant phase; each planula gives rise to a single small polyp. Scyphozoan polyps, or scyphistomae, may produce additional scyphistomae asexually by budding (Fig. 3.7c). At intervals, each produces large numbers of immature medusae, or ephyrae. Ephyrae are formed by strobilation, the horizontal subdivision of the scyphistoma (Fig. 3.7d), and are set free to mature and grow in the plankton. After strobilation the scyphistoma reverts to its normal form and may persist for many years, strobilating at regular intervals (e.g. *Aurelia*, Fig. 3.7d). Some oceanic species (e.g. *Pelagia*) have no polyp; instead, the planula simply matures directly into a medusa. In Cubozoa, each polyp develops into a single medusa, and there is no persistent polyp phase. The planulae of *Chironex fleckeri* settle in mangrove swamps in estuaries, some distance from the open sea, and the young medusae are washed back into the ocean by monsoonal flooding of the estuaries.

Medusae are entirely absent from the class Anthozoa. Gametes are formed in the endoderm of the mesenteries, but there is no recognisable medusa bud or sporosac. Planulae are often brooded; each gives rise directly to a single polyp. In colonial species this becomes the founding member of the colony.

Reproductive system

Cnidaria do not have complex reproductive organs. Gametogenesis, which is usually seasonal, occurs by the differentiation of interstitial cells. In Hydrozoa, gametes are formed in the epidermis and are shed directly into the water. In other Cnidaria the gametes are endodermal; they may be shed into the water through breaks in the body wall or via the gastrovascular cavity and mouth. There are no ducts or specialised glands associated with reproduction. Sexes are usually separate, but some species are hermaphroditic.

Gametes and fertilisation

Eggs may be small with little yolk, or large and yolky. Sperm are always shed into the water, and fertilisation is usually external. In species which brood their eggs, fertilisation occurs at the brooding site, which may be in the gastrovascular cavity or on the outside of the body. Sperm are often attracted to the eggs by highly specific chemicals.

The maturation of the gametes is triggered by environmental factors such as temperature, and their release is triggered by light (sunrise, sunset or the phase of the moon). Many species of Cnidaria may spawn simultaneously in a particular location. While this is commonest in sessile species, dense aggregations of scyphozoan medusae have also been observed spawning. The most spectacular example is the mass spawning of corals on

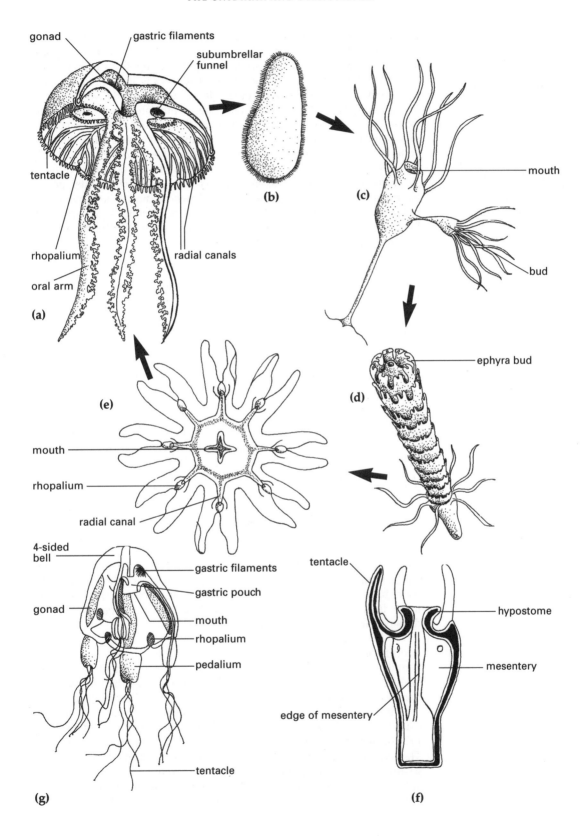

Figure 3.8 (opposite)
Anthozoa. **(a)** Anemone (Actiniaria), showing the pharynx, mesenteries, mesenterial filaments and acontia. **(b)** Structure of a mesenterial filament in transverse section.
(c) Scleractinian coral, showing calcareous skeleton and coenenchyme. **(d)** Gorgonian, showing skeleton made up of a horny axial rod and spicules in the mesogloea (after Pearse *et al.* 1987). **(e)** Alcyonarian soft coral, showing spicular skeleton in the mesogloea.

the Great Barrier Reef. This involves at least 110 species of scleractinian corals, as well as some soft corals and a few other invertebrates which spawn with them. Spawning occurs at night over the third to sixth days after the full moon in spring. Since gametogenesis is stimulated by rising water temperature, spawning occurs in late October or early November on the northern Great Barrier Reef and one month later on southern reefs where the rise in temperature occurs later. During and after mass spawnings (which have also been reported from Japan and some Pacific reefs) the water is cloudy with gametes, zygotes and planulae. Predators can eat only a very small proportion of them, and this may be the adaptive value of synchronised spawning. Recent research has shown that interspecific hybridisation may occur during mass spawning, and that some of the hybrids can settle and metamorphose. It is not known how long such hybrids can survive in the wild. The response of sperm to attractants in the eggs may reduce hybridisation for some species.

Embryonic development

Cleavage is radial, but otherwise quite variable; it may be equal or unequal. In small eggs it is complete and results in the formation of a hollow blastula (coeloblastula), but in large eggs it may be incomplete, with the yolk mass remaining in the centre of the blastula (solid or stereoblastula). Gastrulation — the movement of cells into the interior of the blastula — is also very variable in the Cnidaria. The resulting planula larva is a solid gastrula, consisting of an outer, ciliated ectoderm and an inner mass of endodermal cells, accompanied by yolk in lecithotrophic species.

Larval development

Planulae are elongated (less than 100 μm to about 5 mm long) and have a broad anterior end. They use cilia to swim or creep on the substratum. Many species have an apical organ with a tuft of cilia, which seems to be a sense organ used in selecting settlement sites. Planulae may depend on yolk stores or feed on microplankton. In some species, plankters are caught on a trailing streamer of mucus. Before settlement, the internal mass of endoderm splits to give rise to a gastrovascular cavity lined with gastrodermis.

Planulae may provide for dispersal of species, and always play a crucial role in selecting the substratum on which the polyp will live. In species which brood their eggs, the planulae may spend a very short time in the water between release and settlement. While this decreases their exposure to predators, it tends to lead to their settling close to the parent, limiting dispersal and possibly increasing competition.

Metamorphosis

Planulae are generally competent to settle over a period of some days. Texture and chemical cues have been identified as important in the selection of a settlement site, and planulae may detach and settle again later if the substratum is unsuitable. Initially attachment is by means of mucus, or cnidae, or both. When a suitable site is found, the anterior end of the planula attaches permanently to the substratum and metamorphosis begins. The free (posterior) end develops a mouth, providing an opening into the gastrovascular cavity. Tentacles arise as outpocketings of the body wall; in species with hollow tentacles, they may appear within minutes of attachment, as the gastrovascular cavity is inflated with water. Where present,

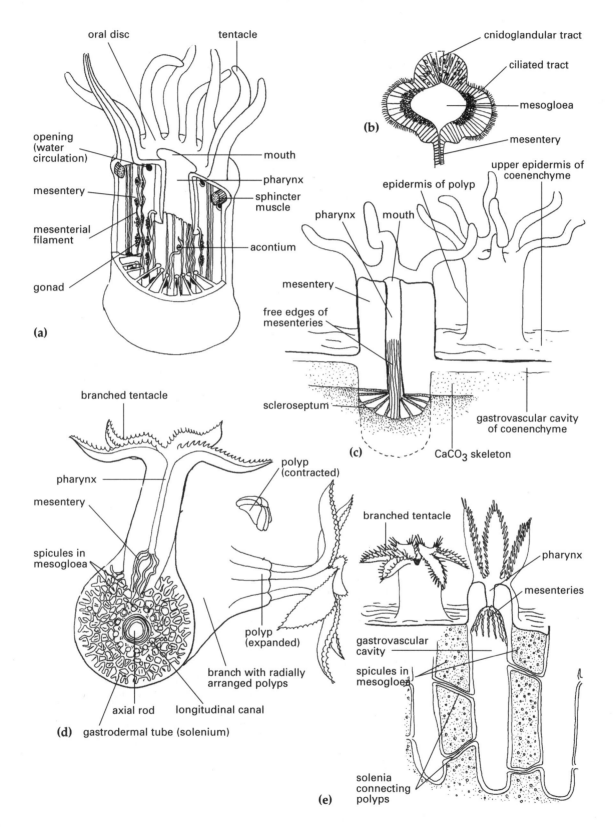

mesenteries develop as ingrowths of the mesogloea and gastrodermis into the gastrovascular cavity. In species with no polypoid form in the life cycle, the planula shortens along the oral–aboral axis and develops directly into a medusa.

In colonial forms, the whole colony is derived by cloning from a single planula. This may develop into a single individual which later buds off other members of the colony, or it may remain undifferentiated and function as a 'growing point' throughout the life of the colony.

Although the planula usually gives rise directly to a polyp or a medusa, some Cnidaria develop to a more advanced juvenile stage before release. The commonest is the actinula, a pear-shaped, ciliated larva with eight tentacles.

Coral reefs

While cnidarians are found at all latitudes and in all marine habitats, they are of special importance in tropical coral reefs. These reefs are biogenic structures composed largely of calcium carbonate, with a relatively thin surface layer of living organisms. The calcium carbonate is laid down by many animal species, which mostly have symbiotic algae, and by free-living algae, so that a more appropriate name is coral–algal reefs. Because light is required for the growth of corals that have symbiotic algae, as well as for the free-living algae, coral reefs grow only in shallow water where there is sufficient light. They will also grow only where the average daily minimum temperature of the water is 19°C or more and the average daily maximum does not exceed 30°C.

Structure of coral reefs

Charles Darwin first classified coral reefs as fringing reefs, barrier reefs and atolls. He proposed that atolls arise from fringing reefs, which originally grow close to the shores of islands; as the island subsides, the fringing reef becomes a barrier reef, separated from the remains of the island by a lagoon. Further subsidence then leads to the disappearance of the island beneath the water, leaving a ring-like atoll surrounding a lagoon.

Modern geological studies have supported this hypothesis for the origin of mid-ocean atolls, but have shown that reefs on continental shelves, such as the Great Barrier Reef on the Queensland shelf, have a very different history. Cycles of sea-level change have meant that the current Queensland continental shelf was below sea level for only about 10% to 20% of the Pleistocene period. During these times of immersion, coral reefs grew on the shelf. Whenever sea levels fell, the reefs died and the limestone platforms they had formed weathered. As the sea level rose again to cover them, these platforms were recolonised by corals and other reef organisms. At each inundation, the shapes of the new reefs were determined by the shapes of the underlying platforms and the effects of currents. The present sites of reefs on the Queensland shelf were exposed until only 8000 to 9000 years ago, so the existing reefs started growing about 8500 years ago, during a period of rising sea levels. The sea level has been steady for the last 6500 years, and most reefs seem to have grown up to surface level by about 4500 years ago. Since then they have been able to expand horizontally to form large reef flats, but not vertically.

Roles of cnidarians in reef formation

Cnidarians with calcium carbonate skeletons, particularly scleractinian corals, form the framework of living coral reefs. Because of the shapes of the corals, this framework is open and has many interstices which shelter other organisms. When corals die, the framework is colonised by algae, by bacteria and by other animals. Some of these bore into the skeleton, weakening it, and the pressure from the weight of organisms growing above the dead material breaks up the framework. Calcium carbonate sands infiltrate the rubble and calcium carbonate precipitates from solution, leading to the formation of limestone, which in turn provides a relatively stable support for the living parts of the reef. Calcium carbonate is laid down by coralline red algae and the green alga *Halimeda*, which have calcified cell walls, by foraminiferans and other protists, and by molluscs, tube worms and other types of animal, as well as by cnidarians. Coralline algae are particularly important because they lay down calcium carbonate as calcite, which is harder than the aragonite produced by corals. Coralline algae dominate the reef rim, where there is heavier wave action than anywhere else on the reef. Much of the carbonate sand which infiltrates reef frameworks comes from *Halimeda*.

Importance of symbiosis in reef growth and productivity

Coral reefs are found in tropical waters, which are usually fairly low in plant nutrients and do not support high rates of photosynthetic productivity; and yet coral reefs are very productive, as can be seen from the abundance and diversity of organisms present on them.

A large proportion of the hard and soft corals, hydroids, gorgonians, zoanthids and other cnidarians found on coral reefs have symbiotic dinoflagellates. Microscopic algae also occur as symbionts in species of foraminiferans, sponges, flatworms and molluscs on reefs. These symbioses contribute to the growth of reefs in a number of ways. They are energetically efficient, because the host animals receive part of their nutrition (see page 43) without using energy to catch and digest food. Plant nutrients, which are scarce in these waters, are recycled efficiently, as the waste nitrogen (ammonia) and phosphate produced by the host are used by the algae. Finally, in symbiotic corals, calcification is faster when the algae are photosynthesising than in the dark. The productivity of coral reefs depends largely on the existence of algal–invertebrate symbioses.

Classification of the phylum Cnidaria

CLASS HYDROZOA
Mesogloea acellular; nematocysts only in the epidermis; gametes develop in the epidermis (or, if gastrodermal, are shed directly into the water, not into the gastrovascular cavity); no mesenteries in the gastrovascular cavity; polyps with the mouth on an elongated hypostome, and no pharynx; medusae with a velum, and relatively small (5 mm to about 60 mm in diameter); symmetry polymerous (with many similar parts) or tetramerous (with structures in multiples of four); life cycles variable, but in most taxa the polyp is the dominant phase; about 2700 species; mostly marine but including all the freshwater Cnidaria. (Orders Hydroida, Hydrocorallina, Siphonophora, etc.)

CLASS SCYPHOZOA

Mesogloea cellular; nematocysts in both the epidermis and the gastrodermis; gametes endodermal; gastrovascular cavity contains mesenteries (except Rhizostomeae); polyps with a hypostome, but no pharynx; medusae with no velum; symmetry tetramerous; medusa the dominant phase, and relatively large (20 mm to 2 m in diameter); each polyp giving rise to many medusae by strobilation; about 200 species, all marine. (Orders Coronatae, Semaeostomeae, Rhizostomeae)

CLASS CUBOZOA

Mesogloea cellular; nematocysts in both epidermis and gastrodermis; gametes endodermal; gastrovascular cavity with mesenteries in the medusa but not the polyp; edge of medusa turned in to form a velum-like structure (the velarium); symmetry tetramerous; bell cuboidal, with four groups of tentacles and four rhopalia at the 'corners'; medusa (20 mm to 250 mm diameter) the dominant phase; each polyp giving rise to a single medusa directly; about 20 species, all marine. (Order Cubomedusae)

CLASS ANTHOZOA

Mesogloea cellular; nematocysts in both epidermis and gastrodermis; mesenteries in gastrovascular cavity; symmetry hexamerous (with structures in multiples of six) or octomerous (eightfold); polyps only; mouth at the inner end of an invaginated pharynx; apex of polyp a flattened oral disc; about 6000 species, all marine.

Sublass Hexacorallia (Zoantharia)

Anthozoa with unbranched tentacles in multiples of six and six primary mesenteries; siphonoglyphs absent or one to many; solitary or colonial. (Orders Actiniaria, Scleractinia, Zoanthidea, Ceriantharia)

Subclass Octocorallia (Alcyonaria)

Anthozoa with pinnately branched tentacles in multiples of eight, and eight primary mesenteries; one siphonoglyph. Almost all colonial. (Orders Alcyonacea, Gorgonacea, Pennatulacea)

Phylum Ctenophora

Diploblastic metazoans with biradial symmetry; tissues but rarely organs; basic body form ovoid; planktonic, with ciliary locomotion; ectodermal and gastrodermal epithelia separated by mesogloea containing muscle cells; single body cavity, mouth and two small anal pores; nervous system a nerve net; no respiratory, circulatory or excretory system; adhesive colloblasts; hermaphrodite sexual reproduction, direct development.

Introduction

All Ctenophora are marine. Most are planktonic, occurring from the surface waters to at least 3000 m. There are about 100 described species, but other undiscovered species probably exist in deep waters. Ctenophores may occur in dense swarms and are sometimes washed up on the shore in large numbers.

Morphology

Ctenophores ('comb jellies') are diploblastic animals with a thick mesogloea which contains immigrant epidermal cells. Unlike cnidarians, ctenophores have true muscle cells rather than epitheliomuscular cells with myofibrils, and these muscle cells develop in the mesogloea. Since muscle is mesodermal in triploblastic animals, some zoologists have suggested that the Ctenophora may be triploblastic, but recent evidence does not support this view (see Chapter 18). The basic form in ctenophores is ovoid or spherical, about 5 mm to 40 mm in diameter, with the mouth at one pole and an aboral apical sense organ at the other. There are eight evenly spaced, longitudinal rows of ciliary combs or ctenes, each connected to the apical organ by a ciliated groove (Fig. 3.9b, d, e). Each ctene consists of a horizontal band of long, partially fused cilia. Two long, highly contractile tentacles with lateral branches (tentillae) arise within tentacular sheaths, into which they can be withdrawn. The mouth opens into a flattened pharynx, which leads to a small stomach and a complex series of gastrovascular canals (Fig. 3.9a). Two of these canals end in anal pores near the apical organ.

Diversity

There are five orders of ctenophores (Fig. 3.9f–i). The Cydippida are conical in form, with a pair of long tentacles. The Beroida are also approximately conical, but lack tentacles. The Lobata are flattened laterally, with large oral lobes and reduced tentacles. The Cestida are ribbon-like, also flattened laterally, and very elongated. The mouth and apical organ are on opposite sides at the centre of the body and there are many short tentacles. The Platyctenida are benthic. The oral–aboral axis in this group is flattened, and the animals creep on the substratum.

Colloblasts

Ctenophora lack cnidae and toxins but have colloblasts, which are extremely sticky. The epidermis of the tentacles (or oral lobes in species without tentacles) consists mainly of colloblasts. Each colloblast has a domed outer surface, with many refractile granules just under the cell membrane, and a stalk inserted into the mesogloea. Each granule is linked to the star-shaped spheroidal body which is embedded in the top of the elongated colloblast nucleus. The spheroidal body is extended into a long spiral filament which coils around the stalk (Fig. 3.9c). When prey touch the colloblasts the granules rupture, releasing the sticky material on to the prey. The spiral filament apparently anchors the colloblast while the prey struggles. Colloblasts are destroyed during feeding and replaced from undifferentiated tissue at the base of the tentacle.

Locomotion

Most species of ctenophores use cilia for locomotion, with little involvement of muscles. The cilia of the comb rows beat towards the aboral pole, so that the animal moves with the mouth forward. Regular waves of beating start at the aboral pole, and are visible because of the iridescence they produce. The beat can be reversed to avoid obstacles.

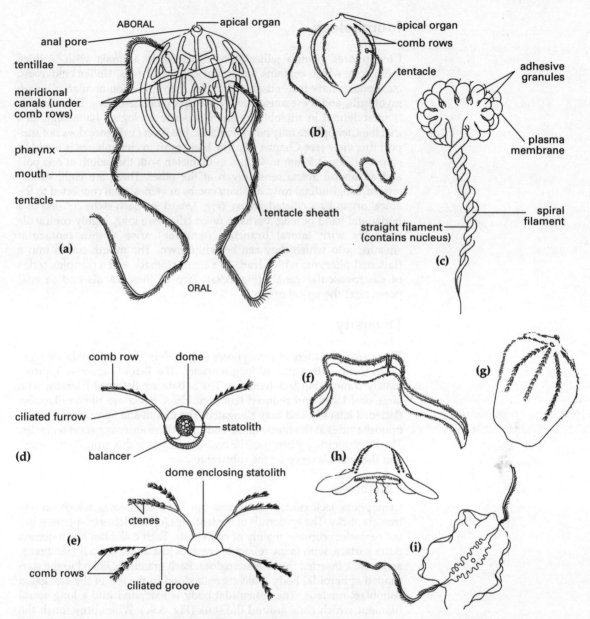

Figure 3.9
Ctenophora. (a) Canal system of *Pleurobrachia* (Cydippida). (b) *Pleurobrachia* in external view. (c) Structure of colloblast. (d, e) Apical organ: (d) lateral view, and (e) apical view. (f–i) Diversity in ctenophores. (f) Cestida. (g) Beroida. (h) Lobata. (i) Platyctenida. (Adapted from various sources.)

Feeding

In cydippids the tentacles, which may be many times the length of the body, are trailed in the water, and any prey encountered stick to the tentillae. Food is ingested by retracting the tentacles and wiping them across the mouth. Ctenophores with short or no tentacles entrap their planktonic prey in mucus on the body surface and transport them to the mouth by ciliary currents.

Bioluminescence

Most ctenophores can produce light biochemically. At night, flashes of blue light, brighter than that of most bioluminescent organisms, are produced in the canals under the comb rows.

Sense organs

The apical organ is a depression which contains a statolith resting on four groups of large cilia, the balancers. The balancers are connected to two ciliary grooves, each running to the top of a comb row (Fig. 3.9d, e). The balancers are the pacemakers for the comb rows, whose beat is controlled by mechanical coupling between adjacent rows rather than by nerve impulses. The pressure of the statolith on a particular balancer increases the beating rate of the two associated comb rows, allowing the animal to control its orientation. The body surface has sensory cells which respond to chemicals, temperature, touch and vibration.

Physiological processes

Digestion

The pharynx secretes digestive enzymes, and the extracellular phase of digestion occurs there. The resulting suspension of particles enters the stomach and is circulated through the gastrovascular canals (Fig. 3.9a). Digestion is completed intracellularly in the cells lining the canals. Indigestible wastes are eliminated through the anal pores and mouth.

Respiration, circulation and excretion

Both the epidermis and the digestive epithelium are used for gas exchange. The gastrovascular canals run throughout the body, providing for the transport of nutrients and oxygenated water, and probably also for excretion.

Nervous system

There is a diffuse subepidermal nerve net, which is more concentrated under the comb rows and around the mouth, and a sparse subgastrodermal net. There are also nerve cells in the mesogloea, including large strands in the tentacles. These synapse with the mesogloeal muscle fibres and control muscular movement.

Reproduction and development

Ctenophores are hermaphroditic. Ovaries and testes differentiate from the endoderm lining the eight meridional canals. The gametes are released through temporary gonopores, and fertilisation is external. Cleavage, which follows a unique biradial pattern, is total and determinate. Gastrulation occurs by invagination or epiboly, and the cydippid larva closely resemble the adult in Cydippida; in other orders, the transformation from cydippid to adult form is gradual.

Classification of the phylum Ctenophora

CLASS TENTACULATA
With tentacles (Orders Cydippida, Lobata, Cestida, Platyctenida)

CLASS NUDA
Without tentacles (Order Beroida)

Chapter 4

The Platyhelminthes, Nemertea, Entoprocta and Gnathostomulida

J.C. Walker and D.T. Anderson

PHYLUM PLATYHELMINTHES

Introduction 59

Diversity 60
   Class Turbellaria 60

Parasitic platyhelminths 61
   Class Trematoda 61
   Class Monogenea 62
   Class Cestoda 62

Relationships within the Platyhelminthes 62

Platyhelminth structure and function 63
   Body wall 63
   Mesenchyme 65
   Feeding and digestion 65
   Osmoregulation 67
   Nervous system and sense organs 68

Reproduction and development 69
   Asexual reproduction in turbellarians 69
   Reproductive systems 70
   Mating 71
   Development 72

Significance of platyhelminths in agriculture and medicine 77
   Turbellaria 77
   Trematodes 77
   Cestodes 78
   Vaccination against helminth parasites 78

Classification of the phylum Platyhelminthes 78

PHYLUM NEMERTEA

Introduction 79

Functional morphology 80

Organ systems 81
   Digestive system 81
   Excretory system 81
   Circulatory system 82
   Nervous system and sense organs 82

Reproduction and development 82

Evolutionary relationships 82

Classification of the phylum Nemertea 83

PHYLUM ENTOPROCTA

Introduction 83

Functional morphology 83

Organ systems 84
   Digestive system 84

Excretory system 84

Nervous system 84

Reproduction and development 84

PHYLUM GNATHOSTOMULIDA

85

Phylum Platyhelminthes

Bilaterally symmetrical, acoelomate, unsegmented, free-living or parasitic worms with a mouth and enteron (absent in some parasitic groups) but no anus; anterior brain and longitudinal nerve cords; protonephridial excretory system; complex hermaphrodite reproductive system; spiral cleavage development in some.

Introduction

Unlike the metazoans described in Chapters 2 and 3, the Platyhelminthes, or flatworms, are bilaterally symmetrical. This is characteristic of motile animals, which have definite anterior and posterior ends, and also have one body surface (ventral) in contact with the substratum and the other (dorsal) uppermost (see Fig. 3.1c, page 30; and Fig. 4.2a). Structures such as eyes are symmetrically arranged on either side of the body. Sense organs and nervous tissues are concentrated at the anterior end. This concentration is called cephalisation.

Unlike the diploblastic Cnidaria and Ctenophora, Platyhelminthes and all of the animals described in the remaining chapters are triploblastic. A third embryonic layer, the mesoderm, lies between the ectoderm and endoderm. This layer substantially increases the options for the development of organs with specific functions, formed by the association of tissues of various kinds. Groups of organs dedicated to the performance of particular functions such as digestion or reproduction are called organ systems. In the Platyhelminthes, mesoderm, as muscle fibres and mesenchyme, fills the spaces between the internal organs (Fig. 4.3). Platyhelminths are thus acoelomate (without a body cavity).

The basic cleavage pattern of platyhelminth eggs is one in which the axes of division are oblique to the egg's polar axis (Fig. 4.1). In addition, because the eggs usually contain moderate amounts of yolk, the early cell divisions are unequal, giving rise to apical micromeres and basal macromeres. The quartette of micromeres from each division of the macromeres come to lie in the junctions between the cells immediately below. The result is a blastula in which the cells are arranged in a spiral pattern. Spiral cleavage, which is characteristic of some free-living platyhelminths and of many other invertebrate phyla, is so uniform that a standard system of nomenclature for the fate of cells is possible. Further details

Figure 4.1
Spiral cleavage. **(a)** Four cells, A–D, following two divisions of the zygote. **(b)** Eight cells, with micromeres 1a–1d lying in the junctions between macromeres 1A–1D. The arrows indicate the origin of each micromere from its parent macromere. **(c)** 33-cell stage viewed from the macromere pole. The macromeres, 3A–3D, have budded off three quartettes of micromeres, and the D macromere, now 4D, has undergone a further division budding off the mesoderm cell 4d. Cells labelled 2d are future ectoderm. Macromeres at this stage are future endoderm.

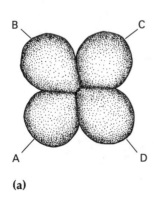

(a)

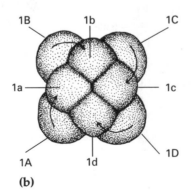

(b)

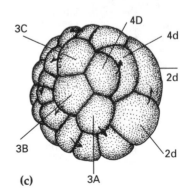

(c)

of this nomenclature are given in the texts listed in the general section of the Bibliography. The quartettes of micromeres are numbered 1a–1d, 2a–2d, 3a–3d and 4a–4d. Individual cells designated in this way often have the same developmental function in several phyla (Fig. 4.1c). The cell 2d is a major contributor to ectoderm, and the cell 4d often gives rise to all of the mesoderm.

Diversity

Class Turbellaria

The structure of platyhelminths determines their diversity in form and habitat. The lack of a circulatory system limits free-living flatworms to relatively small sizes and makes their characteristic flattened shape advantageous for gaseous exchange. The body surface, acting as a permeable respiratory surface, is a site of potential fluid loss. This restricts platyhelminths to environments in which dehydration is unlikely to occur: aquatic and moist terrestrial habitats and as parasites in the body cavities of other animals. Some parasitic flatworms are large.

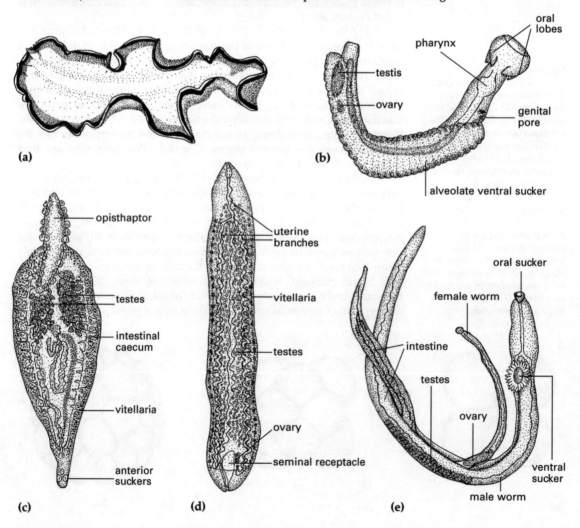

Figure 4.2
(a) Marine polyclad turbellarian, *Pseudoceros bimarginatus*.
(b) Immature aspidobothrean, *Lobatostoma manteri*, from the marine snail *Cerithium moniliferum*. (c) Monogenean, *Pseudothoracocotyla scomberomori*, from the gills of the Queensland mackerel, *Scomberomorus commerson*. (d) Cestodarian, *Austramphilina elongata*, from the freshwater tortoise, *Chelodina longicollis*. (e) Pair of schistosomes: an undescribed species from the eastern pelican, *Pelecanus conspicillatus*.

More than 4500 species of turbellarians have been described. Most are predators or scavengers. Their habitats range from marine to freshwater to moist terrestrial. The majority of aquatic species are benthic, living in sand or mud, under stones, or on aquatic vegetation. Terrestrial planarians are susceptible to desiccation. Most are nocturnal and inhabit shaded, humid situations, particularly in leaf litter and rotting timber. Turbellarians are usually 1 to 2 mm long and 100 to 200 µm in diameter, but some marine polyclads may be 30 to 40 mm and some terrestrial triclads are over 600 mm. Except for the smallest 'microturbellarians' and some land planarians, which are cylindrical, most species are dorso-ventrally flattened. Small tentacles or lateral projections called auricles are often present on the head. Most turbellarians are shades of black, brown or grey, but some marine polyclads and terrestrial planarians are brightly coloured.

Symbiotic relationships exist between turbellarians and other invertebrates. For example, *Bdelloura candida* is a commensal on the gills of horseshoe crabs. Others are true parasites, physiologically dependent on their hosts: a few rhabdocoels live in the digestive system of gastropod and bivalve molluscs, deriving nutrition from their host's tissues. The rhabdocoel *Acholades*, which has no intestinal tract, lives encysted in the tube feet of starfish.

Parasitic platyhelminths

The remaining platyhelminth groups are all parasitic on or in other animals, though their life histories often involve free-living stages, essential for dispersal to new hosts. These groups are the trematodes, the monogeneans and the cestodes.

Class Trematoda

The name Trematoda refers to the cavity in the holdfast organs of these animals (Greek *trema*, a hole). Because of the dorso-ventral flattening of many species, trematodes are called flukes, an old name for flatfish. Most are endoparasitic and have two or three hosts in the life cycle.

Subclass Digenea

Flukes with two or three hosts, the first a mollusc and the last a vertebrate. They usually have two muscular holdfast organs; an anterior oral sucker which surrounds the mouth, and a ventral sucker or acetabulum, which is usually larger (Fig. 4.4a). The name Digenea refers to the occurrence of at least two hosts in the life cycle. The adults occur in the definitive host, normally a vertebrate, and they inhabit virtually any organ in the body. For example, humans may be infected by trematode species which inhabit the intestine, the vascular system or the lungs. Most of the approximately 9000 digenean species have a molluscan intermediate host, usually a gastropod or bivalve.

Subclass Aspidobothrea

Flukes usually having a single host in the life cycle. The ventral sucker is large and is divided by septa into alveoli (Fig. 4.2b). Aspidobothrean trematodes are endoparasitic, ectoparasitic or ectocommensal in or on their hosts, which may be marine or freshwater bivalve or gastropod molluscs, elasmobranch or teleost fish, or turtles.

Class Monogenea

Monogeneans have a single-host life cycle. The most prominent morphological feature is the posterior attachment organ, the opisthaptor (Fig. 4.2c). In addition there is usually an anterior prohaptor, a combination of adhesive glands and small suckers.

Most monogeneans are ectoparasitic on fish, but other vertebrates (frogs, turtles and the hippopotamus), and invertebrates (including copepods and squid) are occasionally infested. Monogeneans are very host-specific and normally occupy a unique site on the host, the structure of the opisthaptor being adapted to suit the form of the host's body.

Class Cestoda

Cestodes are endoparasitic in the gut or coelomic cavity of vertebrates, and usually have life cycles involving two or more hosts. The body normally has an attachment organ, the scolex, followed by a strobila comprising a repeated series of body units called proglottids. Cestodes lack an alimentary tract.

Subclass Cestodaria

This is a small group of unstrobilated cestodes which lack a scolex and have a general resemblance to flukes (Fig. 4.2d). They are endoparasites of fish and turtles.

Subclass Eucestoda

The tapeworms are strobilated cestodes (Fig. 4.7a) with a scolex which attaches to the host's intestinal mucosa. The holdfast structures on the scolex vary from group to group and include rows of hooklets forming a rostellum, muscular suckers, elongate flaps of tissue called bothria, and leaf-like outgrowths called bothridia. Immediately behind the scolex is the neck, an undifferentiated region from which the strobila is proliferated as an anteroposterior sequence of proglottids. Each proglottid contains male and female reproductive organs, immature at the anterior end of the strobila and mature, with eggs in a uterus, posteriorly.

The name Cestoda (from the Latin *cestus*, a girdle), refers to the ribbon-like shape of tapeworms. Around 5000 species are known. Tapeworms range in length from less than 10 mm to more than 20 metres.

Relationships within the Platyhelminthes

Three broad lines of evolution are recognised within the free-living Platyhelminthes: an acoel line, a catenulid line and a line leading to the other orders. The trematodes and cestodes are thought to be a further offshoot of the third group (Box 4.1).

One area of agreement among modern taxonomists is the association of trematodes, monogeneans and cestodes in a single clade, the Neodermata. The basis for this is the replacement of the ciliated larval epidermis by a syncytial neodermis in the adults (see below). Recent studies using the nuclear 18S rRNA gene strongly support a monophyletic origin for the major parasitic taxa. On the basis of ultrastructure and ontogeny, two main clades are recognised within the Neodermata. The first comprises the single class Trematoda (Aspidobothrea and Digenea) but the second clade, the Cercomeromorphae, includes the monogeneans, the cestodes, and related groups. The common feature of the Cercomeromorphae is the possession of a cercomere (a posterior attachment structure bearing hooks).

> **Box 4.1**
> **Relationships within the Platyhelminthes**
>
> Class Turbellaria
>   Order Catenulida   *Stenostomum*
>   Order Acoela   *Convoluta*
>   Order Polycladida   *Pseudoceros, Notoplana*
>   Order Rhabdocoela   *Acholades, Temnocephala*
>   Order Tricladida   *Dugesia, Coenoplana*
>
> 'Clade' Neodermata
> Class Trematoda
>   Subclass Digenea   *Fasciola, Opisthorchis, Schistosoma, Stictodora*
>   Subclass Aspidobothrea   *Lobatostoma*
>
> 'Clade' Cercomeromorphae
> Class Monogenea
>   Subclass Monopisthocotylea   *Pseudothoracocotyla*
>   Subclass Polyopisthocotylea   *Diplozoon*
> Class Cestoda
>   Subclass Cestodaria   *Austramphilina*
>   Subclass Eucestoda   *Anoplotaenia, Echinococcus, Hymenolepis, Taenia*

Platyhelminth structure and function

Body wall
Turbellarians

The turbellarian body wall (Fig. 4.3a) is a ciliated epithelium from 2 to 30 μm thick. The cilia, used in gliding locomotion, are restricted to limited zones on the body surface in some species and may be lacking in ectocommensal forms such as temnocephalids.

The epidermis contains mucus-secreting gland cells which are derived from ectoderm. These cells often have their main body in the underlying mesenchyme, with processes extending between epidermal cells to the body surface. The mucus maintains a moist body surface, lubricating locomotion, and is also used for capturing prey. The epidermis of many turbellar-ians contains rod-shaped bodies called rhabdoids. These are formed by epithelial cells and produce mucus when released from the body surface. Similar rod-shaped bodies called rhabdites are produced by secretory cells in the mesenchyme, below the epidermis. They reach the body surface via intercellular spaces in the epidermis.

The muscles of the turbellarian body, which are unstriated, lie beneath the epidermis and consist of circular fibres, longitudinal fibres and other fibres arranged obliquely. The muscles can act antagonistically or synergistically in changing the body shape.

Trematodes, monogeneans and cestodes

The parasitic Platyhelminthes (trematodes, monogeneans and cestodes) have a distinctive tegument, the neodermis, the nature of which has been clarified through electron microscopy (Fig. 4.3b). The distal cytoplasm of the tegument has an outer membrane and a matrix containing membrane-

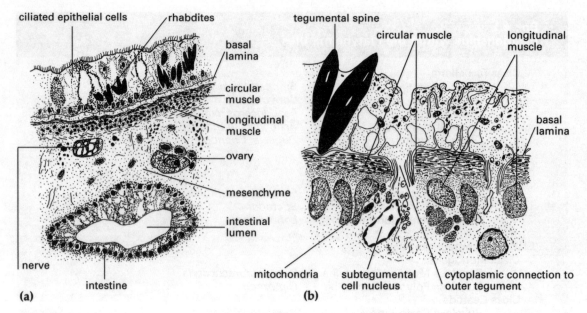

Figure 4.3
(a) Transverse section of dorsal surface of marine polyclad, *Pseudobiceros bedfordi*. (b) Transverse section of dorsal surface of the human blood fluke, *Schistosoma mansoni*.

bound vesicles and mitochondria. From this layer, passing through the basal lamina, are numerous microtubule-lined cytoplasmic connections to tegumental cytons (flask-shaped cell bodies which lie in the mesenchyme). Cytons are the source of the structural elements of the outer tegument. Between the basal lamina and the mesenchyme are bands of circular, longitudinal and oblique muscles similar in structure and function to those of turbellarians.

The distal cytoplasm of the cestode tegument differs from that of trematodes in having a layer of delicate cytoplasmic extensions, or microtriches, similar in form and function to the brush border microvilli of the mammalian intestinal epithelium. The microtriches increase the surface area of the tapeworm body by between two and eleven times, depending on the species of cestode and the region of the strobila. This is a significant adaptation for an animal lacking a gut.

The differentiation of the neodermis has been studied in each of the major groups. The monogenean larval stage is partially covered by ciliated cells arranged in distinct regions separated by cytoplasmic syncytium, both zones being nucleated early in development. The ciliated cells become flattened and their nuclei degenerate, while those of the intervening syncytial areas are lost through the body surface. A basal lamina forms beneath these surface cells; below the lamina, stem cells are found. These produce cytoplasmic processes which connect with the anucleate surface cytoplasm. When the ciliated cells are lost at the end of the free-swimming stage, the discontinuous cytoplasm spreads to form a continuous syncytium, the neodermis.

Digenean larvae (miracidia) have ciliated cells separated by material called ridge cytoplasm. The ciliated epidermal cells are shed at the time of entry into a molluscan host (see below) and the ridge cytoplasm spreads out on the surface. Additional cytoplasmic material is added to the neodermis by cells which differentiate from stem cells lying below the basal lamina. A similar process of differentiation from a nucleated larval epidermis to a typical neodermis occurs in cestodes.

Mesenchyme

The space between the internal organs of platyhelminths is filled with mesenchyme, a cellular connective tissue which supports the organ systems and functions in the transport of metabolites (Fig. 4.4b). There are two main categories of cell: the fixed mesenchymal cells and stem cells. The fixed cells are large, having extensive processes which interdigitate with choroid cells (cells which have fluid filled vacuoles), or with pigmented cells. The stem cells are of two types: neoblasts, and epidermal replacement cells which appear to be derived from neoblasts. Neoblasts, which have a large nucleus and little cytoplasm, are pluripotent or totipotent and function in regeneration following asexual reproduction or trauma. They also act as stem cells for the gonads. The bodies of secretory cells which open at epithelial surfaces may be found in mesenchyme, which is traversed by dorso-ventral and oblique muscle fibres. A filamentous connective tissue consisting of collagen-like fibres lies between the mesenchymal cells. These fibres function as a form of skeleton and act as anchor points for muscle cells.

Studies on trematodes and cestodes have shown that their fixed mesenchymal cells contain large reserves of glycogen, up to 50% by dry weight. These parasites are vulnerable to depletion of glycogen if the host has a low-carbohydrate diet.

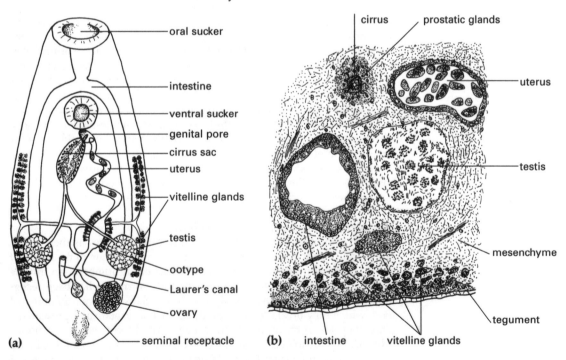

Figure 4.4
(a) Anatomy of a generalised trematode. (b) Transverse section of the human lung fluke, *Paragonimus westermani*.

Feeding and digestion

Although a few turbellarian species feed on plant material such as diatoms and green algae, the majority are predators or scavengers. An extreme exception is the acoel *Convoluta roscoffensis*, which has a symbiotic relationship with the green alga *Tetraselmis*. The young worm ingests algal cells soon after hatching and, as an adult, does not feed at all. It is presumably nourished entirely by its photosynthesising algal symbionts.

The digestive system of acoel turbellarians is a membrane-bound syncytial mass, separated from other tissues by specialised 'wrapping' cells and opening to the outside via a simple, ciliated pharynx. Turbellarians with such a simple tubular pharynx normally feed by ciliary action, sweeping small prey and organic particles into the mouth, which is usually situated mid-ventrally. The syncytium forms only after the ingestion of food in some acoels, and is shed when digestion is complete, leaving a large central cavity. The syncytium is later renewed by fusion of the bordering cells.

Most other turbellarians have an eversible pharynx. That of triclads and polyclads is a folded, or plicate, muscular tube lying within a pharyngeal tube. During feeding the free end of the tube is everted through the mouth. The position of the mouth varies from the middle of the ventral surface to the anterior or posterior end.

Most turbellarians have a sac-like intestine which may have three branches (triclads) or many (polyclads).

Turbellarian digestion is initially extracellular, using enzymes secreted by pharyngeal cells or by gland cells in the intestinal epithelium. The partially digested material is phagocytosed by intestinal cells, in which final digestion occurs.

Trematode redia larvae (see Fig. 4.6) have a muscular pharynx and an intestinal sac, and are able to ingest host tissue and even the germinal sacs of other trematode species. The cells lining the internal surface of the gut have prominent cytoplasmic projections which significantly increase the absorptive surface area. Similar cytoplasmic projections are present on the intestinal epithelium of adult trematodes (Fig. 4.4). The diet of adult flukes depends on their location in the host. Intestinal flukes may feed on material in the host's intestine, liver flukes on bile, and blood flukes on erythrocytes (female *Schistosoma mansoni* ingest 400 000 red cells per hour). The food is taken in through the mouth by a pumping action of the pharynx.

Since the gut of platyhelminths lacks an anus, undigested food must be regurgitated through the mouth. This restriction limits the possibility of the development of regions specialised for particular functions in the intestine.

Because cestodes lack any form of intestine, all nutrients must be absorbed across the tegument, which is specially modified through the development of microtrichs.

Energy metabolism in parasitic helminths

Cestodes and other parasitic helminths use carbohydrates as their major, and possibly only, energy substrate. A lack of carbohydrate in the host diet results in the stunting of worms and a reduction in numbers. An inverse relationship between population density and worm size has been recognised. Parasitic helminths usually consume relatively large amounts of food, which they break down to products such as acetate and propionate. Since they are surrounded by food they have little need for economy. Schistosomes, despite living in a well-oxygenated environment, the blood stream, use an inefficient but rapid anaerobic process, the degradation of glucose to lactate. In general there is a balance between bioenergetic efficiency and metabolic simplicity. Schistosome sporocyst larvae in the snail intermediate host are able to switch between anaerobic and aerobic metabolism, depending on the conditions in the host.

The free-living stages of parasitic helminths usually do not feed, and depend on the energy stores such as glycogen acquired in the previous host.

Osmoregulation

All animals must maintain a relatively constant balance of salts, ions and water in their tissues by a process called osmoregulation. The problem differs according to the environment, and the solutions differ amongst groups. Fresh-water animals must regulate the diffusion of water into their tissues (which are hypertonic to the environment) and conserve salts. In a marine environment the problem is reversed: the need is to conserve water and eliminate salts. Marine animals tend to solve this problem by allowing their tissues to have the same ionic concentration as their environment. Parasitic platyhelminths such as trematodes and cestodes, living in the intestine or body cavities of their hosts, are faced with a similar situation. Animals which, within physiological limits, allow the osmotic potential of their tissues to match that of the environment are called osmoconformers. This is in contrast to osmoregulators, which maintain an osmotic potential which differs from that of their environment.

Despite these differences, the osmoregulatory system of most platyhelminths is structurally similar; a network of tubules running through the mesenchyme, beginning in specialised cells called flame cells and opening to the environment via nephridiopores (Fig. 4.5a). The flame cells and their associated network of tubules are collectively called a protonephridium. The terminal cell of the protonephridium, or cap cell, encloses a number of cilia which extend into a tubular cavity formed by the interdigitation of processes of the cell cytoplasm with that of the adjoining tubule cell (Fig. 4.5b). Under a microscope, the slow beating of the cilia gives the impression of a flickering flame in the mesenchyme; hence the name flame cell.

In the region where the two cells interdigitate, the wall of the tubule has a series of thickened ribs alternating with thin membrane. Between the tubule wall and the cilia there is a ring of microvilli. It is believed that tissue fluid enters the tubule in this region and is driven along the tubule by the beating of the cilia. The microvilli are thought to act as valves controlling the influx of fluid. The movement of fluid along the tubule in response to the beating of the cilia lowers the pressure in the tubule relative to that in the surrounding tissue, causing fluid to enter through the thin-wall region of the cap cell membrane. The fluid entering the tubule must have the same osmolarity as that in the tissues, yet the excreted fluid is hypotonic. It is thought that selective resorption of ions by cells lining the tubule results in the production of urine hypotonic to the internal body fluids. Although the system is thought to function primarily in osmoregulation in turbellarians, its major role in parasitic platyhelminths may be the excretion of waste products of metabolism. Ammonia, urea, amino acids and water-insoluble polypeptides have been found in the protonephridial systems of adult digeneans.

Because of their need to prevent the influx of excess water, freshwater planarians have numerous protonephridia and multiple nephridiopores. The number of protonephridia is usually significantly less in marine species. The excretory system of trematodes differs from that of turbellarians in that the proximal collecting ducts merge and join two main collecting ducts which, in turn, empty into an excretory bladder. The ducts often have fine epithelial processes projecting into the lumen, greatly increasing the internal surface area of the tubule. It has been suggested that these structures are associated with reabsorption. Digenetic flukes usually have a single, posterior excretory pore (Fig. 4.5a) and monogeneans have two anterior pores. Cestodes have protonephridia scattered throughout the

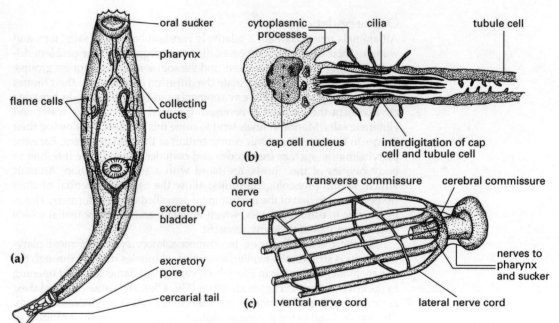

Figure 4.5
(a) Osmoregulatory system of a cercaria from the mudwhelk, *Battilaria australis*. (b) Flame cell of a cercaria of *Cloacitrema narrabeenensis*. (c) Nervous system of a trematode.

body, draining into pairs of dorso-lateral and ventro-lateral collecting vessels which open to the exterior on the posterior margin of the terminal proglottid. The ventro-lateral vessels are usually connected by transverse ducts. Studies of cestode physiology suggest that tapeworms are osmoconformers and that the main function of the protonephridial system is in excretion, not osmoregulation.

One of the many intriguing aspects of trematode physiology is the ability of different stages of the parasitic life history (see below) to respond to totally different environments. A schistosome cercaria, swimming in fresh water, must regulate its osmolarity by actively excreting excess water. As soon as it penetrates the skin of its mammalian host and enters a blood vessel, the larva enters an environment in which it can become an osmoconformer. The miracidia in the eggs laid by the adult female within the host's mesenteric veins must quickly become osmoregulators when they are released into fresh water. If they are successful in finding an appropriate snail host, the intramolluscan stages presumably conform to the osmolarity of their new environment.

Nervous system and sense organs

The simplest turbellarians such as acoels have a nervous system which resembles that of cnidarians. It is a network of nervous tissue lying beneath the surface epithelium but outside the muscles. It is best developed anteriorly and less concentrated posteriorly, where there is some development of a plexus with longitudinal fibres.

The principal nervous system of other turbellarians is situated in the mesenchyme, beneath the subepidermal musculature. In its basic form it consists of cerebral ganglia connected by a broad commissure, and longitudinal cords linked by transverse connections, an arrangement often referred to as ladder-like. There are usually three sets of longitudinal cords (dorsal, lateral and ventral) and there may be many ring commissures joining them (compare Fig. 4.5c). In the more complex turbellarians the elements of the nervous system are separated into motor and sensory pathways. The sensory nerves radiate from the cerebral ganglia to the sense organs of the head. These include tactile receptors, chemoreceptors, photoreceptors, rheoreceptors (which detect water movements over the surface of the body) and statocysts (balance organs).

The distribution of sense organs relates to their function. Thus, benthic turbellarians, which live in an environment generally well illuminated from above, have two photoreceptors in the form of pigment-cup ocelli on the head. Terrestrial triclads such as *Coenoplana coerulea*, which live under leaf litter and other debris, have many pairs of eyes arranged in rows along each side of the body, presumably an adaptation to low light intensities. Tactile receptors are widely distributed but may be concentrated around the mouth and pharynx, especially in predatory species. Chemoreceptors also tend to be concentrated anteriorly, especially on the sides of the head, frequently in ciliated grooves through which water is circulated, thus increasing the volume sampled by each receptor.

Turbellarians which swim or which are interstitial (living in the substratum, between sand grains) are unable to orient themselves to gravity by touch alone. They commonly have statocysts, minute fluid-filled cavities lined with sensory hairs, some of which support a small granule of lime or sand called a statolith. Displacement of the sensory hairs by the statolith stimulates sensory neurones and results in alterations to the animal's orientation. Statocysts are usually closely associated with the cerebral ganglion.

The nervous system of digenean trematodes is similar in layout to that of most turbellarians (Fig. 4.5c). Cerebral ganglia situated posterodorsal to the pharynx give rise to several (usually three) pairs of longitudinal nerve cords running anteriorly and posteriorly. The ventral cords are the most prominent and are united by transverse commissures throughout the length of the worm. Tactile receptors are common on the suckers of flukes.

The main concentration of nervous tissue in cestodes is in the scolex. There is a ring of ganglia from which several nerves arise, the anterior ones innervating the rostellum and other attachment organs. Several major longitudinal nerves arise laterally and extend the full length of the animal. In each proglottid of the tapeworm these lateral nerves have ganglionic swellings from which transverse commissures arise.

In addition to its function in transmitting information via nerve impulses, the nervous system secretes neuropeptides, which can act at sites remote from the point of secretion. Neurosecretory activity in turbellarians and parasitic platyhelminths has a role in a variety of functions, including motility, reproduction and morphogenesis. The secretory cells are usually located in the cerebral ganglion or in the main nerve cords. Neurosecretion is likely to be of great significance for hormone-like control of processes such as growth and development in platyhelminths, because of their lack of a circulatory system, which is the normal route of dispersion of these substances in most animals.

Reproduction and development

Asexual reproduction in turbellarians

Both fresh-water and terrestrial turbellarians may reproduce asexually, mostly by transverse fission. Fresh-water triclads such as *Dugesia* divide behind the pharynx, the original anterior end regenerating a tail and the posterior end growing a new head. This process, which has been studied extensively because of its relevance to wound healing and tissue regeneration in general, involves an interaction between neoblasts and the remaining body tissues. The interaction ensures that the neoblasts regenerate the missing parts of the body and nothing else. The asexual reproduction of trematodes and cestodes is described later because it follows the sexual processes initiated in the adults.

Reproductive systems

Most platyhelminths are hermaphroditic, but normally cross-fertilise. The complexity of the reproductive system varies considerably. Acoel turbellarians lack specific gonads and oviducts, the spermatogonia and oogonia lying free in the mesenchyme. Other species have well-developed gonads linked to complex copulatory organs by a series of ducts modified for diverse functions. The number of gonads varies, some species having many and others only one or two. Normally there are more testes than ovaries. For example, females of *Austrobilharzia terrigalensis*, a schistosome common in seagulls in Australia, have a single ovary, but the males have between twelve and twenty testes. Similarly, each segment of *Anoplotaenia dasyuri*, a tapeworm parasite of the Tasmanian Devil, *Sarcophilus harrisii*, has a single, multilobed ovary and about 200 testes.

Each testis is linked via a fine vas efferens to a common vas deferens, which may expand to form a spermiductal vesicle in which fresh sperm are stored. From there sperm move to the copulatory complex. This includes a seminal vesicle and associated prostatic glands, and the copulatory organ itself. The prostatic glands supply seminal fluid, usually into the seminal vesicle. The copulatory organ may be a cirrus (the end of the male duct, capable of being everted from the body surface) or a penis (the end of the male duct, capable of protruding from the body surface by elongation).

The female gonad of most platyhelminths is separated into two structures, the ovaries, which produce eggs, and the vitelline glands, which produce yolk cells. Ova associated with external yolk cells are called ectolecithal. This is distinct from the situation in which yolk is synthesised within the ovum (entolecithal ova). This is typical of most animals, but amongst the Platyhelminthes entolecithal ova are limited to certain turbellarian groups, including the Acoela and Polycladida. A combined organ producing both ova and yolk cells is called a germovitellarium. Where there are separate vitellaria, as in most turbellarians and flukes, they usually occur as numerous scattered follicles, linked by separate vitelline ducts to a main collecting vessel. Many tapeworms have a single vitellarium in the reproductive system of each proglottid.

Foreign sperm are stored in a copulatory bursa or seminal receptacle, which may be a simple expansion of the oviduct or a discrete sperm storage organ with a separate opening to the exterior. Many digenean flukes and some turbellarians have an additional duct, called Laurer's canal, which arises from the oviduct and opens on the dorsal body surface. It has been presumed that this duct receives the copulatory organ during mating and transfers sperm to the oviduct. However, experiments on three species of eye flukes of the genus *Philophthalmus*, using radiolabelled DNA precursors, have shown that the route of sperm transfer from the outside to the uterine seminal receptacle was via the uterus, not Laurer's canal. The fertilised ova and vitelline cells pass into a compact structure called the ootype and, in response to the secretions of Mehlis' gland into the ootype, the egg shell is formed (Fig. 4.4a).

The structure of the female tract is adapted to the animal's reproductive strategy. There are advantages for free-living turbellarians, which must actively find food and avoid being eaten themselves, in expending relatively less energy producing small numbers of eggs which are deposited in cocoons. For parasitic helminths such as tapeworms or intestinal flukes, the problems are different. Their eggs must escape from the host to the external environment and, in many cases, the larva must find a specific intermediate host. This way of life is one of high risk and low individual survival rates; consequently, parasitic helminths produce large numbers of eggs, thereby ensuring that at least a small proportion will survive. It has been estimated that over 80% of the energy intake of the tapeworm *Hymenolepis diminuta* is expended on processes associated with reproduction. These differences in reproductive strategies are apparent in gross anatomy: the distal oviduct of turbellarians is usually short, while that of cestodes or digenean flukes is long, sometimes branched, and contains large numbers of eggs. Strobilated cestodes have a major reproductive advantage over animals such as free living planarians and parasitic flukes. The single ootype in the reproductive system of these latter animals can process only one egg at a time. The maximum daily rate of egg production for trematodes such as *Fasciola hepatica* is around 25 000. A tapeworm, which may have several thousand proglottids, is able to produce many eggs simultaneously, and production rates of over a million eggs per worm per day are reported.

Although both female and male reproductive systems are usually present, they may not reach maturity at the same time: the animal is said to be protandrous if the testes mature first, and protogynous if the ovaries precede. Many cestodes have protandrous proglottids, the female system becoming mature only as the proglottids are pushed more posteriorly along the body.

Not all platyhelminths are hermaphroditic. The best known gonochoristic (separate sexes) group is the schistosomes, which are digenean flukes which live in blood vessels in vertebrate hosts (Fig. 4.2e). The male schistosome has a broad, flat body, the edges of which are folded to form a ventral groove, the gynecophoric canal, in which the thread-like female is held. Sex is chromosomally determined in the zygote, females being heterogametic (ZW) and males homogametic (ZZ), but studies on schistosome biology indicate that males must be present for normal growth and sexual differentiation of the females.

Mating

Despite being hermaphroditic, most platyhelminths mate by cross fertilisation. Free-living worms pair with the male gonopore of one aligned with the female pore of its mate. Each worm's copulatory organ is everted by hydrostatic pressure and inserted into the female atrium of the mate, where sperm are deposited.

Observations on the mating behaviour of parasitic species are difficult to achieve. Some experiments using radiolabelled DNA precursors have been performed on philophthalmid flukes, from under the eyelids of birds. In these studies cross fertilisation was found to be the most common method of insemination.

Although the proglottids of a single tapeworm are capable of cross insemination, in some species (including the dog tapeworm, *Echinococcus granulosus*) the usual method of fertilisation is self-insemination, in which the cirrus is inserted into the vagina of the same proglottid.

A small number of turbellarian species mate by a process called hypodermic impregnation. The male copulatory organ pierces the body wall of the mate and sperm are released into the mesenchyme, through which they migrate to fertilise the ova. This fertilisation mechanism also occurs in some monogeneans and cestodes.

Development
Turbellarian development
Early development includes typical spiral cleavage (Fig. 4.1) in some acoels and polyclads with entolecithal ova. Mesoderm proliferates from several blastomeres, one of which is the 4d cell. The presence of yolk cells around the ovum of ectolecithal turbellarians significantly modifies cleavage, making it difficult to trace cell fates during development. The same is true of development in trematodes, monogeneans and cestodes, all of which have ectolecithal ova.

The embryonic development of most turbellarians is direct, producing complete but minute worms. A few entolecithal polyclad species produce a planktonic Müller's larva which swims by means of eight ventral, ciliated lobes for several days before settling and metamorphosing into an adult worm.

Trematode development
A trematode egg develops and hatches as a ciliated larva, the miracidium, which infects the molluscan intermediate host. Miracidia are usually released from the eggs in water and swim until they find a suitable mollusc. Occasionally the egg containing the miracidium is eaten by a snail and the larva is released inside the host. At the time of infection of the mollusc, the ciliated epithelium is shed and the syncytial neodermis is formed. The sequence of developmental stages in the mollusc (Fig. 4.6) is highly variable, but generally the miracidium transforms into a sporocyst within which a redial generation proliferates, and cercariae (the larvae infective to the next host) develop in the rediae. Sporocysts are sac-like bodies with no gut (Fig. 4.6b-2). Rediae are similar in shape but have a mouth, muscular pharynx and blind-ending gut, and are capable of ingesting their host's tissues (Fig. 4.6b-4). The cercariae escape from sporocysts or rediae and leave the mollusc prior to infecting the next host. They are tailed larvae and can swim (Fig. 4.6b-5).

A proliferating series of intramolluscan stages is a fundamental feature of digenean biology, increasing parasite numbers and allowing significant flexibility in life history patterns. If external environmental conditions are unfavourable for survival of cercariae, rediae of trematodes such as the sheep liver fluke, *Fasciola hepatica*, may switch from producing cercariae to generating more rediae.

A typical trematode larval sequence is exhibited by *Stictodora lari*, a small fluke which occurs in the intestine of fish-eating birds such as cormorants, gulls and pelicans (Fig. 4.6). The minute eggs passed in the bird's faeces are eaten by *Battilaria australis*, a common mudwhelk of southern Australia. The miracidium escapes from the egg and transforms into a branched sporocyst in the snail's rectal blood sinus. Large numbers of rediae are produced in the sporocyst, and these migrate to the sinuses between the digestive gland and gonad. They ultimately completely destroy the gonad tissue. Cercariae with a pair of eyespots and a finned tail are released from the rediae and escape from the snail. These actively swimming larvae penetrate the body of small estuarine fish, lose their tail and form a cyst in the musculature. The encysted larvae, now called metacercariae, develop into adult worms in the intestine of birds (and occasionally cats) which eat the fish. This life history is similar to that of human liver flukes such as *Opisthorchis sinensis*.

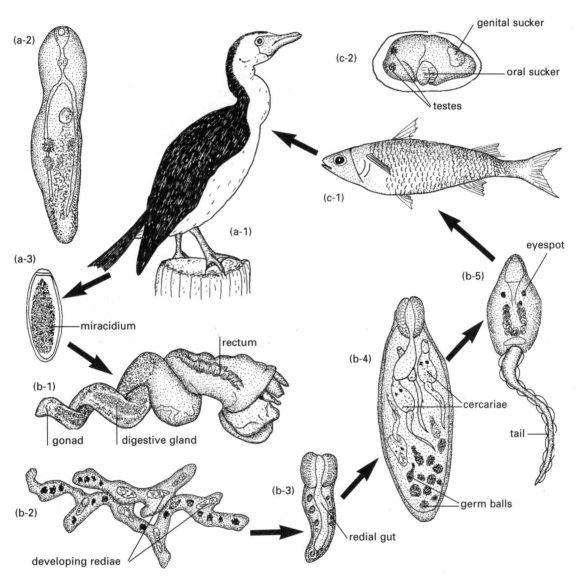

Figure 4.6
Life history of the trematode *Stictodora lari*. **(a-1)** Definitive host, the little pied cormorant, *Phalacrocorax melanoleucos*. **(a-2)** Adult fluke from the intestine of a cormorant. **(a–3)** Egg, passed in bird's faeces. **(b-1)** First intermediate host, the mudwhelk, *Battilaria australis*, shell removed. **(b-2)** Sporocyst from the rectal blood sinus of *Battilaria australis*. **(b-3)** Immature redia from blood spaces around the snail's digestive gland and gonad. **(b-4)** Mature redia containing cercariae.
(b-5) Free-swimming cercaria. **(c-1)** Second intermediate host, the sea mullet, *Mugil cephalus*. **(c-2)** Metacercaria from the muscles of a mullet.

Host reactions to trematode larvae

Molluscs, like all animals, have defence mechanisms which protect them against colonisation by other organisms. The basis of these protective responses is the ability to recognise foreign material as non-self and mount a cellular reaction which kills and removes the invader. For example, the miracidia of the human blood fluke *Schistosoma mansoni* elicit no reaction in the normal intermediate host snail *Biomphalaria glabrata*, but are unable to develop further in the Australian planorbid snail *Isidorella newcombi*. Miracidia invading this snail species are quickly killed and destroyed by a cellular reaction.

A similar reaction is mounted to germinal sacs which are themselves infected and not developing normally. A haplosporidian parasite, *Urosporidium* sp., occasionally infects the germinal sacs of *Stictodora lari* in *Battilaria australis*. The mudwhelk, which normally tolerates these trematode parasites, responds with a prominent cellular reaction to the protistan-infected trematodes. There is good evidence supporting the concept that trematode germinal sacs actively control the defences of susceptible molluscan hosts, though the exact mechanism is not understood.

Austrobilharzia terrigalensis, a schistosome of birds with larval stages in the same estuarine whelk, *Battilaria australis*, is unusual in that it never occurs by itself; there is always another species of trematode present. Presumably *A. terrigalensis* requires the presence of the other trematode in order to avoid the mollusc's defences. In *Planaxis sulcatus*, a gastropod common in northern Australia, including islands of the Great Barrier Reef, *A. terrigalensis* is normally found by itself. A possible explanation for the different behaviour of this trematode in the two snail species is that *A. terrigalensis* entered the Australian region from the north and has had time to adapt to *Planaxis* and its defence mechanisms, but its association with *Battilaria* has not been long enough for this adaptation to occur. In time it may develop the ability to control the defences of *B. australis* on its own and not have to rely on the presence of another parasite in the snail.

Monogenean development

The ciliated larvae of monogeneans are called oncomiracidia. They have a well-developed opisthaptor which is used in attaching to a new host. Those which parasitise fish swim actively, though because of the vast difference between the swimming speed of fish and oncomiracidia, eggs are usually released by adult worms when the fish hosts are less active or grouped for activities such as spawning. The transformation from the ciliated larval epidermis to the syncytial neodermis occurs at the time of attachment to the new host, and further development is a simple metamorphosis to the adult.

Cestode development

The initial infection of intermediate hosts by cestodes is frequently via the ingestion of the egg, which contains the oncosphere. Occasionally the larva escapes from the egg as a free swimming, ciliated coracidium and is eaten by an invertebrate such as a copepod. The cestodarian *Austramphilina elongata* has a ciliated larva called a lycophore, which penetrates the exoskeleton of juvenile crayfish, *Cherax destructor*.

The next larval stage of cestodes is a six-hooked (hexacanth) larva, which migrates to a site amongst the viscera of the invertebrate or vertebrate host. Here the hexacanth larva typically transforms into a metacestode, which grows into an adult tapeworm — sometimes in the same host,

THE PLATYHELMINTHES, NEMERTEA, ENTOPROCTA AND GNATHOSTOMULIDA

but more commonly following one or more transfers to new hosts, usually through predation of one host by the next.

Cestode species have a great variety of larval stages, sometimes including asexual reproduction. One of the best studied is the hydatid cyst, the larval stage of the dog tapeworm *Echinococcus granulosus* (Fig. 4.7d). The intermediate host is most commonly a herbivore, especially sheep, but infection can also occur in humans. The oncosphere larva is released from the egg in the small intestine, passes through the intestinal wall and is carried in the blood stream to another tissue, often the liver, where it settles and begins to grow. The cyst is a fluid-filled sphere, lined internally with a germinal epithelium. This germinal layer buds off daughter cysts into the cavity of the mother cyst. Within these cysts the germinal epithelium produces protoscolices (Fig. 4.7e), each of which has the capacity to mature into an adult tapeworm if eaten by a dog (the definitive host). A large cyst may contain thousands of infective protoscolices, so a canid which has regular access to offal from infected intermediate hosts may harbour large numbers of adult worms: as many as 400000 have been found in a single dog. Interestingly, each protoscolex also has the capacity to develop into another hydatid cyst if it is released into the body cavity of the intermediate host.

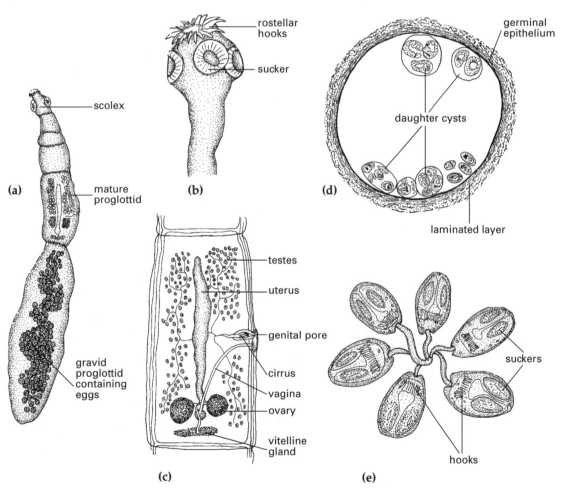

Figure 4.7
(a) Adult *Echinococcus granulosus* from the intestine of a dog. (b) Scolex of the pork tapeworm, *Taenia solium*. (c) Mature proglottid of a taeniid tapeworm. (d) Hydatid cyst from the viscera of an intermediate host. (e) Group of protoscolices from a hydatid cyst.

Box 4.2
Diagrammatic summary of some of the diverse life histories of platyhelminths.

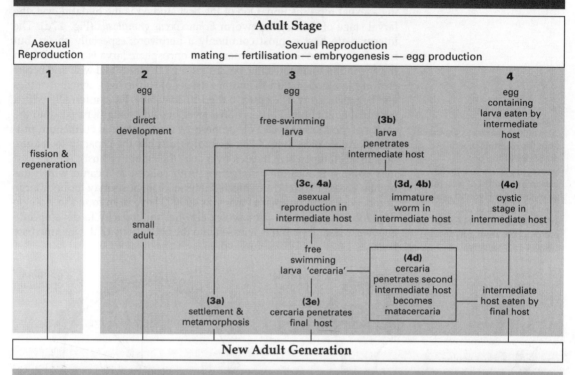

Diagrammatic summary of some of the diverse life histories of platyhelminths. Representatives of each major group in the phylum are used to illustrate the sequence of stages passed through in progressing from one adult generation to the next.

1 Triclad planarians such as *Coenoplana coerulea* often reproduce by binary fission, each half regenerating the missing body structures (1).

2 Sexual reproduction of most triclad turbellarians involves direct development from the embryo into a small worm, which grows to maturity (2).

3.1 Marine polyclads such as *Pseudoceros bimarginatus* reproduce sexually. Their eggs release a planktonic Müller's larva (3) which settles down to the substratum and metamorphoses into an adult worm (3a).

3.2 Eggs of the digenetic trematode, *Schistosoma mansoni* are passed in human faeces. In fresh water they release a miracidium (3) which finds and penetrates a specific snail host (*Biomphalaria*) (3b). In the snail the parasite reproduces asexually (3c), producing cercariae which escape from the mollusc and enter human skin directly (3e).

3.3 Eggs of the cestodarian *Austramphilina elongata*, a parasite in the body cavities of long-necked turtles, *Chelodina longicollis*, hatch in fresh water, releasing a ciliated lycophore larva (3b) which penetrates the exoskeleton of juvenile crayfish, *Cherax destructor* (3d). In this host the worm develops into an immature adult which matures if the crayfish is eaten by a turtle.

3.4 Eggs produced by monogenetic trematodes such as *Pseudothoracocotyla scomberomori* hatch when released into water (3), liberating a free-swimming oncomiracidium. If this larva comes into contact with the correct fish species, it attaches itself and metamorphoses into an adult worm (3a).

4.1 Adults of the aspidobothrean trematode *Lobatostoma manteri* parasitise fish (*Trachinotus blochi*). The parasite eggs are passed in the host's faeces. Eggs containing the cotylocidium larva (4) are eaten by marine snails (*Cerithium moniliferum*) in which an immature adult worm develops (4b). The snails are eaten by the fish.

4.2 Eggs of the Chinese liver fluke, *Opisthorchis sinensis*, a digenetic trematode, are eaten by the snail host (4), in which the released miracidium initiates asexual reproduction (4a) which produces cercariae. These penetrate the body of fresh-water fish and encyst as metacercariae (4d). The adult worms develop in the bile ducts of humans who eat the fish raw.

4.3 The eggs of the eucestode *Anoplotaenia dasyuri*, a parasite of the Tasmanian Devil, *Sarcophilus harrisii*, are passed in the host's faeces and ingested by a herbivorous intermediate host, usually a wallaby (4). The eggs hatch in the intermediate host, releasing the metacestode which develops into a cysticercus in viscera such as the heart or lungs (4c). The maturation of the worm relies on predation of the intermediate host by the final host.

In Australia there is an alternative cycle of transmission of *Echinococcus granulosus* involving wild dogs (such as dingoes) and macropod marsupials, including swamp wallabies. In macropods the lungs are the most commonly infected organ. The large number of hydatid cysts present often causes significant pathology. This damage to the lungs of the wallabies makes infected animals more susceptible to predation by dingoes, thus ensuring transmission of the parasite.

The life history patterns of a range of representative platyhelminths are summarised in Box 4.2.

Significance of platyhelminths in agriculture and medicine

Turbellaria

In the 1960s a triclad flatworm from New Zealand, *Artioposthia triangulata*, was introduced to Great Britain and Ireland, where it has become a serious pest because it feeds on earthworms. Although the introduction appears to have been accidental, probably by way of plants imported in containers, it highlights the danger of the release of any species into new localities.

Trematodes

The sheep liver fluke, *Fasciola hepatica*, is a major pathogen of domesticated sheep and cattle and occasionally humans, and has a cosmopolitan distribution. In Australia and New Zealand the molluscan intermediate host is *Austropeplea tomentosa*, a small amphibious snail found in shallow streams or in moist, shaded areas on pasture. The cercariae of *Fasciola hepatica* encyst on vegetation and are ingested by grazing ruminants. Human infection usually results from eating watercress harvested from wet areas inhabited by the snail host. In the mammalian host the young worms migrate through the tissue of the liver to the bile ducts, often causing significant damage. The adult flukes reside in the gall bladder. Fascioliasis in sheep and cattle causes reduced weight gain and low milk yield. Infected livers are rejected at abattoirs in many countries, resulting in considerable economic loss to primary producers.

Amongst many trematode species which infect humans the most important are the blood flukes or schistosomes. Two forms of infection exist: intestinal schistosomiasis caused by *Schistosoma mansoni*, *S. japonicum* and *S. mekongi*, and urinary schistosomiasis caused by *S. haematobium*. It is estimated that at least 200 million people are infected in 74 countries. Most of the infections (and those of greatest severity) occur in young children, because they have frequent contact with waters in which the molluscan intermediate hosts live. The cercariae released from the snail hosts penetrate the skin and migrate via the heart and lungs to the liver, where they mature into adult worms. From there they move to the mesenteric veins of the intestine and the veins surrounding the urinary bladder. Schistosomiasis results from the reaction by the human host to eggs which become trapped in tissues instead of escaping to the environment in faeces or urine. As more and more eggs become trapped in organs such as the bladder wall, ureters, intestinal wall and liver, fibrous tissue replaces the normal structures and the functions are severely affected.

We saw earlier that the schistosome *Austrobilharzia terrigalensis* parasitises seabirds, particularly the Silver Gull, *Larus novaehollandiae*, and passes through its larval stages in the mudwhelk, *Battilaria australis*. If the cercariae of this or other bird schistosomes penetrate human skin, they are killed by a cellular immune response, producing a form of dermatitis known as cercarial dermatitis or swimmers' itch. In some countries this is a significant problem for people working in rice fields.

Cestodes

Tapeworms, especially species of the genera *Echinococcus* and *Taenia*, are important as agents of human disease and of economic loss in animal husbandry. The pork tapeworm, *Taenia solium* (Fig. 4.7b), is a parasite of the human intestinal tract, with the larval stage (cysticercus) in pigs. Humans pass the infection to pigs via egg-carrying proglottids voided with faeces and become infected by eating pork containing cysticerci.

It was mentioned earlier that humans can acquire hydatid infections by ingesting the eggs of *Echinococcus granulosus* passed in dog faeces. In humans the hydatid cysts develop mainly in liver or lungs, but may be found in almost any tissue. In a site where they have room to grow, hydatid cysts may be unnoticed for many years, but in other situations they cause debilitating disease. Each cyst is full of highly antigenic fluid which slowly leaks out, sensitising the host. One particular complication is death due to anaphylactic shock following cyst rupture and the release of a large volume of the fluid. Effective control of *E. granulosus* involves preventing farm dogs from having access to offal from the usual intermediate host, sheep. An impediment to control in Australia is the existence of a sylvatic cycle of infection involving dingoes and other wild dogs, and macropods. In some regions sheep are infected by eggs left on pasture by dogs which live in adjacent bushland.

The cysts of *E. granulosus* have a single cavity (that is, they are unilocular) and are surrounded by a thick capsule deposited by the host. A different species, *E. multilocularis*, has cysts which are multilocular and not delimited by host tissue. These cysts metastasise throughout the body like a malignant tumour, and the disease caused by *E. multilocularis* has a high rate of mortality.

Vaccination against helminth parasites

The possibility of preventing disease by vaccination is always an attractive option. It has been a goal for those working on human schistosomiasis for many years. Initial results using live, irradiated cercariae to protect mice against further infection were extremely encouraging. The current focus of research is on the development of defined antigen vaccines using the techniques of molecular biology. Unfortunately the results so far have not been promising. This is in sharp contrast to those obtained with recombinant antigen vaccines against cestodes. Oncospheres have been shown to be a potent source of host-protective antigens for tapeworms of the family Taeniidae. Two recombinant antigen vaccines against the larval stages of *Taenia ovis* and *Echinococcus granulosus* have been shown to induce levels of protection as high as 98% in sheep.

Classification of the phylum Platyhelminthes

The following outline of the classification of the phylum is based on the widely accepted scheme of Ulrich Ehlers. The traditional classes and orders are used, although Ehlers and others have demonstrated that the characters used to demarcate the class Turbellaria (acoelomate Bilateria, free-living, with a ciliated epidermis) are shared with other animals, and thus are not definitive. The 'Turbellaria' are paraphyletic (only the major orders are listed here). The name is still useful within the context of the phylum, however, provided that its use does not necessarily imply any phylogenetic relationship. Some details of the major, well-defined groups of platyhelminths are given.

CLASS TURBELLARIA
Free-living flatworms (some commensal or parasitic); ciliated surface epithelium.
Order Catenulida — Simple pharynx and gut; unpaired protonephridial system; dorsal male genital pore; aciliary spermatozoa; elongate fresh-water and marine species.
Order Acoela — Lacking intestine and protonephridia; small species of marine and brackish-water sediments; some planktonic or symbiotic.
Order Polycladida — Many-branched intestine; mostly benthic marine species of moderate size; some pelagic, some symbiotic.
Order Rhabdocoela — Bulbous pharynx; marine and freshwater species; some symbionts on or in other invertebrates. (Suborder Temnocephalida with anterior tentacles and posterior adhesive organs is ectocommensal on freshwater invertebrates such as decapod crustaceans and is abundant in Australia and New Zealand.)
Order Tricladida — Three-branched intestine; mostly free-living marine, freshwater and terrestrial species (includes the familiar planarians).

CLASS TREMATODA
Parasitic flatworms with a ventral sucker.
Subclass Digenea
Trematodes with at least two hosts in the life cycle, the first a mollusc.
Subclass Aspidobothrea
Trematodes with a single host in the life cycle.

CLASS MONOGENEA
Parasitic flatworms with a posterior attachment organ (opisthaptor) and a single host in the life cycle.
Subclass Monopisthocotylea
Monogeneans with a simple opisthaptor.
Subclass Polyopisthocotylea
Monogoneans with a complex opisthaptor with multiple suckers.

CLASS CESTODA
Parasitic flatworms without an alimentary canal.
Subclass Cestodaria
Cestodes with a trematode-like body form.
Subclass Eucestoda
Cestodes with an anterior attachment region (scolex) and a strobila of proglottids.

Phylum Nemertea

Bilaterally symmetrical, possibly acoelomate, unsegmented worms; gut with mouth and anus; anterior brain and longitudinal nerve cords; protonephridial excretory system; simple circulatory system; dorsal eversible proboscis; separate sexes, sac-like gonads; spiral cleavage development.

Introduction

The phylum Nemertea contains about 900 species of unsegmented worms, ranging in length from a few millimetres to several metres. The body is usually more or less cylindrical (Fig. 4.8a), but may be flattened dorsoventrally. Most nemerteans are benthic marine animals, including a number that are commensal in the gill chambers of crabs, bivalves and ascidians. *Carcinonemertes* feeds on the eggs carried by female crabs. Some species of nemerteans are planktonic in deep ocean waters, and a few occur in fresh water and moist terrestrial habitats. Nemerteans have left no fossil record.

Functional morphology

The characteristic feature of nemerteans is a dorsal, eversible proboscis (Fig. 4.8a), situated in a fluid-filled cavity above the gut. The proboscis cavity is lined by epithelium and is formally recognised as a coelomic space, the rhynchocoel (Fig. 4.8b, c), but there is no general body cavity. The internal organs are surrounded by mesenchyme, as in Platyhelminthes. The body surface, which lacks a cuticle, is covered by a ciliated, glandular epithelium with rhabdites, resting on a thickened basement membrane, the dermis. Thick layers of muscle lie beneath the dermis. The arrangement of the layers differs in the subclasses. The Palaeonemertea have striated muscle. An external layer of circular muscle is followed by longitudinal muscle, then sometimes by an inner circular layer; nerve cords in palaeonemerteans are located in the dermis or in the longitudinal muscle layer. Heteronemertea (Fig. 4.8b) have smooth muscle. An outer longitudinal muscle layer is followed by circular muscle, then by an inner longitudinal layer; nerve cords are located in the outer longitudinal muscle layer. Hoplonemertea also have smooth muscle. An external layer of circular muscle is followed by longitudinal muscle, as in palaeonemerteans, but the nerve cords are located internal to the muscle layers. These structural differences are not associated with distinctive patterns of movement and locomotion.

Nemerteans, in spite of their well-developed musculature, are restricted to slow ciliary crawling and some slow, muscular bending and peristaltic burrowing. These limitations on movement are associated with their acoelomate construction and simple nervous system; yet all nemerteans are carnivores, feeding on active prey such as annelids and crustaceans. This capability stems from the action of the proboscis as a prey-capture mechanism. The proboscis is a tubular infolding of the body wall. When infolded, it is lined by a continuation of the epithelium of the body surface, surrounded by muscle layers arranged in the reverse sequence to those of the body wall. Further layers of circular and longitudinal muscles form the wall of the proboscis cavity. Muscle contractions pressing on the fluid in the rhynchocoel cause a rapid eversion of the proboscis, which coils around the prey. The latter is then drawn to the mouth and ingested. In hoplonemerteans, the proboscis carries calcified stylets which pierce the prey during capture. Infolding of the extended proboscis is brought about by a long retractor muscle running between the tip of the proboscis and the posterior wall of the proboscis cavity.

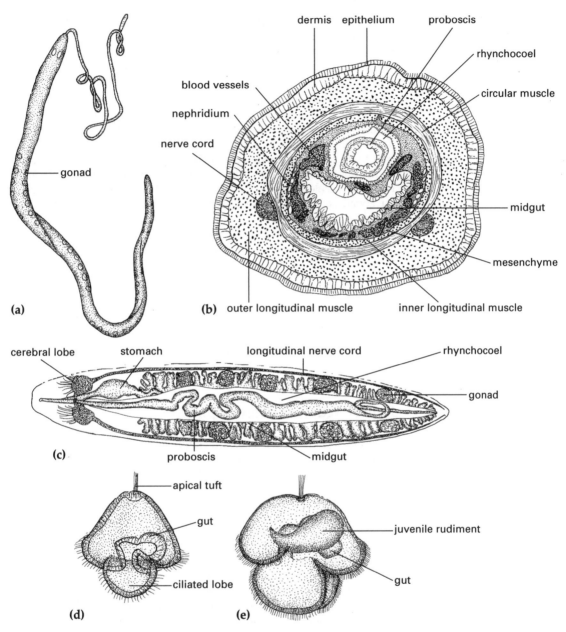

Figure 4.8
(a) The heteronemertean *Bennettiella insularis*.
(b) Transverse section of *Bennettiella*. (c) Frontal longitudinal section through *Amphiporus*. (d) Pilidium larva.
(e) Older pilidium with developing juvenile rudiment.
(a, b after Gibson 1981; c after Nielsen 1995; d, e after Gibson 1972.)

Organ systems

Digestive system
The gut begins with a short ectodermal foregut (buccal cavity, oesophagus and stomach), followed by a long, tubular midgut with numerous paired lateral diverticula. This ends posteriorly in an ectodermal rectum and anus.

Excretory system
Nemerteans, like platyhelminths, have protonephridia with flame cells. The number of protonephridia varies in different species from one pair to

several thousand pairs. The flame cells are often closely associated with the circulatory system. The protonephridia function in excretion, ionic regulation and osmoregulation.

Circulatory system

Nemerteans have a circulatory system in the mesenchyme. Two main longitudinal channels are linked by anterior and posterior transverse vessels and a network of smaller vessels (Fig. 4.8b). The system is lined by a mesodermal epithelium and is thus definable as a coelomic space, in contrast to the typical unlined haemocoelic blood space of other invertebrates. Numerous cells float in the blood. Unlike the coelomic body cavity in the recognised coelomate phyla, the nemertean blood system does not contain the gonads, and there are no coelomoducts opening to the exterior.

Nervous system and sense organs

The nemertean central nervous system (Fig. 4.8b, c) begins with two pairs of cerebral lobes, dorsal and ventral, surrounding the anterior entrance to the proboscis, above the gut. From the ventral cerebral lobes, a pair of lateral longitudinal nerve cords extends to the posterior end of the body. The nerve cords lie either in the dermis or in the underlying muscle layers. Numerous transverse commissures connect the two nerve cords. There is also a nerve plexus beneath the surface epithelium. The simple sense organs include ciliated pits and pigmented ocelli on the anterior region.

Reproduction and development

Sexes are separate and gonads are simple and temporary, developing as paired rows of bilateral sacs in the mesenchyme (Fig. 4.8c). Each gonad sac is lined by a mesodermal epithelium from which gametes are proliferated. Simple temporary gonoducts develop from ectodermal invaginations. The gametes are usually spawned freely, with external fertilisation, but some species deposit their eggs in gelatinous masses and a few are viviparous. Cleavage is spiral, but the details of embryonic development are still not clear. At least some of the mesoderm develops from blastomere 4d, but other cells are also involved.

Palaeonemerteans and hoplonemerteans develop directly to a juvenile worm, often with a ciliated, lecithotrophic, worm-like planktonic phase, but heteronemerteans develop through a planktotrophic pilidium larva (Fig. 4.8d) which appears to be a modified trochophore. The adult rudiments (Fig. 4.8e) are formed in a highly specialised way from localised cell discs invaginated from the surface of the larva, and there is a complex metamorphosis to the benthic juvenile stage.

Evolutionary relationships

Embryological and molecular evidence links the Nemertea with the coelomate phyla Sipuncula, Mollusca and Annelida, but the relationships among these groups are not clear. The blood spaces in nemerteans cannot be related developmentally to the coelom of the other phyla and may have evolved independently (see Chapter 18 for further discussion).

Classification of the phylum Nemertea

The taxonomy of nemerteans is difficult, relying mainly on internal differences detectable only by histology, and much remains to be learned. The Australasian nemertean fauna is plentiful but still poorly known. Only a few species have been described and identified. The phylum is divided into two classes.

CLASS ANOPLA
Mouth below and behind cerebral ganglia; separate proboscis pore anteriorly.
Order Palaeonemertea — with direct development.
Order Heteronemertea — with a pilidium larva.

CLASS ENOPLA
Mouth in front of cerebral ganglia and usually combined with proboscis pore as a common opening.
Order Hoplonemertea — proboscis armed with stylets. The freshwater nemerteans (*Prostoma*) and terrestrial nemerteans (*Geonemertes*) are part of this group.
Order Bdellonemertea — proboscis secondarily unarmed; a single genus, *Malacobdella*, commensal in the mantle cavity of bivalve molluscs.

Phylum Entoprocta

Bilaterally symmetrical, acoelomate, unsegmented sessile animals, usually with multiple zooids, having an apical ring of tentacles surrounding both mouth and anus; U-shaped gut; protonephridial excretory system; hermaphrodite reproductive system; spiral cleavage development.

Introduction

The Entoprocta comprise about 150 species of small, acoelomate, sessile ciliary feeders (Fig. 4.9a) attached to rocks, shells and the surfaces of other benthic invertebrates. A few are solitary, but most are branching forms with budding growth and multiple zooids. *Pedicellina* is a common example. Almost all genera are marine, only *Urnatella* inhabiting fresh water. Jurassic fossil entoprocts are known.

Functional morphology

The zooid (Fig. 4.9b) has a globular body or calyx with a flattened apical surface, the atrium, surrounded by a ring of ciliated tentacles. Opposite the atrium is a basal, cylindrical stalk. The body surface is covered with a thin cuticle, containing chitin, overlying a ciliated surface epithelium. Thin longitudinal muscles act to bend the tentacles and to contract and bend the stalk. The tentacles carry lateral rows of compound cilia which draw a water current on to the frontal faces of the tentacles, and frontal tracts of smaller cilia which carry filtered particles to the mouth situated in the midline of the atrium.

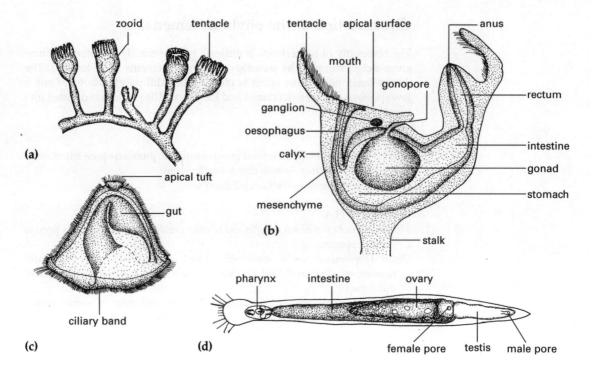

Figure 4.9
(a) The entoproct *Pedicellina*. (b) Zooid of *Pedicellina*, general structure. (c) Planktonic larva of the entoproct *Electra* (after Nielsen 1995). (d) Gnathostomulid anatomy (after Sterrer 1982).

Organ systems

The spaces between the internal organs are filled by mesenchyme. There are some irregular, fluid-filled channels through which fluid can move between the stalk and the body, but no definitive coelom can be identified.

Digestive system
The gut (Fig. 4.9b) is U-shaped, with a narrow oesophagus, swollen stomach, short intestine and rectum, opening at an anus which also lies within the ring of tentacles.

Excretory system
The excretory system consists of a pair of protonephridia flanking the gut, with a common nephridiopore in the midline just behind the mouth.

Nervous system
The nervous system is centred on a bilobed ganglion between the mouth and anus. Peripheral nerves passing to the tentacles and stalk control ciliary beat and contractile movements.

Reproduction and development

Entoprocts are hermaphrodites. The gonads are a pair of sacs (Fig. 4.9b) adjacent to the protonephridia, each with a simple duct opening to a common gonopore behind the nephridiopore. Sperm are shed freely, but eggs are retained and fertilised in the ovaries. The embryos are extruded in a jelly which adheres within the ring of tentacles, where development is completed and the larvae hatch. The embryos of *Loxostomella* develop a placental connection with the surface of the parental zooid; the freshwater *Urnatella gracilis* is viviparous.

Development proceeds through spiral cleavage and formation of mesoderm from the cell 4d. The larva (Fig. 4.9c) is a type of planktotrophic trochophore, but has a free-swimming period of only a few hours. The postero-ventral surface of the larva develops precociously as an attachment organ, by which the larva adheres before metamorphosis to the zooid form.

Phylum Gnathostomulida

Bilaterally symmetrical, acoelomate, unsegmented marine worms of interstitial habitats; gut with mouth and jawed pharynx, anus vestigial; no excretory system; hermaphrodite reproduction system; spiral cleavage development.

The Gnathostomulida are a group of very small, unsegmented acoelomate worms of interstitial habitats, first described in 1956. About 80 species are known. The body (Fig. 4.9d) is cylindrical, with a slightly enlarged head and tapering tail. The external surface is ciliated, and the animals glide by metachronal ciliary beating. Muscles are striated but weakly developed. The tubular gut has a muscular pharynx with cuticular jaws, but no anus. A sparse mesenchyme lies between the internal organs, which include two to five pairs of protonephridia. Gnathostomulids are hermaphrodites, with a well-differentiated reproductive system. An anterior ovary and sperm storage sac is followed by one or more pairs of testes and a posterior copulatory organ. Copulation leads to internal fertilisation and the release of a single, relatively large egg by the rupturing of the body wall. Development begins with spiral cleavage and is direct.

Gnathostomulids have a unique combination of structural and functional features, and their relationship to other phyla of spiral cleavage worms is unclear.

Chapter 5

The Aschelminthes

W.L. Nicholas

	Introduction 87	

PHYLUM NEMATODA	Functional morphology 88
	    Locomotion 88
	    Cuticle and epidermis 88
	    Musculature 90
	    Digestive system 90
	    Secretory/excretory system 92
	    Nervous system 94
	    Reproductive organs 94
	
	Sex 95
	    Genetics, developmental biology
	    and *Caenorhabditis elegans*
	    96
		Embryonic development 96
		Ecology 98
		    Marine nematodes 98
		    Freshwater nematodes 99
		    Soil nematodes 99
		    Plant-feeding nematodes 99
		    Parasitism and other
		        associations with
		        invertebrates 100
		    Parasites of vertebrates 101

PHYLUM NEMATOMORPHA	105	
PHYLUM GASTROTRICHA	105	
PHYLUM KINORHYNCHA	106	
PHYLUM LORICIFERA	108	
PHYLUM ROTIFERA	Morphology 109 Reproduction and development 110 Ecology 110	
PHYLUM ACANTHOCEPHALA	Morphology 111 Reproduction 112 Development and life cycle 114	Classification of the Aschelminthes 114

Introduction

Most Aschelminthes are microscopic, worm-like, aquatic animals of dubious evolutionary relationship, unknown to the general public so that they have no common names. Some, however, are quite large parasites that have been known from antiquity, usually described as roundworms. The composition of the Aschelminthes is as much a question of convenience as of presumed phylogeny, and there is no consensus amongst zoologists as to which groups should be included or whether the Aschelminthes constitutes a single phylum (see Chapter 18 for further discussion). Specialists in the various groups consider each to be of phylum rank, so that the Aschelminthes becomes a superphylum of doubtful phylogeny. It is controversial, though convenient, to include the Acanthocephala, relatively large parasitic worms, with the Aschelminthes. There are grounds for considering that the Acanthocephala are closer to the Rotifera than to the other Aschelminthes.

The phyla included in the Aschelminthes, and the major subgroups within each phylum, are listed in the classification at the end of the chapter. The various groups are of very different sizes. The Nematoda are overwhelmingly the most important in number of species and ecological significance, and as parasites of humans, animals and plants. They dominate the marine meiobenthos (the microscopic inhabitants of marine sediments) and play very significant roles in soils. The Rotifera comprises many species, which are prominent in fresh waters. The Kinorhyncha and Loricifera each contain only a few species of minute marine animals. The more numerous Gastrotricha are also minute and aquatic. The small number of Nematomorpha are parasites of arthropods, and may be quite large. The Acanthocephala are all parasitic.

Aschelminths are bilaterally symmetrical, triploblastic, unsegmented animals (except Kinorhyncha), with a body cavity that is not a coelom. There is an epidermis which also gives rise to the nervous system. The epidermis secretes an external cuticle (which is moulted) in all phyla except Rotifera and Acanthocephala. There is a gut with a mouth, an anus and an absorptive mid-region with an endodermal lining (except that Acanthocephala have no gut). Most of the musculature is mesodermal. The Aschelminthes are often characterised as pseudocoelomates. A body cavity develops within the mesodermal tissues, but is not considered to be a coelom because it is not lined by an epithelium. The cavity may in some be a remnant of the embryonic blastocoel. The body may be annulated, or divided into distinct regions, but except in the Kinorhyncha there is no serial repetition of internal organs, so the body is not metamerically segmented. Bilateral symmetry may be overlaid by radial symmetries or be barely discernible, as in the Acanthocephala.

Phylum Nematoda

Unsegmented worms with layered cuticle, moulted periodically; mouth surrounded by six lips; dorsal and ventral nerve cords in epidermis; excretory system of renette cells or tubules; separate sexes, tubular gonads and gonoducts.

About 15 000 species of nematodes have been described, but this is undoubtedly a small fraction of the total number of species. Estimates of

500 000 to one million species have been authoritatively made. Nematodes can be found everywhere that life can be supported, although they are absent from plankton except as parasites. Nematodes are abundant in marine and freshwater sediments and in the soil, where they depend on a film of water for activity. Their capacity to survive periodic drying and extreme cold permits them to inhabit deserts and polar regions. Many species are parasitic in plants and animals, and most are microscopic. There are two classes, the Adenophorea and Secernentea. In the Adenophorea the 'excretory organ', a misleading name, is a glandular renette cell that opens by an anterior ventral pore. Adenophorea usually possess numerous epidermal glands, as well as caudal glands that open on the tip of the tail, but lack phasmids (paired caudal sensilla). Males usually have two testes. The Secernentea have lateral 'excretory' canals opening by an anterior ventral pore, and paired phasmids, but lack epidermal glands and caudal glands. Males almost always have a single testis.

Functional morphology

The appearance and internal anatomy of typical free-living nematodes is illustrated in Figure 5.1. The symmetry of the head, and transverse sections through the pharyngeal and intestinal regions, are shown in Fig. 5.2.

Locomotion

The body form of most nematodes (Fig. 5.1a, d) is adapted to progression by sinuous body waves which is efficient for moving through soil, mud or the tissues of plants and other animals but inefficient for swimming. This accounts for the absence of nematodes from plankton. Generally the body is a smooth or annulated cylinder, tapered at both ends. The body waves are produced by the sequential contraction of longitudinal muscles (Fig. 5.2f), opposed by the flexible but relatively inextensible cuticle. The repertoire of body movements is stereotyped, primarily as forward or backward body waves, or curling and uncurling. Locomotion comes from the forces exerted by the body waves on the substratum: surface tension facilitates crawling through thin films of liquid. The common adoption of this means of locomotory propulsion, and the underlying mechanical principles, account for the typical body form and uniform anatomical structure that characterises most nematodes. Although swimming is possible using very rapid body waves, it is energetically expensive because much of the work is wasted in turbulent water movement. However, there are exceptions. Some minute marine nematodes crawl on tube-feet, some plant parasites become globular and sedentary, and a few other soil nematodes are strongly annulated and progress more like earthworms.

Cuticle and epidermis

The epidermis (hypodermis) secretes a cuticle that is moulted four times before a nematode becomes adult. The cuticle often bears cuticular hairs (setae), spines, or annulations of very variable coarseness, and may have longitudinal ridges (alae). The cuticle must be flexible enough to bend with body waves, or when coiling, but not crimp or buckle when the longitudinal muscles contract. Muscle contraction acts on the cuticle through fine fibres that cross the epidermal cells. Nematodes generally maintain a higher hydrostatic pressure within their tissues than that in the external medium, and the cuticle must also resist any tendency of the body to

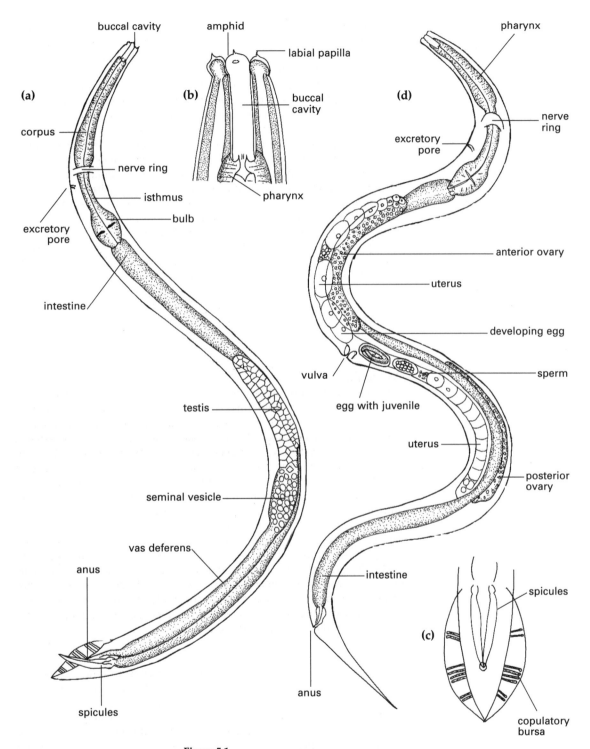

Figure 5.1
Anatomy of Rhabditidae (Secernentea), soil-inhabiting nematodes. **(a–c)** *Mesorhabditis* sp.: (a) Male. (b) Buccal cavity. (c) Copulatory bursa in ventral view. **(d)** *Caenorhabditis elegans*, a protandrous hermaphrodite.

Figure 5.2 (opposite)
Nematode anatomy.
(a) Arrangement of cephalic sensilla and lips on the head of *Caenorhabditis elegans* (Secernentea). **(b)** Head of *Ceramonema carinatum* (Adenophorea). **(c)** Transverse section through the pharyngeal region of a small, free-living nematode. **(d)** Section through the intestinal region of the same nematode. **(e)** Ultrastructure of the nematode sensillum.
(f) Transverse section through the intestinal region of *Ascaris suum*.

expand due to osmotic forces. The structure of the cuticle of parasitic nematodes often changes after moulting to suit different habitats. The functions of the cuticle explain its complex and diverse structure when examined with an electron microscope.

Generally, the nematode cuticle can be resolved into four basic layers. Externally there is a very thin trilaminar epicuticle, 6–40 nm thick. On the outer surface there is usually a tenuous surface coat of glycoprotein, secreted by epidermal glands. The three underlying layers, of variable but greater thickness, are composed of collagens stabilised by disulfhide bonds and other cross-linkages. The outer cortical layer is relatively rigid and amorphous. The median layer is less rigid, but often contains dense rigid rods, bars or plates. A more uniform basal layer is often fibrous, sometimes with spaced radial bars. In large nematodes there is often a basket-work of helically arranged collagen fibres making opposing 55–65° angles to the long axis of the body. In some, such helical fibres underlie the outer cortical layer. Chitin is not present in the cuticle but is found in the egg shell. The buccal cavity, pharynx, rectum, vulva and excretory duct secrete a cuticle of simpler structure that must also be shed when the animal moults.

At moulting, the inner cuticle is resorbed before a new epicuticle forms, after which the succeeding layers are secreted. Between moults the epidermis is very thin, except at four points around the circumference of the body where it thickens to form four longitudinal cords in which the epidermal cell nuclei are located.

Musculature

The longitudinal body-wall muscles, used in locomotion, are divided into four quadrants by the epidermal cords (Fig. 5.2d,f). There are no circular body-wall muscles. Each muscle cell contains a contractile zone and a region with a nucleus and other organelles, from which a process extends to the nervous system to form synapses. At the molecular level, the contractile mechanism involves the same classes of proteins that are found in other animals — thick filaments of myosin and paramyosin and thin filaments of actin, tropomyosin and troponin. In nematodes the body wall muscles are obliquely striated. The sarcomeres are staggered relative to the long axis of the body so that, unlike the more familiar pattern found in vertebrate striated muscles with perpendicular z-bands, the z-bodies are obliquely arranged, making a small angle of about 6° to the long axis of the body. This arrangement, also found in some other invertebrates, maintains the force of contraction when muscles lengthen as the body flexes. Thin fibrils connect the muscle cells through the epidermis to the cuticle. Other muscles, with single cross-striated sarcomeres, operate the mouth parts, pharynx, rectum and accessory sexual organs.

Digestive system

The buccal cavity (Fig. 5.1b) is lined by cuticle that is continuous with that of the pharynx. The cuticle may form teeth, jaws, stylets, or a simple triradiate slot (Fig. 5.2b). A great variety of different buccal structures can be found in nematodes that otherwise closely resemble one another in form. The food and habit of a species can often be deduced by examining the buccal cavity. Examples are shown in Figure 5.3. In a few nematodes there is no buccal cavity, the mouth opening directly into the pharynx.

The pharynx (Fig. 5.1d) is an autonomous muscular organ that pumps food through the non-muscular intestine. Its muscular-epithelial cells

THE ASCHELMINTHES

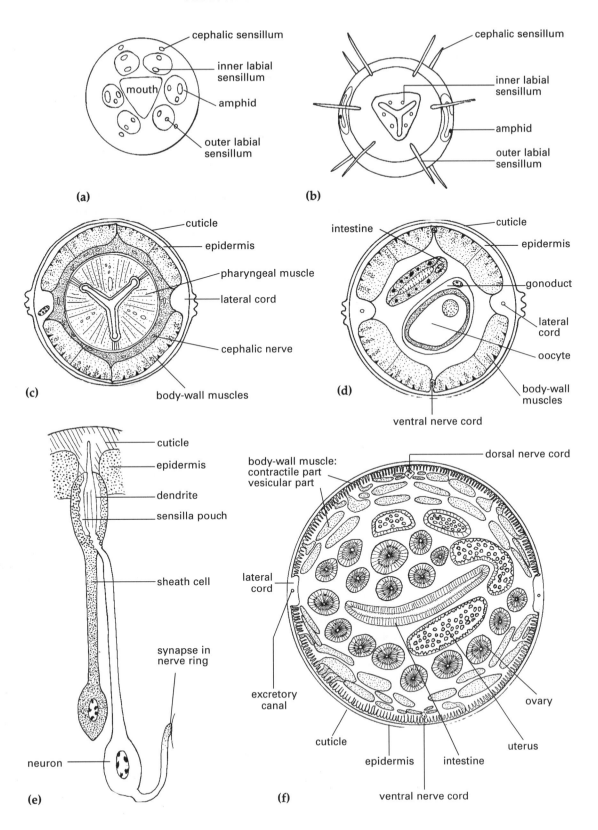

Figure 5.3
Heads and buccal cavities of representative Adenophorea from (a–f) marine habitats and (g–j) terrestrial and freshwater habitats. The odontostyle is used by *Paralongidorus* to feed on plant tissues, but the stylet of *Dorylaimus* is used with an omnivorous diet.

contain radial muscle filaments (Fig. 5.2c) that extend from its cuticular lining to the basal lamina that separates the pharynx from the pseudocoel. Typically the pharyngeal lumen is triangular in cross-section (Fig. 5.2c), and the elasticity of its cuticular walls opposes the contraction of the myofilaments in providing the pumping mechanism. The cuticular lining, secreted by the muscular-epithelial cells, often forms valves, sometimes toothed, as in the Rhabditidae, to prevent the regurgitation of food. Often its basic cylindrical form is modified by one or more localised muscular swellings known as pharyngeal bulbs. In rhabditids a muscular corpus (Fig. 5.1a) leads into a narrow non-muscular isthmus (surrounded by the nerve ring) and then an expanded muscular bulb. The pharynx contains intrinsic glands with ducts that open into the lumen, whose number and arrangement varies in different nematode taxa. Neurons within the pharynx coordinate pumping with minimal inhibitory control from the nerve ring. Food passes from the pharynx to the intestine via a multicellular valve.

The intestine (Fig. 5.2d) is a straight, short, non-muscular tube, sometimes with extrinsic muscles. It is the only part of the gut not lined by cuticle, and its absorptive lining usually forms microvilli. In small nematodes there are only two or four cells in cross section, but in larger parasites there may be hundreds (Fig. 5.2f). The cells also serve as storage organs with protein or lipid inclusions. The intestine is followed by a short cuticle-lined rectum opening close to the hind end of the body, which terminates in a tail of variable length. In females the rectum opens at the anus, but in males it usually opens together with the gonoduct, forming a cloaca. Paired rectal glands open into the rectum.

Secretory/excretory system

Nematodes do not possess flame cells, nephridia or any of the osmoregulatory organs typical of other invertebrates. The so-called excretory organs differ fundamentally in the two classes. In Adenophorea a prominent, glandular renette cell lies ventral to the anterior region of the alimentary canal, with a duct opening by a pore somewhere between the nerve ring and the lips. Though clearly not excretory, its function is unclear. In some marine nematodes the renette cell may secrete mucus used in feeding or tube construction. In Secernentea, canals lie within the lateral epidermal cords (Fig. 5.2f). In its most complete development the canal system forms an H, with the horizontal bar representing a ventral canal linking the two lateral canals and opening by a short cuticular duct at a pore ventral to the nerve ring (Fig. 5.1d). In some nematodes the system appears to secrete enzymes. In others, for example, the free-living larvae of parasitic Strongylida, a pulsating ampulla proximal to the pore plays an osmoregulatory function in hypotonic surroundings. In still others, its function is unknown.

Nematodes generally maintain the hydrostatic pressure in their tissues above that of their surroundings, and many of the characteristics of nematode anatomy can be explained as consequences of this state. The influx of water in hypotonic media is counteracted by the resistance of the cuticle to stretching and by body-wall muscle tone. Some free-living nematodes can tolerate a wide range of osmotic pressures. *Rhabditis marina*, for example, can tolerate distilled water or double-strength sea water for long periods. The situation in other marine nematodes is less clear. When immersed in fresh water they usually swell and become paralysed. After short exposures this is reversible.

THE ASCHELMINTHES

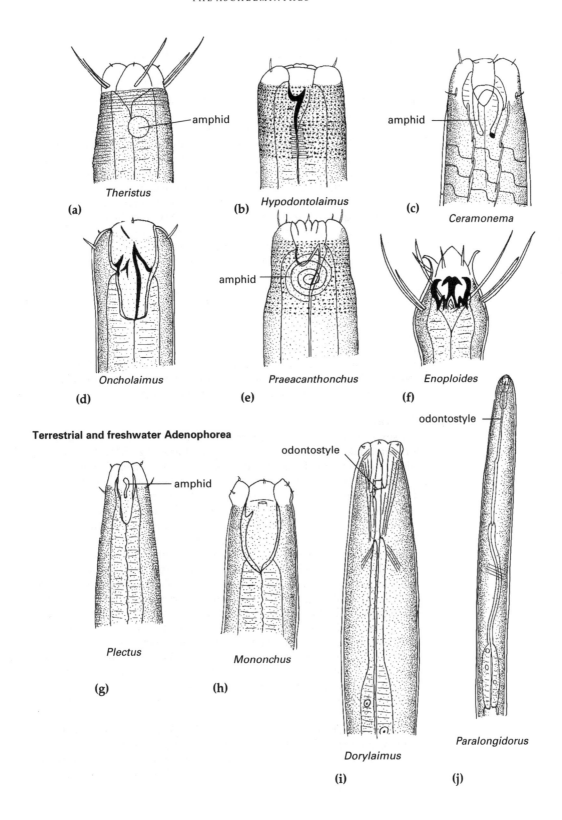

Terrestrial and freshwater Adenophorea

Nervous system

The nervous system of the free-living nematode *Caenorhabditis elegans* is known in greater anatomical detail than that of any other animal. The much larger size of the parasitic *Ascaris suum*, which has a similar neuroanatomy, has facilitated investigations of electrophysiology and chemical neurotransmission.

A group of ganglia form a nerve ring around the pharynx (Fig. 5.1d), constituting the central nervous system. Males have 381 neurons, plus 92 glia supporting cells; the hermaphrodites have 302 and 56 respectively. A ventral ganglionated motor nerve runs from the nerve ring to the hind end of the body within the ventral epidermal cord (Fig. 5.2d, f). Commissures carry nerve processes from the ventral nerve cord to form a dorsal motor nerve in the dorsal epidermal cord (Fig. 5.2f). Six cephalic nerves, arising from neurons in nerve ring ganglia, innervate the cephalic sensilla (Fig. 5.2a, b). Nematodes are unusual in that, instead of axons from motor neurons reaching the muscle cells to form synapses, the muscle cells extend processes to the motor nerves, either in the motor nerve cords or, if anterior to the nerve ring, to nerve ring ganglia.

Most nematode sensilla are derived from modified cilia. The amphids at the sides of the head (Figs 5.2 and 5.3) are chemosensory organs that are particularly well developed in Adenophorea, in which they are often larger in males than females, with the detection of female sexual attractants as one of their functions. The basic organisation of cephalic sensilla is shown in Fig. 5.2e. Some aquatic nematodes have paired photoreceptors (ocelli) on the pharynx, with a simple lens and a pigment cup.

Reproductive organs

The sexes are separate. Gamete formation, maturation, fertilisation and egg formation takes place within tubular gonoducts (Figs 5.1a, d, 5.2f) lying within the pseudocoel. The gonoducts are formed by an epithelium, in part with glandular and contractile capabilities. Females usually have two gonoducts, but sometimes only one, opening to the exterior by a vulva that may be located anywhere between the head and the anus, depending on the species. Male Secernentea usually have only a single testis and duct (Fig. 5.1a), but Adenophorea may possess one or two testes, which join a single sperm duct. In females special muscles, ovojectors, lie near the vulva. In males the sperm duct leads into a short ejaculatory duct that opens with the anus to form a cloaca. An adjacent epidermal pouch secretes two hard cuticular spicules (Fig. 5.1c) that are used to prise open the vulva during copulation, facilitating the injection of sperm. There are accessory glands, and a system of protractor and retractor muscles operates the spicules. In some nematodes two cuticular flaps alongside the cloaca form the bursa, with intrinsic muscle bands, which can envelop the vulva during copulation (Fig. 5.1c). Sensilla are present within the spicules, around the cloaca, and often in Adenophorea as a ventral pre-cloacal row. These accessory sexual organs can take many forms and are of primary importance in nematode taxonomy.

The ovary, at the tip of the gonoduct (Fig. 5.1d), generates oocytes by mitosis, and these move through a very short region where they enter meiotic prophase, then a region where stored sperm fertilise the oocytes and meiosis is completed. Fertilised eggs pass through an oviduct to the uterus, where an egg shell is laid down and embryonic development commences.

Nematode egg shells are multilayered, the egg secreting a thin vitelline membrane, a chitin layer, and an inner lipoidal layer. The uterus may secrete additional protein outer layers. In some nematodes that parasitise plants and animals, the inner lipoidal layer may make the egg a dormant infective stage, resistant to drying and toxic chemicals until, in response to specific stimuli, the juvenile secretes hatching enzymes. *Ascaris suum* eggs can, for example, develop viable juveniles in formalin or sulfuric acid.

Spermatogenesis follows a similar pattern to oogenesis, with the testis at the tip of the gonoduct. The gonoduct becomes successively a seminal vesicle, vas deferens, and ejaculatory duct (Fig. 5.1a). Mitosis produces spermatogonia that enter a short region of meiosis, leading to primary and secondary spermatocytes, and finally spermatids. Nematode sperm are unusual in several ways. There is no flagellum, no acrosome and, with the exception of some Enoplida, no nuclear membrane. The chromatin becomes a very compact body. Nematode sperm develop a pseudopodium and become amoeboid in the female uterus. Motility does not depend, as it does in other animals, on actin, but on a unique nematode sperm protein.

Sex

Nematodes reproduce sexually, with many variations, but there is no asexual reproduction, and one egg can give rise to only one adult. This is in striking contrast to some other parasitic invertebrates such as the Platyhelminthes. Typically, nematode males copulate with females and fertilisation is internal. Small nematodes produce 200 to 300 eggs, but parasites often many more. *Ascaris suum* can produce 200 000 eggs a day for many months.

Soil-inhabiting nematodes commonly exhibit a form of hermaphroditism in which individuals with female morphology produce first sperm and then oocytes in the same gonad, followed by self-fertilisation. Rare males may occur in the same population that are capable of inseminating the hermaphrodites. Female and hermaphrodite nematodes possess XX sex chromosomes, males XO or XY sex chromosomes. In spermatogenesis, meiosis typically gives rise to equal numbers of X and O spermatozoa, yielding progeny with equal numbers of males and females.

Parthenogenesis is not uncommon, resulting either from failure to form bivalents at meiosis I or by the restitution of diploidy due to the failure of disjunction. Some plant parasites reproduce parthenogenetically, without males present, until environmental conditions become adverse, whereupon some juveniles develop as functional males. Anomalies in some genera of plant parasites have led to polyploidy, so that a series of related species with differing multiples of the presumably original diploid chromosome number have arisen.

Heterogamy, or alternation of generations, is found in nematodes among plant, insect and vertebrate parasites. A parasitic generation multiplies within the host, either parthenogenetically as in *Strongyloides* (parasites of mammals including humans), or hermaphroditically as in *Rhabdias* (reptile and amphibian parasites), but their eggs then develop into sexually reproducing free-living males and females. Their progeny can infect new potential hosts. Heterogamy also occurs in nematodes infecting insects and plants, but without a free-living generation.

Genetics, developmental biology and *Caenorhabditis elegans*

The small nematode *Caenorhabditis elegans* has proved particularly valuable for research in developmental genetics and molecular biology, a role it shares with the fruit fly *Drosophila*. Both species are used for this purpose in many laboratories world-wide. It is significant that the total genetic information (that is, the size of the genome) is much smaller than that of mammals and is intermediate between that of insects and bacteria (Table 5.1).

Table 5.1 A comparison of the genetic information of insects and bacteria.

Organism	Base pairs in DNA	Estimated number of genes
Bacterium: *E. coli*	4.2×10^6	2 350
Nematode: *C. elegans*	8.0×10^7	15 000
Fly: *Drosophila*	1.4×10^8	12 000
Human	3.3×10^9	125 000

C. elegans is usually cultured on agar in Petri dishes, together with the bacterium *Escherichia coli* as food. As a self-fertilising hermaphrodite it can develop from egg to reproductive adult in about 85 hours. The position and differentiation of every one of its 959 cells has been followed under the microscope as the egg develops into the adult.

The developmental fate of all of the cells of *C. elegans* can be predicted from the hierarchical sequence of cell divisions which produces them, i.e. the cell lineage. In this respect the development of *C. elegans* is much more predictable than in most animals. The fate of a cell is partly an expression of its intrinsic genetic program and partly the result of interaction with adjacent cells. As with other animals, chemical messengers passing between cells are important in determining cell differentiation. Genetically programmed cell death also plays a part in development. Of great interest is the frequency of homologous gene sequences controlling similar developmental processes in *C. elegans* and other animals, as with homeobox genes. This suggests, as mentioned in Chapter 2, that important developmental processes originated very early in metazoan evolution and have been conserved in higher animals such as insects and mammals.

Embryonic development

Nematode eggs divide to produce an elongated ball of cells, become tadpole shaped, then rapidly elongate to form a mobile, fully formed juvenile worm curled up within the egg shell (Fig. 5.4a). Some species are ovoviviparous, hatching in the uterus. The juveniles often moult within the egg before hatching.

Embryonic development follows a fixed pattern. Initial cleavage divisions of the egg give rise to stem cells called AB, C, EMS, D and P (Fig. 5.4a,b). AB will give rise to epidermis, neurons, muscles, part of the pharynx and other structural cells. C will also give rise to epidermis, neurons and muscles. EMS divides into E (the progenitor of the intestine) and MS (which gives rise to muscle cells, pharyngeal cells, gonoduct cells and neurons). D gives rise to muscle cells. The germ cell line separates from the other stem cells by four successive divisions of the P cell line. The P4 cell divides into two gonad precursor cells that will not divide again until after the juvenile has hatched. Often one of the pair fails to develop, so that only a single testis or ovary forms. In gastrulation, beginning at about the 100 cell stage, the E, P, MS and D cell descendants move successively within the

Figure 5.4
Embryonic development of *Caenorhabditis elegans*.
(a) Development of fertilised egg up to the stage at oviposition. (b) Cell divisions during cleavage and developmental fates of the cells (after Wood 1988).

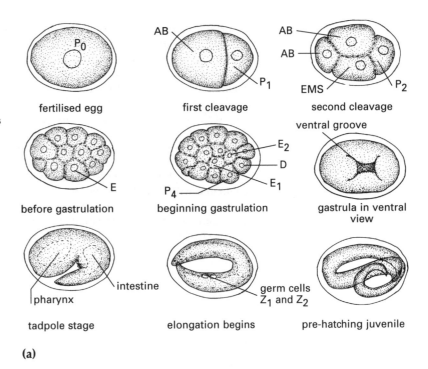

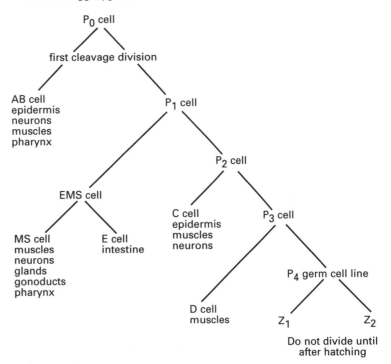

embryo from the ventral surface (Fig. 5.4a). The stomodaeum, precursor of the pharynx, and proctodaeum, precursor of the rectum, are separate incursions. Elongation of the embryo following gastrulation is caused by the contraction of circumferentially arranged epidermal actin filaments and maintained by cuticle formation.

In *C. elegans* there are about 550 cells at hatching, 811 surviving cells in the adult hermaphrodite and 971 in the adult male, excluding the gonads which produce one to two thousand gametes. In the hermaphrodite 113 cells are programmed to die, 111 in the male. In the larger parasitic nematodes, cell division produces many more cells during post-embryonic development, for example 50 000 body-wall muscle cells in *Ascaris suum*, compared with 95 in *C. elegans*.

The gonads usually develop during the last juvenile stage and the sexual organs are only fully apparent after the fourth moult. In many parasitic nematodes there is a resting infective stage, usually the third juvenile stage. In parasitic nematodes the cuticle and mouth parts may change with moulting to accommodate differences in life style. In some terrestrial nematodes, such as *C. elegans*, juveniles can enter a resting 'dauer' larval stage. Starvation or pheromones induce dauer formation in *C. elegans*, and development resumes with the presence of food.

Ecology

Marine nematodes

Nearly all marine nematodes belong to the class Adenophorea and are taxonomically and structurally more diverse than terrestrial or parasitic nematodes. They are the most numerous Metazoa in the marine meiobenthos, which consists of nematodes, copepods, ostracods, acarines, turbellarians, gastrotrichs, kinorhynchs and the early larval stages of some of the larger invertebrates. Though much smaller in size than the macrobenthos, the contribution of nematodes to marine ecosystems is important because of their great numbers, very high metabolic rates and potentially high reproductive rates. Nematodes generally constitute between 60% and 90% of the individuals and most of the meiobenthic biomass in the sediments of estuaries, beaches and oceans, their dominance reaching over 90% in the deepest ocean depths. The populations are also very diverse. Some examples are given in Table 5.2.

Table 5.2 Density and diversity of nematodes in various habitats.

Habitat	Region	Density per 10 cm^2	No. species
Beach sand	Australia	100–300	56
Mangrove mud	Australia	100–500	40
Shallow sea	North Sea	250–500	478
Deep sea	North Atlantic	90–250	300

The nematode fauna includes bacterial feeders, algal feeders (especially diatom feeders) omnivores and predators. Each of these categories has mouth parts specialised for their particular food, some examples of which are illustrated in Figure 5.3a–f. Nematodes are food for small fish, shrimps and crustaceans. The meiobenthos, including the nematodes, are typically aerobic and therefore restricted to the top few centimetres of sediment, but some nematodes are among the very few Metazoa to penetrate the deeper anoxic sediments. Here not only is oxygen lacking, but toxic hydrogen sulfide, an inhibitor of cytochrome oxidase, accumulates through bacterial action. Some marine nematodes have their body surface clothed in symbiotic sulfur bacteria.

Freshwater nematodes

The class Secernentea predominates over the Adenophorea in fresh water, as it does in the soil (Fig. 5.3g–j). Nematodes are very important members of the benthic fauna of all kinds of fresh waters, but they are taxonomically much less rich and occur in lower densities than in the sea. They include bacterial feeders, algal feeders, omnivores, predators, and species with a stylet for feeding on plants and fungi.

As in marine ecosystems, nematodes are among the few metazoans to inhabit anoxic sediments underlying fresh waters and to tolerate an accumulation of hydrogen sulfide. Thermal stratification makes such anoxic conditions a frequent (though usually seasonal) occurrence in lakes.

Soil nematodes

Nematodes are present in all types of soil, usually at densities of hundreds or thousands per square metre. They can be roughly classified as bacterial feeders, fungal feeders, plant-root feeders, omnivores and predators. They exert an important influence on the decomposition of organic matter in the soil through grazing on bacteria and fungi. They are attacked by endoparasitic fungi and nematode-trapping fungi, some of which produce sticky traps or constricting loops for the purpose. Soil insects and some specialised families of soil mites feed on nematodes. Nematodes reach higher densities on agricultural land, where plant-feeding nematodes become more numerous.

For activity, nematodes require at least a film of water on soil particles, but their ability to enter cryptobiosis enables some species to inhabit the driest of deserts, polar regions and extreme environments of all kinds. Some persist for many years in an anhydrobiotic state until it rains, when they complete their life cycle and reproduce in a matter of days. Nematodes are among the few animals surviving on the Antarctic mainland; some recover from freezing and thawing, while others are physiologically adapted to supercooling without freezing.

Plant-feeding nematodes

Many soil nematodes feed on plants. Tylenchida and Aphelenchida possess a hollow mouth stylet that acts like a hypodermic syringe (Fig. 5.5b–d). This stylet enables the nematode to pump digestive enzymes into plant cells and ingest the cell content. Some species enter the plant tissues, migrating through the stem and causing galling and distortion. Some Tylenchida are specialised for parasitism. The female becomes a swollen cyst attached to a root. In *Meloidogyne* (Fig. 5.5d,e), salivary secretions induce giant syncytial cells to form in the root, on which the immobile female feeds. The swollen dry cysts of *Heterodera* and *Globodera* (containing juvenile worms) and the dry egg masses of *Meloidogyne* can survive in soil for years. Infective juveniles emerge when environmental conditions are appropriate. In some this may arise when chemical signals emanate from

INVERTEBRATE ZOOLOGY

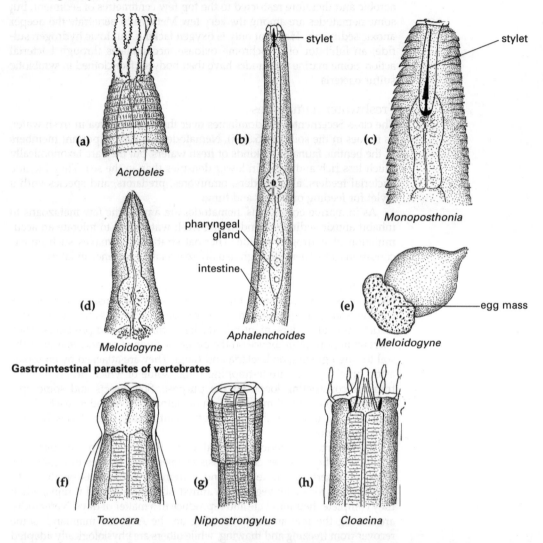

Figure 5.5
Head and pharyngeal regions of various soil-inhabiting and parasitic Secernentea. **(a–d)**: *Acrobeles* (a) feeds on bacteria; *Aphelenchoides* (b), *Monoposthonia* (c) and *Meloidogyne* (d) feed on plant cells, which they puncture with a stylet. **(e)** The swollen body of the adult female of *Meloidogyne* with an extruded gelatinous egg-mass. **(f)** *Toxocara canis* from the intestine of dogs. **(g)** *Nippostrongylus brasiliensis* from the intestine of rats. **(h)** *Cloacina* sp. from the stomach of kangaroos.

the roots of suitable plant hosts. Dorylaimid nematodes possess a different kind of hollow stylet that may be used to feed on plant or animal tissue. Some dorylaimids are important agricultural and horticultural pests because they transmit pathogenic plant viruses, carried from plant to plant in the pharynx. Plant-feeding nematodes can cause economic losses through the failure or reduced yields of all the major agricultural and horticultural crops. Chemical control is expensive and often not very effective.

Parasitism and other associations with invertebrates

Many kinds of nematodes invade the bodies or cling to the body surfaces of other invertebrates of all kinds. Often this provides a means of dispersal, so that, for example, bacteria-feeding nematodes can reach new sources of food on dung beetles or flies. This can lead to facultative or obligate pathogenic associations. Some nematode invaders remain immature until the host dies, then feed on the decomposing cadaver. In *Steinernema* and

Heterorhabditis, specific symbiotic bacteria are carried by infective juveniles within the intestine. When the nematode penetrates the haemocoel of the insect host, the bacteria are released, killing the host. The nematodes reproduce in the cadaver, feeding on the bacteria and liberating infective larvae into the soil. These two 'entomogenous' genera can be cultured cheaply in great numbers and have been exploited commercially as biological insecticides with a wide range of insect hosts.

Nematodes in the orders Tylenchida and Aphelenchida, which typically use a stylet to feed on plant tissues, also include many insect parasites. Generally, males are free-living, but after copulation the female invades a suitable insect host, usually a larval stage. Here she becomes greatly swollen with eggs, and releases juveniles in the adult insect that reach the exterior via the reproductive organs or gut. In some genera there is a second parasitic generation within the insect host before the juveniles are released.

The Mermithida parasitise many kinds of invertebrates, both terrestrial and marine. In this large group it is usually the first juvenile stage that invades the host, though in some the eggs must be swallowed by an intermediate host which must then be eaten by the final host. The juveniles grow within the haemocoel, often becoming much longer than the host, sometimes reaching many centimetres in large hosts. The intestine becomes a blind food-storage organ and nourishment is obtained through the body surface. Eventually the nematodes escape, killing the host, to reproduce sexually as free-living males and females and deposit eggs.

The Oxyurida includes many species that inhabit the alimentary canal of insects, often apparently harmlessly. All stages occur in one host, except the infective eggs which are voided with the faeces. The juvenile stages of many vertebrate parasites use invertebrates as obligate or facultative intermediate hosts in their life cycles.

Parasites of vertebrates

Every vertebrate is the potential host of many different kinds of nematodes, which often parasitise a restricted set of related species. One or another species invades every organ system. Consequently there is a huge number and variety of nematode parasites (Fig. 5.5f–h), often with very complicated and differing life cycles, of which only a few can be briefly dealt with here. The great majority belong to the class Secernentea. The life cycle often includes extensive migrations in the host's tissues. Juvenile nematodes may make similar migrations through the tissues of unsuitable hosts, even though the life cycle is aborted. Migrating juveniles of parasites of domestic or wild animals are a cause of human disease, as, for example, when human blindness is caused by the migrating juveniles of the dog roundworm, *Toxocara canis* (Fig. 5.5f). Most nematodes are less than 2 mm long, but some parasitic nematodes are much larger — up to 500 mm or more, and with greatly enhanced egg production, as already noted in *Ascaris suum*.

The Rhabdiasidae and Strongyloididae are interesting in that free-living, sexually reproducing generations which feed on bacteria alternate with parasitic females, either hermaphroditic or parthenogenetic. The parthenogenetic female of *Strongyloides stercoralis* inhabits the intestinal mucosa of humans, dogs and other animals. Its eggs are passed in the faeces and may develop into a free-living generation whose progeny will develop into parasitic females, or the eggs may develop directly into infective juveniles that give rise to parasitic females. Infective juveniles penetrate the skin, and debilitating infections may persist for many years through autoinfection.

The Strongylida, distinguished by the possession of a characteristic copulatory male bursa, also show a transitional state between free-living and parasitism. Typically the eggs hatch into free-living larvae that feed on bacteria in the host's faeces. The third juvenile stage is an infective stage; the fourth stage and adult males and females are parasitic. Adults usually inhabit some region of the gut, but may inhabit the respiratory system or any organ from which eggs can be voided. The infective juveniles of parasites of terrestrial carnivores usually invade the host by skin penetration, while those of grazing animals are usually taken up by ingestion. Two human parasites, the hookworms *Necator americana* and *Ancylostoma duodenale*, can cause serious human disease through blood loss. They penetrate the skin and, following migration through the lungs, reach the small intestine, where they grasp the mucosa with teeth and feed on blood.

The distribution of Strongylida is determined by the tolerance of environmental conditions by the juvenile stages. With human hookworms this means the moist tropics or subtropics, but other Strongylida are adapted in one way or another to colder or drier climates. Some can develop to the infective stage before hatching, and others may invade intermediate hosts such as snails or earthworms and await accidental ingestion by a suitable host. Many Strongylida are economically very significant parasites of domestic animals.

The Oxyurida are parasites of invertebrates and vertebrates, which usually become infected by ingesting eggs passed in another individual's faeces. All stages inhabit the host's gut. Many appear to feed on the gut contents with little effect on the host. The human pin worm *Enterobius vermicularis*, the most universal human parasite, lives in the hind gut. The parasitic female must crawl out through the anus to burst and release eggs on the perianal skin. This can cause transitory inflammation, itching and scratching. Eggs, which do not survive long, are passed rapidly from host to host, often on the fingers.

The Ascaridida are large nematodes inhabiting the intestines of vertebrates. Often there are one or two intermediate hosts in which juveniles migrate to deeper body tissues. *Ascaris lumbricoides* and *Ascaris suum* have only one host, human or pig respectively, and infection is through ingesting the eggs. These eggs, passed in the faeces, survive for years in the soil and are impervious to disinfectants. The juveniles migrate through the liver and lungs to return to the intestine in which they hatched. It seems that the parasite behaves as though the same individual acts as intermediate and definitive host.

The Spirurida have two hosts in the life cycle, and the adult parasites may occur in many different tissues. *Dracunculus medinensis* is a specific human parasite found in the drier parts of Africa and southern Asia. The adult female, up to 80 cm long, lies under the skin, producing an ulcer through which bursts of microscopic juveniles are released whenever the skin is immersed in water. The intermediate host is a copepod crustacean (water-flea). People become infected by drinking water containing infected copepods. Infection is often an annual event when people congregate for washing and drinking water around wells.

Within the Spirurida, the Filarioidea are the most ubiquitous vertebrate parasites, infesting all classes of terrestrial vertebrates. The thread-like adults inhabit various body tissues, often connective tissue, where the

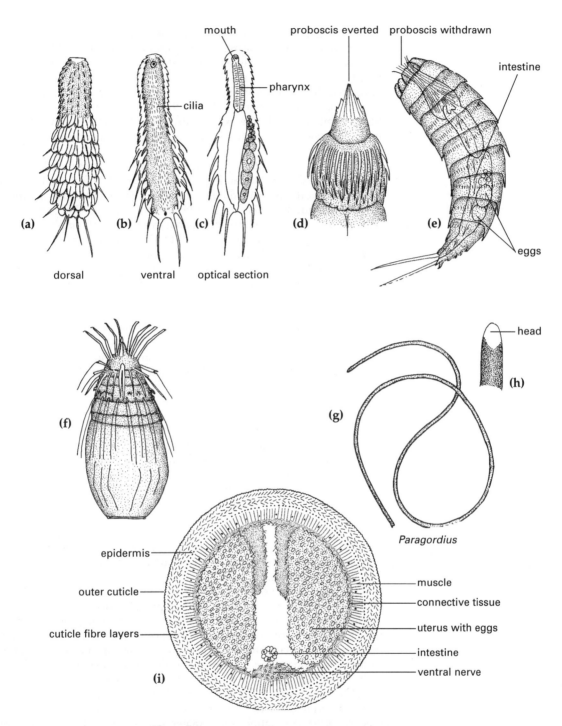

Figure 5.6
(**a–c**) Gastrotricha: an undescribed species of Chaetonotida from sandy beaches. (a) Dorsal view. (b) Ventral view. (c) Optical section. (**d, e**) Kinorhyncha: *Echinoderes* sp., a species with a retractable proboscis, from mangrove mud. (**f**) Loricifera: *Pliciloricus habilis* (after Kristensen and Shirayama 1988). (**g–i**) Nematomorpha: *Paragordius* sp., found in rivers and streams. (g) Adult. (h) Head. (i) Transverse section.

female gives birth to microscopic juveniles (microfilariae) at a very early stage of development. The microfilariae find their way to the blood system or, less often, enter the dermal lymphatic vessels. Intermediate hosts are blood-sucking insects, mites or ticks in whose tissues the microfilariae develop into infective juveniles. The juveniles escape from the arthropod mouth parts during feeding and enter the wound. In many species, microfilariae concentrate in the skin capillaries at the time of day when the insect intermediate host feeds, accumulating in deeper tissues at other times.

The seven species that parasitise humans are transmitted by mosquitoes, black-flies or midges. They often successfully infect whole populations in tropical regions, although some species do not cause significant disease. The adults of *Onchocerca volvulus* congregate in skin nodules, but only the microfilariae, which accumulate in sub-dermal lymphatics, cause serious human disease. Chronic inflammation damages the skin, eye lens and retina, leading to blindness. The disease, known as onchocerciasis or river blindness, is associated with rivers and streams because the insect hosts, black-flies (Simuliidae), breed in flowing water. It occurs in western and central Africa, Yemen, Central America and northern South America.

Wuchereria bancrofti and *Brugia malayi* inhabit the lymphatic nodes of humans. *B. malayi* also parasitises dogs, monkeys and other animals in southern Asia, but *W. bancrofti* is a purely human parasite, with a worldwide tropical distribution. These Filarioidea are transmitted by particular species of mosquito. They cause chronic lymphatic inflammation, which may lead to complete blockage, resulting over many years in hideous deformities known as elephantiasis, in which legs, arms, testicles or breasts become enormously distended with accumulated lymph. *Dirofilaria immitis* (heartworm), a parasite of dogs that is also transmitted by certain mosquitoes, inhabits the chambers of the heart and great veins, eventually leading to cardiac failure. *Dirofilaria roemeri* has a peculiar predilection for the knee joint of kangaroos, and is transmitted by tabanid flies.

The distribution of Filarioidea depends on the behaviour of the arthropod host, and on sufficient ambient temperatures to permit the microfilariae to develop to the infective stage at a time when the host may still take another meal of blood or lymph. Where conditions are right, Filarioidea frequently saturate the vertebrate host population, often without causing significant symptoms.

One order of the nematode class Adenophora, the Trichocephalida, has evolved as vertebrate parasites. Trichocephalids possess proportionately narrow heads and cervical regions, often with a very narrow pharyngeal tube, and large extrinsic pharyngeal gland cells. *Trichinella spiralis* is exceptional in its very large range of host species, such as humans, pigs, rats, wolves, bears (including polar bears), walruses and many others. The life cycle is entirely parasitic, the worms being passed from host to host through the ingestion of juveniles encysted in the muscles of prey or carrion. The adults, about 1.5 mm long, develop rapidly in the intestinal mucosa. After copulation, the females release up to 1500 microscopic juveniles that will encyst within muscle fibres over several weeks, where they develop to infective juveniles that can remain viable for years. The ingestion of poorly cooked pork, which can hold a million larvae in a single bite, can produce a fatal human infection through massive muscle damage.

Phylum Nematomorpha

Unsegmented long worms with layered cuticle, moulted periodically; gut reduced, non-functional; cerebral ganglion and midventral epidermal nerve cord; no excretory system; separate sexes, simple gonads; juveniles parasitic in arthropods.

The major group of Nematomorpha is the Gordioidea or bootlace worms, which are parasites in the body cavity of terrestrial arthropods, usually crickets, grasshoppers or beetles. They are often much longer than their hosts, in which they must coil up, killing the host to become free-living adults. The adults typically enter fresh water to copulate and deposit their eggs. Long, thin, darkly pigmented, slowly undulating worms sometimes appear in drinking water in summer.

Adult Gordioidea (Fig. 5.6g–i) possess a brown pigmented, tough cuticle of helically wound, oppositely orientated layers of collagen fibres. Anteriorly there is an unpigmented 'head', the calotte. Externally a thin lipoprotein epicuticle is present. The epidermis thickens midventrally to form a ventral cord holding a ventral nerve cord. Anteriorly the nerve cord joins a cerebral ganglion. Within the epidermis there is a layer of longitudinal muscle cells. The body cavity, a pseudocoel, is filled with stellate mesenchymal cells. Paired gonads run the length of the body to open, separately or together, into a posterior cloaca. In the female there is a glandular uterus. Eggs are fertilised internally, following copulation, by spermatozoa held in a seminal receptacle. The adult gut is not functional. A vestigial mouth, short occluded pharynx and narrow intestine lead to a cuticle-lined posterior cloaca. Eggs hatch to release a minute larval stage equipped to invade an arthropod host (Fig. 5.7a). *Nectonema*, the single genus of a second nematomorph group Nectonematoidea, is a tissue parasite in the juvenile stage of marine Crustacea, becoming free-living and planktonic when sexually mature. The cuticle of *Nectonema* possesses bristles. There is a capacious body cavity, a dorsal and ventral nerve cord, a minute mouth, slender pharynx and intestine, but no anus. It may be that in both Gordioidea and Nectonemoidea the gut is functional in the larval stages.

Phylum Gastrotricha

Unsegmented, with terminal adhesive tubes; spiny, often plated, layered cuticle, moulted periodically; cerebral ganglion and lateral nerve cords; protonephridial excretory system; hermaphrodite or parthenogenetic, with simple gonads.

The Gastrotricha (Fig. 5.6a–c) are very common in the benthic fauna of marine and freshwater habitats, but are often overlooked because of their minute size, usually less than 0.5 mm. There are two classes. The Macrodasyida are marine, inhabiting the interstitial water of sandy and muddy sediments. The Chaetonotida are common in fresh-water, often gliding among algae. Their cephalic region is slightly swollen and their hind end is forked.

Figure 5.7 (opposite)
Nematomorpha: **(a)** The infective larva of *Gordius* (after Dorier 1930). **(b–g)** Rotifera: **(b)** *Rotaria* (Digononta, Bdelloidea), common among pond weed. **(c)** Section through the mastax (after Beauchamp 1965). **(d)** *Keratella* (Monogononta), a free-swimming rotifer. **(e)** Sculptured lorica of *Keratella*. **(f, g)** *Floscularia* (Monogononta), a sessile rotifer.

The dorsal and lateral body surfaces are covered by spines, scales or warts. The ventral surface has cilia by which the animal glides smoothly. On the head there are often tufts of feeding and sensory cilia. Macrodasyids have adhesive tubules in rows laterally, and at the hind end. In chaetonotids the tubules may be restricted to the terminal forks. The epidermis secretes a multilayered cuticle. One or more trilaminar membranes, 7–10 nm thick, form an exocuticle. The endocuticle is much thicker and has a complex ultrastructure. The epidermis contains ciliated cells and gland cells, including the glandular adhesive tubes. Some of the cilia are modified as sensory receptors. A layer of circular muscle underlies the epidermis, external to bands of longitudinal muscle cells. A narrow pseudocoel lies between the body wall and the alimentary canal.

A cuticular buccal cavity leads into a muscular pharynx (Fig. 5.6c), with a triradiate lumen, followed by a straight intestine, and a short rectum and anus. The pharynx is composed of muscular-epithelial cells with radial filaments. The pharynx and rectum are lined by cuticle. The intestinal cells, which have microvilli, are secretory and absorptive. Circular and longitudinal muscles surround the alimentary canal. The brain surrounds the pharynx and gives rise to cephalic nerves and two posteriorly directed lateral nerve cords. Photoreceptors may be present and there are paired cephalic pits that are presumed to be chemoreceptors. Freshwater Chaetonotida may have a pair of protonephridia beginning with ciliated flame cells and ending with pores to the exterior.

Macrodasyids are oviparous, cross-fertilising hermaphrodites. Paired gonads, each acting as both ovary and testis, lie within the pseudocoel. Accessory sexual organs are varied. Generally a penis passes sperm to a female seminal receptacle, but insemination may sometimes be hypodermic. Spermatozoa may or may not possess a flagellum. The fertilised eggs develop within the pseudocoel. Chaetonotids are usually parthenogenetic, lacking male sexual organs. Females produce only a few large yolky eggs (Fig. 5.6c). There is no larval stage.

Phylum Kinorhyncha

Segmented, with layered cuticle of articulated segmental plates, moulted periodically; epidermal circumpharyngeal nerve ring and longitudinal nerve cords; protonephridial excretory system; separate sexes, tubular gonads.

Kinorhynchs (sometimes named Echinodera) are minute, spiny, segmented marine animals, generally less than 1 mm long, with about 150 described species. They are found in marine mud and sand from intertidal coastal beaches to the ocean depths. A species of *Echinoderes* is common in the surface mud of Australian estuarine mangroves (Fig. 5.6d,e).

Segmentation involves the cuticle, musculature and nerve ganglia, dividing the body into 13 segments or zonites. The epidermis secretes a chitinous cuticle, forming hard cuticular plates and flexible intersegmental joints. It is moulted repeatedly during growth. Numerous backwardly pointed spines and finer setae arise from the segments, with usually two or three longer spines on the terminal segment. There are also glandular adhesive tubes. The first segment, the head, carries 80 to 90 hollow spines

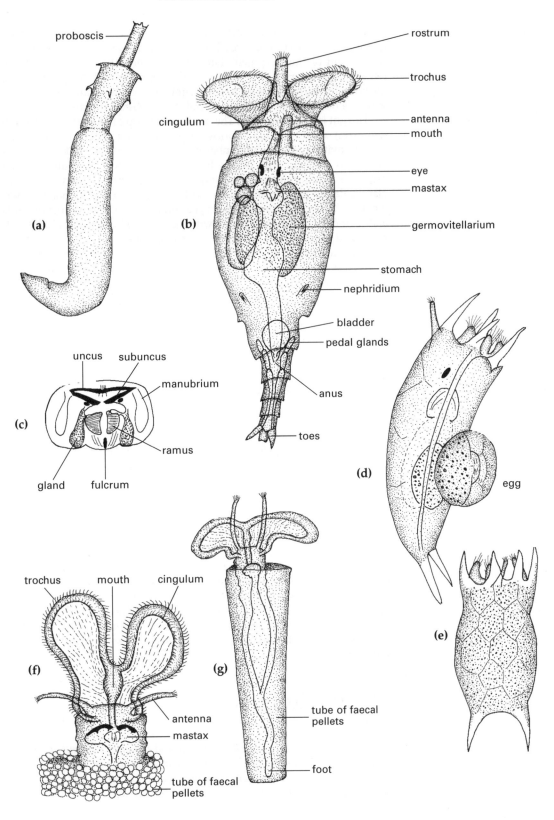

(scalids) in several rows, articulated at their bases. These serve locomotory and sensory functions. The first segment can be retracted into the second and or third segment. During locomotion the head is repeatedly retracted and everted so that the scalids sweep through an arc, thrusting the body forward. The third to 13th segments form the trunk, each usually with one dorsal tergal and two ventral sternal cuticular plates.

The gut is a straight tube. The mouth, at the tip of an evertible oral cone (Fig. 5.6d), is surrounded by forward-pointing oral stylets. It leads into a buccal cavity and muscular pharynx, lined by cuticle, followed by an absorptive mid-gut with microvilli, and a rectum lined by cuticle. Radial muscles surround the pharynx, and there are extrinsic salivary and gastric glands. Paired protonephridia on the 10th segment open by pores on segment 11. There is a circum-pharyngeal brain, paired ventral nerve cords, and segmental ganglia associated with the epidermis. A simple ocellus with a lens and pigment cup is present in some species.

Sexes are separate. Paired sac-like ovaries lead to oviducts opening between segments 12 and 13. On each oviduct there is a seminal receptacle, suggesting that fertilisation is internal following copulation. Paired testes discharge banana-shaped flagellate sperm through ducts opening on segment 13. Males possess cuticular rods that presumably facilitate copulation. Eggs are yolky; juveniles hatch at an early developmental stage and segments are added posteriorly as the worm grows.

Phylum Loricifera

Unsegmented, microscopic, in marine interstitial habitats; retractable head with scalids; plated cuticle, moulted periodically; dorsal cerebral ganglion; protonephridial excretory system; separate sexes, simple gonads.

The Loricifera were unknown until 1946, when specimens were discovered in subtidal sediments in the North Sea. Since then about 50 species have been described from marine sediments world-wide (Fig. 5.6f). They had escaped notice because of their minute size (less than 250 nm), transparency and difficulty of recovery from marine sediments. The body consists of a mouth cone, head, neck, thorax and abdomen. The head and neck bear many rows of appendages of variable form (scalids). Successive rows are paddle shaped, club shaped or spiny, each articulated at its base. The mouth cone, head and neck can be withdrawn into the abdomen, which is encased with stronger, more rigid cuticular plates forming a protective case or lorica. Loricifera have separate sexes and a free-living larval stage.

Phylum Rotifera

Unsegmented, with anterior ciliary corona and posterior foot; cuticle secreted within epidermis, not moulted; pharynx modified as mastax; cerebral ganglion and longitudinal nerve tracts; protonephridial excretory system; separate sexes, tubular gonads and gonoducts; spiral cleavage.

Rotifers are microscopic animals that occur in great numbers in freshwater lakes and ponds. Many are free-swimming members of freshwater plankton, while others attach themselves to objects such as submerged plants and are sessile species. Some occur in brackish water, but few are truly marine. Some inhabit soils or bryophytes, but depend for activity on a film of moisture. Because they can tolerate drying or freezing, rotifers can colonise temporary pools and polar regions. A few are parasitic.

Over 2000 species have been described, but no doubt this is an incomplete list. There are three classes. The Monogononta, a wholly freshwater group, reproduce both parthenogenetically and sexually. Both sexes have a single gonad. Males are dwarfs, usually lacking a gut. The Digononta, which mostly inhabit fresh water, reproduce parthenogenetically with a paired female gonad; but never produce males. The marine order Seisonacea, consisting of relatively few species, are epizoic on crustacea. They have a reduced corona and paired gonads.

Morphology

The body is divisible into a cephalic region, a trunk and a foot (Fig. 5.7b). The cephalic region has cilia, forming a corona that sweeps suspended food particles into the mouth and also acts as a swimming organ. The corona takes many different forms. Often two ciliated bands surround the cephalic region, as a trochus anterior to the mouth and a cingulum posterior to it (Fig. 5.7f). The trochus surrounds a non-ciliated apical field. In the Digononta the trochus is divided into two discs raised on pedestals (Fig. 5.7b), while the cingulum remains a single post-oral band. The synchronised beating of the cilia, circulating particles in opposite directions, gives the illusion of two rotating wheels, hence the name of the phylum. The cilia are often coalesced to form beating membranelles or more rigid spines. A large paired gland, the retrocerebral organ, opens on the apical region.

The trunk is generally broader than the cephalic region and its syncytial epidermis may be internally reinforced to form a lorica, into which the cephalic region and foot can be instantly withdrawn when the animal is disturbed. The epidermis does not secrete a cuticle but has a gelatinous surface coat. The skeletal structure lies within the syncytial epidermis. The foot is usually narrower than the trunk (sometimes reduced and often annulated), and ends in one to four toes on which mucus-secreting glands open (Fig. 5.7b). Their secretions enable the rotifer to adhere to objects, either temporarily as in crawling species or permanently as in sessile species (Fig. 5.7f,g). Crawling by looping along, alternately attaching the head and foot, is an alternative to swimming in some rotifers.

The ventrally located mouth, often with a ciliated buccal tube, leads into a muscular pharynx. The walls of the pharynx support a relatively massive complex organ for mechanically breaking up food, the mastax (Fig. 5.7c), which is characteristic of the phylum. The structure of the mastax varies with the nature of the food. It is formed from hardened articulating pieces ('jaws' or trophi) that may be used to grind, crush or tear. Food leaving the mastax passes through a ciliated oesophagus to the sac-like stomach and then the intestine, the latter often ciliated. Paired salivary glands open into the pharynx, and paired gastric glands open into the stomach. The gonoducts and the excretory ducts open into the intestine, which empties to the exterior through a dorsal cloaca at the hind end of the trunk.

There are no longitudinal or circular body wall muscle layers, but discrete muscles lie within the pseudocoel and there are extrinsic muscles associated with the alimentary canal. The osmoregulatory system consists of two coiled, tubular protonephridia (each with several flame cells) that open into the cloaca, often via a bladder.

The nervous system is bilaterally symmetrical, with a brain in the cephalic region and several ganglionated nerve cords, of which two ventral cords are the most prominent. Rotifers often have two or more photoreceptors (ocelli) formed from modified cilia, each with a lens and a red pigment cup. There may be a pair associated with the brain, two laterally on the corona and two in the apical field. Ocelli are absent in sessile rotifers. There are other sensilla formed from cilia, possibly with a tactile function, and ciliated pits on the apical field which may be chemoreceptors. Paired lateral antennae and a dorsal antenna also occupy the apical field.

Females possess either one combined ovary and yolk producing organ, the germovitellarium (Fig. 5.7b), or two germovitellaria. Eggs pass from the germovitellarium through an oviduct to the cloaca. In males (where they occur) a sac-like testis passes flagellated sperm through a ciliated sperm duct to a penis. Because most males lack a functional gut, the cloaca is replaced with a male gonopore. Paired glands are associated with the sperm duct.

Reproduction and development

During copulation in rotifers, the females are usually impregnated hypodermically by the male's penis. Reproduction is often parthenogenetic; in Digononta it is exclusively so. The Seisonacea have separate sexes, with females being inseminated by males. In the Monogononta, reproduction is more complex, giving rise to three forms of ova. For most of the year amictic females lay diploid eggs that develop parthenogenetically into more amictic females. Seasonally, or in deteriorating conditions, females become mictic females that lay haploid eggs. Unfertilised eggs develop into males, while fertilised eggs produce a resistant shell (Fig. 5.7d,e) and become dormant until appropriately stimulated, when they hatch as a new generation of females.

Embryonic development follows a spiral cleavage pattern. Nuclear division is completed early in development and many tissues become syncytial by the loss of intercellular plasma membranes. There is no larval stage.

Ecology

Rotifers feed on bacteria, cyanobacteria, algae and protozoans, protruding the mastax jaws to catch their prey. Some are specialists that can multiply very rapidly in response to abundant food. Others are generalists that usually have a slower reproduction. Planktonic species that must swim continuously are often globular and clothed in spines or rods that aid flotation. They may show seasonal changes in body form as well as in reproduction. Rotifers, in turn, are food for larger zooplankton such as copepods.

One euryhaline estuarine genus, *Brachionus*, has been exploited commercially in aquaculture as food for the earliest larval stages of cultivated prawns and fish. It can be produced in vast numbers in tanks, feeding on cultured algae such as *Chlorella*.

The production of resistant eggs by bdelloid rotifers allows them to survive unfavourable seasons and to exploit temporary pools, ponds or damp soil following rain. It facilitates their cosmopolitan distribution. Moss and lichens are a favoured habitat. Some rotifers can survive freezing or drying as adults for years in a cryptobiotic state. Such species are found in polar regions in melt-water. It is notable that rotifers and nematodes, both with species capable of cryptobiosis, have exploited the same transiently favourable habitats, feeding on microorganisms.

Some rotifers, including the Seisonacea, live on the surface of other invertebrates or inhabit their gills; some live within the gut or coelom of invertebrate hosts, while others live within algal cells.

Phylum Acanthocephala

Unsegmented worms parasitic in vertebrates; retractile hooked proboscis; no general cuticle; syncytial tegument with lacunar system; gut absent; reduced nervous system; protonephridial excretory system, usually absent; separate sexes, saccular gonads, tubular gonoducts; complex parasitic life cycles.

The Acanthocephala are all parasitic throughout the life cycle, with only a shelled egg occurring free in the environment. There is no trace of a gut at any stage and all nutrition must be absorbed through the body surface. Over 800 species have been described. The adults inhabit the intestines of all classes of vertebrates, the first or only intermediate host being an arthropod. Acanthocephala have received less attention than other parasitic helminths because they are of little medical and only minor veterinary importance. One species, *Moniliformis moniliformis*, which normally parasitises cockroaches and rats in Sydney, occasionally occurs in humans but does little harm. Another very large species, *Macranthorhynchus hirudinaceus*, up to 700 mm long, with a world-wide distribution, parasitises free-range pigs and can infect humans. Most Acanthocephala are less than 10 mm long.

Morphology

Acanthocephala (Fig. 5.8a, b) possess a distinctive proboscis, armed with rows of recurved hooks. The proboscis can be inverted into a muscular pouch within the main body, the proboscis receptacle, by a retractor muscle. The everted proboscis anchors the adult worm to the intestine of its host, becoming deeply embedded. It also provides a limited means of locomotion necessary for copulation. The body consists of a presoma and trunk. The presoma comprises the proboscis, proboscis receptacle, cerebral ganglion, lemnisci and muscles (Fig. 5.8b). The trunk encloses muscles, ligaments, reproductive and excretory organs (rarely present) and a large pseudocoel, which is functionally isolated from the smaller pseudocoel of the presoma. The body surface often has numerous spines, smaller than those on the proboscis.

The acanthocephalan body does not secrete a cuticle. The structure equivalent to the epidermis of other animals, called the tegument (Fig. 5.8h), is syncytial. Its structure can only be satisfactorily resolved using the

Figure 5.8 (opposite)
Acanthocephala: The anatomy of adult *Moniliformis moniliformis* from the intestine of rats, and its development stages from cockroaches. **(a)** Entire male. **(b)** Proboscis and associated structures. **(c)** Embryo from female pseudocoel. **(d)** Shelled, fully developed egg at the time of oviposition. **(e)** Acanthor larva hatched from egg in the gut of an intermediate host. **(f)** Acanthella larva from the tissues of an intermediate host. **(g)** Infective cystacanth larva from the tissues of an intermediate host. **(h)** Section through the body wall (tegument) of adult worm. **(i)** Ultrastructure of body surface.

electron microscope. On the outer surface (Fig. 5.8i) there is a tenuous layer of glycoprotein (the glycocalyx) which is important in nutrition and is resistant to host enzymes. Underlying the glycocalyx is a plasma membrane. Beneath the plasma membrane lies a striped layer made up of structural protein bars, perpendicular to the membrane, with vacuolated cytoplasm in between (Fig. 5.8i). Further down lies the much thicker layer containing a feltwork of collagen fibres. Deeper still the fibres are predominantly radially aligned. Endoplasmic reticulum, mitochondria, glycogen, lipid and other vesicles of various kinds are interspersed amongst the fibres. There are a few large nuclei, which may be dendritic or fragmented according to the species. No cell division occurs as the tegument grows.

Two long protrusions of the presomal tegument extend deeply into the trunk pseudocoel as the lemnisci (Fig. 5.8b). An interconnecting network of fluid-filled tubes, without lining or pumps (the lacunar system, Fig. 5.8h), permeates the tegument. Circumferential wider tubes and major longitudinal tubes, running the length of the trunk, interconnect the system; the different arrangements characterise the taxonomic classes. Beneath the tegument lie successively a basement lamina, a single layer of circular muscles and a single layer of longitudinal muscles. The muscle cells are also penetrated by an interconnecting network of fluid-filled channels. Some acanthocephalans have paired protonephridia, but most do not have excretory organs. The protonephridia consist of bundles of flame cells, a reservoir, and ciliated ducts opening via the gonoducts.

The central nervous system takes the form of a large ganglion on the ventral base of the proboscis receptacle, which gives rise to a number of nerves. Two ganglia lie adjacent to the posterior genitalia. Nerves are also associated with what are presumed to be sensory receptors at the tip of the proboscis and on the neck and genitalia.

Reproduction

Acanthocephala have separate sexes. Reproduction is exclusively sexual, with internal fertilisation following copulation. The reproductive organs are associated with a thin ligament strand running through the middle of the pseudocoel from the base of the receptacle to the hind end. Two thin-walled sacs enclose the ligament and gonads, filling the pseudocoel; these often rupture during development.

In males, two testes lie one behind the other on the ligament (rarely only one), followed posteriorly by one to eight prominent cement glands. There is a cup-shaped copulatory bursa at the posterior end of males, with a penis at its centre, that can enclose the female gonopore but is usually retracted within the male trunk. During copulation the products of the cement glands are discharged via ducts through the vas deferens and penis. The cement seals the female vagina.

In females the ovary fragments into numerous bodies, sometimes called ovarian balls, that float freely in the ligament sacs (or where these rupture, in the pseudocoel). Spermatozoa fertilise the oocytes in the ovarian balls, and the fertilised eggs are released into the pseudocoel to complete embryonic development to the acanthor stage. A complicated organ called the uterine bell, on the ligament at the posterior end of the pseudocoel, engulfs the eggs and sorts them according to shape and size. Shelled acanthors (Fig. 5.8d) are passed to the oviduct and expelled from the gonopore. Immature eggs are returned to the pseudocoel.

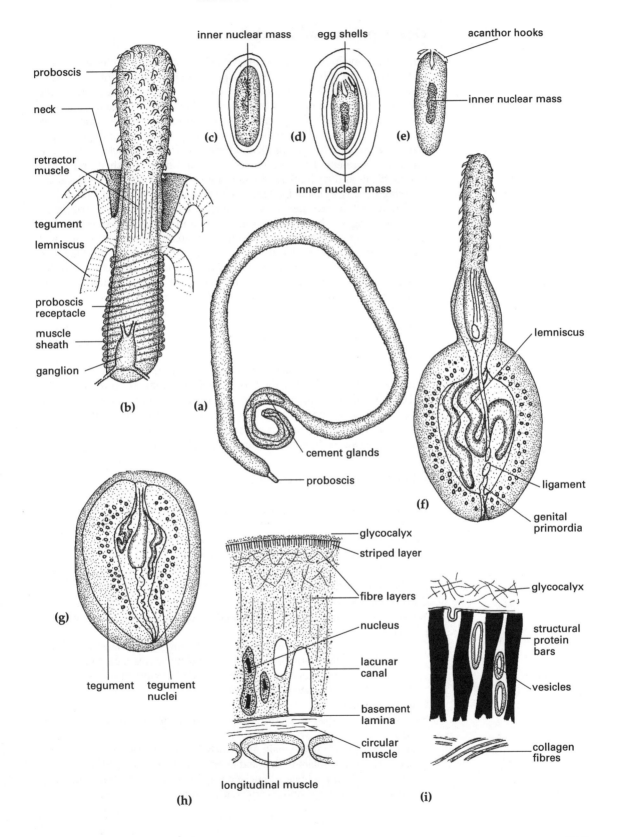

Development and life cycle

Embryonic development in acanthocephalans can be interpreted as a modified form of the spiral cleavage development characteristic of some other invertebrate phyla. The migration of most of the nuclei to the centre of the embryo to form an inner nuclear mass (Fig. 5.8c) can be considered as gastrulation, leading to the separation of the mesoderm, with no trace of any rudimentary alimentary canal ever developing. A number of large peripheral nuclei remain uncondensed, and these will form the nuclei of the adult syncytial tegument. The embryo secretes a multilayered shell and forms muscle fibrils and six hooks at the anterior end (Fig. 5.8d). This is the acanthor stage, ready to leave the female and be passed in the host's faeces.

The acanthor becomes active if swallowed by a potential arthropod host, escaping to the intestine (Fig. 5.8e) and using its hooks to claw through the intestine to the body cavity. Here in an appropriate host it will increase considerably in size and develop through the acanthella stage (Fig. 5.8f). As the acanthella grows, the inner nuclear mass subdivides and differentiates into a series of primordia of the adult organs. Many of the tissues remain as syncytia as they develop, but the muscles form as discrete cells. When the adult structure has developed in miniature, the proboscis inverts and becomes invested by a membrane, forming the cystacanth (Fig. 5.8g). The cystacanth will establish itself in the intestine of a definitive host if swallowed.

In some species the cystacanths, if ingested by an animal other than the definitive host, may invade the host's tissue and remain dormant but are still capable of infecting the definitive host. Such paratenic hosts widen the range of possible definitive hosts so that, for example, a species with an aquatic crustacean as an intermediate host may parasitise a fish-eating bird via a crustacean-feeding fish, acting as a paratenic host.

Classification of the Aschelminthes

All aschelminths are bilaterally symmetrical pseudocoelomates having a gut with mouth and anus (or gut absent), but lacking a circulatory system. Development is direct, with no ciliated larval stage. The characteristic features of each phylum were summarised earlier in the chapter.

PHYLUM NEMATODA
Class Adenophorea
Nematodes with post-labial amphids; caudal phasmids absent; renette excretory system.
Orders — Chromadorida (mostly marine), Monhysterida (mostly marine), Enoplida (mostly marine), Dorylaimida (mostly terrestrial), Mononchida (freshwater and terrestrial), Mermithida (invertebrate parasites), Trichocephalida (vertebrate parasites).
Class Secernentea
Nematodes with labial amphids; caudal phasmids present; tubular excretory system.
Orders — Rhabditida (terrestrial), Aphelenchida (plant feeders and insect parasites), Tylenchida (plant feeders and insect parasites), Strongylida (vertebrate parasites), Oxyurida (invertebrate and vertebrate parasites), Ascaridida (vertebrate parasites), Spirurida (vertebrate parasites).

PHYLUM NEMATOMORPHA
Superfamilies — Nectonemoidea (marine), Gordioidea (freshwater and terrestrial).

PHYLUM GASTROTRICHA
Orders — Macrodasyida (mostly marine), Chaetonotida (mostly freshwater).

PHYLUM KINORHYNCHA (marine)

PHYLUM LORICIFERA (marine)

PHYLUM ROTIFERA
Class Monogononta
Freshwater, swimming, creeping or sessile; single germovitellarium; usually females and transient males.
Class Digononta
Mostly freshwater, some terrestrial, swimming or attached; paired germovitellaria; parthenogenetic females only.
Class Seisonacea
Marine, epizoic on Crustacea; reduced corona; sexual reproduction.

PHYLUM ACANTHOCEPHALA (vertebrate parasites)
Class Archiacanthocephala
Class Palaeacanthocephala
Class Eoacanthocephala

Chapter 6

The Sipuncula and Priapula

D.T. Anderson

	The coelom *117*	
PHYLUM SIPUNCULA	Introduction *117*	Nervous system and sense organs *118*
	Functional morphology *117*	Reproduction and development *118*

	Organ systems *118*
	    Coelomic system *118*
	    Digestive system *118*
	    Excretory system *118*
		Classification of the phylum Sipuncula *119*

PHYLUM PRIAPULA	Introduction *120*	
	Functional morphology *120*	

	Organ systems *120*
	    Digestive system *120*
	    Nervous system *120*
	    Excretory and reproductive systems *121*	

The coelom

The animal groups which have evolved a complex anatomical and functional organisation and a concomitant complexity of behaviour patterns and ways of life all share a distinctive type of body cavity, the coelom. We saw in Chapter 1 that a coelom is a fluid-filled cavity between the body wall and the internal organs, lined by a coelomic epithelium or peritoneum. Current opinion is still divided on whether a coelom evolved only once, or several times in different lines of animal evolution (see Chapter 18). In all cases, the coelomic cavity allows flexible and independent movements of the body wall and gut wall, acts as a hydrostatic skeleton generating shape changes during these movements, accommodates the development of organs, and provides a storage space for gametes and other products of internal activity.

This chapter describes a group of marine worms, the phylum Sipuncula, in which the coelomic organisation is shown in its basic form, despite certain peculiarities of structure and function. The phylum Priapula, a small group of superficially similar worms, is also described.

Phylum Sipuncula

Bilaterally symmetrical, coelomate, unsegmented marine worms with anterior introvert; U-shaped gut with mouth and anus; brain and ventral nerve cord; nephridial excretory system; no circulatory system; separate sexes, coelomic gonads; spiral cleavage development.

Introduction

Sipunculans are a small phylum, comprising about 220 living species of bilaterally symmetrical, unsegmented, coelomate worms 2–200 mm long. They live among rocks or in sand or gravel, intertidally and subtidally, and are sedentary deposit feeders. *Phascolosoma* is a typical Australian genus. Sipunculans have left no fossil record.

Functional morphology

The cylindrical body of a sipunculan (Fig. 6.1a) is swollen posteriorly and thin anteriorly. The anterior part, or introvert, can be inturned, head first, into the larger posterior part, in an action of protective withdrawal. When everted, the anterior end extends a ring of small, ciliated, food-collecting tentacles (Fig. 6.1c). The tubular gut is U-shaped (Fig. 6.1b), passing from the anterior mouth to the posterior end of the body before looping forwards to a dorsal anus at the base of the introvert.

The external surface is covered with a thick cuticle, composed of cross-layered collagen fibres with interspersed microvilli. There is no chitin. A thin layer of cuticle continues over the tentacles. Beneath the cuticle is a surface epithelium, followed by a collagenous dermis and layers of circular and longitudinal muscle (Fig. 6.1d). In generalised sipunculans the longitudinal muscles comprise a continuous layer, but in many species the muscles are in distinct bands separated by extensions of the coelom. The muscle layers of the body wall work in conjunction with the coelomic fluid, which acts as a hydrostatic skeleton, to generate peristaltic burrowing movements and extension of the introvert. Introversion is brought about by specialised retractor muscles running between the base of the tentacles and the body wall posteriorly (Fig. 6.1b).

Organ systems

Coelomic system

The main coelom extends throughout the body but is separated anteriorly from a small tentacle coelom. All internal surfaces are covered by a thin coelomic peritoneum. Coelomocytes proliferated from the peritoneum float in the coelomic fluid. The tentacle coelom encircles the oesophagus, with branches into the tentacles and into one or two compensation sacs extending along the oesophagus dorsally and ventrally. The sacs are components of a hydraulic system by which the tentacles are extended and contracted. The ciliated tentacles collect food particles from the surrounding sediment and also have a role as a respiratory surface. The coelomic system is the only fluid circulation in sipunculans, which lack a separate blood system. The coelomocytes contain haemerythrin.

Digestive system

The gut begins with a long oesophagus, extending through the introvert, and continues as a spiral intestine which loops back on itself and ends in a short rectum opening at the anus.

Excretory system

Near the anus, small nephridiopores open ventro-laterally from a pair of tubular nephridia (Fig. 6.1b, d) attached to the body wall on either side of the nerve cord. The inner ends of the nephridial tubes are expanded as large, ciliated funnels which separate gametes from other coelomic cell types during spawning. The role of the nephridia as gonoducts in sipunculans is obvious, but how they function in excretion, osmoregulation and ionic balance is not clear.

Nervous system and sense organs

The central nervous system of sipunculans comprises a small, bilobed brain lying dorsally in the head, connected by circumoesophageal commissures to a long, thin ventral nerve cord (Fig. 6.1b, d). The branches of the peripheral nervous system are very fine and are generally not visible macroscopically. The head has a pair of ciliated sensory pits and a pair of simple ocelli.

Reproduction and development

The gonad in sipunculans is a ventral, bilobed organ suspended in the coelom by a mesentery. Sexes are separate. In females, the oocytes are released from the gonad early in previtellogenesis and complete the remainder of their growth and vitellogenesis in the coelom. Male spermatocytes similarly undergo spermiogenesis in the coelomic fluid. In ripe animals, the coelom is packed with gametes. Most species spawn their eggs or sperm freely into the water, where fertilisation occurs. Spiral cleavage leads to an embryogenesis in which mesoderm is proliferated as paired, ventro-

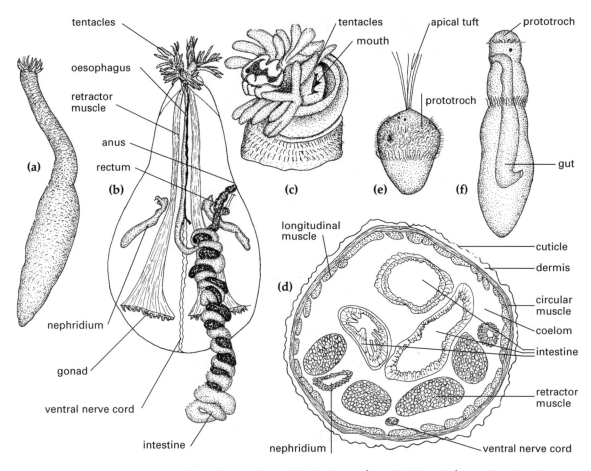

Figure 6.1
Sipuncula. **(a)** *Phascolosoma*. **(b)** General anatomy of a sipunculan. **(c)** Mouth and tentacles of *Phascolosoma*. **(d)** Transverse section. **(e)** Trochophore larva. **(f)** Pelagosphaera larva. (a–c after Rice 1993; d after Anderson 1996; e, f after Rice 1967).

lateral mesodermal bands from the cell 4d, and the coelom develops as splits within the mesodermal bands (schizocoely), later merging into a single cavity. Most species develop through a planktonic, lecithotrophic trochophore larva (Fig. 6.1e), although a few have direct development to the juvenile stage. The trochophore may metamorphose directly into a juvenile, but the majority of sipunculan species develop through a more complex planktonic larval stage called a pelagosphaera (Fig. 6.1f). Depending on egg size, the pelagosphaera may continue as a lecithotrophic stage or may become planktotrophic, with a functional gut and a ciliated feeding organ around the mouth. The larvae appear to be carnivorous, feeding on small zooplankton. The nephridia develop during the pelagosphaera stage. Metamorphosis, which occurs after several months of larval growth in the plankton, involves loss of the larval cilia and settlement as a sedentary juvenile.

Classification of the phylum Sipuncula

CLASS SIPUNCULIDA
Ring of tentacles around the mouth (e.g. *Sipunculus*).

CLASS PHASCOLOSOMIDA
Ring of tentacles dorsal to the mouth (e.g. *Phascolosoma*).

Phylum Priapula

Bilaterally symmetrical, pseudocoelomate unsegmented marine worms with anterior introvert; thin cuticle moulted periodically; gut with mouth and anus; epidermal nervous system; protonephridial excretory system; no circulatory system; separate sexes.

Introduction

The priapulans are a very small group in the modern fauna, comprising only about 17 species of unsegmented, burrowing marine worms, but have a long history. Priapulans similar to modern forms are common as fossils in the Cambrian Burgess Shale deposits of Canada. Modern forms such as *Priapulus* and *Halicryptus* occur mainly in colder waters. Despite the superficial similarities between priapulans and sipunculans, the priapulans are currently recognised as pseudocoelomates with affinities to kinorhynchs and loriciferans (see Chapters 5 and 18).

Functional morphology

As in sipunculans, the body comprises a tubular trunk and anterior introvert, but the latter is swollen and is covered with short, recurved spines. In *Priapulus* (see Fig. 1.1i) the posterior end of the trunk carries a pair of vesiculate caudal appendages, possibly respiratory. At the body surface is a chitinous cuticle, secreted by a surface epithelium. Layers of circular and longitudinal muscle lie beneath the epithelium. The eversion of the introvert is caused by the contraction of the body wall muscles, with the fluid in the body cavity acting as a hydrostatic skeleton as it does in sipunculids. Retraction is due to a double ring of retractor muscles. Repeated eversions and retractions of the introvert are used, together with the peristaltic action of the body wall, in burrowing through the sediment.

Organ systems

The body cavity of priapulans has been interpreted as a coelom, but recent studies have shown that it lacks a lining peritoneum and is thus a type of pseudocoel. Amoebocytes containing haemerythrin are present in the fluid.

Digestive system
The terminal mouth opens into a muscular pharynx ringed by rows of cuticular teeth. The eversion of the pharynx, a separate action from the eversion of the introvert, is used in feeding. Larger species such as *Priapulus* are carnivores, grasping annelids and other prey; small interstitial species ingest bacteria and other microorganisms. The long, tubular midgut ends in a short, cuticle-lined rectum opening at a posterior anus.

Nervous system
The central nervous system begins with an anterior ring in the surface epithelium around the mouth. A ventral nerve cord, also superficial in position, extends to the posterior end.

Excretory and reproductive systems

The body cavity contains a pair of large protonephridia (with solenocytes) and a pair of gonadal sacs. Both discharge through a pair of common ducts opening on either side of the anus. Sexes are separate. In larger priapulids, free spawning occurs and fertilisation is external. Development is little known, but appears to begin with a form of radial cleavage. A burrowing, feeding 'larva' with a ring of longitudinal chitinous shields on the trunk precedes metamorphosis to the juvenile form. There is no larval ciliation or planktonic phase.

Chapter 7

The Mollusca

J.M. Healy

PHYLUM MOLLUSCA

Major divisions of the phylum 123

Basic molluscan features 124

Class Gastropoda 125
  Basic gastropod features 125
  Torsion 127
  Gastropod classification 127
  Gastropod diversity 127
  Evolutionary history of the gastropods 133
  Physiological processes 133
  Reproduction and development 137

Class Cephalopoda 139
  Basic cephalopod features 139
  Cephalopod classification 140
  Cephalopod diversity 141
  Locomotion 141
  Evolutionary history 142
  Physiological processes 143
  Reproduction and development 147

Class Bivalvia 148
  Basic bivalve features 148
  Bivalve diversity 150
  Form and function 152
  Locomotion 154
  Evolutionary history 154
  Physiological processes 155
  Reproduction and development 156

Class Scaphopoda 157
  Basic scaphopod features 157
  Scaphopod diversity 157
  Form and function 158
  Evolutionary history 159
  Physiological processes 159
  Reproduction and development 160

Class Monoplacophora 160
  Basic monoplacophoran features 160
  Monoplacophoran diversity 161
  Form and function 161
  Evolutionary history 162
  Physiological processes 162
  Reproduction and development 163

Class Polyplacophora 163
  Basic polyplacophoran features 163
  Polyplacophoran diversity 163
  Form and function 164
  Evolutionary history 165
  Physiological processes 165
  Reproduction and development 166

Class Aplacophora 167
  Basic aplacophoran features 167
  Form and function 167
  Evolutionary history 168
  Physiological processes 168
  Reproduction and development 170

Classification of the phylum Mollusca 171

Phylum Mollusca

Bilaterally symmetrical, unsegmented coelomates with reduced coelom; haemocoelic body cavity (expanded circulatory system); head, foot and visceral mass; dorsal mantle secreting calcareous spicules or shell (sometimes lost); gut with buccal mass and radula (sometimes lost); nervous system with brain, pedal ganglia and visceral ganglia; coelomic kidneys; coelomic gonads; separate sexes or hermaphrodite; spiral cleavage development.

Major divisions of the phylum

The phylum Mollusca constitutes one of the most widespread and conspicuous of the invertebrate groups, with 50 000–60 000 species, including such familiar animals as snails, oysters and octopods. Molluscs are also among the oldest of the bilaterians, with a confirmed record of shelled forms dating back to the early Cambrian.

Seven extant classes of molluscs are recognised (Fig. 7.1), the three largest being the Gastropoda (marine, freshwater and terrestrial snails and slugs: about 40 000 species), Bivalvia (clams, oysters, scallops etc.; about 8000 species) and Cephalopoda (*Nautilus*, cuttlefish, squid, octopods; about 600 species). The remaining four classes, although containing fewer species, are no less significant in terms of the evolution of the phylum. These are the Scaphopoda (tooth or tusk shells; 300–500 species), Monoplacophora (small, deep sea molluscs with cap-shaped shells; 20 species), Polyplacophora (chitons, with a shell composed of eight plates; 500–600 species) and Aplacophora (vermiform molluscs with an epidermis covered in needle-like spicules; about 250 species).

Figure 7.1
Classes of the Mollusca (after Morton and Yonge 1964 and Salvini-Plawen and Steiner 1996).

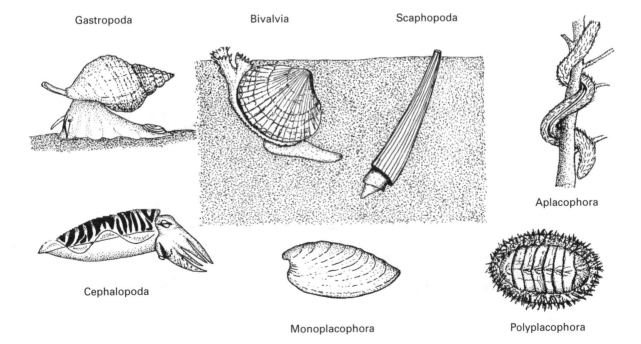

Basic molluscan features

The morphological diversity of the molluscs (Fig. 7.1) makes it difficult to give a simple definition of the phylum. In general the molluscan body is divisible into a head, a ventral foot, a dorsal visceral mass and a spicule- or shell-secreting integument called the mantle. The head is well developed in gastropods and cephalopods and equipped with eyes, but in most classes the head is small or poorly differentiated. The foot is the primary locomotory organ in most molluscs, used either for surface creeping by muscular and ciliary action on a mucous trail, or in burrowing. In species that are attached to a substratum or are sedentary, the foot may be reduced or absent.

The haemocoel forms the major body cavity of molluscs, usually in the form of several large, connected sinuses. Haemocyanin is the chief oxygen-carrying blood pigment, although a number of species have haemoglobin. A heart of variable complexity is usually present. A coelomic space is represented by the pericardium, kidneys and gonads, but its homology with the coelom of other coelomates such as Sipunculida and Annelida is not clear.

Part of the molluscan mantle overhangs the rest of the body, forming a recess (the mantle cavity) which is open to the environment. In aquatic molluscs the mantle cavity acts as the exit point for faecal and kidney waste, and also contains the ctenidia or gills and often a chemosensory epithelium termed the osphradium. In terrestrial gastropods ctenidia are usually absent and the mantle cavity is converted into a pulmonary chamber.

One of the most distinctive molluscan features is the buccal mass (Fig. 7.3i). This consists of a radula, a ribbon-like structure composed of numerous rows of chitinous teeth; a cartilaginous, radula-supporting odontophore; and often chitinous jaws and a chemosensory subradular organ. Complex muscles surround these structures. A buccal mass is absent in bivalves, most of which feed via ctenidial filtration (see below) or by the use of palps, siphons or tentaculate structures.

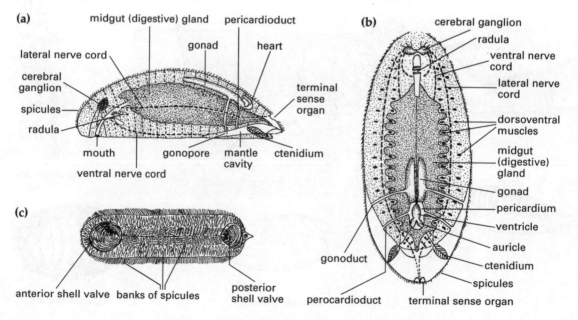

Figure 7.2
(a, b) Reconstruction of an early spiculate mollusc: (a) Lateral view. (b) Dorsal view (after Salvini-Plawen 1972, 1981, 1985). (c) Reconstruction of fossil mollusc-like coeloscleritophoran (affinities unknown), showing cap-shaped valves and spicules on the integument (after Conway-Morris and Peel 1992).

True metamerism in organ and muscle arrangement is absent in molluscs, despite the repetition of ctenidia, muscle scars and renal pores in monoplacophorans (Fig. 7.15) and polyplacophorans (Fig. 7.16). Development proceeds via spiral cleavage, and in marine and some freshwater species often involves larval stages such as the ciliated trochophore and/or the veliger. Direct development has evolved in representatives of all classes, especially the Gastropoda and Cephalopoda.

The earliest molluscs (Fig. 7.2a, b) were probably small (1–2 mm in length), with a dorsally chitinised integument lacking a calcareous shell but possibly having embedded spicules, a ventral head-foot with an anterior mouth and toothed radula, a posterior mantle cavity with paired ctenidia, a simple gut with stomach, digestive gland and intestine, and paired kidneys and gonads. The muscular foot was equipped with mucus glands and cilia for surface gliding. Animals of this type may have been dioecious or hermaphroditic. Fertilisation probably occurred either internally, with sperm transferred in spermatophores or by a penis, or within the mantle cavity. Intriguing lower Cambrian fossils such as the spiculose worm-shaped Halkieriidae (Fig. 7.2c) appear to offer connecting links with the Mollusca, Brachiopoda and Annelida (see Chapter 18).

Class Gastropoda

Basic gastropod features

Gastropods (snails and slugs) form the largest and most diverse molluscan class and the only one to have successfully colonised terrestrial habitats. Typically they possess a calcareous, conical, spiral shell and are asymmetrical (Fig. 7.3c). The shell has become greatly reduced in a number of groups and is absent in marine and terrestrial 'slugs', resulting in external bilateral symmetry. Most gastropods have a well-developed head with two cephalic tentacles, each associated with an eye, and a buccal mass containing the odontophore, radula and often jaws (Fig. 7.3i). The number and morphology of radular teeth vary considerably within the class, reflecting taxonomic relationships and food substratum preferences. The foot is usually prominent and used for crawling (Fig. 7.3a), typically on a mucous trail, but is greatly reduced in sedentary forms, modified for swimming in some pelagic groups and enlarged in burrowers. A chitinous operculum, sometimes strengthened with calcium, is carried on the postero-dorsal surface of the foot in many gastropods (Fig. 7.3c). Most commonly it is used to seal the shell aperture, but in a few groups such as the stromb and spider shells it assists with movement. In conispiral gastropods a strap-like columellar muscle enables rapid retraction of the animal into the shell. The mantle cavity is anterior and contains one or a pair of ctenidia and often a chemosensory osphradium and hypobranchial gland (Fig. 7.3c, j). Primitively the ctenidia are paired and bipectinate, with two rows of leaflets on a central axis, but in many groups only a single, monopectinate ctenidium remains. A ctenidium is absent in most terrestrial species (Fig. 7.7k), but in some marine and freshwater forms, secondary gills may be present (Fig. 7.6f).

Gastropods may be dioecious or hermaphroditic and occasionally parthenogenetic. Fertilisation can occur in the surrounding water, the mantle cavity or internally. Internally fertilising groups use a penis or spermatophores (or both), and several have more than one type of sperm. Development proceeds via lecithotrophic or planktotrophic stages (planktonic trochophore, often followed by a veliger), or may be direct.

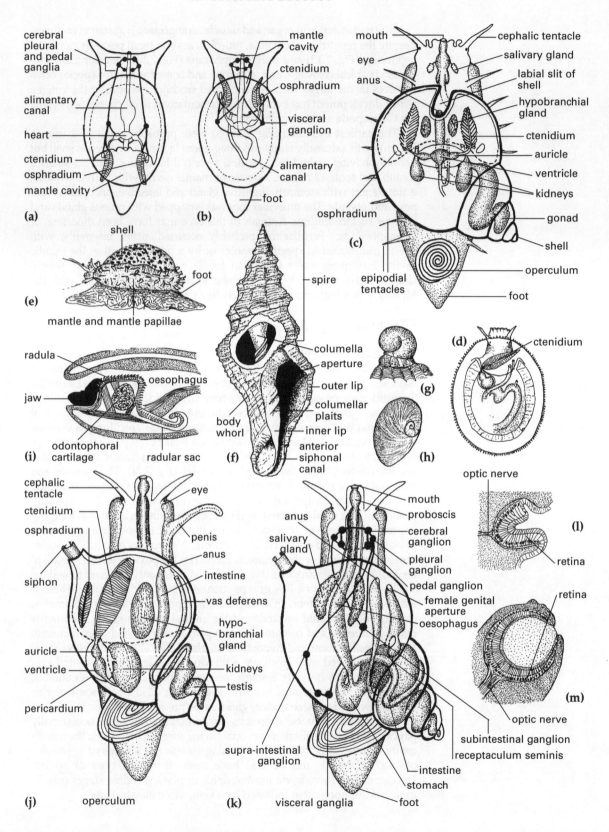

Figure 7.3 (opposite)
(a) Pre-torsional gastropod with posterior mantle cavity, symmetrical alimentary canal and nervous system.
(b) Post-torsional gastropod with posterior mantle cavity, twisted alimentary canal and nervous system. **(c)** Vetigastropod prosobranch, showing paired bipectinate ctenidia, osphradia, kidneys, heart auricles and slit in shell. **(d)** Acmaeid limpet with single bipectinate ctenidium. **(e)** Cowrie (*Cypraea*). **(f)** General features of prosobranch shell. **(g)** Protoconch (larval shell) of volute. **(h)** Operculum of *Littorina*. **(i)** Relationship of the radula to the jaw in a gastropod buccal mass. **(j)** Caenogastropod prosobranch, showing single monopectinate ctenidium, osphradium, hypo-branchial glands, kidney and heart auricle. **(k)** Caenogastropod nervous and alimentary system. **(l, m)** Eyes of patellogastropod (*Patella*) and vetigastropod (*Trochus*). (a, b after Hyman 1967; c, j, k after Ivanov 1940; d after Graham 1985; e after Abbott 1962; f, g after Cox 1960; h after Fretter and Graham 1962; l, m after Hilger 1885.)

Torsion

It is primarily the phenomenon of torsion which distinguishes the class Gastropoda from all other molluscs. Torsion occurs at the larval (veliger) stage and involves differential growth of the retractor muscles. This results in the relocation of the mantle cavity, and the organs it contains, from a posterior position overhanging the foot to an anterior position partly overhanging the head (Fig. 7.3 a, b). The mantle cavity and its associated organs, the alimentary tract and certain nerves and ganglia are twisted during torsion. Torsion must be distinguished from shell coiling, which is a completely independent phenomenon and is not limited to gastropods. There are cephalopods and monoplacophorans with coiled shells. Explanations as to why torsion evolved centre largely on a perceived protection of the head and foot by withdrawal into the anterior mantle cavity. Anterior repositioning of the mantle cavity also enables the ctenidia and osphradia to function better. Fouling of the mantle cavity by faecal and nephridial waste, which was probably an initial problem, was solved through the evolution of either a slit (Fig. 7.3c) or lines of small holes in the shell directly over the mantle cavity, as in many vetigastropods such as the abalones and slit shells, or through an increase in the angularity of the water current washing through the mantle cavity, giving a transverse rather than a ventrodorsal flow (Fig. 7.3j).

Gastropod classification

The morphology and arrangement of the ctenidia and nervous system, the structure of the heart, the shape of the foot, and radular morphology have all been used to designate higher groupings within the Gastropoda. Thiele (1929–1931) divided the main group of marine snails, the Prosobranchia, into three orders: Archaeogastropoda (including groups such as the true limpets, abalones, trochoideans, fissurelloidean limpets and neritids); Mesogastropoda (with many superfamilies including the cowries, littorinid periwinkles, tritons, spider shells, sand and mud creepers) and the Neogastropoda (including murex and coral shells, cone shells, augers and buccinid whelks). This system of the Prosobranchia proved popular for over 40 years. Cox (1960), in *Treatise on Invertebrate Paleontology*, suggested uniting the Mesogastropoda and Neogastropoda into a single unit, the Caenogastropoda, but this was largely ignored until recently. In the late 1970s and 1980s, studies of gastropod classification and phylogeny entered a new phase in which comparative anatomy was complemented by ultrastructural studies of several organs or cell types, including excretory cells, mantle epithelium, osphradia and spermatozoa. Osphradial and sperm ultrastructure, in particular, have provided much new insight into higher level systematics and phylogeny.

Recent classifications of the Gastropoda, such as those proposed by Haszprunar (1988) and Ponder and Lindberg (1997), herald a new trend towards the introduction of molecular (DNA, RNA, proteins) data, combined with a cladistic approach (relationships based on shared advanced features). The classification adopted in this chapter reflects some of the more important changes to occur in the last 10 years.

Gastropod diversity

Among the Mollusca, the Gastropoda are unmatched in terms of morphological and ecological diversity, exploiting almost every aquatic and terrestrial habitat. The success is largely explained by the supreme adaptability of the basic gastropod body plan. Perhaps the best way to illustrate the morphological diversity of the Gastropoda is to briefly review the subclasses Prosobranchia and Heterobranchia, both of which are distributed worldwide, including a strong representation in the Australian fauna.

Subclass Prosobranchia

Prosobranch gastropods exhibit complete torsion, with the mantle cavity in an anterior position. They have separate sexes. Prosobranchs constitute the most basal members of the class Gastropoda, with fossils known from the mid-Cambrian.

Among the former 'archaeogastropod' assemblage, three chief orders are now recognised, the Patellogastropoda or true marine limpets (Fig. 7.4a), the the more diverse Vetigastropoda (Fig. 7.4b–f) and the Neritimorpha (Fig. 7.4g, h). The variety of species falling within these groups is summarised in Box 7.1. A fourth order, the Cocculiniformia, comprises deep-water gastropods with limpet-like shells that feed on organic debris.

Figure 7.4
Prosobranch diversity.
(a) Acmaeidae (*Acmaea*).
(b) Pleurotomaridae (*Pleurotomaria*). (c) Trochidae (*Trochus*). (d, e) Fissurellidae:
(d) *Fissurella*. (e) *Scutus*.
(f) Haliotidae (*Haliotis*).
(g) Neritidae. (h) *Nerita*, showing apertural teeth.
(a, d–f, h after Wilson 1993; b, g after Abbott 1962).

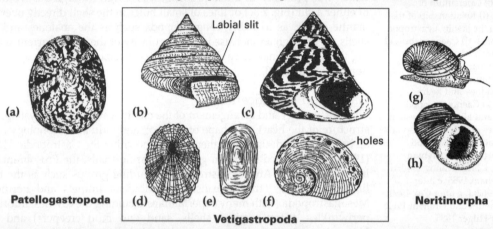

Box 7.1
Basal prosobranch gastropods

ORDER PATELLOGASTROPODA
True marine limpets with cap shell (Fig. 7.4a); intertidal and shallow water browsers of algae on rocky surfaces.

ORDER VETIGASTROPODA
Including the following superfamilies:
Pleurotomarioidea (slit shells, Fig. 7.4b) — deep-sea browsers on sponges.
Haliotoidea (abalones, Fig. 7.4f) — shallow water, limpet-liked algal grazers, with the shell slit converted into a series of holes; commercially important.
Trochoidea (trochus and turban shells, Fig. 7.4c) — shallow to deep-water algal or sponge grazers; some commercially important.
Fissurelloidea (keyhole and shield limpets, Fig. 7.4d, e) — intertidal and shallow water algal browsers.

ORDER NERITIMORPHA
Neritoidea (Nerites, Fig. 7.4g, h) — algal grazers, including shallow-water marine, freshwater and terrestrial families.

The former 'mesogastropods' and 'neogastropods' are now grouped together as the order Caenogastropoda. This is a highly diverse and complex group of snails that has radiated into many ways of life (see Box 7.2). Three suborders are distinguished: the Architaenioglossa (Fig. 7.5d) which live in freshwater and terrestrial habitats; the Neotaenioglossa, a very diverse range of primarily marine snails (Fig. 7.5a–c, e–m); and the exclusively carnivorous, marine Neogastropoda (Fig. 7.5n–w). All caenogastropods have a functionally advanced respiratory system, with a single monopectinate ctenidium on the left of the mantle cavity and a left–right respiratory current.

Subclass Heterobranchia
Heterobranch gastropods exhibit various changes in the respiratory system, including replacement of ctenidia by other forms of gill, or modification of the mantle cavity as a lung. Most heterobranchs are hermaphrodite. One superorder of marine heterobranchs, the Allogastropoda (Fig. 7.6a–d),

---

Box 7.2
Advanced prosobranch gastropods

ORDER CAENOGASTROPODA
Suborder Architaenioglossa
   Freshwater grazers (superfamily Ampullaroidea, Fig. 5d), including the operculate pond snails, families Ampullariidae, Viviparidae.
   Terrestrial grazers (superfamily Cyclophoroidea).

Suborder Neotaenioglossa
Including the superfamilies:
   Cerithioidea (Fig. 7.5b, g) — deposit-feeding and grazing sand- and mud-creeping cerithiids, potamidids and batillariids; and the periwinkle-like planaxids of rocky shores.
   Naticoidea (Figs 7.5g, 7.8h) — carnivorous sand snails.
   Cypraeoidea (Fig. 7.5m) — cowries, sponge grazers.
   Stromboidea — spider shells, algal grazers.
   Janthinoidea (Fig. 7.5i, j) — including the coral- and anemone-associated wentle traps (Epitoniidae); and the floating violet snails (Janthinidae), which hang upside down on a bubble raft at the ocean surface.
   Littorinoidea (Fig. 7.5f) — true periwinkles of rocky shores, algal grazers.
   Rissooidea — microscopic, ubiquitous algal grazers and detritivores.
   Tonnoidea (Fig. 7.5k, l) — carnivorous triton, helmet and tun shells.
   Vermetoidea (Fig. 7.5a) — worm shells, uncoiled, sessile suspension feeders.

Suborder Neogastropoda
Including the superfamilies:
   Conoidea (Fig. 7.5s–w) — cone shells (Conidae), venomous predators of worms, molluscs and fish; also auger shells (Terebridae) and turrid shells (Turridae), worm predators.
   Cancellaroidea — suctorial predators.
   Muricoidea (Fig. 7.5n, o, q, r) — mainly scavengers and predators, such as the murex shells (Muricidae), the colourful volutes (Volutidae) and the largest living gastropod, *Syrinx aruanus* (Turbinellidae) from northern Australia, New Guinea and Indonesia.

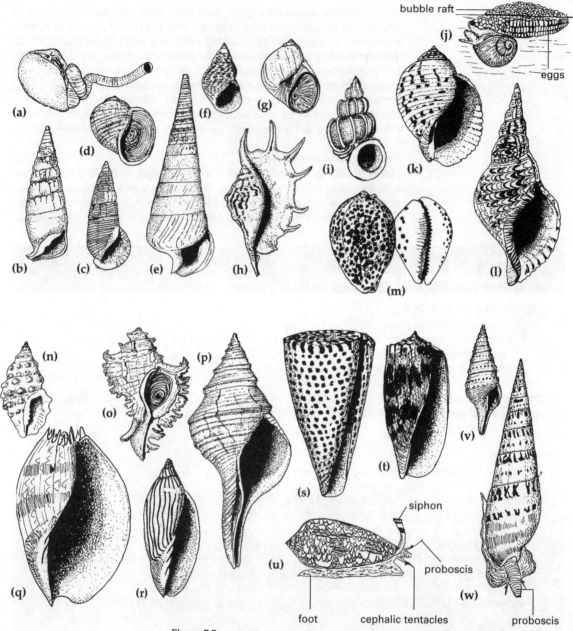

Figure 7.5
Caenogastropod diversity: **(a)** Vermetoidea (worm shells). **(b, c)** Cerithioidea (sand and mud creepers). **(d)** Freshwater Ampullarioidea. **(e)** Campaniloidea. **(f)** Littorinoidea (periwinkles). **(g)** Naticoidea (sand snails). **(h)** Stromboidea (spider shells and allies). **(i, j)** Janthinoidea (wentle traps and floating violet snails). **(k, l)** Tonnoidea: (k) tun shell, (l) triton shell. **(m)** Cypraeoidea (cowries). **(n, o)** Muricoidea (murex shells). **(p)** Muricoidea: Turbinellidae (*Syrinx aruanus*). **(q, r)** Muricoidea (volutes). **(s–w)** Conoidea: (s, u) Conidae, including (s) worm eaters (*Conus litteratus*), (t) fish eaters (*Conus geographus*) and (u) mollusc eaters (*Conus textile*). (v) Turridae (*Gemmula*). (w) Terebridae (*Terebra*). (a, c, f–h, k–o, q, t, v, w after Wilson 1993, 1994); d after Abbott 1989; i, u, w after Abbott 1962; j after Fraenkel 1927.)

THE MOLLUSCA

retains many prosobranch features, but the other two superorders, Opisthobranchia and Pulmonata, show distinct modifications.

Superorder Opisthobranchia

This superorder includes a number of familiar marine gastropods common around the Australian coastline, from mud flats to coral reefs. A trend exists within the Opisthobranchia towards reduction and loss of the shell. This is evident even within a single order, as in the Cephalaspidea ('bubble shells'), where the shell is prominent and thick in some families (e.g. Acteonidae) but thin and fragile in others (e.g. the Hydatinidae) (Fig. 7.6e). In the orders Sacoglossa, Anaspidea (sea hares), Notaspidea (side-gilled slugs) and Nudibranchia (nudibranchs), the substantial or total loss of the shell has resulted in many morphologically bizarre forms. This is especially evident in the abundant, widespread and often colourful nudibranchs, many of which have secondary derived 'gills' posteriorly (Doridimorpha) or club or frond-like cerata structures arising from the body wall (Aeolidiida) (Fig. 7.6f, g). Nudibranchs of the family Glaucidae are epipelagic, using lateral cerata for flotation or swimming, and graze on the tentacles of drifting colonies of cnidarians such as *Velella* and *Physalia*. These and other aeolidoidean nudibranchs are able to ingest cnidocysts and transfer them intact to their own cerata for use in defence.

Many opisthobranchs have chemical defences. Sea hares emit a chemically noxious ink if disturbed. Nudibranchs of the genus *Phyllidea* are renowned for the acrid secretion they exude from their warty mantle epithelium. This provides a very effective defence against predation, which certain reef flatworms have managed to take advantage of by mimicking the colour pattern of *Phyllidea*.

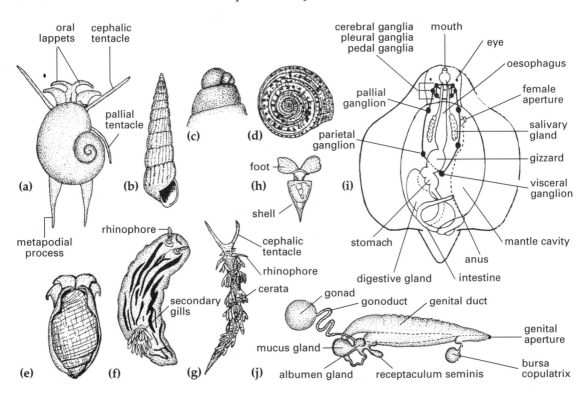

Figure 7.6
Gastropoda. Heterobranchia. (a–d) Superorder Allogastropoda. (a) Valvatoidea (*Cornirostra*). (b) Pyramidelloidea (*Turbonilla*). (c) Protoconch of pyramidellid. (d) Architectonicoidea (sundial shells). (e–j) Superorder Opisthobranchia. (e) Cephalaspid (*Acteon*). (f, g) Nudibranchs: (f) *Hypselodoris*, showing rosette of secondary gills. (g) *Tularia*, with club-shaped dorsal cerata. (h) Swimming shelled pteropod (*Clio*). (i) Basic anatomy of advanced opisthobranch, including detorted nervous system. (j) Reproductive system of *Aplysia*. (a after Ponder; b after Fretter and Graham 1962; c after Cox 1960; e after Barnes 1980; f, g after Willan and Coleman 1984; h after Cooke; i after Ivanov 1940; j after Fretter 1946.)

Superorder Pulmonata

Pulmonates share many anatomical features with the Opisthobranchia, such as a similar arrangement of ganglia (Figs 7.6, 7.7j) and a complex and usually simultaneously hermaphroditic reproductive system (Figs 7.6j, 7.7i). Probably these two groups arose from a common ancestral source. Three major orders are recognised: (1) the Systellommatophora, including the amphibious marine slugs (Onchidoidea) from mangroves (Fig. 7.7e); (2) the Basommatophora, including the amphibious marine groups.

Siphonarioidea (Fig. 7.7b, from rocky shores, with radially-ribbed, limpet-like shells) and Amphiboloidea (Fig. 7.7c, from mangroves and estuarine areas, with a globular shell, and an operculum), and the freshwater Lymnaeoidea and Planorboidea, both well represented in Australian rivers

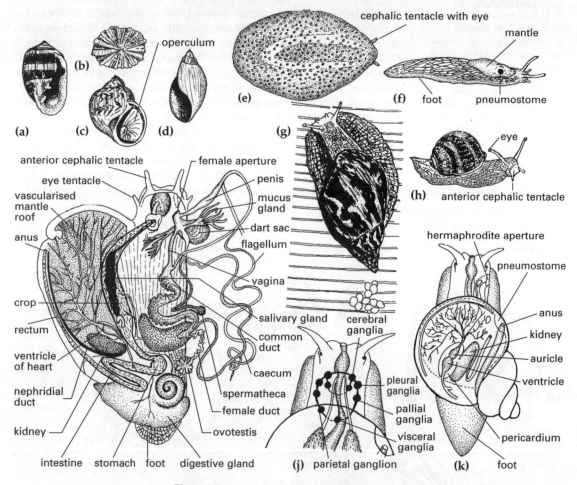

Figure 7.7
Gastropoda, Superorder Pulmonata. **(a–d)** Basommatophorans: (a) *Cassidula* (Ellobioidea). (b) *Siphonaria* (Siphonarioidea). (c) *Salinator* (Amphiboloidea). (d) *Physa* (Physoidea). **(e)** A marine slug, *Onchidium* (Systellommatophora). **(f–h)** Stylommatophorans: (f) *Arion* (terrestrial slug). (g) *Achatina* (giant African land snail). (h) *Helix* (garden snail, Helicoidea). **(i)** Internal anatomy of *Helix*. **(j)** Concentration of ganglia in a generalised pulmonate. **(k)** Mantle cavity converted to pulmonary chamber. (a, b after Abbott 1989; c, d after Hubendick 1978; e after Smith and Kershaw 1979; f, h after Hyman 1967; g after Watson 1985; i after Grove and Newell 1969; j, k after Ivanov 1940.)

and creeks; (3) the Stylommatophora or land snails, including the Rhytidoidea (the large forest snails of Australia such as *Hedleyella* and *Pedinogyra*), Limacoidea and Arionoidea (Fig. 7.7f, slugs) and Helicoidea (Fig. 7.7g, h, garden snails).

Various introduced helicoidean species, such as *Helix aspersa* and *Theba pisana* are important pest species in Australia and many other parts of the world. The giant African land snail, *Achatina fulica* (Achatinoidea, Fig. 7.7g), is a major agricultural threat throughout South-East Asia and New Guinea and has occasionally caused concern in northern Queensland when specimens have been detected.

Evolutionary history of the gastropods

Although the precise origin of the Gastropoda remains uncertain, anatomical and palaeontological evidence suggests that they are closely related to the Monoplacophora and probably arose from them some time in the early Cambrian. The stratigraphic record confirms the appearance of marine prosobranchs, mainly vetigastropods such as the Pleurotomarioidea (slit shells and allies) and trochoideans, in the early Palaeozoic. Patellogastropods and the Neritimorpha appear later in the Palaeozoic, followed by the first caenogastropods. Many of the living caenogastropod superfamilies first appeared during the Cretaceous. Heterobranch origins are less certain, but the available evidence suggests that allogastropods, opisthobranchs and pulmonates were all present in the late Palaeozoic.

Physiological processes
Respiration and circulation

Primitively, gastropods possessed a pair of ciliated, bipectinate ctenidia housed in the anterior mantle cavity. Within the subclass Prosobranchia a gradual reduction and simplification of the ctenidial complex has occurred. The primitive condition can still be observed in most vetigastropods (slit shells, abalones, keyhole and shield 'limpets' and trochus and turban shells) (Fig. 7.3c). In most neritimorphans (see Box 7.1) the right hand ctenidium is lost, while that on the left retains the bipectinate structure. In patellogastropod limpets a single bipectinate ctenidium is retained on the left in the family Acmaeidae (Fig. 7.3d) but is lost in the family Patellidae, the gill leaflets of patellid limpets being secondarily derived structures. In caenogastropods the single, left ctenidium has become monopectinate (Fig. 7.3j), and even this is lost in terrestrial groups (Cyclophoroidea), with the mantle cavity becoming a pulmonary chamber. In the subclass Heterobranchia, true ctenidia are absent. Secondary gills that have developed in various ways are either housed within the mantle cavity or, as in many nudibranchs, are completely exposed (Fig. 7.6f). In most pulmonates the mantle cavity has been converted to a pulmonary chamber, the mantle edge being fused to the head and penetrated by a contractile aperture called the pneumostome (Fig. 7.7f, i, k). Surprisingly, in freshwater pulmonates the pulmonary chamber may be periodically filled with water, so that air-breathing can be used only intermittently. Secondary gills are present in some marine pulmonate 'limpets'.

The mantle edge in many aquatic gastropods is well served by blood sinuses, indicating that this region plays an important support role in respiratory exchange. Haemocyanin is the primary blood pigment, appearing colourless when bound to oxygen but blue upon release of oxygen. The heart features two auricles and a ventricle in basal prosobranchs (patellogastropods, vetigastropods, neritomorphans; Fig. 7.3c) but only one auricle and ventricle in caenogastropod prosobranchs and heterobranchs (Figs 7.3j, 7.7i, k).

Feeding and digestion
Gastropods consume a wide range of foods, including algae, detritus, plankton, a variety of cnidarians, sponges, polychaetes, crustaceans, other molluscs, carrion, and even live fish and body fluids. In most cases feeding involves direct rasping by the radular teeth on the food substratum followed by the transfer of dislodged particles to the mouth. Enzymes for extracellular digestion are produced by the salivary glands, oesophageal glands (if present) and digestive gland. Most extracellular digestion takes place in the stomach. The digestive gland is the principal site for intracellular digestion. In several pulmonates (chiefly stylommatophorans) the crop is enlarged for food storage, and extracellular digestion is effected by bacterial cellulases. In detrital or algal grazers, including many prosobranchs such as trochids, abalones, patellogastropods, nerites, sand and mud creepers, spider shells and littorinids, and in some sessile marine prosobranchs that are ciliary and mucus trap feeders, the mucus string which traps ingested particles is drawn into the stomach by either a revolving mucous style or a crystalline style containing enzymes (Fig. 7.8a). The style is located in a style sac attached to the stomach and is rotated by ciliary action.

Several types of radular pattern are seen in gastropods. Within the Prosobranchia there are five major types:
1. rhipidoglossate, with fine, numerous, often elongate teeth, including many laterals and marginals, in vetigastropods and neritimorphs;
2. docoglossate, with cusps often coarse and mineralised, in patellogastropods (Fig. 7.8c);
3. taenioglossate, with usually seven teeth per row including one pair of laterals and two pairs of marginals, in many caenogastropods (Fig. 7.8b);
4. stenoglossate, with the central tooth usually broad and multicusped and often one pair of laterals, in many carnivorous neogastropods (Fig. 7.8d); and
5. toxoglossate, with isolated, hollow teeth, injected (usually with venom) through the proboscis (Fig. 7.8f,). Toxoglossan radulae are found in the Conoidea (Fig. 7.8e), a group including cone shells, augers and turrids (Fig. 7.5s–w).

In predatory caenogastropods the radula is first used to drill a circular, bevelled hole through the shell of the prey, usually another mollusc, and is then inserted with the proboscis to rasp out the flesh (Fig. 7.8g). Secretions (acidic or non-acidic) from glands in the foot are often used to weaken the shell fabric during drilling. Many neogastropods also scavenge on dead crustaceans, fish and other molluscs. The transformation of the marginal teeth in toxoglossans into highly effective poison darts is perhaps the strangest use of the radula within the Mollusca. The most famous of these gastropods are the beautiful but often highly dangerous cone shells. Conidae usually swallow their prey whole, using the extensible proboscis. Most of the 400–500 species (Fig. 7.5s–u) are either exclusively worm-eaters (narrow aperture) or mollusc eaters (wider aperture), but cone shells with very wide apertures are hunters of fish (Fig. 7.8f) and their venoms have been responsible for several human fatalities in the Indo–Pacific region. Mollusc-eating species may also cause injury and occasionally death. Cone shell toxins are some of the most active biocompounds known and are now the focus of intense pharmaceutical research in Australia and elsewhere.

The caenogastropods of the superfamily Eulimoidea show a complete gradation from free-living species to those which implant themselves into the surface of echinoderms, causing a gall of host tissue to develop around them. The culmination of this range is the bizarre *Enteroxenos*, a worm-shaped, endoparasitic sac containing only gonads and embryos and absorbing haemolymph from its holothurian host.

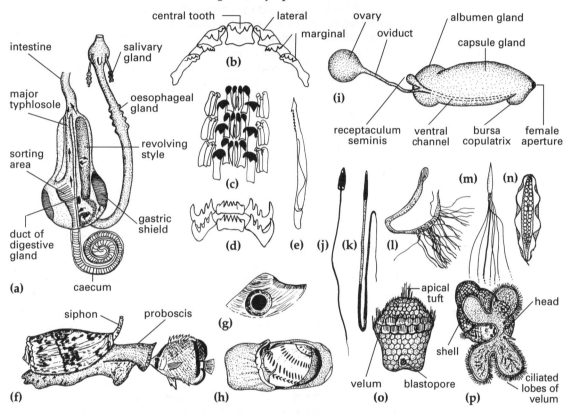

Figure 7.8

Gastropoda. (a) Alimentary system of a trochid vetigastropod. (b) Transverse row of teeth from taenioglossan radula, with seven teeth per row (e.g. *Littorina*). (c) Transverse row of teeth from a docoglossan radula (e.g. *Patella*). (d) Transverse row of teeth from a stenoglossan radula (e.g. the neogastropod *Buccinum*). (e) Harpoon-like radula tooth of cone-shell toxoglossan radula. (f) *Conus geographus* with the proboscis expanded to engulf envenomed fish. (g) Drill-hole left by sand snail (Naticidae). (h) Sand snail (*Natica*). (i) Female reproductive tract of caenogastropod *Littorina*. (j–n) spermatozoa of prosobranchs. (j) Vetigastropod sperm. (k) Caenogastropod fertile sperm (euspermatozoon). (l) Caenogastropod infertile sperm (paraspermatozoon) bearing tufts of attached euspermatozoa. (m) Multitailed parasperm of caenogastropod mudwhelk. (n) Paraspermatozoon of a spider shell (Strombidae), showing the undulating wings. (o) Trochophore of a patel-logastropod. (p) Veliger of a caenogastropod. (a after Owen 1996; b, c–e, g, i, p after Fretter and Graham 1962; d, e after Lankester; f after Coleman; h after Healy 1988; m after Healy and Jamieson 1981; n after Tochimoto 1967; o after Cox 1960.)

In contrast to the Bivalvia, the use of ctenidia for filter feeding is very rare in gastropods, occurring only in species of the sessile caenogastropod families Calyptraeidae, Struthiolariidae, Turritellidae and Siliquariidae. The cemented worm shells (Vermetidae, Fig. 7.5a) unwind mucous nets into the water to trap suspended food or plankton, then periodically haul in the net using the radula.

Within the Heterobranchia, radulae can range from having a few teeth per transverse row, as in many opisthobranchs, to over 50 or more per row in pulmonates. The shapes of the teeth are extremely variable, particularly among the Allogastropoda and Opisthobranchia, which consume a wide range of foods such as algae, detritus, plant sap, cnidarians, polychaetes and sometimes other molluscs. Pyramidellid allogastropods (Fig. 7.6b, c) lack a radula and employ the proboscis and a sharp stylet to pierce the body wall of bivalves or tube-dwelling polychaetes and suck up body fluids such as haemolymph. Pulmonates (Fig. 7.7) mainly feed on algae or detritus in aquatic habitats and on living or rotting leaves in terrestrial situations. Some land pulmonates are carnivorous, feeding on earthworms or on other pulmonates.

Excretion and ionic regulation

The morphology and physiological capabilities of the excretory organs of gastropods vary considerably between groups. Left and right kidneys are present in the most basal marine prosobranchs, such as the Patellogastropoda (true limpets) and many Vetigastropoda (trochoids, haliotoids, etc.) (Fig. 7.3c). In these forms the left kidney regulates ionic levels within the blood, while the right is concerned with the extraction and release of nitrogenous waste. In all other gastropods the right kidney is either reduced to a vestige and partially incorporated into the gonoduct (e.g. Neritimorpha) or absent (most other groups such as Caenogastropoda, Heterobranchia) (Figs 7.3j, 7.7i). In such cases the left kidney is responsible both for removing nitrogenous waste and regulating the ionic concentration of the blood. In freshwater gastropods the kidney must be capable of very efficient extraction of salts from the urine in order to maintain a favourable internal osmotic environment. The resulting filtrate is often extremely hypotonic.

Aquatic gastropods usually release their nitrogenous waste as ammonia, via the excretory pore(s) into the mantle cavity, from where it is flushed out by water currents. In terrestrial gastropods such as the land prosobranchs and most pulmonates, water retention is at a premium and waste is normally released as uric acid, conveyed outside the mantle cavity by an elongate ureter. Water loss in terrestrial forms is further minimised by closure of the shell aperture via an operculum (prosobranchs) or a mucous epiphragm (pulmonates), and behavioural modifications such as nocturnal activity, dry season burrowing and aestivation.

Nervous system and sense organs

In prosobranchs the nervous system usually shows evidence of torsion in the form of crossed nerves between the pleural, parietal and visceral ganglia (Fig. 7.3b). This is also present in basal heterobranchs but not in the majority of more advanced heterobranchs (most opisthobranchs and pulmonates), due to concentration of the cerebral and cerebropedal ganglia and shortening of the commissures giving the so-called de-torted nervous system (Figs 7.6i, 7.7j). Despite sometimes marked cephalisation, a complex, highly organised brain comparable to that of coleoid cephalopods has never evolved in gastropods.

The most important sense organs of gastropods are the eyes, cephalic tentacles, statocysts and osphradium. Eyes usually occur on the cephalic tentacles but are reduced or absent in several burrowing forms (e.g. prosobranchs of the family Naticidae, and several shelled opisthobranchs) and in some pelagic groups (e.g. violet snails). Often each eye is borne on a short optic peduncle (Fig. 7.3e, j). In stromb and spider shells the optic peduncle is longer than the cephalic tentacle and the eyes are terminal and coloured. The prolific land pulmonate group Stylommatophora is distinguished by having invaginable eye peduncles as well as invaginable cephalic tentacles (Fig. 7.7 f–h). Gastropod eyes are extremely varied in their complexity, ranging from the open retinal cup of patellogastropods (Fig. 7.3l), to semi-enclosed flask-like retinae in some vetigastropods such as abalones and trochids (Fig. 7.3m), to the complex eyes of caenogastropods and heterobranchs, featuring inner and outer corneal layers, a spherical lens and a thick retina. Pallial eyes are also known in some marine caenogastropods, notably among the mud and sand creepers of the Cerithioidea. In addition, some mangrove pulmonate slugs (Onchidiidae) exhibit numerous light-sensitive ocelli on their dorsal (mantle) surface.

Cephalic tentacles are believed to have both tactile and chemosensory capabilities and are usually highly retractile. Patellogastropods, many vetigastropods and some caenogastropods may also have mantle edge or epipodial tentacles (Fig. 7.3c–e).

The gastropod osphradium is primarily chemosensory and is of particular importance for detecting food, especially the carrion and prey of carnivorous species. In the basal prosobranchs (Box 7.1) two osphradia are present, one associated with each ctenidium (Fig. 7.3c). Only one occurs in caenogastropods and many heterobranchs (Fig. 7.3j). In terrestrial gastropods the osphradium is vestigial or absent.

Cerebrally innervated statocysts containing one to many calcareous statoliths are positioned near the pedal ganglia in most gastropods, and presumably assist in righting movements and general body orientation.

Reproduction and development

Prosobranchs typically have separate sexes, but there are several examples of protandric hermaphroditism (e.g. violet snails — Janthinoidea), parthenogenesis (e.g. many freshwater cerithioideans) and a few cases of simultaneous hermaphroditism. By contrast the Heterobranchia are almost exclusively hermaphrodites (usually simultaneous, sometimes protandric), with only a few known instances of dioecism occurring in certain interstitial opisthobranchs. The reproductive systems of internally fertilising groups vary considerably in their morphology, being least complex in dioecious prosobranchs and most complex in simultaneous hermaphroditic heterobranchs. In all gastropods the gonad lies close to the digestive gland, normally within the coil of the shell (Fig. 7.3c, j, k) . A gonoduct (vas deferens, oviduct or hermaphrodite duct) proceeds from the gonad to the mantle cavity where, in externally fertilising species, gametes are released (usually via the kidney ducts) to the sea water. In internally fertilising gastropods, the gonoduct within the mantle cavity may be open or closed, and is associated with a variety of accessory glands and ducts for storing and packaging eggs and sperm for release (egg-capsule gland, albumen gland, prostatic gland, spermatophore-forming organ, seminal vesicle) or for receiving sperm (seminal receptacle) (Figs 7.6j, 7.7i).

In most heterobranchs each animal has both male and female systems, and an albumen gland is normally associated with the gonad. Land snails (stylommatophoran pulmonates) (Fig. 7.7i) have extraordinarily complex reproductive systems; some of these snails, such as the garden snails (Helicoidea), exchange calcareous 'love darts' during copulation, presumably as a stimulatory adjunct. Sperm transfer in most internally fertilising gastropods is via a penis, spermatophores, or both. Giant infertile sperm are also used for fertile sperm transfer in some marine caenogastropods (Fig. 7.8l).

Spermatozoa

Gastropods exhibit perhaps one of the most diverse ranges of sperm morphologies in the animal kingdom. Within the Prosobranchia, it is also necessary to differentiate between the genetically viable, fertilising sperm or euspermatozoa and one or more types of genetically non-viable sperm or paraspermatozoa. Among prosobranchs the following types of euspermatozoa are encountered:

1. the simple 'classic' aquasperm of externally fertilising groups such as patellogastropods, haliotids and trochoideans. Aquasperm are characterised by a large, conical acrosome, a short nucleus, a short midpiece containing paired centrioles surrounded by a ring of mitochondria, and finally the flagellum (Fig. 7.8j);
2. filiform introsperm (some Trochoidea, all Neritimorpha, all Caenogastropoda), characterised by a conical acrosome, a rod-like nucleus, an elongate midpiece with a flagellar axoneme surrounded by a mitochondrial sheath and often glycogen deposits, and a short end piece (terminal portion of axoneme) (Fig. 7.8k). Eusperm intermediate between these two morphological types are prevalent in many small trochoideans, and seem to be associated with fertilisation within the mantle cavity.

Paraspermatozoa are found in most neritimorphs and in the majority of caenogastropods. Their most likely general function is nutritive or stimulatory support of euspermatozoa. Generally, paraspermatozoa are characterised by the absence or reduction of the nucleus and usually the presence of numerous axonemes and dense vesicles. Some have multiple external tails (as in sand creepers and mudwhelks; Fig. 7.8m), or undulating lateral wings (as in spider shells; Fig. 7.8n). Some paraspermatozoa may be up to 1 mm long and bear hundreds of attached euspermatozoa, (Fig. 7.8l). The functional role of such paraspermatozoan-euspermatozoa associations may be to help transport and to prevent premature dispersal of the euspermatozoa.

Heterobranchs produce only euspermatozoa, characterised by a small, rounded acrosomal vesicle, typically a helical nucleus, and always a very elongate and complex helical midpiece. The longest recorded molluscan sperm belongs to a pulmonate land snail, the helicoidean *Pleurodonte acuta* (length 1.75 mm). The sperm of many gastropods can remain viable in the seminal receptacle for long periods. In land snails the sperm may be viable for up to 2 years, and undoubtedly this is an important factor in the rapid spread of introduced species, the individuals being able to lay fertilised eggs long after mating has occurred.

Spermatophores

Although most prosobranchs either liberate sperm directly into the surrounding water (marine only) or have a penis to transfer sperm (marine and freshwater), a number of taxa utilise spermatophores in their reproduction. For example spermatophores are typical of the large and diverse superfamily Cerithioidea. In the uncoiled, sessile worm-shells (Vermetidae) spermatophores are released into the water column by males and subsequently collected by females in their mucous feeding 'nets'. In neritimorph prosobranchs, some marine opisthobranchs and amphibious and terrestrial pulmonates, spermatophores are used in conjunction with, rather than instead of, a penis during copulation. Spermatophores prevent premature dispersal of spermatozoa, and maintain their viability.

Eggs, fertilisation and development

Fertilisation in gastropods may occur outside the mantle cavity (ect-aquatic fertilisation; e.g. patellogastropods, abalones, many trochoideans), within the mantle cavity (ent-aquatic fertilisation: e.g. some trochoideans) or within the reproductive tract of the recipient snail (internal fertilisation; e.g. a few trochoideans, all neritimorph and caenogastropod prosobranchs, all heterobranchs). The fertilised eggs emerging from the oviduct are encapsulated by the oviducal gland, then covered in a jelly-like substance. In neriti-morphs, caenogastropods and heterobranchs, capsules containing from one to several fertilised eggs are laid in a cluster. Most commonly the capsules are flexible. In several groups of stylommatophoran pulmonates, single eggs with thick calcareous shells are produced, some reaching over 20 mm in length. The predatory sand snails or Naticoidea incorporate eggs into either collar-shaped sand masses or a curved, voluminous mass of jelly.

Development proceeds via spiral cleavage in all gastropods. Eggs hatch as trochophores (Fig. 7.8o) in prosobranchs such as abalones and many patellogastropods, then progress to the veliger stage (Fig. 7.8p) in the plankton. In many trochoideans, and in neritimorphs, several caenogastropods and some marine opisthobranchs, the veliger stage is reached before hatching. Many caenogastropods and heterobranchs undergo direct development. Planktonic and direct development may occur in representatives of the same family or even the same genus.

Class Cephalopoda

Basic cephalopod features

Cephalopods are primarily swimming molluscs. They include such well-known groups as cuttlefish, squid and octopods and the chambered nautiluses, as well as several important fossil groups including the ammonoids, straight-shelled 'nautiloids' and squid-like belemnoids.

Several features help to define the Cephalopoda (Figs. 7.9–7.11). The animals are bilaterally symmetrical. The shell usually has numerous internal chambers (Fig. 7.9c), but is most commonly internalised and reduced in living forms. The head is well developed, with complex eyes, cerebral ganglia organised as a discrete brain enclosed in a cartilaginous casing, and a buccal apparatus containing a chitinous two-jawed beak, radula and odontophore. Most species are carnivorous, but some deep sea forms are probably detritivores. Prehensile, muscular appendages called arms or tentacles surround the mouth and assist in food capture, mating and locomotion. All cephalopods have 8–10 of these tentacles (usually with suckers) except *Nautilus*, which has up to 90 (Figs 7.9b, c). With few exceptions, cephalopods are active swimmers, using reverse jet propulsion generated through muscular contraction of the funnel and/or mantle wall and often undulating movements of lateral fins or an interbranchial web. Benthic octopods can swim but move primarily by using the sucker-bearing arms.

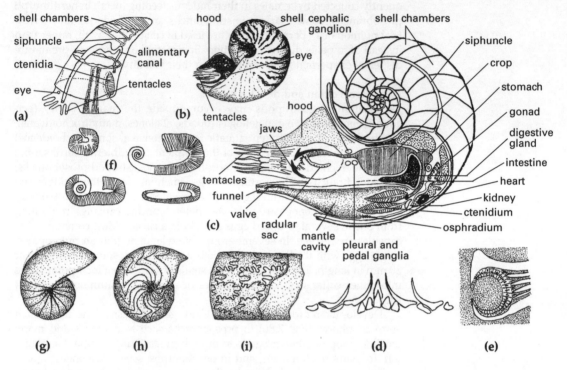

Figure 7.9
Cephalopoda. (a) Reconstruction of a late Cambrian cephalopod, *Plectronoceras*. (b) *Nautilus*, in swimming position. (c) Internal structure of *Nautilus*. (d) Row of teeth from a *Nautilus* radula. (e) Eye of *Nautilus*. (f) Uncoiling and coiling of ammonoid shells. (g) Shell of coiled nautiloid showing simple septal sutures. (h,i) Increase in the complexity of the septal suture in ammonoids. (a after Yochelson *et al*. 1973; b after Abbott 1962; c after Naef 1928; d after Keferstein 1865; e after House 1988; f after d'Orbigny 1840–55; g–i after Barnes 1980.)

Bipectinate ctenidia are retained within the mantle cavity (Figs 7.9c, 7.10b) but lack cilia. *Nautilus* has two pairs, but only a single pair is present in other extant cephalopods. Osphradia are present only in *Nautilus*. The circulatory system consists of open haemocoelic sinuses in *Nautilus*, but shows a secondary condition of closed vessels in coleoids. Heart morphology is variable; in coleoids a pair of additional branchial hearts is associated with the ctenidia.

Sexes are separate. Fertilisation usually occurs within the mantle cavity or internally. A single, typically dorsal gonad is present, with one vas deferens in males and one or two oviducts in females. Spermatozoa are packed in complex, elongate spermatophores which are transferred from males to females by a modified arm (or a group of arms in *Nautilus*) or by a penis. Eggs are yolky and laid in individual capsules. Copulation is often preceded by complex courtship rituals which sometimes involve rapid attractant and warning colour changes in the skin. Development is direct, without a trochophore or veliger stages.

Cephalopod classification
Extant cephalopods are divided into two subclasses. The Nautiloida contains the five or six living species of the genus *Nautilus*. There is an external

chambered shell with simple sutures and an axial siphuncle, two pairs of ctenidia, kidneys and heart auricles, blood circulation through sinuses, no ink sac, a divided funnel, and up to 90 suckerless arms (Fig. 7.9b, c). The Coleoida contains all the remaining members of the class, including the Orders Sepioida (cuttlefish), Spirulida (the ram's horn shell), Sepiolida (dumpling 'squids'), Teuthoida (true squid), Vampyromorpha (vampire squid) and Octopoda. In coleoids there is a reduced internal shell or no shell, one pair of ctenidia, kidneys and auricles, blood circulatory system via enclosed vessels assisted by branchial hearts, an ink sac typically present, a unitary funnel and 8–10 usually sucker-bearing arms (Figs 7.10, 7.11). The wholly extinct Ammonoida, characterised by their conspicuous septal suture marks, comprise a further subclass (Fig. 7.9f, h, i).

Cephalopod diversity

With the exception of the few surviving species of *Nautilus*, all living cephalopods have an internalised, reduced shell or have lost the shell, but numerous and widely distributed fossil shells have made us aware of the importance of the shelled nautiloids and ammonoids in Palaeozoic and Mesozoic seas. Living *Nautilus* provides some indication of the mode of life of these forms, but the exceptionally large size (3-metre diameter in some ammonoids; 2-metre-long shells in some straight-shelled 'nautiloids') and/or unwieldy shell shapes of several extinct groups suggest that many may have been relatively quiescent deposit or planktonic feeders.

The diversity among the modern Cephalopoda derives largely from increased swimming mobility, or in the case of many octopods, a combination of benthic and swimming mobility. Dominant among the coleoids are the nektonic teuthid squids (300 or more species) and the largely benthic octopods (up to 200 species). All groups of living cephalopods occur in Australian waters, including the cosmopolitan ram's horn squid (*Spirula spirula*, Fig. 7.10f) with its internal, coiled and chambered shell, and the much rarer and even stranger relative of the octopods, the vampire squid (*Vampyroteuthis infernalis*; Fig. 7.11a) of the Vampyromorpha.

Locomotion

In general, cephalopods utilise jet propulsion as the primary means of locomotion. This is achieved through muscular contraction of the funnel and/or mantle cavity combined and the use of the funnel as a means of concentrating and directing the streaming water. In most groups, lateral fins help to stabilise the body during rapid motion, change direction, and effect hovering manoeuvres. In forms with a calcareous shell, active regulation of gas and liquids within the shell chambers by associated tissues leads to very fine buoyancy control. Although often portrayed as a cumbersome animal, *Nautilus* (Fig. 7.9b, c) is capable of surprising speed when necessary, all derived through funnel-generated jetting. The externally shelled cephalopods are limited in their vertical distribution by shell strength. The maximum depth to which they can dive without imploding is about 800 metres.

Cuttlefish are more effective swimmers than nautilids because of body streamlining, reduction and internalising of the shell, and the conversion of the mantle cavity into a highly muscular, contractile space (see Fig. 7.10a, b). Teuthid squids are among the fastest of all aquatic animals, and some can temporarily leave the water and 'glide' like flying fish. The shell is reduced from that seen in cuttlefish to a fusiform, organic structure — the chitinous gladius or 'pen' — which gives body support with minimal weight (Fig. 7.10h). Cartilaginous rods also provide internal support for fins and ctenidia. The loss of the chambered shell as a buoyancy regulator through shell reduction and internalisation (Fig. 7.11o) has been compensated in true squids and some swimming octopods by the incorporation of ammonia solution within the body tissues. Being less dense than sea water, the ammonia helps maintain neutral or slight positive buoyancy. This saves considerable energy for nektonic species, which must either move over large distances vertically in search of their food or maintain a stable position within the water column for long periods. Most octopods spend much of their time on the seabed, using the arms for crawling, but they are all capable of rapid swimming when required (e.g. to escape predators). Several deep sea forms can slowly descend through the water using the expanded arm web as a parachute.

Most coleoids are equipped with an ink sac which produces quantities of brown-black, opaque ink containing melanin and (in some cases) a mucus-like substance. When disturbed, particularly when being pursued by predators, coleoids release the ink from the ink sac via the rectum and mantle cavity into the water jet. The resulting cloud then acts as a diversion while the cephalopod escapes. The sometimes noxious chemical content of the ink may further assist the escape by disabling the would-be attacker. In deep-sea species, especially certain octopods, the ink sac may be absent, but in others the ink may be released with bioluminescent compounds or bacteria, creating a glowing or even temporarily blinding phantom image.

Evolutionary history

Despite the fact that cephalopods have left an impressive fossil record from the Late Cambrian to the Recent, the origin of the group and its relationships to other molluscan classes are still uncertain. Cephalopods probably arose from monoplacophorans with dorso-ventrally elongate, internally septate shells. The earliest fossils generally accepted as belonging to the Cephalopoda are those of the genus *Plectronoceras* (Fig. 7.9a) which show the characteristic siphuncular hole in each septum. Ammonoids probably arose from nautiloids in the middle Palaeozoic and their shells subsequently became coiled, straight, or in some cases even irregular in shape (Fig. 7.9f). Ammonoids can usually be differentiated from nautiloids by their more complex suture marks (where the septum contacts the interior surface of the shell), which initially consisted of wide undulations but in Cretaceous forms became folded and refolded, often into intricate patterns (Fig. 7.9h, i). Rare fossils also indicate that the ammononid radula possessed 7 teeth per row, like extant Coleoidea, rather than 13 per row as in nautiloids (Fig. 7.9d). As a group the ammonoids were extraordinarily successful and are abundant in fossil-bearing marine rocks, especially those representing shallow-water communities. With the eventual extinction of the Ammonoidea during the late Cretaceous and a concurrent decline in the nautiloids, cephalopod evolution proceeded along the path of rapid mobility, pioneered during the late Palaeozoic and throughout the Mesozoic by the belemnoids. In these squid-like animals, believed to be the first of the Coleoidea, the body was streamlined and the shell reduced, straightened and internalised (Fig. 7.10h). The various lineages of modern coleoids developed in the Cretaceous and early Tertiary, with further shell reductions.

THE MOLLUSCA

Physiological processes
Respiration and circulation

In all cephalopods respiratory water currents across the ctenidial surfaces are generated through rapid, rhythmic contraction of the mantle. The ctenidia in *Nautilus* are attached basally to the mantle, while in coleoids the ctenidia are attached both basally and dorsally. In *Nautilus* the blood moves to and from the ctenidia and other organs via large sinuses throughout the haemocoel, with discrete vessels being present only near the heart and under the integument. By contrast, the organs of coleoids are served by an efficient, almost fully closed circulatory system composed of: a main heart (ventricle and two auricles); two accessory branchial hearts (one attached

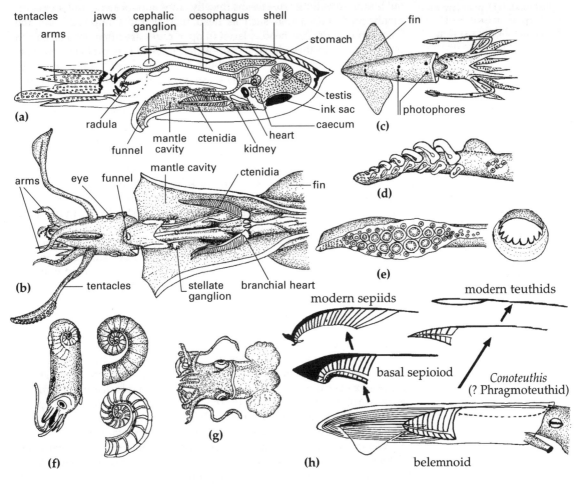

Figure 7.10
Cephalopoda. **(a)** Principal features of sepiid anatomy. **(b)** Teuthid squid, opened mid-ventrally. **(c)** Distribution of photophores on body of teuthid squid. **(d)** Arm hooks of teuthid squid. **(e)** Sucker rings of teuthid squid. **(f)** *Spirula*; whole and sectioned shells shown at right. **(g)** *Heteroteuthis*, a sepiolid. **(h)** Evolutionary changes in coleoid shells: from extinct belemnites, via the extinct sepioid *Belosepia*, to modern sepiids (left), and via extinct phragmoteuthids (e.g. *Conoteuthis*) to teuthids (right). (a after Naef 1928; b after Barnes 1980; c after Herring 1977; d, e after Okutani *et al.* 1987; f after Cooke 1895; g after *Fauna of Australia* 1997; h after Morton and Yonge 1964).

Figure 7.11 (opposite)
Cephalopods: **(a)** *Vampyroteuthis* (Vampyromorpha). **(b, c)** Male (left) and female (right) of *Argonauta* (Octopoda).
(d) Deep-sea male octopod, *Bathypolypus*, showing expanded tip of hectocotylised arm.
(e) Transverse row of teeth from the radula of *Octopus*. **(f)** The relation of the radula to the horny upper and lower jaws in the buccal mass of a cephalopod. **(g)** Upper jaw with the sharp, sclerotised 'beak'.
(h) Teuthid digestive system.
(i) Basic features of the nervous system of a sepiid, including giant fibres. **(j)** Octopod eye.
(k) Blood flow from the heart through ctenidia and renal organs. **(l)** Spermatophore of a teuthid (*Loligo*).
(m) Spermatophore of an octopod (*Eledone*). **(n)** Teuthid (*Loligo*) laying egg capsules in a cluster. **(o)** Possible steps in the transition from ciliary water circulation in monoplacophoran ancestors (1) to a chamber pump in nautiloids (2–4), leading to a mantle pump in modern coleoids (5). (a after Pickford 1950; b after Müller 1853; c after Yonge 1960; d after Verrill; e after Nixon 1988; f after Bidder 1966; g after Clarke and Maddock 1988; h after Bidder 1950; i after Hillig 1912; j after Wells 1962; k after Morton and Yonge 1964; l after Pierce 1950; m after Fort 1941; o after House 1988.)

to each ctenidium); vessels to and from the head, viscera and mantle cavity; and finally terminal capillaries (Figs 7.10b, 7.11k). The branchial hearts help to maintain a high blood flow, especially during periods of activity such as hunting or escaping, when musculature demands increased oxygen availability. The presence of the blood pigment haemocyanin greatly facilitates the uptake of oxygen from the respiratory current.

Feeding and digestion

With the exception of certain detritus-feeding octopods and squids from the deep sea, all cephalopods are active carnivores. The prey taken depends on the size, speed and arm dexterity of the hunter. *Nautilus* typically seeks out bottom-dwelling crustaceans (mostly crabs and lobsters), but also commonly feeds on carrion such as dead fish. In the Philippines and the Solomon Islands, fish heads placed in open cages are often used as bait to trap *Nautilus*. In these largely deep-water animals it is probably not surprising that chemosensory capabilities (the paired osphradia, absent in other cephalopods) and the tactile sense (the numerous small arms lacking suckers) are more important than sight.

Crustaceans are also the primary food of cuttlefish and dumpling squids, whereas octopods take both crustaceans and molluscs. In these three groups and the fish-hunting teuthid squids, sight is usually more important than chemosensory or tactile senses for locating and capturing prey, at least within the photic zone. The pair of elongate 'tentacles' of cuttlefish, dumpling squid and teuthid squid are at first extended at high speed to clasp the prey, then retracted, bringing the prey towards the arms and the mouth, to be torn by the beak and then ingested by the radula (Fig. 7.11f, g). Suckers and chitinous sucker hooks form the basis of this clasping ability (Fig. 7.10d, e). Shallow-water octopods generally cover crustaceans with the arms and web, then immobilise them with a neurotoxin that is secreted from the posterior pair of salivary glands, administered with the aid of the beak. Octopods lack hooks or sucker rings. Molluscs may be pulled apart using the arms (bivalve prey) or poisoned after the radula has rasped a hole through the shell (bivalve or gastropod prey). Piles of drilled and discarded mollusc shells are often seen around octopus lairs. Blue-ringed octopods (*Hapalochlaena*), which inhabit the eastern and northern coastlines of Australia and several other areas of the Indo-Pacific, are capable of inflicting fatal bites in humans. The colour pattern of numerous blue rings over the body intensifies when the animal is disturbed, providing a warning signal to intruders. The radula is often vestigial in deep-sea cirrate octopods, and it is absent in the ram's horn squid *Spirula spirula*, which feeds on drifting detritus or small nektonic crustaceans.

Well-preserved fossils indicate that extinct cephalopods used beaks and radulae similar to those of modern forms. The cephalopod beak complex is composed of a tanned chitin-protein (with calcareous tips in *Nautilus*) and is controlled by powerful muscles which give the beak its shearing and tearing capabilities (Fig. 7.11f, g).

Having entered the mouth and been shredded by the radula, food is subjected to enzymes (chiefly proteases), firstly from the anterior salivary glands and secondly from the digestive gland emptying into the stomach (Figs 7.9c, 7.10a, 7.11h). In *Nautilus* the digestive gland is a single organ, but in coleoids it is usually divided into a small (pancreatic) and large (hepatic) component. The stomach functions principally to mix food and enzymes, then passes its contents to the large caecum for sorting by cil-

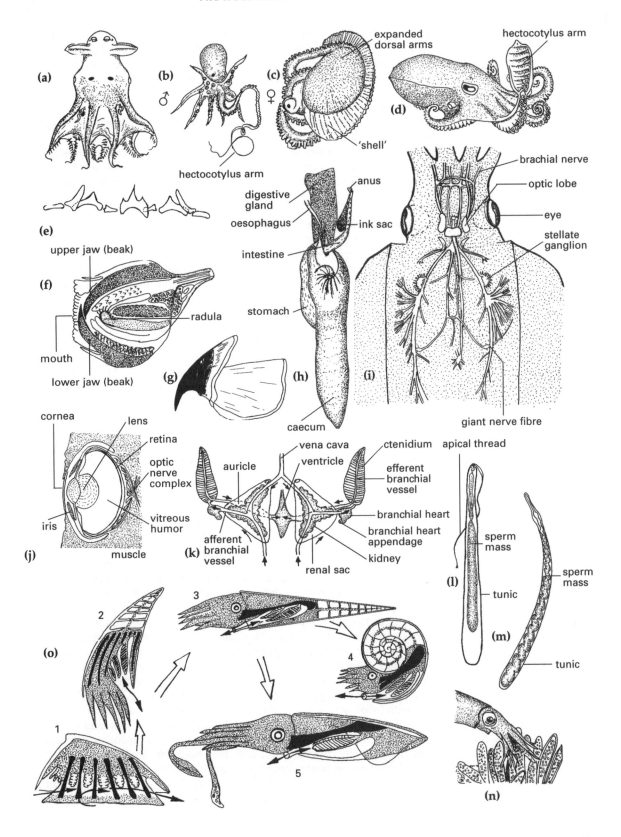

iated leaflets. The subsequent absorption of nutrients occurs either within the hepatic portion of the digestive gland (most cephalopods) or in the caecum itself (teuthid squids). Indigestible remains are discharged via the intestine and anus, which opens into the mantle cavity.

Excretion

The kidneys of all cephalopods are large, saccular structures into which project several out-pouches from the traversing afferent branchial veins. Four kidney are present in *Nautilus*, but only two in coleoids. Wastes are secreted via the vein out-pouches, with additional wastes secreted from peri-cardial glands (in *Nautilus*) or branchial heart glands (in coleoids) (Fig. 7.11k).

Nervous system and sense organs
Ganglia and nerve fibres

Cephalopods represent the pinnacle of neuronal development among invertebrates and have complex, adaptable behaviour patterns. Although *Nautilus* still shows clear evidence of unfused ganglia, in coleoids a definite brain is recognisable, shielded by a cartilaginous casing and made up of a supraoesophageal cerebral ganglia and suboesophageal pedal and brachial ganglia (Fig. 7.11i). Statocysts positioned close to the brain help to maintain orientation in all cephalopods, especially in swimming groups such as *Nautilus*, cuttlefish and teuthid squid and in the vertically 'parachuting' deep-sea octopods. Experimental studies have demonstrated the ability of cephalopods, especially octopods, to learn from experience, but they retain the acquired information or skills for only a limited time.

Cuttlefish and teuthid squid have long, thick nerves containing giant fibres (Fig. 7.11i), whose function is to coordinate a simultaneous contraction of the mantle musculature during jet propulsion, giving a rapid escape reaction. Such giant fibres have been extensively used in the experimental investigation of nerve cell chemistry and axonal transmission.

Eyes, chromatophores and photophores

Cephalopod eyes are complex structures, ranging from the pinhole open eyes of *Nautilus* (Fig. 7.9e) to the more complex eyes of coleoids, with an iris, lens, retina and often a cornea (Fig. 7.11j). The visual capability of such eyes varies from simple detection of light in *Nautilus* and some deep sea octopods to the ability to discriminate between shapes and patterns in shallow water coleoids. Opinions differ as to whether a defined image is formed. What appears certain is that pattern and colour are detected.

Perhaps the most impressive ability of coleoids is the almost instantaneous translation of visual input into camouflage and communicative colour patterns involving chromatophores. Chromatophores within the mantle epithelium are used for camouflage against predators and for mating and defence. Through nerve-directed stimulation from the brain, the pigment-containing chromatophore cells can enlarge (making the skin darker) or diminish (making skin paler), through the contraction or relaxation of associated myofibrils. A combination of differently coloured chromatophores and silvery, light-reflecting iridocytes in the skin result in colour displays that are often very complex. During mating, hormones are believed to play a significant role in maintaining warning or attracting colour patterns. The use of chromatophores is essentially limited to those coleoids which spend a considerable proportion of their time in the photic zone. The remarkable speed at which colour change can be effected is unmatched in the animal kingdom.

Photophores or 'light organs' in cephalopods are subcutaneous tissue patches which in darkness are capable of emitting light (bioluminescence), sometimes in precisely timed pulses or rhythms. These structures are chiefly associated with oceanic cephalopods (teuthid squid and octopods), especially those which can live at great depths. Their main function is probably signalling to other individuals, but other functions such as prey attraction are also likely. Light, whether generated through the action of luminescent bacteria living subcutaneously or produced by cells within the cephalopod's skin, involves the activity of the enzyme luciferase. In cephalopods, photophores occur in various places, depending on the species (arms, tentacle tips, around the mouth or most commonly dorso-laterally and around the eyes) and are usually arranged according to a species-specific pattern (Fig. 7.10c).

Reproduction and development

In view of their advanced motor-neural development, it is not surprising that cephalopods often exhibit complex courtship displays and copulatory activity. Throughout the class the sexes are separate, the gonad is single, egg development is direct, and males package sperm in slender, usually highly structured spermatophores. Males and females can often be distinguished by external, secondary sexual characters: notably the presence in males of a structure for transferring spermatophores during copulation. This is either a modified arm (the hectocotylus) or a penis (Fig. 7.11b, d).

Spermatozoa

Spermatozoa vary markedly in their structure within the Cephalopoda, reflecting in part the taxonomic affinities of each group and also the environment of fertilisation. The classic ectaquasperm, associated with dispersive release of eggs and sperm into sea water, does not occur in this group. The sperm of most decabrach cephalopods have a conical acrosome, curved nucleus, mitochondrial 'spur' and a flagellum. In octopods, the acrosome is corkscrew-shaped and the nucleus and flagellar regions are elongate. Like most other molluscs, cephalopods do not exhibit sperm dimorphism.

Spermatophores

Spermatophores are produced in the seminal vesicle and stored in a pouch, the Needham's sac, within the mantle cavity. Typically, cephalopod spermatophores are composed of an outer investment enclosing (posteriorly) an elongate sperm mass and (anteriorly) a cement body, ejaculatory body and apical cap (Fig. 7.11l, m). The removal of the cap during spermatophore transfer at copulation results in eversion of the ejaculatory body and cement body (which anchors the spermatophore) and the sperm. Spermatophore size ranges from about 5–10 mm to over a metre in the giant squid *Architeuthis*. Like the sperm, the spermatophores differ in structure between the various members of the class. In *Nautilus*, Sepioida, Sepiolida, Teuthida, Spirulida and the Vampyromorpha, the spermatophores are rod-shaped and slightly wider posteriorly than anteriorly, with a long thread issuing from the apical cap (Fig. 7.11l); in incirrate octopods (e.g. *Octopus*) the spermatophores are very elongate and the sperm are often organised into a coiled rope (Fig. 7.11m); in cirrate octopods (e.g. *Opisthoteuthis*) the spermatophores are little more than small, structureless sacs containing sperm.

Figure 7.12 (opposite)
Bivalvia: **(a)** Forces closing and opening the shell. **(b)** Interior of left valve of dimyarian bivalve, with two adductor scars. **(c)** Transverse section through the edge of a living bivalve, showing the relationship of the shell to the mantle tissues. **(d)** Internal anatomy of a heterodont. **(e)** Direction of water flow in a heterodont. **(f–h)** Transverse sections through the mantle cavity of: (f) a protobranchiate, (g) a lamellibranchiate, and (h) a septibranchiate. **(i)** Detail of protobranchiate ctenidium. **(j)** Detail of filibranch–lamellibranch ctenidium. **(k)** Detail of tissue connection between lamellae in a eulamellibranch ctenidium. **(l)** Internal anatomy of a monomyarian bivalve (*Pecten* or scallop), with single adductor muscle. **(m)** Basic organisation of nervous system in a bivalve. (a, b, d, e–g after Barnes 1980; c after Kennedy et al. 1969; i–k after Cox 1969; l after Pierce 1950; m after Buchsbaum 1948.)

Courtship and copulation

In most cases, copulation in cephalopods is preceded by some form of courtship. Especially in cuttlefish and squid, males use chromatophore patterning to attract females and also, having found a partner, to ward off rival males. The extraordinary colour changes which occur during the courtship of many cephalopods are only possible because of their advanced sensory capabilities. In deep-sea squid and octopods, the positioning and flashing sequence of photophores on the arms probably fulfil the same display function during courtship as the chromatophores of shallow-water species.

During copulation, the male grasps the female in order to transfer spermatophores. The transfer is achieved using the hectocotylus (in sepioids, sepiolids, most teuthoids, incirrate octopods such as *Octopus*, Fig. 7. 11b, d), or a penis (e.g. some teuthoids, *Vampyroteuthis* and cirrate octopods), or in *Nautilus*, the spadix, a structure consisting of four arms collectively acting as the intromittent organ. Females may be grasped dorsally (as in some cirrate octopods), posteriorly (as in some incirrate octopods), or commonly head-on (in sepioids and teuthoids). In the last case the encounter may leave deep gashes in the skin, resulting from the firm grip of the rasp-like sucker hooks from the arms of the paired animals. Mass spawnings are typical of oceanic squids and several films have vividly captured the frenzied activity and eventual adult mortality associated with such spectacular events. Pronounced sexual dimorphism occurs in pelagic octopods, especially in the Argonautidae or paper nautiluses. Female argonauts are many times the size of the male. They construct a thin, highly sculptured egg case in the shape of a nautiloid or ammonoid shell, using their flattened dorsal pair of arms (Fig. 7.11c). The long, whip-shaped hectocotylus of the tiny male (Fig. 7.11b) eventually becomes detached to enter the mantle cavity of the female. Occasionally hundreds of the 10–20 cm diameter 'shells' (sometimes with eggs still inside) may be washed up on southern Australian beaches, especially along the Victorian and Tasmanian coasts.

Fertilisation

Fertilisation in cephalopods may occur outside the mantle cavity (ectaquatic fertilisation; e.g. sepioids, some teuthoids), within the mantle cavity (ent-aquatic fertilisation; e.g. some teuthoids, many octopods) or within the oviduct or ovary (internal fertilisation; e.g. the octopod *Eledone*). In the first two cases, eggs emerging from the oviduct are encapsulated by the oviducal gland, then often covered by a jelly-like substance by the nidamental gland (Fig. 7.11n).

Development and brooding

In all cephalopods the eggs are yolky and development is direct, without any veliger stage. The degree of parental care varies widely, ranging from no care at all in *Nautilus*, most sepioids, teuthoids and many deep sea octopods, through to constant guarding and aeration of the deposited egg mass by the female in shallow water octopods. The young hatch as fully developed juveniles.

Class Bivalvia

Basic bivalve features

As the class name indicates, bivalve molluscs have a shell with two distinct valves (Fig. 7.12). These are composed of a complex fabric of layered calcium

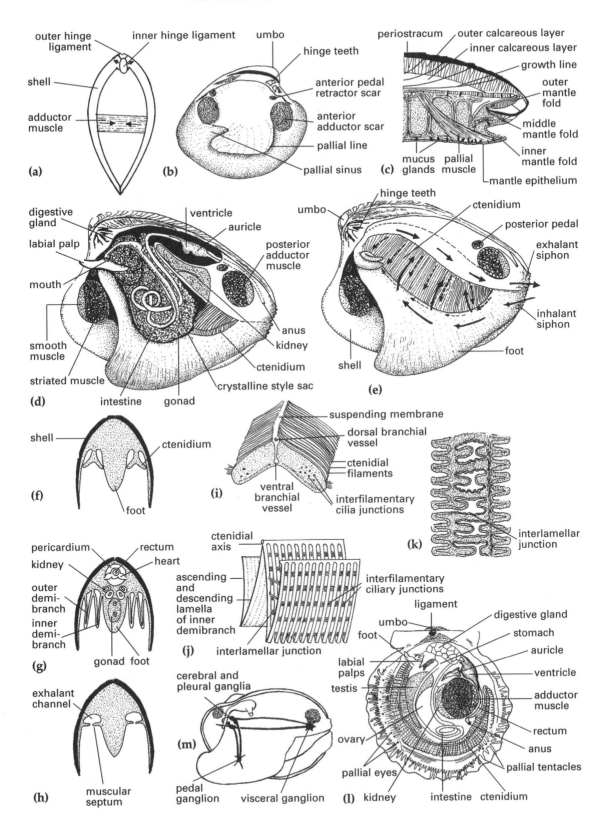

carbonate and shell protein and are articulated dorsally by the shell hinge complex, which typically contains cardinal and lateral teeth and a conchiolin ligament. The body is bilaterally symmetrical and strongly compressed laterally, with a vestigial head lacking cephalic tentacles, eyes and radula; the paired mantle folds are fused to varying degrees ventrally and developed posteriorly as inhalant and exhalant siphons (Fig. 7.13a). The foot is well developed in burrowing bivalves (Fig. 7.13d, m, o) but reduced, vestigial or absent in forms cemented or fixed by a byssus to the substratum. Ctenidia (Fig. 7.12e) are well developed and are used in respiration and often food collection. Bivalves are dioecious or hermaphroditic, with a single gonad situated near the digestive gland. Fertilisation is wholly external or takes place within the mantle cavity. The larvae are planktonic and planktotrophic, and pass through trochophore and veliger stages before settlement.

Bivalve diversity

The Bivalvia have proved to be the most taxonomically challenging of all molluscan classes. This is reflected in the numerous classifications that have been proposed over the last 150 years, each one stressing different discriminators for higher levels (hinge teeth, stomach morphology, ctenidial substructure, shape of muscle scars etc.). No one scheme has proven acceptable to both palaeontologists and neontologists, with the result that the most recent classifications retain a mixture of names from various systems. Six subclasses are usually recognised: the Palaeotaxodonta; Cryptodonta; Pteriomorphia (marine mussels, pearl and rock oysters and allies, scallops and spiny oysters and allies); Paleoheterodonta (marine trigonioids and freshwater mussels); Heterodonta (venus and surf clams and their allies, true cockles and giant clams, tellinoideans, mud-burrowing myoids and wood-eating shipworms) and Anomalodesmata (thracioideans, clavagelloidean 'watering pots', and deep-sea septibranchs). Examples of all of these are shown in Fig. 7.13. The scheme by Boss (1982) may be consulted for detailed diagnoses of all taxa to the family level.

The greatest diversity occurs in the subclass Pteriomorphia (epibenthic, usually attached forms) and in the sub-class Heterodonta (shallow to deep burrowing, mostly unattached forms).

Subclass Palaeotaxodonta
This subclass is believed to contain the most primitive bivalves; it includes the shallow to deep-sea Nuculoida and the earliest recorded fossil bivalves, the lower Cambrian genera *Pojetaia* and *Fordilla*. Several nuculoids are found in Australian waters, some species being very common. The shells are typically triangular, glossy externally and pearly internally, with many small teeth along the hinge line (taxodont dentition) (Fig. 7.13a, b).

Subclass Cryptodonta
This class, containing the 'date clams' or Solemyoida, shares a similar ctenidial structure with the Palaeotaxodonta, but nuculoids and solemyoids differ in many features such as the hinge teeth (vestigial in solemyoids), gut (reduced or absent in solemyoids) and shell form. The Australian fauna contains only a few species of *Solemya*, both showing characteristic elongate shells with a periostracum extending considerably beyond the shell margins (Fig. 7.13c).

Subclass Pteriomorphia

Among the Pteriomorphia, families such as the hammer-oysters (Malleidae) and pen shells (Pinnidae) have an almost infaunal existence, using byssal threads to attach themselves to buried stones or shells, while others such as the file shells (Limidae, Fig. 7.13j) and many scallops (Pectinidae, Fig. 7.12l) exhibit a crude form of jet propelled swimming, using adductor muscles to flap the valves. The spiny oysters (Spondylidae, Fig. 7.13h), although related to the scallops, are usually cemented and possess many projecting spines on each valve. Marine mussels (Mytiloida, Fig. 7.13f) may be epifaunal and are anchored by a byssus or even burrow into coral rock. Ark shells and their allies (Arcoida, Fig. 7.13d, e) are frequently byssally attached and are abundant from mudflats to depths of over 200 metres. They are characterised by radially ribbed shells and often a hairy periostracum. Various pteriomorphian families form the basis of substantial fisheries, including pearls and pearl shell (Pteriidae, northern Australia, Fig. 7.13g), scallops (Pectinidae, southern Australia, Fig. 7.12l) and oysters (Ostreidae, mostly southern and eastern Australia, Fig. 7.13i).

Subclass Paleoheterodonta

The world's only surviving species of the marine paleoheterodont order Trigonioida are restricted to shallow waters around the Australian coastline. The five or six living species of *Neotrigonia* exhibit a pearly shell interior and strongly grooved (schizodont) teeth (Fig. 7.13k). They provide a unique window on a group of bivalves which dominated shallow water in late Palaeozoic and Mesozoic communities. Trigonioids gradually became supplanted by more adaptable heterodont groups such as the true cockles (Cardiidae) and venus shells (Veneridae). The other order of paleoheterodonts, the Unionoida or freshwater mussels (Fig. 7.13l), occurs world-wide and is particularly well represented in Australian rivers, streams and standing bodies of water by the Gondwanan family Hyriidae. Unionoids are often long-lived, some species reaching ages of more than 80 years and taking up to 20 years to reach sexual maturity. These bivalves are particularly sensitive to pollution, and populations are used as indicators of freshwater environmental health.

Subclass Heterodonta

The Heterodonta are the most abundant and widespread subclass of the Bivalvia. They extend back to the Palaeozoic, with groups such as the still extant Lucinoidea. Heterodonts underwent a significant late Mesozoic radiation into a large range of sediments and habitats to become the dominant infaunal bivalve subclass. Some heterodonts, such as the true cockles and giant clams (Tridacnoidea, Fig. 7.13p), represent independent adaptations to an epibenthic life style. Heterodonts have exploited a wide variety of habitats, including mangroves (trough clams or Mactroida), submerged wood (shipworms or Teredinidae, Fig. 7.13s), the high-energy and constantly moving sands of ocean beaches (pipis such as the Donacidae (Fig. 7.13o) and freshwater streams and ponds (the ubiquitous Corbiculoida). The venus shells (Veneroida, Fig. 7.13q), myid and basket clams (Myoida, Fig. 7.13m) and tellins (Tellinoida, Fig. 7.13n) are abundant inhabitants of sand and sand–mud bottoms. A number of heterodont bivalves have become commensal associates of tube-dwelling cnidarians and burrowing crustaceans.

Figure 7.13 (opposite)
Bivalvia: **(a)** Palaeotaxodonta: *Yoldia* (Nuculanidae) showing feeding proboscides.
(b) Taxodont dentition of *Ledella* (Nuculanidae). **(c)** Cryptodonta: shell of *Solemya*.
(d–j) Pteriomorphia: (d) *Arca* (Arcidae). (e) *Glycymeris* (Glycymerididae). (f) *Mytilus* (Mytilidae). (g) pearl oyster, *Pteria* (Pteriidae). (h) *Spondylus*, spiny oyster, cemented to the substratum. (i) Shell of *Crassostrea* (Ostreidae), a rock oyster. (j) Shell of *Lima* (Limidae), a swimming file shell. **(k, l)** Paleoheterodonta: (k) Shell and schizodont teeth of *Neotrigonia* (Trigonioida). (l) Shell of *Cucumerio* (Unionoida, a freshwater mussel. **(m–s)** Heterodonta: (m) The clam *Mya* (Myoida). (n, o) Tellinoids: (n) Tellinidae (deep burrowers). (o) Donacidae (surf-zone bivalves). (p) *Tridacna* (giant clams) with exposed mantle tissue patterned by endosymbiotic zooxanthellae. (q) Veneridae (*Pitar*, venus shells). (r) Solenidae (Solen, deep-burrowing razor shells). (s) Teredinidae (*Teredo*, shipworms) in wood, with detail of shell valve. **(t)** Anomalodesmata (Clavagellidae with large secondary shell attached to tiny original valves. (a after Yonge 1939; b after Puri 1969c, d, f, k, m, p after Cox 1969; e, h, k, n, q after Lamprell and Whitehead 1992; g, j, p, r, t (right) after Abbott 1962; o after Barnes 1980; s after Lane 1961; t (left) after Sieverts 1934.)

Subclass Anomalodesmata

This small but probably very old subclass shows remarkable morphological differences between families and superfamilies in the order Thracioida, including the mangrove-burrowing Laternulidae, which have inflated, very thin shells, and the strange 'watering pot shells' or Clavagellidae, which have tiny valves and a long, secondary, shelly tube (Fig. 7.13t). The carnivorous, largely deep-sea order Septibranchia, whose members lack ctenidia, is usually included within the Anomalodesmata, although their affinities are still open to question.

Form and function in bivalves
Shell morphology

Bivalve shells are extremely varied in shape, microstructure, sculpture and colour, reflecting the wide range of habitats exploited by the class (Fig. 7.13). In most groups the two valves are approximately equal in size, but in cemented taxa such as the true oysters (Ostreoidea), spiny oysters (Spondylidae) and jewel box shells (Chamoidea) the attached lower valve is almost always the larger. An organic periostracum composed of conchiolin (a scleroprotein) forms the outermost layer, which is usually thin and sometimes eroded but can be thick in highly acidic environments (mangrove muds, etc.). Underlying the periostracum are the outer and inner shell layers composed mainly of calcium carbonate, usually with varying amounts of conchiolin in the shell fabric (Fig. 7.12c). In several groups of bivalves the inner layer may consist of nacre, a special form of aragonite in which the foliate crystals are interleaved with conchiolin to give a 'mother-of-pearl' lustre (e.g. nuculoids, pterioids or pearl shells, trigonioids, marine mytiloid and freshwater unionoid mussels). Both the periostracum and outer shell layer are secreted by the mantle edge, whereas the inner shell layer is secreted by the mantle surface. Pearls are produced when irritants such as sand grains become trapped between the inner shell layer and the mantle and are encrusted by nacre.

In all species the valves are hinged dorsally by an elastic external ligament composed of conchiolin and usually have some system of interlocking teeth and sockets positioned along the hinge line (Fig. 7.12a, b). The initial post-larval portion of each valve, termed the umbo, normally projects slightly beyond the hingeline. Several patterns of hinge teeth organisation are recognised, the most common being taxodont (several small teeth of similar size) and heterodont (larger cardinal teeth near the umbo, long laterals on either side). Taxodont dentition occurs in the nuculoids and independently in some members of the subclass Pteriomorphia (ark shells and allies or Arcoida; Fig. 7.13e). Heterodont dentition is by far the commonest pattern within the Bivalvia, although the shape of the cardinal and lateral teeth may differ profoundly between taxa.

On the inside of each valve, marks on the shell surface indicate the shape and extent of many soft parts. The anterior and posterior adductor scars mark the position of the transverse muscles associated with the opening and closing of the valves (Figs 7.12b, 7.13k, n, q). When contracted, the adductors stretch the external ligament and compress any internal ligament that may be present (Fig. 7.12a). When the adductors relax, the tension stored in the ligament(s), causes the valves to open. Scars for the pedal protractor and retractor muscles are also often visible, but may merge with the adductor scars.

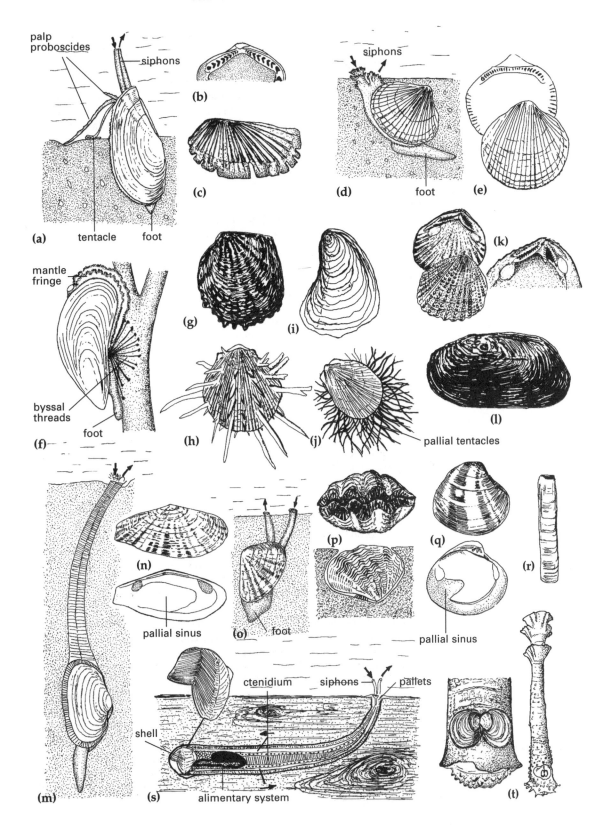

Most unattached bivalves retain anterior and posterior adductor muscles and are therefore said to be dimyarian. The scars may be equal (isomyarian) or of different size (anisomyarian). In cemented forms only the posterior adductor is retained (the monomyarian condition), but it is enlarged and shifted closer to the centre of the valve. Many bivalve groups which produce organic threads (a byssus, Fig. 7.13f) to attach themselves to the substratum (marine mussels, pearl shells, pen shells) are strongly anisomyarian — with a tiny anterior scar still visible and a large, centred posterior scar. The anisomyarian condition clearly indicates how monomyarian bivalves have been derived from dimyarian ancestors.

Locomotion

Bivalves are broadly divisible into burrowers and surface dwellers. In burrowing forms movement usually involves the muscular extension of the foot into the sediment by pedal protractors, followed by the haemocoelic dilation of the tip of the foot for anchorage, and finally the use of pedal retractors to draw the animal deeper into the sediment. Rapid burrowers usually have smooth shells to reduce drag (Fig. 7.13o, r), especially in high-energy habitats such as ocean beaches where constant burrowing may be a necessity. Heavy sculpture such as ribs and lamellae in some species are important in maintaining the position within the sediment. Siphons for inhalant and exhalant water currents are almost always well developed in burrowing bivalves and may be separate (Fig. 7.13d, o), partially fused (Fig. 7.13s) or fused (Fig. 7.13a, m). Siphonal length is regulated by protractor and retractor muscles, but in some sedentary deep burrowers the siphonal length may be constant.

In the majority of surface-dwelling forms (cemented, byssally attached or free) the foot is reduced or absent (Figs 7.12l, 7.13f). In oysters the newly settled post-veliger larva has a well-developed foot, which it uses to crawl along the substratum surface in search of a final place for cementation. This foot is lost during metamorphosis.

Evolutionary history

Bivalves probably arose either from early monoplacophorans or via the extinct Rostroconchia. The earliest recognisable bivalves comprise the Lower Cambrian genera *Pojetaia* from Australia and Greenland and *Fordilla* from North America, Europe and Siberia, both assigned to the Palaeotaxodonta. Most of the diversity observed in the living Bivalvia relates to the exploitation of new habitats made possible by the evolution of the lamellibranch-style ctenidium, and in the deep burrowers by mantle fusion and the development of elongate siphons. During the late Palaeozoic and throughout the Mesozoic era, several surface-dwelling groups appeared, including byssally attached and valve-cemented pteriomorphians and heterodonts. The byssus is thought to have evolved as a means of assisting larval settlement, but subsequently was retained into adulthood as a primary anchoring mechanism. Clustering of certain pteriomorphians such as marine mussels, rock oysters and pearl shells, often on unstable sediments, is possible only because of the retention of the byssus. This crowding facilitates free spawning and rapid larval dispersal, and also improved chances of settlement by offering a firm substratum for the attachment of a new generation. It is therefore not surprising that several 'pest' bivalves are clustering species.

Physiological processes
Respiration and circulation

As in most other molluscs, ciliated ctenidia form the principal organs of respiratory exchange in bivalves. In the carnivorous deep-sea septibranchs the mantle epithelium fulfils this function, the ctenidia being replaced by muscular septa which generate water currents. Respiration and nutrition are intimately linked in most bivalves because the ctenidia have become specialised for filter feeding. There are two main types of ctenidia. In the primitive nuculoids and solemyoids the ctenidia (Fig. 7.12f, i) have a main axis containing afferent and efferent blood vessels and two rows of filaments — the protobranchiate condition. Scattered ciliary tufts provide stabilising cross-links between the filaments, and from the filaments to the mantle surface.

Most protobranchs are deposit feeders. The ctenidia are respiratory and their cilia only generate respiratory currents and remove fouling sediment. In other bivalves the ctenidia have become larger in size and W-shaped in cross-section, due to the elongation and folding of the filaments on both sides of the axis into V-shaped structures called demibranchs (Fig. 7.12 g, j). The folding greatly increases the surface area available for respiration and for filtering particulate material from the inhalant current. This more complex organ is called the lamellibranch ctenidium, of which three levels of development are recognised: the filibranch grade, with filaments connected by tufts of cilia, as in mytilid mussels, trigonioids and scallops (Fig. 7.12j); the pseudolamellibranch grade, with filaments connected by scattered tissue bridges as in ostreid rock oysters; and the eulamellibranch grade, with filaments connected extensively by tissue bridges as in the unionid freshwater mussels and the heterodonts (Fig. 7.12k). Tissue connections between the demibranch and the mantle or visceral mass stabilise and preserve the physical integrity of the entire lamellibranch complex.

The heart is dorsal in most bivalves, with a median ventricle and paired lateral auricles. Usually the ventricle is traversed by the intestine (Fig. 7.12d, e). In marine ark shells such as *Anadara trapezia*, haemoglobin occurs as a respiratory pigment in the blood. Other bivalves appear to lack respiratory pigments. The mantle probably provides an additional surface for respiration in all bivalves, especially in juveniles or small-sized species in which the ratio of surface area to volume is favourable.

Feeding and digestion

The lack of a radula greatly limits the types of food available to bivalves. Nuculoids are deposit feeders, using a pair of tentaculate structures termed proboscides to pass mucus-trapped particles by means of ciliary action to a pair of oral palps for sorting and transfer to the mouth (Fig. 7.13a). With the advent of filter feeding using the lamellibranch ctenidium, bivalves were free to exploit a range of substrata as infaunal and epifaunal animals. Some deep burrowing groups such as the tellins have returned to deposit feeding, utilising a greatly elongated inhalant siphon.

Filter feeding in lamellibranchs involves trapping particles (sediment, plankton, organic debris) on the ctenidial filaments and transferring the particles to food grooves which occupy the bend in each demibranch. The labial palps sort particles which are too large for ingestion, for immediate removal as pseudofaeces by periodic contractions of the mantle cavity. Mucus-entrapped particles pass to the mouth for ingestion and are wound into the stomach on a mucus string by the rotary action of a crystalline style.

Not all bivalves are deposit or filter feeders. Deep sea septibranchs use a muscular septum within the mantle cavity, together with the inhalant siphon, to suck in small animals on or immediately above the substratum surface. These and certain species of the scallop family Propeamussiidae constitute the only carnivorous Bivalvia. Equally unique is the life style of the Indo–Pacific giant clams (Tridacnidae) which lie ventral side upwards, lodged among living coral (Fig. 7.13p). Nutrition is derived from filter feeding and also from phagocytosis of symbiotic zooxanthellae which live in the sunlight-exposed mantle and siphonal tissues. The zooxanthellae give these tissues their bright and often varied colouration. Date clams (Solemyoida) and heterodonts of the family Lucinidae obtain their nutrition through the activity of endosymbiotic bacteria on the ctenidia, which live on sulfides from the sediment. Shipworm bivalves (Teredinidae, Fig. 7.13s) have for many centuries damaged or destroyed boat hulls and wharf pylons. Wood pulp generated by the chisel-like action of the valves is converted into sugars with the aid of bacterial cellulases plus cellulases produced by the digestive gland of the shipworm. Ctenidial filter feeding is still necessary for obtaining protein and may in fact be the only source of nutrition in some teredinids.

Excretion and ionic regulation
Bivalves possess a pair of kidneys postero-dorsally between the gonad and the heart (Fig. 7.12d, l). Each kidney is folded into an upper and lower portion, which connect, respectively, with the pericardial sac (where fluids are received) and the mantle cavity (where wastes are released via an excretory pore). In most bivalves the main function of the kidneys is to remove nitrogenous waste, but in freshwater and estuarine species they also maintain the solute content of the body fluids in the face of severe osmotic stress.

Nervous system and sense organs
Even in the absence of a discernible head, bivalves usually retain a pair of cerebropleural ganglia, with connections leading to the visceral ganglia dorsally and to the pedal ganglia ventrally (Fig. 7.12m). Most of the sensory capabilities of bivalves relate to predator avoidance. Pressure or vibration-sensitive papillae or short tentacles are often associated with exposed ends of the siphons, the mantle edge, or both. The swimming file shells (Limidae) have particularly long and colourful mantle-edge tentacles (Fig. 7.13j). Several groups of epibenthic bivalves have numerous light-sensitive ocelli within the mantle edge. In certain groups such as the scallops (Pectinidae, Fig. 7.12l) and the related spiny oysters (Spondylidae) the ocelli may be developed as true eyes with a lens, cornea and retina. Some bivalves possess a chemosensory epithelium within the mantle cavity, probably analogous to the osphradium of gastropods and *Nautilus*. Exudates from starfish have been experimentally shown to induce an escape response in surface-dwelling cardiid cockles and scallops.

Reproduction and development
In all bivalves, the sperm at some stage come into contact with the surrounding water; internal fertilisation and the development of complex reproductive tracts are unknown in the class. For this reason, it is not surprising that sperm morphology in the Bivalvia is less diverse than in the Gastropoda or the Cephalopoda. However, even within the confines of the aquasperm type, bivalves exhibit much variation in the length and sub-

structure of the acrosomal complex, the nuclear length, mitochondrial number, and sometimes the positioning of the centrioles. Some subclasses can be differentiated by acrosomal features, such as the multiple acrosomal vesicles of the Paleoheterodonta or the so-called 'temporary acrosome' of the Anomalodesmata. Similarly, features that are diagnostic of orders, superfamilies, families and genera can often be discerned, a fact which offers useful insights into the difficult world of bivalve phylogeny and taxonomy.

Fertilisation in bivalves occurs either in the surrounding sea water or within the mantle cavity. Free spawning occurs widely among the Bivalvia, including some of the most successful groups (for example rock oysters, pearl oysters, marine mussels, scallops, giant clams and many venerid clams). Ent-aquatic fertilisation is usually associated with brooding, which is relatively widespread within the Bivalvia: for example, in freshwater groups (freshwater unionoid mussels, sphaeriid pea clams), in small species and in several deep-sea forms (e.g. septibranchs). Brooding also occurs in a number of rock oysters (Ostreoidea).

Development in most marine bivalves proceeds via planktonic trochophore and veliger stages. In brooding species the trochophore and often the veliger stage are passed within the egg. Freshwater mussels exhibit a semiparasitic larval stage (glochidium) which attaches to fish gills for 3–6 weeks then detaches to begin benthic life.

Class Scaphopoda

Basic scaphopod features
Scaphopods are infaunal, bilaterally symmetrical molluscs, easily recognised by the presence of a curved tubular shell which is open at both ends (Fig. 7.14b). The mantle is also tubular, and ctenidia are absent. The head is reduced and lacks eyes, but contains a well-developed buccal mass featuring a strong radula (five teeth per transverse row, rarely seven, Fig. 7.14h), an odontophore and a subradular organ. Numerous (30–300) tentacle-like structures called captacula surround the base of the head and are involved in the collection of food, which includes a wide range of interstitial fauna (Fig. 7.14a, b). The prominent foot is used for burrowing, and for expelling water during respiration and gametes during spawning. Scoop-like epipodial lobes (order Dentaliida) or a broad pedal disc (order Gadilida) also assist in burrowing (Fig. 7.14c–e). With rare exceptions, scaphopods have separate sexes. The gonad is single and dorsal and fertilisation occurs externally or (more uncommonly) within the mantle cavity. Larvae are planktonic and lecithotrophic (Fig. 7.14f,g).

Scaphopod diversity
Despite often being considered a minor class within the Mollusca, the Scaphopoda comprise almost as many species as the Cephalopoda or the Polyplacophora. They often make up a significant component of benthic communities from the littoral and the continental shelf down to abyssal depths over 6000 metres, suggesting that their ecological importance is generally underestimated. Australian waters support a diverse scaphopod fauna (107 species), largely because the continent borders three oceanic systems (the Indian, Pacific and Southern Oceans).

INVERTEBRATE ZOOLOGY

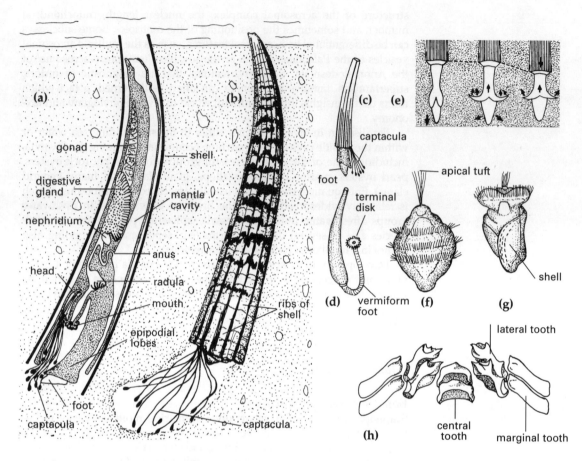

Figure 7.14
Scaphopoda. (a) Dentaliid in shell. (b) Animal with extended captacula. (c) External features of order Dentaliida. (d) External features of Order Gadilida. (e) Shape changes and directions of movement of foot during burrowing in a dentaliid. (f, g) Trochophore and pre-settlement larva of a dentaliid. (h) Tooth row of radula of a dentaliid. (a after Lacaze-Duthiers 1856 and several other sources; c, d after Palmer 1974; e after Trueman 1968; f–h after Lacaze-Duthiers 1856.

Form and function
Shell morphology

Scaphopod shells (see Fig. 7.14b–d) consist of an outer chitinous layer (the periostracum), and three shelly (aragonitic) layers, including an inner layer with a concentric substructure directly sheathing the mantle. As the animal grows, new shell is added at the anterior aperture, while posteriorly the apex is enlarged by shell resorption or natural wear. This permits a greater flow of water for respiration, the removal of waste products and the release of the gametes. The posterior mantle may also secrete a shell pipe (sometimes termed the secondary shell).

Locomotion

Scaphopods inhabit a wide variety of sediments, ranging from coarse sand or coral rubble to clayey mud, with the majority of species inhabiting fine sand or sandy mud. Most Dentaliida inhabit only the upper few centimetres of the substratum. Members of the Gadilida, however, have been observed in captivity to burrow more than 300 mm below the substratum surface. Scaphopods burrow by means of the foot. In the Dentaliida the foot is conical and associated with a pair of scoop-shaped epipodial lobes. Burrowing occurs through muscular extension of the foot and epipodial lobes, followed by expansion of the epipodial lobes to grip the substratum, then muscular contraction which draws the animal further into the substratum (Fig. 7.14e). The burrowing process in Gadilida is less well known, although an expanded, terminal disk on the vermiform foot clearly performs a similar anchoring function to the epipodial lobes of the Dentaliida (Fig. 7.14d). Unlike the Dentaliida, the Gadilida have a large pedal haemocoel, weakly developed pedal longitudinal muscles, and three pairs of pedal retractor muscles — an arrangement which highlights the role of hydraulic forces in these animals.

Evolutionary history

The earliest known scaphopod fossil is *Rhytiodentalium kentuckyensis*, a smooth-shelled dentaliid from the Middle Ordovician of the USA. Current opinion favours either a common origin for scaphopods and the extinct Palaeozoic rostroconchs or derivation of scaphopods directly from rostroconchs. The order Gadilida arose in the Mesozoic. Of the extant classes, the Scaphopoda appear to be closest to the Bivalvia.

Physiological processes
Respiration and circulation

Lacking ctenidia or secondary gills, Scaphopoda use the mantle surface for respiratory exchange (Fig. 7.14a). The circulatory system has no heart. Blood sinuses are associated with the foot, pallial cavity, viscera and anus. Ciliary tracts on the mantle epithelium draw in water via the posterior (narrow) shell aperture, which lies close to the substratum. The contraction of the foot expels the water via the same shell aperture. Little is known of the physiology of scaphopod respiration: it is presumably aerobic in most species, but the habit in some gadilids of burrowing well below the substratum surface suggests a capacity for anaerobic respiration.

Feeding and digestion

During feeding, a cavity in the sediment is first created by the foot in the vicinity of the anterior shell aperture. Food particles are collected from the sediment by the retractile captacula and transferred to the proboscis lips and mouth either by ciliary movement along the captacula or, more rarely, directly by the captacular tips. Detritus may also be passed from the foot groove to the mouth. Movements of the foot play a significant role in maintaining the flow of fresh detrital material from the substratum. The diet typically consists of foraminiferans and/or detritus, but bivalve spat, ostracods, invertebrate eggs and diatoms are also sometimes ingested. Food is stored in the buccal pouch and then partially crushed by the jaw and radular apparatus, before being passed to the stomach and ultimately the digestive gland. Faecal and nitrogenous wastes are expelled from the mantle cavity via the posterior shell aperture.

The scaphopod radula usually has five teeth per transverse row. Some degree of mineral coating of the radular teeth, usually iron based, appears to be typical of the Scaphopoda, as it is in the Polyplacophora and in certain patellid gastropods. Presumably this enhances the strength and durability of the teeth.

Excretion

The paired kidneys connect via separate pores to the mantle cavity. Presumably in deep-burrowing gadilids the animals return periodically to the substratum surface to release wastes. Pumping of nitrogenous products directly into the surrounding sediment is also possible.

Nervous system and sense organs
The configuration of the scaphopod nervous system is generally similar to that of bivalves, with paired ganglia associated with the head, foot and visceral mass. In the absence of eyes, scaphopods largely rely on tactile and chemosensory input. Statocysts lie close to the pedal ganglia and may help to maintain correct orientation of the animal within the substratum. The tips of the captacula may have some chemoreceptive function in the selection of food items.

Reproduction and development
Sexes are usually separate in the Scaphopoda. In both sexes the gonad is a single elongate organ composed of many interconnected acini lying in contact with the digestive gland.

Scaphopod sperm are of the uniflagellate aquasperm type and are similar to those of many prosobranch gastropods, bivalves and monoplacophorans. Scaphopod oocytes are small and numerous, but are large and yolky in some gadilid species.

Fertilisation usually occurs in the sea water. Eggs and sperm are released into the mantle cavity via a temporary connection with the right kidney, then expelled either through the narrow posterior shell aperture (in Dentaliida) or via the anterior shell aperture (in Gadilida). A pumping action generated by repeatedly extending and withdrawing the foot into the shell helps expel gametes.

Following spiral cleavage, development proceeds via a short-lived (3–5 days) lecithotrophic trochophore larva (Fig. 7.14f) in which the main locomotory cilia are arranged primarily as equatorial bands. These bands later shift anteriorly as the mantle edge and curved shell surface fuse ventrally to form a tube (Fig. 7.14g). After settlement the ciliary bands and apical tuft are shed. The larval shell is usually swollen and is only rarely observed attached to the juvenile shell.

Class Monoplacophora

Basic monoplacophoran features
Living members of this largely extinct class (Fig. 7.15) are bilaterally symmetrical, with a broad, calcareous shell covered in a fine periostracum, and also with 5–8 pairs of pedal retractor muscle scars. The head is of moderate to large size, but it is indistinct from the rest of the body. It features a prominent mouth associated with an anterio-lateral velar lobe and usually two clumps of short post-oral tentacles. There is a well-developed buccal mass containing a radula and odontophore. Eyes and cephalic tentacles are absent. The broad foot is surrounded by a spacious, almost circular mantle cavity which contains several pairs of monopectinate ctenidia but lacks osphradia. The terminal anus and lateral excretory pores all empty into the mantle cavity (Fig. 7.15b, c). The sexes are separate except in the tiny (<1 mm) species *Micropilina arntzi*, which is a hermaphrodite brooder. One or two pairs of gonads are present. Fertilisation probably occurs in the mantle cavity, but may be external in some species.

A little over 40 years ago monoplacophorans were known only from Cambrian to Devonian fossils characterised by a cap or horn-shaped shell with 5–8 pairs of muscle scars. In 1957 Henning Lemche announced in the journal *Nature* the discovery of a living monoplacophoran collected from deep water off the Costa Rican coast. This species, *Neopilina galatheae*, sub-

THE MOLLUSCA

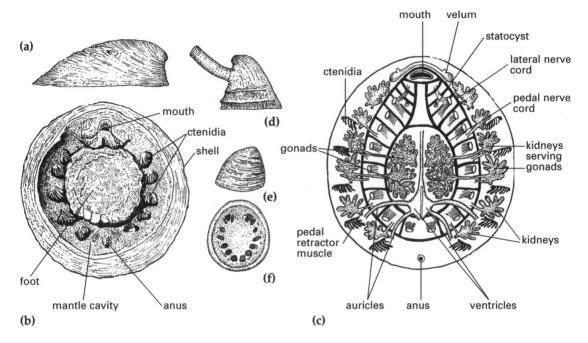

Figure 7.15
(a) Lateral view of the shell of *Neopilina*. (b) Ventral view of *Neopilina*. (c) External and internal features of *Neopilina*. (d) Shell of a Middle Cambrian monoplacophoran, *Yochelcionella daleki*. (e, f) External and internal views of a Cambrian monoplacophoran, showing multiple paired muscle scars. (a, b, c after Lemche and Wingstrand 1959; d after Runnegar and Jell 1976; e, f after Knight and Yochelson 1960.)

sequently became the focal point for discussions of monoplacophoran relationships to other molluscs. Lemche claimed that his new species showed clear indications of metamerism in the repetition of ctenidia, nephridia, muscle scars and nerve connectives. At first the metameric organisation of neopilinids was widely accepted and used as evidence of molluscan connections with annelids, but in recent reevaluations of the anatomy of extant monoplacophorans this view has been either challenged or rejected.

Monoplacophoran diversity

Monoplacophorans are distributed widely through the world's ocean systems. They have not been recorded from Australian waters, but species have been taken north of New Zealand (*Micropilina tangaroa*, at depths of over 1200 m) and most recently off Antarctica (*Micropilina arntzi* and *Laevipilina antarctica*). Twenty living species from seven genera are now known, all belonging to the single family Neopilinidae. They are mostly small (less than 10 mm shell length) abyssal animals. In contrast, the Palaeozoic forms are often of moderate size (30 mm up), have thicker shells, and are associated with shallow water communities.

Form and function

Throughout the Mollusca the presence of a cap-like shell is frequently correlated with life on a hard substratum, often in a high-energy environment. The first living monoplacophorans to be dredged, however, appeared to be associated with a muddy clay bottom. Lemche suggested that *Neopilina galatheae* lives upside down in the sediment (foot upwards) and subsists through filter feeding. But recent work has shown that stones and shells often form part of the substratum where monoplacophorans are found, providing a suitable surface for a crawling mollusc. Gut contents indicate that monoplacophorans are detritivores, most likely ingesting sediment and organic particles that settle on the bottom. *Laevipilina antarctica* has large numbers of bacteria between the microvilli of some epidermal cells and inside others (bacteriocytes). Such bacteria are thought to exist symbiotically and assist in the nutrition of the monoplacophoran.

Evolutionary history

All available fossil evidence indicates the importance of the Monoplacophora in the Palaeozoic. The class, which arose some time in the early Cambrian, is the most likely source for the Gastropoda, Cephalopoda and possibly (via the extinct Rostroconchia) the Scaphopoda and Bivalvia. The close correspondence of shell and muscle scar features in fossil and living forms, and the distinctive anatomical traits of species studied histologically, allows the class to be defined unambiguously and distinguished from other molluscs (see Fig. 7.15e, f). Recent reassessments of anatomy in the family Neopilinidae have shown that the numbers of ctenidia, nephridia, muscle scars and nerve connectives do not closely correlate with each other, thereby providing strong evidence against metameric organisation. The rapid decline of the Monoplacophora after the Devonian is possibly associated with the expansion of the Gastropoda during the later Palaeozoic.

Physiological processes

Respiration and circulation

Although it could be assumed that gaseous exchange occurs via the paired ctenidia, it has been suggested that the blood-filled kidneys, which also border the mantle cavity, are the primary respiratory surface in monoplacophorans. In all probability the ctenidia, mantle, kidneys and other exposed epithelial surfaces are all involved in gas exchange. The heart usually consists of two parallel, longitudinally aligned ventricles each giving rise to two elongate auricles (Fig. 7.15c). With the exception of venous and arterial vessels associated with the heart, the blood is conducted via haemocoelic sinuses throughout the body. In *Micropilina* the heart is replaced functionally by horizontal muscles.

Feeding and digestion

Monoplacophorans apparently subsist on detritus that has settled in the substratum surface. The unusually large oesophageal pouches have often been mistakenly referred to as coelomic pouches. After particles are ingested via the mouth and radula, they are passed to the stomach where they are subjected to the action of enzymes released from the style. Most digestion appears to be extracellular. Faecal material is voided via the terminal anus.

Excretion

Three to six pairs of well-developed kidneys are present in monoplacophorans, positioned between the ctenidia (Fig. 7.15c). The kidneys are not connected to each other or to the pericardial sac, but open separately into the mantle cavity.

Nervous system and sense organs

The nervous system consists of anteriorly located, paired ganglia (cerebral, pleural and pedal), circum-oral connectives, and two circular nerve cords (one associated with the broad foot, the other with the mantle). Both cords are linked by connectives (Fig. 7.15c). The principal sense organs appear to be the post-oral tentacles, probably chemosensory, and statocysts.

Reproduction and development

Monoplacophorans have one or two pairs of laterally positioned gonads (Fig. 7.15c). Eggs and sperm are usually released via the kidneys into the mantle cavity. Sperm are of the uniflagellate aquasperm type in *Neopilina* and *Laevipilina*, indicating that these gametes at least at some stage come into contact with sea water. Because of the generally small size of monoplacophorans, it seems likely that fertilisation takes place in the mantle cavity rather than in the water column. Unfortunately nothing is known of sperm morphology or sperm transfer in the oviducal brooder *Micropilina arntzi*. Embryos of this species increase in size during brooding, presumably through some mechanism for nutrient transfer from the adult, and eventually emerge as crawling juveniles.

Class Polyplacophora

Basic polyplacophoran features

Polyplacophorans (chitons) form a small but distinct group of bilaterally symmetrical marine molluscs characterised by a dorso-ventrally compressed, mainly calcareous shell composed of eight curved, slightly overlapping plates. The plates are embedded at their margins in an externally bristled, scaly or spiculose region of the mantle termed the girdle (Fig. 7.16a), which holds the entire shell complex together and may even cover it completely. The head is poorly differentiated from the foot and lacks eyes and cephalic tentacles, but possesses a well-developed mouth, oral palps and a buccal mass (Fig. 7.16e). Radular teeth, 17 per transverse row, have their cusps mineralised, reducing damage from scraping the rocky substratum (Fig. 7.16c). Working in conjunction with the girdle and shell, the broad, muscular foot clamps the animal firmly to the rock substratum. The mantle cavity is divided into two channels, one on either side of the foot, with each channel containing a few to many bipectinate ctenidia. The sexes are usually separate. Fertilisation occurs either externally or within the confines of the mantle cavity. Hermaphroditism is found in some members of the order Acanthochitona. Typically, development proceeds via spiral cleavage, a planktonic trochophore larva and direct metamophosis (Fig. 7.16h, i). Brooding within the mantle cavity is common, and viviparity has been recorded in one species.

Polyplacophoran diversity

The class contains three extant orders: the primitive Lepidopleurina, with a few pairs of ctenidia in the posterior region of the mantle cavity and with weak or no valve articulation; and the more advanced Ischnochitonina and Acanthochitonina, both with numerous ctenidia throughout the mantle cavity and usually strong valve articulation. All are represented in the Australian fauna. World-wide there are 500–600 species in the class, some families containing over 100 species each. In Australian waters the class is better represented in the coastal and sublittoral habitats of the southern states such as Victoria and South Australia.

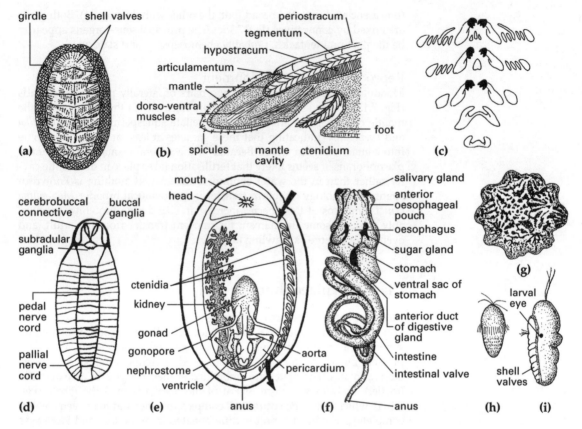

Figure 7.16
(a) *Chiton olivaceus*: dorsal view.
(b) Sectional view of girdle, shell valve and mantle cavity of a chiton. (c) Sequence of changes in a row of radular teeth of a chiton, beginning as a single folded structure and maturing into a complex series of differentiated teeth.
(d) Nervous system of a chiton.
(e) Circulatory and respiratory systems in a chiton. (f) Digestive system of a chiton. (g) Mature oocyte of *Lepidochitona*, showing sculpturing of 'hull'.
(h, i) Trochophore and presettlement larva of a chiton.
(a after Salvini-Plawen 1971; b after Barnes 1980; c after Sirenko and Minichev 1975; e, g after *Fauna of Australia* 1997; f after Fretter 1937; h after Christiansen 1954; i after Grave 1932.)

Form and function

Like limpets, chitons primarily inhabit rocky shores, where microalgal sources are plentiful and dissolved oxygen levels are high. Several species live sublittorally, usually to depths of around 50 m, and species of the primitive family Lepidopleuridae have been recorded from over 7000 m. The basic form of chitons appears to have remained unchanged throughout their fossil record, and apart from differences in the numbers of gills and degree of mantle envelopment of the shell, most living species show little morphological development.

The shell plates are complex structures and are composed of a thin dorsal periostracum, often eroded away or encrusted, an underlying tegmentum composed of conchiolin and calcium carbonate, and a basal hypostracum of aragonitic calcium carbonate (Fig. 7.16 a, b). The hypostracum protrudes anteriorly to form the articulating surface of each valve. Usually a substantial portion of the shell is visible externally, but in several species the girdle portion of the mantle may cover a large proportion of the valves, or in some cases sheath the entire shell (as in *Cryptochiton*). A powerful suction is generated through the interaction of the mantle girdle and foot. As the dorso-ventral muscles of the foot contract, the inner rim of the girdle is lifted up, thereby creating a partial vacuum within the mantle cavity. The ability of chitons to roll up into a ball if dislodged affords protection from predators while the animal 'rights' itself. Once mature, chitons do not move large distances. Many species develop feeding home-ranges similar to those of limpets.

Evolutionary history

Although the earliest undisputed chitons date from the Ordovician, some Late Cambrian fossils (*Matthevia*) have recently been attributed to this class. Anatomical and molecular studies of extant polyplacophorans also indicate an early origin for the group, probably appearing after the Aplacophora but before the Monoplacophora and Gastropoda. The shell similarities of fossil chitons to living species suggest that they inhabited hard substrata and well-oxygenated water. The chance of fossilisation in such environments, particularly of fully articulated specimens, is low, and this is reflected in the rarity of fossil chitons.

At present there is no evidence to indicate that the Polyplacophora gave rise to other major groups of molluscs. The class appears to be most closely related to the Aplacophora, in spite of several significant differences in body organisation. The presence in chitons of a multivalved shell and numerous mantle spicules or bristles invites comparison with the worm-like Lower Cambrian Halkieriidae, with body-wall spicules or setae and a cap-shaped shell at anterior and posterior extremities (see Fig. 7.2c). However, marked dissimilarities in the substructure of spicules and shells between these two groups seems to rule out any relationship between them.

Physiological processes

Respiration and circulation

In all chitons, respiratory exchange occurs through multiple pairs of bipectinate ctenidia in the mantle cavity (Fig. 7.16b, e). Their linear arrangement and angular, downward orientation effectively creates separate inhalant and exhalant channels within the mantle cavity, particularly when the girdle lies in contact with the substratum. Water is drawn anteriorly by the ctenidial cilia, then progresses posterio-dorsally over the ctenidia before exiting as a single exhalant stream at the posterior extremity of the mantle cavity. The two anterior inhalant apertures and the posterior aperture are created by localised raising of the girdle.

The polyplacophoran heart is located posteriorly and consists of two auricles and a ventricle enclosed within an extensive pericardium (Fig. 7.16e). Oxygenated blood flows from the ctenidia to the auricles and is then pumped anteriorly by the ventricle. Experimental evidence suggests that blood circulation throughout the body is very slow. The principal blood pigment is haemocyanin.

Feeding and digestion

Almost all chitons live by scraping algae from rocks with the radula. A few deeper-water species feed on sponges and members of at least one Pacific genus, *Placiphorella*, actively trap small crustaceans and polychaetes using an enlarged frilled head. When food is detected on the substratum using the chemosensory subradular organ, the radula is extended. Particles retrieved on the radular teeth are trapped in mucus within the mouth. As they are passed by ciliary action along the oesophagus they are subjected to amylytic enzymes from the oesophageal gland or 'sugar gland'. The food is then further digested by stomach and digestive gland enzymes before the remains continue along the convoluted intestine (Fig. 7.16f). There is no style. A valve separating the anterior from posterior sections of the intestine helps to maintain adequate periods for nutrient absorption and regulate the flow of waste. Faecal pellets entering the mantle cavity via the anus are flushed out posteriorly by the exhalant respiratory current.

Excretion and ionic regulation
The paired kidneys are long, often highly ramified organs emptying posteriorly within the mantle cavity (Fig. 7.16e). No chiton species has evolved the ability to tolerate estuarine levels of salinity, and hence the group has never invaded freshwater habitats.

Nervous system and sense organs
Chitons possess a simple nervous system which reflects the overall bilateral symmetry of the body and an emphasis on the foot rather than the head. The dominant elements of the system are the longitudinally orientated, paired pedal nerve cords and paired pallio-visceral nerve cords (Fig. 7.16d). These are linked anteriorly by a cephalic nerve and more posteriorly by several smaller transverse nerves, forming a ladder-like network. Ganglia are poorly developed or absent. Chitons are unique in possessing hundreds or thousands of multicellular sensory organs, called aesthetes, within the surface layer of the tegmentum of each valve. Aesthetes are each housed in a curved tube and are connected together by a network of small tubes. These structures appear to have a light-sensing function, probably in phototactic responses. Some aesthetes may be developed as small 'eyes', each with a lens, vitreous layer and primitive 'retina' of photosensitive cells.

Reproduction and development
Sexes are almost always separate in the Polyplacophora, the exceptions being a few species of *Lepidochitona* which are simultaneous hermaphrodites. There are no obvious external characters which identify the sex of the individual, with the occasional exception of colour differences. Although only a single, dorsal gonad is present in most chitons, two closely apposed gonads have been noted in species of *Nuttallochiton* and *Notochiton*, suggesting that the single condition is derived. The gonad empties into the mantle cavity via paired gonoducts opening anteriorly to the excretory pores.

Spermatozoa
Chiton sperm fall into two structural categories: Lepidopleurina type, with an elongate well developed acrosome, short nucleus, posterior mitochondria and flagellum; and Ischnochitonina and Acanthochitonina type, with a minuscule acrosome, tear-drop shaped nucleus (thread-like anteriorly), lateral or posterior mitochondria and flagellum. For many years it was believed that sperm of the second type lacked an acrosome, but recent studies have demonstrated that the tiny acrosome fuses with the surface of the egg, as in other molluscs. Experimental studies have shown that chiton sperm have a positive chemotactic response to mature eggs.

Eggs and fertilisation
The eggs of chitons are enclosed in an inner vitelline layer and an outer layer called the hull. Typically the hull has a complex surface sculpture that is characteristic for each species (Fig. 7.16g), but in species of the primitive genus *Lepidopleura* it is smooth-surfaced. Spines and cupules, formed through the activity of follicular cells in the gonad, have been observed in

several different arrangements, and it has been suggested that these elements, by increasing the surface area, may be important in keeping eggs suspended within the water column before and after fertilisation.

Fertilisation in chitons usually takes place in the sea water after males and females have released their mature gametes via the exhalant respiratory current. A few males usually initiate spawning of surrounding females by releasing sperm, the effects of which can extend for several metres. Spawning is greatly facilitated by the clustering habit of chitons and will normally only occur in calm water. In brooding species sperm are released by the male close to the inhalant current of the female, then drawn into the mantle cavity of the female, where fertilisation occurs. Embryos are retained in the mantle cavity for varying lengths of time depending upon the species. In hermaphroditic species of *Lepidochitona*, self fertilisation occurs within the mantle cavity and the resulting embryos are brooded.

Development

As in other molluscs, fertilised chiton eggs undergo typical spiral cleavage. The trochophore is the only larval stage in the Polyplacophora. After breaking free from the enclosing hull, the trochophore (Fig. 7.16h) swims for a few hours to days. Single-celled larval 'eyes', a ciliated prototroch and an apical tuft are initially present. Before settlement, seven of the eventual eight plates become visible on the future dorsal side of the elongating trochophore (Fig. 7.16i). On settling, the animal undergoes dorso-ventral compression and loses its larval features. In brooding species and the viviparous *Callistochiton viviparus* from Chile, the trochophore stage develops within the hull.

Class Aplacophora

Basic aplacophoran features

The Aplacophora embraces two distinct groups of oblong to vermiform, bilaterally symmetrical marine molluscs, the Neomeniomorpha and Caudofoveata. In these animals there is no shell, but the mantle is studded with numerous calcareous needle or scale-like spicules (Fig. 7.17). In addition to the unusual body shape and presence of spicules, all aplacophorans exhibit a poorly differentiated head with a cerebral ganglion, mouth and usually a radula; a tetraneural nervous system (paired lateral and ventral nerve cords); a foregut gland; a central digestive gland; dorsal elongate gonad(s); a dorsal heart with an elongate ventricle and two auricles; and a posteriorly positioned mantle cavity. Although ctenidia are present only in caudofoveates, secondary gills are sometimes developed in Neomeniomorpha. There are no kidneys. Wastes are excreted via epithelial papillae or possibly through the gonopericardial duct.

Form and function

Aplacophorans are typically small animals (3–30 mm length) and usually inhabit deep water (200 to over 5000 m) although some species are between 100 and 300 mm long and inhabit waters less than 20 m deep. The habits of the two subclasses are very different. Neomeniomorpha occur either on the surface of the substratum (sand or mud) or on colonial cnidarians, and glide on a mucous trail, using cilia that line the pedal groove (Fig. 7.17c). Body wall musculature is employed chiefly to gain purchase around branches of the host hydroid on which the animal feeds. Caudofoveates are

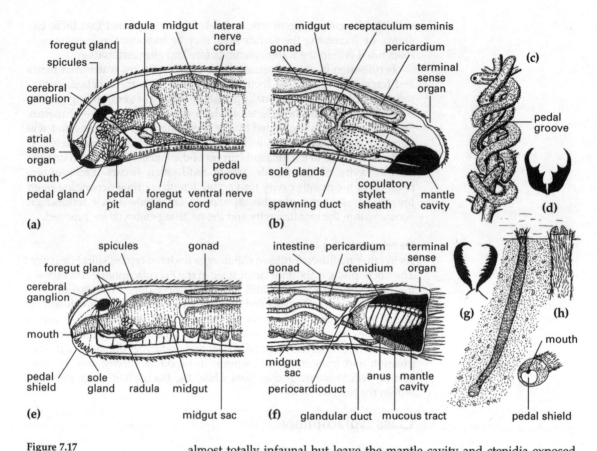

Figure 7.17
(a, b) Anatomy of anterior and posterior regions of a neomeniomorph. (c) Two neomeniomorphs entwined on a gorgonian coral. (d) One row of radula of a neomeniomorph, showing two cusped teeth. (e, f) Anatomy of anterior and posterior regions of a caudofoveate. (g) One row of radula of a caudofoveate, showing two serrated teeth. (h) Caudofoveate in the living position, with ctenidia exposed above the substratum, and views of anterior and posterior ends. (a, b, e after Salvini-Plawen 1985; c after Salvini-Plawen 1971; f after Salvini-Plawen 1972, 1975; g after Salvini-Plawen and Nopp 1974; h after Salvini-Plawen 1985.)

almost totally infaunal but leave the mantle cavity and ctenidia exposed slightly above the substratum (Fig. 7.17h). They feed on detritus and microfauna in the sediment. Both subclasses occur in Australian waters.

Evolutionary history

Although often dismissed as a highly specialised group in older literature, there is now almost universal acceptance that aplacophorans were derived at a very early stage in the history of the phylum Mollusca and that their worm-shaped forms reflect the anatomy of the earliest molluscs. This view is supported by recent cladistic analyses of comparative anatomy and independently by molecular data (see Fig. 7.18). Aplacophorans have some specialisations, such as the reduction or loss of the foot and abandonment of marked dorso-ventral compression with the adoption of a worm shape, and the loss of the ctenidia and the radula in some species, but these are correlated with habitat and diet. Anatomically the Aplacophora are most closely related to the Polyplacophora among the extant classes. As with the Polyplacophora, there are resemblances to the Lower Cambrian spiculate Halkieriidae (see Fig. 7.2c).

Physiological processes
Respiration and circulation

Due to their preference for deep-sea habits, it is perhaps not surprising that little is known about respiration in aplacophorans. Gas exchange presumably takes place in the posterior mantle cavity, either at the surface of the

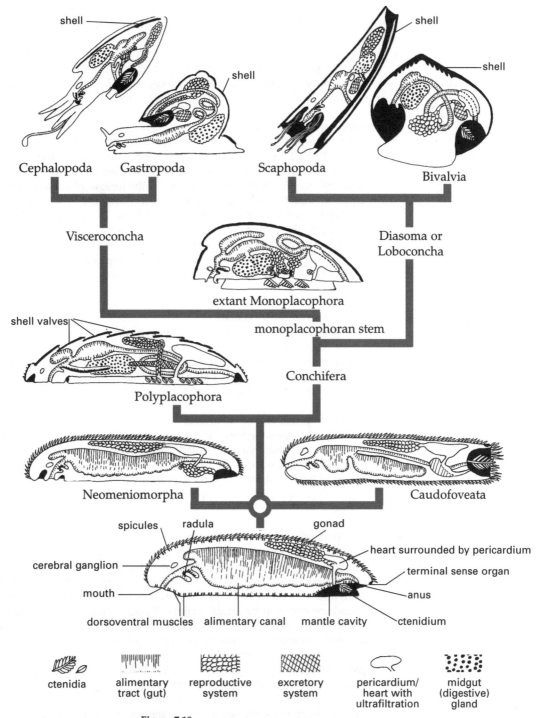

Figure 7.18
Diagram summarising body organisation and evolutionary relationships among the classes of extant Mollusca, and their probable derivation from an aplacophoran-like stem mollusc. A radula is present in all groups except the Bivalvia. In this reconstruction the aplacophoran subclasses Neomeniomorpha and Caudofoveata are treated as classes. (After Salvini-Plawen and Steiner 1996.)

pair of ctenidia (caudofoveates Fig. 7.17f, h) or through the wall of the mantle cavity or secondary gills (Neomeniomorpha). In Neomeniomorpha the surface of the pedal pit and pedal groove, and (if present) the epidermal papillae, may act as supplementary respiratory surfaces. The aplacophoran heart consists of an oblong ventricle and two poorly defined auricles. The remainder of the circulation is an open haemocoel.

Feeding and digestion
A radula is present in most aplacophorans, usually with two teeth per row (Fig. 7.17d, g), but is absent in some Neomeniomorpha. Both subclasses have a foregut gland which secretes digestive enzymes (Fig. 7.17 a, e). In some Neomeniomorpha the secretions of the foregut gland are sometimes emptied directly onto the cnidarian prey, and the resulting digested material is sucked up into the buccal cavity. In most cases, however, gouging by the radula moves food into the buccal cavity where it is processed with the foregut gland enzymes, and later by enzymes produced by the midgut. Digestion in aplacophorans is largely extracellular, but particles may also be phagocytosed by the midgut wall and digested intracellularly.

Excretion and ionic regulation
No special organs for excretion or ionic regulation are present in aplacophorans. In Neomeniomorpha, nitrogenous waste is believed to be concentrated in amoebocytes which migrate to the dorsal papillae and are lost through the abrasion of the papillae on the substratum. The gonopericardial duct might also excrete wastes in caudofoveates.

Nervous system and sense organs
In both subclasses the nervous system is simplified but exhibits the tetraneural pattern observed in polyplacophorans and monoplacophorans, with a cerebral ganglion and paired ventral (pedal) and lateral nerves (Fig. 7.17a, b, e, f). Other than a small sensory epithelium at the postero-dorsal extremity Fig. 7.17b, f), aplacophorans appear to lack external tactile or light-sensitive organs.

Reproduction and development
Gonads are dorsal and paired in Neomeniomorpha but are usually fused into a single gonad in most caudofoveates (Fig. 7.17a, b, e, f). Caudofoveates have separate sexes, but in Neomeniomorpha simultaneous hermaphroditism is evident in the production of oocytes in the median part of the gonad and sperm in the lateral sections. Neomeniomorpha often have copulatory stylets, and their gonoducts empty directly into the mantle cavity. Internal fertilisation seems to be typical of Neomeniomorpha. Caudofoveates always lack copulatory structures and release gametes via a complex pathway beginning with transfer via ciliary ducts, first to the pericardium and then into the mantle cavity.

The sperm of neomeniomorphans are similar in many respects to those of caenogastropods, having a conical acrosome, a rod-shaped nucleus, an elongate mitochondrial midpiece and glycogen piece and a short, free tail section. Such features indicate that fertilisation is internal in these aplacophorans. In contrast, caudofoveate fertilisation occurs either within the mantle cavity or outside the animal, and the sperm have a small acrosome, short nucleus, simple midpiece, and tail.

Aplacophoran eggs are typically large and yolky. Development (examined only in Neomeniomorpha) proceeds via spiral cleavage to a planktonic trochophore. Granular or scale-like spicules and the pedal groove are usually visible in the settlement stages. Although it has been claimed that a linear series of transverse plaques similar to those of chitons occurred during embryogenesis in *Nematomenia*, this has never been confirmed.

Classification of the phylum Mollusca

The relationships between the seven molluscan classes are summarised in Figure 7.18. Within each class, the definition and arrangement of groups is currently the subject of intense review, particularly in the class Gastropoda, where numerous higher-level taxa have been introduced or redefined as new anatomical data come to hand. Some authors have even suggested abandoning formal Linnean categories above the superfamily level, and applying names only to strictly monophyletic clades. The following system attempts to incorporate some of the more widely accepted changes to traditional schemes. Brief definitions are provided of the subphyla, classes and subclasses. The major orders in each class are listed, together with some of the superfamilies that are commonly encountered in molluscan studies.

SUBPHYLUM CONCHIFERA
Shell a single calcareous structure, sometimes restricted to the larval stage.

CLASS GASTROPODA
Single, conispiral shell, secondarily cap-shaped; reduced or absent in some; torsion during development; mantle cavity anterior; head with radula and jaws.
Subclass Prosobranchia
Mainly marine, some freshwater, a few terrestrial; shell and operculum usually well developed; mantle cavity typically with ctenidia; sexes usually separate. (Orders Patellogastropoda (Patelloidea, Acmaeoidea); Vetigastropoda (Haliotoidea, Trochoidea, Fissurelloidea); Neritimorpha (Neritoidea); Cocculiniformia; Caenogastropoda (Littorinoidea, Rissooidea, Cypraeoidea, Stromboidea, Cerithioidea, Naticoidea, Tonnoidea, Janthinoidea, Muricoidea, Conoidea))
Subclass Heterobranchia
Marine, freshwater, terrestrial; shell often thin, reduced or absent; operculum only in basal forms; true ctenidia absent; hermaphroditic.
Superorder Allogastropoda — Mainly marine; retaining well developed shell and operculum.
Superorder Opisthobranchia — Marine; shell and operculum usually reduced or absent; mantle cavity anterior to posterior or absent; secondary gills usually present. (Orders Cephalaspidea, Anaspidea, Notaspidea, Nudibranchia, Gymnosomata, Thecosomata)
Superorder Pulmonata — Mainly terrestrial or freshwater, a few marine; shell usually present, thin, absent in slugs; operculum absent; mantle cavity a pulmonary chamber. (Orders Basommatophora (marine Siphonarioidea, Ellobioidea and Amphiboloidea, freshwater Lymnaeoidea and Planorboidea); Systellommatophora; Stylommatophora (e.g. Helicoidea, the garden snails and allies, and Limacoidea, the terrestrial slugs))

CLASS CEPHALOPODA
Marine; shell external and chambered in nautiloids, usually internal, reduced or absent; head with complex eyes, radula, beak-like jaws; mouth surrounded by grasping tentacles; swimming via jet propulsion.
Subclass Nautiloida
Shell external, with internal chambers linked by siphuncular tube; up to 90 tentacles without suckers; 2 pairs of ctenidia.
Subclass Coleoida
Shell internal, reduced, flattened; 8–10 arms or tentacles, with suckers; 1 pair of ctenidia.

Superorder Decabrachia — With 10 arms, 2 as long tentacles; suckers with horny rings and often hooks; nidamental glands. (Orders Sepioida, Spirulida, Sepiolida, Teuthida)

Superorder Octopodiformes — With 8 arms; suckers without horny rings or hooks; nidamental glands absent. (Orders Vampyromorpha, Octopoda)

CLASS BIVALVIA
Marine and freshwater; shell of 2 lateral valves articulated dorsally by ligament and usually with hinge teeth; body and mantle cavity laterally compressed; head reduced, lacking radula and jaws; 1 pair of ctenidia, often enlarged.
Subclass Palaeotaxodonta
Marine; shell small, triangular, interior nacreous; hinge teeth multiple and linearly arranged; protobranchiate. (Order Nuculoida)
Subclass Cryptodonta
Marine; shell small to moderate, thin, interior non-nacreous; hinge lacking teeth; protobranchiate. (Order Solemyoida)
Subclass Pteriomorphia
Marine and freshwater; shell of moderate to large size, thin, with compressed valves; frequently anchored by byssal threads or valve cementation; foot reduced; filibranchiate. (Orders Arcoida, Mytiloida, Pterioida, Limoida, Ostreoida)
Subclass Paleoheterodonta
Marine and freshwater; shell of moderate to large size, externally ribbed or smooth; hinge teeth radial, often grooved; periostracum well developed; filibranchiate or eulamellibranchiate. (Orders Trigonioida, Unionioda)
Subclass Heterodonta
Marine and freshwater; shell small to large, externally sculptured or smooth; with cardinal and lateral hinge teeth; siphons often well developed; eulamellibranchiate. (Orders Veneroida, Cardioida, Myoida, Tellinoida, Teredinoida, etc.)
Subclass Anomalodesmata
Marine; shell varied in shape, but usually thin, with weak hinge dentition; eulamellibranchiate or septibranchiate. (Orders Thracioida, Septibranchia)

CLASS SCAPHOPODA
Marine; shell tubular, anterior aperture wider than posterior; head reduced, lacking eyes, but with radula, jaws and retractile captacula; mantle cavity tubular, lacking ctenidia. (Orders Dentaliida, Gadilida)

CLASS MONOPLACOPHORA
Marine; shell cap-shaped; head reduced, lacking eyes, but with radula and velar lobes; mantle cavity circular; multiple pairs of ctenidia, kidneys and gonads.

SUBPHYLUM ACULIFERA
Mantle epidermis with embedded spicules; shell (if present) composed of 8 plates.

CLASS POLYPLACOPHORA
Marine; shell of 8 plates; spicule embedded mantle margin; head reduced, lacking eyes but with well-defined mouth and radula; mantle cavity bilateral, with a few to numerous pairs of ctenidia; aesthetes often associated with shell plates. (Orders Lepidopleurina, Ischnochitonina, Acanthochitonina)

CLASS APLACOPHORA
Marine; body oblong to vermiform, lacking well defined head; radula and jaws present; shell absent; mantle with numerous calcareous spicules.
Subclass Neomeniomorpha
Ctenidia absent; animals usually associated with hydroids; foot long, narrow; copulatory stylet posteriorly.
Subclass Caudofoveata
1 pair of ctenidia in mantle cavity; burrowing, orientated vertically; foot forming anterior pedal shield; no copulatory stylet.

Chapter 8

The Annelida

G. Rouse

PHYLUM ANNELIDA

Introduction 175

Basic annelid organisation 176
  Coelom 177
  Metamerism 177
  Chaetae 177

Annelid diversity 178
  Polychaetes and clitellates 178
  Unusual annelid or probable annelid groups 182

Annelid structure and function 183
  Body wall 183
  Nervous system 186
    Sense organs 186
  Circulation and respiratory structures 188

Buccal organs 189
Feeding methods 191
Nephridia and excretion 194
Reproduction and development 195
  Asexual reproduction in the Polychaeta 195
  Sexual reproduction in the Polychaeta 196
  Sexual reproduction in the Clitellata 200

Fossil annelids 201

Classification of the phylum Annelida 201

Phylum Annelida

Bilaterally symmetrical, metamerically segmented coelomates; thin external cuticle, not moulted; paired, epidermal chaetal bundles; head with prostomium and peristomium; brain and double ventral nerve cord; protonephridial or nephridial excretory system; closed circulatory system with dorsal pumping vessel; coelomic gonads, coelomoduct gonoducts; separate sexes or hermaphrodite; spiral cleavage development; trochophore larva (often suppressed).

Introduction

The phylum Annelida contains more than 17 000 described species. No doubt there are many still to be discovered and named. Lamarck (1802) first used the term 'Annélides' when naming a group of organisms from the broad taxon Vermes erected by Linnaeus in the mid-eighteenth century. The name Annélides was based on the Latin word *anellus*, meaning a little ring, in reference to the presence of ring-like segments. Although the name Annelida has continued to be used, there are now suggestions that it may not define a monophyletic group and should be abandoned. Through most of this century, the Annelida have been split into three classes; the Archiannelida, Polychaeta, and Clitellata. Many texts do not use the name Clitellata and instead use two classes for the group, the Oligochaeta (earthworms and close relatives) and Hirudinea (leeches).

The class Archiannelida was erected for a group of generally minute annelids that were presumed to be primitive because of their simple body-structure. This simple morphology is now regarded as secondarily related to the interstitial habitat of the animals. Most of the archiannelidan families have now been transferred to the Polychaeta. This limits the number of commonly accepted annelid classes to two, the Polychaeta and Clitellata.

The membership of the taxon Polychaeta and the subdivision of the group has remained unstable. Eighty-three families of polychaetes are currently recognised. Until recently the system most commonly used for classifying polychaete families into higher groups was derived from Quatrefages (1866), who split the Polychaeta into the Errantia (wandering forms) and Sedentaria (sedentary forms) according to whether they were mobile or lived in tubes or burrows. This classification was supplanted in the 1960s and 1970 by ones which split the Polychaeta into as many as 22 orders with no real linkage between them. This situation was also unsatisfactory and gave no useful insight into polychaete evolution. A recent cladistic analysis of the Annelida and other groups has resulted in a new classification of the Polychaeta which is used in this chapter (Box 8.1 and Fig. 8.15).

The taxon Clitellata is considered to be monophyletic, based on the common presence of a clitellum, the organisation of the reproductive system, and features of sperm ultrastructure. The current consensus is that the Clitellata is the sister group to the Polychaeta, but new evidence suggests that the clitellates may actually fall within the Polychaeta. If this shown to be the case then the classification of the Annelida will have to be altered again. However, in this chapter the Polychaeta and Clitellata will be referred to as classes of the phylum Annelida.

Figure 8.1
(a) Correlation between body regions of a larva and an adult annelid. (b) Ultrastructure of hooked chaeta (uncinus) of serpulid polychaete, *Spirorbis spirorbis*. (c) Body wall of the clitellate *Pheretima*, showing invaginated epidermal cells (chaetoblasts) that form the chaetal follicle. (d) Segment of a generalised annelid. (a after Nielsen 1995; b from Bartolemaeus 1995; c after Barnes 1974.)

Basic annelid organisation

Annelid worms consist of three basic regions, with the majority of the body comprised of a region of repeated units called segments. Each segment is, in principle, limited by septa dividing it from neighbouring segments. A segment usually carries parapodia (in polychaetes) and chaetae, in addition to various segmentally arranged internal organs. The only parts of the annelid body that are not segmental are the head and a terminal post-segmental region called the pygidium. The head is comprised of two units, the prostomium and the peristomium. The prostomium usually contains the brain and carries sensory appendages. The peristomium is the region surrounding the mouth. The postsegmental pygidium includes the zone from which new segments are proliferated forwards during growth (Fig. 8.1a).

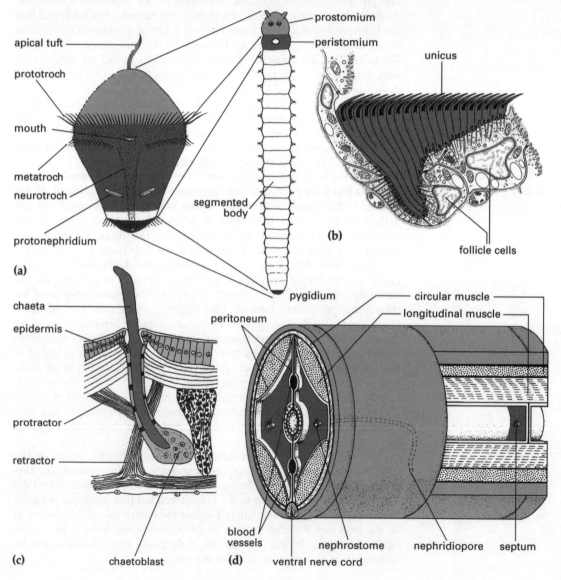

Coelom

A schizocoelic coelom is formed when solid mesoderm hollows out and develops spaces within (Figs 8.1d, 8.6a). Although a coelom is strictly defined as having a peritoneal lining over all surfaces, there are numerous examples of polychaetes that have the coelom directly in contact with the longitudinal muscle layer. The coelom is invaded by tissue in certain polychaetes and members of the Hirudinea, but this is regarded as a secondary phenomenon.

The perceived advantages of a coelom, mentioned at the beginning of Chapter 6 are:
1. The gut is separated from the body wall and, with its own musculature, can move independently.
2. The coelom provides a space where gametes can mature and where nutrients and waste products can be moved or stored.
3. The coelom provides a hydrostatic skeleton upon which muscles can act.

The evolution of the coelom introduced some physiological problems, related to the enlargement of the body and the distance of the gut from the outer surface. The combination of these factors means that the transportation of respiratory gases and nutrients cannot be accomplished by diffusion alone, as in Platyhelminthes. Most taxa with a coelom therefore have a circulatory system, though this is not always so, as we have seen with the Sipuncula. A number of annelid groups also rely on movement of the coelomic contents to facilitate gas exchange.

Metamerism

The serially repeated segmentation seen in annelids is known as metamerism. Each segment is a unit which in most species is isolated from adjacent segments by membranous septa. In addition to the septa in each segment there are usually dorsal and ventral mesenteries that separate a bilateral pair of coelomic compartments. Segments are formed sequentially in annelids and are established during development from paired mesodermal growth zones that originate from the 4d micromere. These growth zones are located at the posterior end of the body (Fig. 8.1a). Some authors have suggested that the first three segments are formed simultaneously in many annelids, but this has been shown to be incorrect. In reality the three segments are formed rapidly in sequence. Many annelids, such as leeches and myzostomids, have a fixed number of segments, but in many others segments continue to be added throughout life. Structures such as locomotory, excretory, and respiratory organs are generally repeated in each segment. However, all segments are basically united by the digestive, vascular, muscular and nervous systems and so have little autonomy. The septa separating each segment can be virtually complete, as in most clitellates, but in many polychaetes the septa are incomplete or even absent, resulting in coelomic continuity along much of the body.

Chaetae

Chaetae (also called setae) are bundles of chitinous, thin-walled cylinders held together by sclerotinised protein. They are produced by a microvillar border of certain invaginated epidermal cells and so can be defined as cuticular structures that develop within epidermal follicles (Fig. 8.1b, c). Chaetal ultrastructure is similar in all cases but there is a considerable diversity of form. Chaetae can be long thin filaments or stout multihooked

(a) (b) (c)

Figure 8.2
(a) Tufted chaetae of polynoid polychaete larva. (b) Compound chaetae of chrysopetalid polychaete larva. (c) Neuropodium of sabellid polychaete, *Fabricinuda*, with a row of uncini.

structures (Fig. 8.2a,c). In some cases they form compound structures with ligaments (Fig. 8.2b). Apart from annelids, chaetae are found in the Echiura (see later in this chapter) and the Brachiopoda (Chapter 15). It is possible that these groups fall within the Annelida and that chaetae have only evolved once. Most authors reject the idea that brachiopods have any annelid affinities, though molecular evidence does support this idea (see Chapter 18 for further discussion).

In the Echiura, paired ventral chaetae are present anteriorly and one or more rings of chaetae may be present posteriorly. The presence of paired dorsal and ventral bundles of chaetae has been considered a synapomorphy for the Annelida by some authors. However, many polychaetes have only a single, usually ventral, paired segmental chaetal bundle, and the plesiomorphic annelid condition with regard to chaetal distribution is as yet uncertain. Chaetae have been lost in leeches and a few polychaete groups.

Annelid diversity

Polychaetes and clitellates

Though uncommon in fresh water and virtually absent from the land, polychaetes occupy every marine habitat, their life styles ranging from inhabiting the sediments of the deepest oceans to swimming freely and preying on plankton. Polychaetes vary greatly in form (Figs 8.3a–c, 8.4) and range in size as adults from a fraction of a millimetre to well over 3 metres in length. Polychaetes are a dominant component of the macrobenthos of many marine communities in terms of numbers of individuals and species, and in some areas they dominate the biomass.

The most commonly observed polychaetes are the more errant forms that live amongst algae or under stones, or crawl over sediment (Fig. 8.3b, c). They tend to have well-developed eyes and sensory appendages (Fig. 8.3b, c). Many polychaetes are burrowers in sand or mud; some dig continuously through the sediment, while others live in permanent burrows or tubes that they secrete or construct from gathered materials. Those that live in permanent tubes are basically sessile and have special food-gathering appendages that can be projected from the tube (Fig. 8.3a). These appendages either collect food from the surrounding surfaces or filter it directly from the water, and are often the only parts visible. They tend to be brightly coloured, resembling flowers or anemones or even spaghetti (Fig. 8.4a). A variety of polychaetes can be found living a permanently pelagic existence. These forms all appear to be derived from benthic crawl-

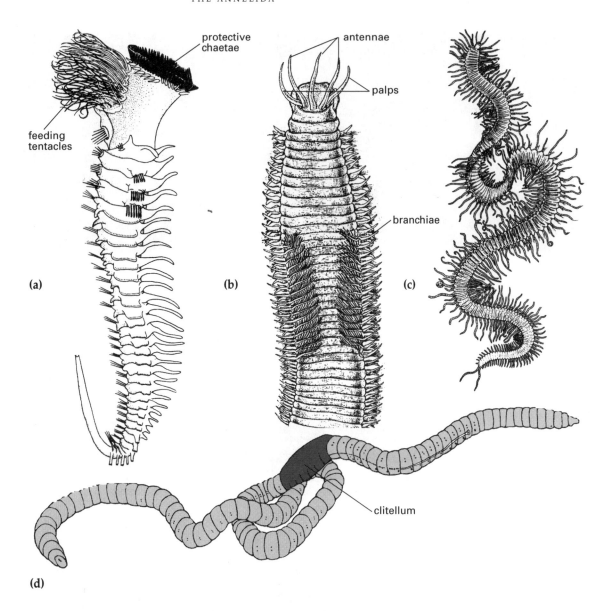

Figure 8.3
(a) Lateral view of *Lygamis indicus* (Sabellariidae).
(b) Dorsal view of *Marphysa disjuncta* (Eunicidae). (c) Dorsal view of *Trypanosyllis zebra* (Syllidae). (d) The clitellate *Lumbricus terrestris*. (a after Day 1967; b after Hartman 1961; c from Barnes 1974; d from Sherman and Sherman 1972.)

ing forms; they can have very well-developed eyes and be active predators on other plankton (Fig. 8.9d).

While less diverse morphologically than the Polychaeta, the Clitellata (Fig. 8.3d) are commonly found in marine, freshwater and terrestrial environments. They are divided into more than 42 families. Clitellates range in size as adults from less than a one millimetre (many freshwater and marine forms) to well over 2 metres (as in the giant Gippsland earthworm, *Megascolides australis*, found in Victoria). Almost all clitellates, except predatory or parasitic leeches, are burrowing detritus feeders and very few form any sort of permanent tube.

Leeches (Hirudinea) are a carnivorous group of clitellate annelids (Fig. 8.5c) that almost certainly have evolved from detritus-feeding clitellate ancestors. They are hermaphroditic and have a cocoon-secreting clitellum,

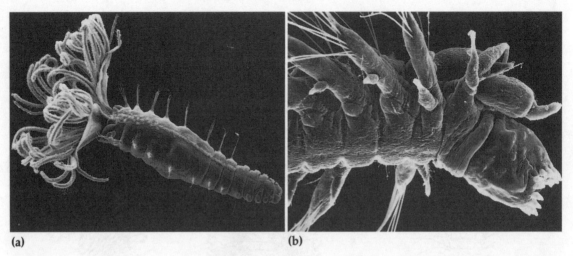

Figure 8.4
(a) Scanning electron microscope (SEM) photograph of small sabellid polychaete *Amphicorina androgyne*. (b) SEM of phyllodocid polychaete, with the muscular axial pharynx everted.

but generally show coelomic reduction and a loss of chaetae. Although leeches are more commonly thought of as blood-sucking parasites, they exploit a wide range of feeding habits and life histories. In fresh water, macrophagous leeches that prey on invertebrates are more common in terms of number of species and abundance than species that feed on blood. In marine systems the reverse occurs. Most marine leeches are ectoparasitic on fishes. Which taxon comprises the sister group of the leeches within the Clitellata is still the subject of debate, so that their taxonomic status is uncertain (Box 8.1).

Box 8.1
The annelids and related groups

PHYLUM ANNELIDA
Class Polychaeta (almost all marine; a few freshwater)
  'Clade' Scolecida (mainly burrowing polychaetes)
  'Clade' Palpata
    'Clade' Aciculata (errant, burrowing and planktonic polychaetes)
      'Clade' Phyllodocida (including Myzostomida)
      'Clade' Eunicida
    'Clade' Canalipalpata (burrowing and tubicolous polychaetes)
      'Clade' Sabellida (including Pogonophora)
      'Clade' Spionida
      'Clade' Terebellida

Class Clitellata (marine, freshwater, terrestrial)
  Subclass Tubificata (marine, freshwater and terrestrial)

  Subclass Lumbriculata (freshwater)

  Subclass Diplotesticulata (freshwater and terrestrial)

  'Clade' Hirudinea (leeches)
    'Clade' Branchiobdellida (freshwater, on crayfish)
    'Clade' Acanthobdellida (aquatic, on salmonid fish)
    'Clade' Euhirudinea (marine, freshwater and terrestrial)

PHYLUM(?) ECHIURA (marine)

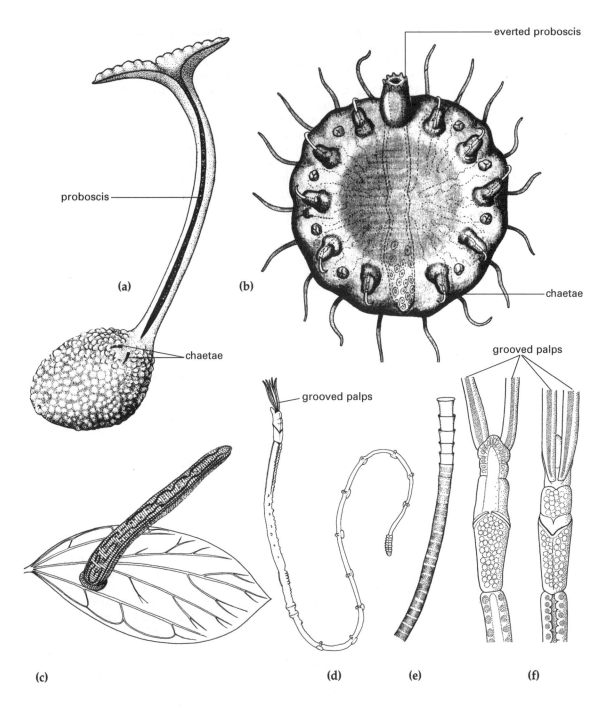

Figure 8.5
(a) Ventral view of female echiurid, *Bonellia viridis*. (b) Ventral view of a myzostomid polychaete, showing the characteristic hooked chaetae. (c) A blood-sucking terrestrial leech from southern Asia, *Haemadipsa*. (d) Generalised siboglinid (= Pogonophora) removed from the tube. (e) Anterior portion of the tube of a siboglinid. (f) Ventral and dorsal views of *Siboglinoides dibrachia*. (a after Fretter and Graham 1976; b, c from Barnes 1974; d, e from Southward 1993; f from Ivanov 1963.)

Unusual annelid or probable annelid groups
Myzostomida

The Myzostomida is an enigmatic group of 'worms' whose members are obligate commensals or parasites of echinoderms, mainly crinoids (Fig. 8.5b). While historically they have been regarded as closely related to groups such as trematode flatworms and tardigrades, current opinion now places them in the Annelida. A recent suggestion is that myzostomes are derived from polychaetes with a muscularised axial pharynx, and constitute a family in the Phyllodocida (see Box 8.1). This is based on assumptions about the homology of certain features such as the structure of the pharynx, the nature of the parapodia and the occurrence of specialised chaetae called acicula, also found in polychaetes.

Pogonophora

The Pogonophora is another enigmatic group (Fig. 8.5d, e, f) the systematic placement of which has been a problem since the first species was described in 1914. They can reach 1.5 m in length, as in the Vestimentifera from deep-sea hydrothermal vents. The lack of a recognisable gut resulted in controversy over the dorsal and ventral orientation and hence whether they were 'deuterostomes' or 'protostomes'. Studies on pogonophoran development have now shown that the nerve cord is ventral, so that they would seem to be protostome. There has also been doubt as to whether the group should be classified as one phylum (Pogonophora) or two (Pogonophora and Vestimentifera), but a recent reassessment of their morphology indicated that the Pogonophora are actually annelids. Features such as the distinctive chaetae and excretory systems were found to be similar to certain polychaetes. A series of cladistic morphological analyses, focused on the Polychaeta, all place the Pogonophora as derived polychaetes, close to sabellids and serpulids (feather-duster worms, Box 8.1 and Fig. 8.4). The ranking of the Pogonophora as a phylum is incorrect. Their taxonomic status as derived polychaetes means they should revert to the original name and status, family Siboglinidae, and be placed in the polychaete group Sabellida.

Phylum(?) Echiura

The echiurids were originally considered to be annelids by early authorities such as Cuvier and Lamarck, before being placed with the sipunculids and priapulids in the phylum Gephyrea. The Gephyrea was eliminated as a taxon at the end of the last century and the echiurids were again placed in the Annelida until Newby proposed a separate phylum, Echiura, based on a detailed embryological study of *Urechis caupo*. His proposal is retained provisionally here (Box 8.1 and Fig. 8.5).

Echiurids are unsegmented marine coelomates. Two features identifying the Echiura are the flattened or grooved proboscis, formed from the prostomium, and the anal sacs. The presence of chaetae in echiurids indicates that they are closely related to annelids. Echiurids burrow in soft sediments or live under rocks, and feed on detritus by projecting the extensible proboscis (Fig. 8.5a). The lack of segmentation in echiurids is usually considered a primitive feature rather than a loss, but this is now being more closely assessed and the group may well prove to fall once again within the Annelida.

Annelid structure and function

Body wall
The annelids, like the Echiura and Sipuncula, have a body covered by an external cuticle which is not moulted. Epidermal microvilli secrete a network of fibres that are in part collagenous and also contain scleroprotein. Chaetae are also cuticular structures, but contain large amounts of chitin. The epidermis is usually a columnar epithelium, ciliated in certain areas of the body. Beneath the epidermis and its basal lamina lies a layer of circular muscle. The circular muscle layer forms a nearly continuous sheath around the body, except in polychaetes with well-developed parapodia. Beneath the circular muscle layer lie thick longitudinal muscles. In virtually all annelids these are present as four distinct bands (Fig. 8.6a). In addition to the circular and longitudinal muscle layers there can be series of 'oblique' muscle fibres that join the ventral area of the body with the mid-lateral region. The innermost body wall component in most, but not all annelids, is a thin peritoneal layer lining the coelom.

Parapodia
Parapodia are unjointed extensions of the body wall (Fig. 8.6), found in nearly all polychaetes but absent in clitellates and echiurans. Parapodia are equipped with musculature derived mainly from the circular muscle layer and usually carry chaetae. They may be supported internally by one or more large, thick chaetae called acicula. Parapodia vary in structure but basically can be considered to consist of two elements; a dorsal notopodium and a ventral neuropodium (Fig. 8.6a). In addition to various kinds of chaetae, the notopodia and neuropodia can also carry a variety of cirri and gills (Fig. 8.6c, e, f). They are most elaborate in actively crawling or swimming forms, in which they form large fleshy lobes that act as paddles (Fig. 8.6a). Parapodia of burrowing or tubicolous polychaetes may be no more than slightly raised ridges carrying hooked chaetae called uncini (Figs 8.2c, 8.6b, d).

Coelom
As we have seen, the coelom in many annelids is organised as a series of compartments divided by intersegmental septa (Fig. 8.1d). This compartmentalisation means that if the worm is damaged or severed the coelomic contents will be lost from only a few segments and locomotion can be maintained. In other annelids there may be only a few septa dividing the coelom. Under these circumstances much more coelomic fluid is lost with injury and locomotion is severely affected. However, there are advantages in having large coelomic compartments, particularly for burrowing polychaetes. A number of annelid groups, particularly many leeches and a number of small polychaetes, have little or no coelomic space in the body. It appears that this is a secondary phenomenon in each group. In leeches the role of the coelom in locomotion has been replaced by a sucker-based mode of movement. Of the polychaetes with a small adult size, those with no coelom tend to use ciliary movement.

Locomotion
In the errant polychaete groups (such as the family Nereididae), septation tends to be complete, giving segments of a constant volume that are basi-

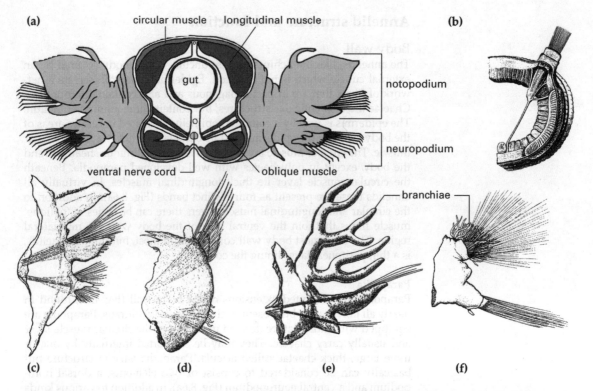

Figure 8.6
(a) Transverse section of a nereidid polychaete showing body wall, coelom and parapodia. (b) Parapodium of *Arenicola marina* (Arenicolidae); neuropodia have hooked chaetae that project on a low ridge. (c) Parapodium of *Haploscoloplos elongatus* (Orbiniidae). (d) Parapodium of *Polydora neocardalia* (Spionidae). (e) Parapodium of *Scoloplos dendrobranchus* (Orbiniidae). (f) Parapodium of *Pareurythoe americana* (Amphinomidae). (a from Barnes 1974; b after Ashworth 1912; c–f after Hartman 1951, 1957, 1961.)

cally independent units. These groups can show several patterns of locomotion, from slow crawling to rapid swimming. For example, in *Nereis diversicolor* the longitudinal muscles on each side of a given segment act out of phase with each other; when a parapodium moves forwards the one on the other side moves backwards and when the longitudinal muscles on one side contract those on the other relax (and are stretched). This results in waves of lateral undulations passing forward along the body (Fig. 8.7a). The parapodia come into contact with the substratum when they are at the crest of each wave (Fig. 8.7b). When crawling slowly the wavelength of the undulations is relatively short (four to eight segments) with a small amplitude. For greater crawling speed the worm increases both the wavelength and amplitude of the undulations.

In burrowing and tube-dwelling annelids both the circular and longitudinal muscle bands tend to be well developed and to act in concert, producing peristaltic contractions. Earthworms are excellent burrowers and are also capable of crawling (Fig. 8.7c). Both activities are achieved through the passage of alternating waves of circular muscle contraction and longitudinal muscle contraction along the body. The circular muscle contractions make the segments long and narrow, so that the region of the body in which they are occurring extends forwards. The longitudinal muscle contractions make the segments short and fat, providing anchor points for further extension. Polychaetes that live in tubes also tend to have complete septa, limited parapodia and movement that is essentially peristaltic, as in earthworms. The parapodia often carry uncini (Fig. 8.2c) that can dig into the side of the tubes. These can then act as anchors, allowing the worm to withdraw rapidly into the tube through contraction of the longitudinal muscle bundles.

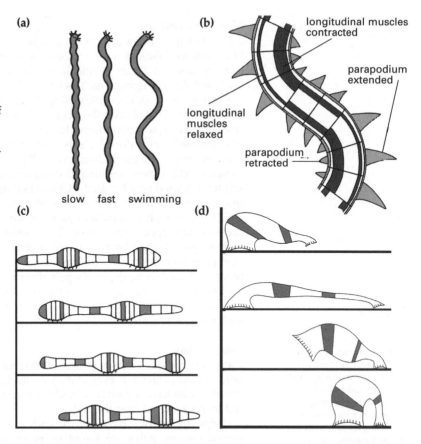

Figure 8.7
(a) *Nereis*: slow crawling, fast crawling and swimming. In slow crawling the interval between segments at the same phase of contraction is usually 6–8 segments, in rapid crawling 11–15. Swimming has a phase of 40. (b) Slow crawling in *Nereis*, showing patterns of longitudinal muscle contraction. (c) Crawling in an earthworm, *Lumbricus* (from Russell-Hunter 1979). (d) Crawling by looping in a leech (Hirudinea).

Many burrowing polychaetes, in contrast, have little septation between segments and the coelomic contents are free to move from one part of the body to another. This allows for a form of burrowing that is energetically less efficient than that of earthworms in soil, but very effective in aquatic muds and sands (*Arenicola* is an example). Burrowing is accomplished by contracting the circular muscles of the posterior region of the body, which forces coelomic fluid into the anterior region, causing it to swell. The more posterior segments then contract via longitudinal muscles and are drawn forward to the anchored anterior segments. Following this, the head is pushed forward and the proboscis everted, allowing the animal to further deepen the burrow. *Arenicola* is only capable of maintaining cycles of such activity for a short time and is largely sedentary in its burrow. Annelids with complete septation can sustain burrowing or crawling for much longer.

Leeches, with their reduced coelom and lack of chaetae, cannot move like other annelids. They are not burrowers and either crawl over surfaces or swim. Anterior and posterior suckers work in concert with the circular and longitudinal musculature. In order to crawl, the leech attaches its posterior sucker and contracts much of its circular musculature (Fig. 8.7d). This makes the body much thinner and extends it forwards. The anterior sucker is then attached, the posterior one released, and the longitudinal muscles are contracted. This draws the body forward. By repeating these steps a leech can move rapidly.

Nervous system

Annelids have a brain or cerebral ganglion in the prostomium (Fig. 8.8a). The brain morphology varies: mobile active forms such as *Australonereis* (Nereididae) and *Eunice* (Eunicidae) have the most complex brains (Fig. 8.8c), and deposit-feeders such as *Ctenodrilus* (Ctenodrilidae) have simple brains with little differentiation (Fig. 8.8a). This variation in brain morphology is correlated with the degree of sensory input the brain receives, since burrowing sedentary forms tend to have few sensory appendages. In all cases the brain is dorsal and is connected to the ventral nerve cord by two circumo-esophageal connectives (Fig. 8.8a–c). In polychaetes with complex brains there are three distinct divisions: a forebrain, mid-brain and hind brain (Fig. 8.8c). In the Clitellata (Fig 8.8b) and polychaetes with simple brains there are no obvious subdivisions.

The forebrain has palpal and buccal centres and the anterior roots of the circum-oesophageal connectives. The mid-brain has antennal and optic centres and the posterior roots of the circum-oesophageal connectives. The hind brain includes centres for the nuchal organs (see below). The ventral nerve cord, usually made up of a pair of cords that are bound together, runs the length of the body. It varies in thickness and dilates into a ganglion in each segment, from which pairs of segmental nerves pass out to the body wall, muscles and gut.

Sense organs

There are six major kinds of sensory structures among the Annelida. These include palps, antennae, eyes, statocysts, nuchal organs and lateral organs. Annelids also have a variety of epidermal sensory cells responsive to light and touch scattered over the body.

Palps are of two basic forms, both innervated from the forebrain. Ventral sensory palps are found in members of the polychaete groups Phyllodocida and Eunicida. In most cases they are tapering or digitiform

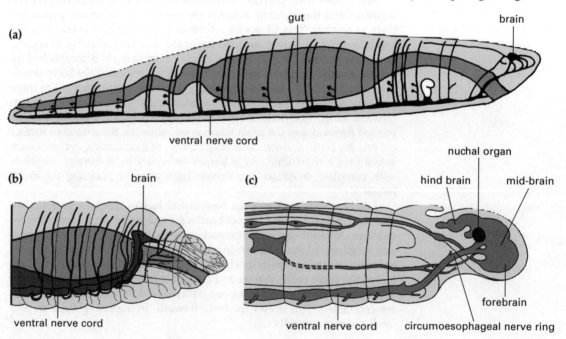

Figure 8.8
(a) Nervous system of *Ctenodrilus serratus* (Ctenodrilidae). (b) Anterior part of nervous system of *C. serratus*, showing simple brain. (c) Lateral view of *Eunice* (Eunicidae), showing the anterior nervous system, including a brain with three distinct regions. (a, b after Gelder and Palmer 1976.)

and relatively short compared to grooved palps (Figs 8.3b, 8.4b). Grooved palps are found in a large number of polychaetes and generally have a feeding function (see page 191), but presumably also serve a tactile role. Antennae are found only in the Phyllodocida and Eunicida. They are always located on the prostomium and probably have a largely tactile role. They are innervated by the midbrain. Three antennae are usually present, forming a lateral pair and a single median antenna. Most antennae are simple, tapering or digitiform, but they may be articulated. The lateral antennae may be located at or near the frontal edge of the prostomium, whereas the median antenna is usually located behind the frontal margin (Fig. 8.3b). Statocysts, which act as gravity sensors, are found in a range of polychaetes, usually burrowing or tubicolous forms. There may be only a single pair, as in many sabellids, or more than 20 pairs, as in orbiniids. They are always located dorsally in the anterior part of the body and are usually innervated from the circum-oesophageal connectives. Statocysts may be simple open pits in the epidermis or deep invaginations that connect to the outside via ciliated canals. In some cases they are subepidermal and have no outside connection. The space within the statocyst contains either sand grains or special hard secretions called statoliths. These fall against receptor cells that line the lumen of the statocyst, providing stimuli that allow the worm to orient itself.

Nuchal organs are paired ciliated structures which are generally innervated directly from the posterior part of the brain. They are usually assumed to have a chemosensory role, but this has not been demonstrated. Nuchal organs are present only in polychaetes. They may be ciliated patches or eversible folded or finger-shaped structures.

Most annelids have some type of photoreceptor or eyes, lacking only in a few burrowing forms. Many clitellates have individual photoreceptor cells spread over the body, but in general eyes are located in the prostomium. The complexity of eyes varies from simple pigmented cups or ocelli, through well-developed camera-type eyes, to compound eyes analogous to those found in arthropods. Ocelli occur in a wide range of polychaete families. They can be as simple as two cells — a sensory cell and a pigmented support cell (Fig. 8.9a). Other forms of ocelli are more complex but still may be composed of only a few tens of cells. They probably perceive information about light direction and intensity. In certain families of the Phyllodocida, particularly the swimming predatory Alciopidae (Fig. 8.9c, d), the eyes are probably capable of forming an image. The eyes of alciopids can be up to 1 mm across and so large that they protrude laterally from the head and press into the brain. They consist of a primary retina containing thousands of cells, a secondary retina overlain by a lens, and other accessory structures (Fig. 8.9c).

Compound eyes, which evolved independently of those of arthropods, are found on the radiolar crown of some sabellids and serpulids. The numerous compound eyes are arranged along the radioles or at the extreme tips. Each compound eye consists of up to 50 separate units called ommatidia (Fig. 8.9b). Each ommatidium is composed of three cells: one receptor cell forming a ciliary receptive segment and two pigment cells surrounding an extracellular lens (crystalline cone). The eyes cannot form images but are extremely sensitive to visual motion because, in concert, they cover such a wide area. Any motion sensed by the eyes will trigger retraction into the tube.

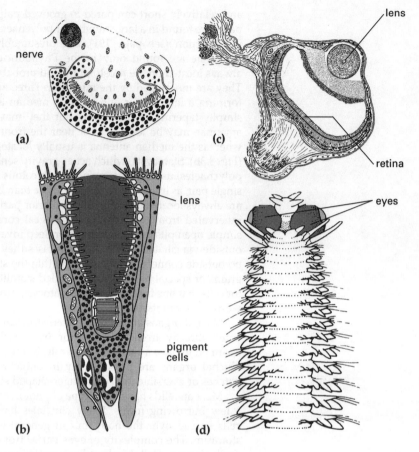

Figure 8.9
(a) Simple ocellus of *Armandia brevis* (Opheliidae).
(b) Ommatidium from a compound eye of *Sabella melanostigmata*. (c) Longitudinal section through an eye of *Vanadis tagensis* (Alciopidae).
(d) Dorsal view of *Plotohelmis capitata* (Alciopidae). (a from Mill 1978; b from Nilsson 1994; c after Hermans and Eakin 1974; d after Day 1967.)

Circulation and respiratory structures

The closed circulatory system of most annelids (many polychaetes and clitellates), and echiurids consists of dorsal and ventral longitudinal vessels, linked by smaller vessels, capillary beds and gut lacunae. The blood flows anteriorly in the dorsal vessel and towards the pygidium on the ventral side. The ventral vessel lies directly beneath the gut and has branches that supply the body wall muscles and the epidermis and (if present) the branchiae. The blood sent to these places is passed to the dorsal vessel by lateral vessels in each segment. From the dorsal vessel the blood flows around the gut to the ventral vessel (Fig. 8.10b). Blood flow in annelids depends on contractions of the dorsal vessel and on movements of the body wall. There are specialised pumping organs or 'hearts' in some polychaetes, usually tube-dwelling forms. Many branchiae are also contractile. This closed circulatory system is reduced or absent in leeches, where it may be replaced by coelomic canals. A limited circulatory system in which some of the major blood-vessels are present but the distal capillary vessels are missing is found in a number of polychaete families (e.g. phyllodocids). A circulatory system is absent in many small polychaetes such as protodrilids, and also in glycerids, goniadids and capitellids.

In some annelids, such as earthworms, the body surface functions in gas exchange. But in many polychaetes and some clitellates, elaborate branchiae (gills) are developed as extensions of the body wall and contain vascular loops and epidermal capillaries. These vastly increase the surface area available for gas exchange. Most branchiae are segmental structures repeated along part of the the body. They tend to be associated with parapodia, and have a wide variety of shapes (Fig. 8.6e, f). They can be simple and cirriform or broad flat sheets. In some cases the branchiae are elaborately branched or spirally arranged filaments. In tube-dwellers or burrowers they tend to be at the anterior or posterior end, usually where there is maximum water flow (Fig. 8.3b). Terebellids have paired branchiae on the dorsal part of the first three segments with a surface area equivalent to one-third of the body wall.

The radiolar crown is the primary site of respiration in the tubicolous Sabellidae and Serpulidae (Fig. 8.4a), though there is some evidence that auxiliary respiration is achieved by generating water currents through the tube. The blood moves from the abdomen anteriorly along a sinus surrounding the alimentary canal. Anteriorly it passes through dorsal, transverse, and circum-oesophageal vessels to a ventral vessel which carries blood posteriorly. There are a series of 'peripheral' blood vessels with blind ends. One of these is the vessel that supplies the radiolar crown (Fig. 8.10a). Contractile myoepithelial cells surround the radiolar crown blood vessel, which rhythmically fills and empties when the crown is extended into the water. Blood returning from the branchial crown enters the ventral vessel and travels posteriorly.

The blood of most annelids does not contain cells, although they have been recorded in a number of polychaete families. Groups that lack a circulatory system, such as the Capitellidae and Glyceridae, have special cells in the coelom that contain a form of haemoglobin. Most other annelids have respiratory pigments dissolved in the blood, while many small species lack pigments completely. The respiratory pigments of annelids are red haemoglobins or green chlorocruorins. These pigments enable annelids to draw about 50–60% of the oxygen from the water that comes into contact with the body or branchiae.

Buccal organs

The initial mouth opening (or buccal organ) in larval annelids may give rise to a variety of structures in the adults. The structure of buccal organs was used as a basic criterion for grouping polychaete families in an influential

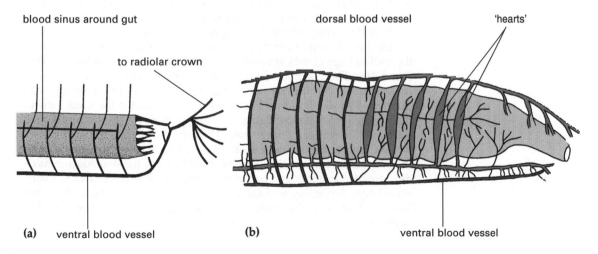

Figure 8.10
(a) Anterior circular system of *Pomatoceros triqueter* (Serpulidae). (b) Anterior circular system of *Lumbricus terrestris*. (a after Thomas 1940; b after Edwards and Lofty 1972.)

study by Dales. His system has been modified for use here. Five types of buccal organ can be distinguished (Fig. 8.11):
1. muscular axial pharynx,
2. simple axial proboscis,
3. ventral buccal organ, which may be simple or have well developed musculature,
4. buccal organ absent or occluded, and
5. dorsal buccal organ.

In many polychaetes the buccal organ is an eversible pharynx with thickened, muscular walls on all sides, referred to as a muscular axial pharynx (Fig. 8.11a). This pharynx may be retracted into a sheath. The external opening seen when the pharynx is fully retracted is often termed the mouth, but it is really the opening to the pharynx-sheath. The mouth is located at the tip of the pharynx when fully everted. The mouth may be surrounded by terminal papillae, or by jaws. The jaws have various forms and are usually paired. Glycerids have a long, muscular axial pharynx, covered externally with papillae and tipped with four jaws. Other groups, such as the Hesionidae and Phyllodocidae, lack jaws and rely on the muscular wall of the pharynx to grasp their prey.

A sac-like eversible pharynx is present in certain taxa (e.g. arenicolids, some maldanids and opheliids) and is everted by fluid pressure from the coelom (Fig. 8.11b). There is no particular development of musculature or glands; this type of pharynx is referred to as a simple axial proboscis. These worms tend to have reduced septa in the anterior part of the body. This allows the contraction of the posterior part of the body to exert considerable force on the buccal apparatus, because of the free movement of the coelomic contents. The proboscis is retracted by muscles associated with a thickened first septum and connected to the proboscis.

A variable, but often complex set of folds, musculature and glands is present on the ventral side of the pharynx in many polychaetes and is usually referred to as a ventral buccal organ (Fig. 8.11c). This is the most common form of buccal organ in polychaetes. Ventral buccal organs may be simple eversible muscular pads, as in sand or mud-eating groups such as the Orbiniidae and Opheliidae. In the Eunicida, the ventral and lateral walls of the involuted buccal organ are muscular and the lining is sclerotised into a varying number of jaw pieces. The jaws are separated into a pair of ventral mandibles and two or more pairs of lateral maxillae. These can be everted for grasping and tearing prey (Fig. 8.11d).

The buccal cavity in many polychaetes lacks obvious differentiation of the wall as large glands or additional muscular layers, and is not eversible (Fig. 8.11e). This condition (buccal organ absent) is found in several filter-feeding groups, such as the Sabellidae and Serpulidae. In the Siboglinidae (formerly Pogonophora) the buccal organ is occluded in the adults, where the digestive tract is closed anteriorly and posteriorly. A mouth and anus are present during early development but are closed off as the larva grows. In the Clitellata there is far less diversity in buccal organ structure than in the Polychaeta. In most clitellates the dorsal wall of the stomodaeum has a differentiated muscularised pad that is often eversible; this is used in a variety of ways, depending on the feeding mode (Fig. 8.11f). In the leech groups there can be an eversible proboscis or a muscular pharynx armed with two or three teeth.

THE ANNELIDA

Figure 8.11
(a–e) Longitudinal sections through anterior ends of (a) *Aphrodite aculeata* (Aphroditidae) with muscularised axial pharynx and jaws, (b) *Abarenicola vagabunda* (Arenicolidae) with simple axial pharynx, (c) an orbiniid showing ventral buccal organ, and (d) *Marphysa sanguinea* (Eunicidae) ventral muscularised pharynx and jaws. (e) Longitudinal section through the anterior end of *Pomatoceros triqueter* (Serpulidae) showing the absence of a buccal organ. **(f)** Action of the dorsal pad-like pharynx of the clitellate, *Aulophorus*. (a–d after Dales 1962, 1963; e after Thomas 1940; f from Barnes 1974.)

Accessory feeding structures

Where the buccal organ is simpler (ventral or simple axial) or absent, other structures facilitate feeding. These are often called palps, tentacles, or a radiolar crown. Recent studies have suggested that all of these structures can be regarded as homologous and referred to as grooved palps.

Grooved palps have ciliated longitudinal paths, often located in a groove, giving each palp a U-shaped or V-shaped cross-section. A single pair of grooved palps is present in many polychaetes (e.g. spionids). In terebellids, trichobranchids, ampharetids and pectinariids there are multiple grooved palps. In the sabellids and serpulids the grooved palps form the prostomial radiolar crown (Fig. 8.4a). The Siboglinidae have peristomial grooved palps. These structures have a feeding role in providing symbiotic bacteria within the worm with their nutritional requirements. The bacteria, in turn, supply the worm with energy (see below).

Feeding methods

The diversity of feeding structures in annelids can be classified into several categories according to their function. These categories are not indicative of phylogenetic relationships, but are useful for ecological purposes. Each category can contain taxa that have different buccal organs.

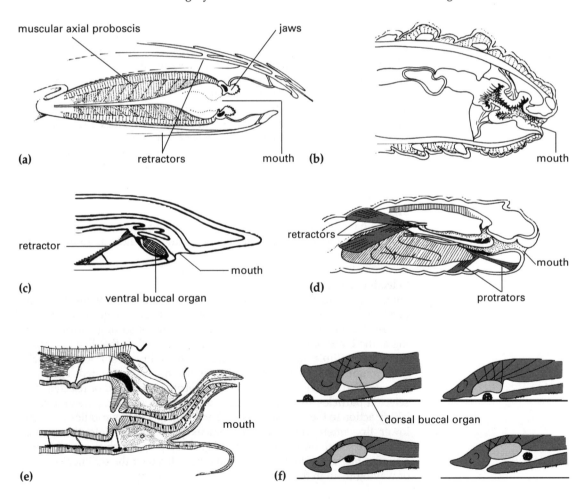

Raptorial feeders

Raptorial feeders use their buccal apparatus, usually an eversible muscular ventral or muscular axial organ, to seize their food (Fig. 8.4a). The two groups of polychaetes that have evolved jaws (Eunicida and Phyllodocida) use them to seize live animals (carnivores), tear off pieces of algae (herbivores), or grasp dead and decaying matter (scavengers). Some of the raptorial groups, such as the Phyllodocidae, appear to have lost their jaws, and use their eversible proboscis to grasp prey. Members of the eunicid family Onuphidae, such as *Australonuphis*, are well known from Australian beaches. They are large burrowing polychaetes that emerge where waves break to seek their prey. They have powerful jaws and feed on small to medium-sized animals. *Australonereis* (Nereididae), an errant polychaete which builds a temporary tube in predominantly sandy habitats in sheltered habitats on the Australian east coast, feeds on algae using powerful jaws attached to the eversible pharynx. Some clitellates can evert the dorsal pad which has mucus glands that stick the prey, usually other worms, to the pad for ingestion. Although leeches are more commonly thought of in relation to vertebrates, many prey on invertebrates, including other annelids.

Non-selective deposit feeders

Many types of annelid eat mud, sand or soil. Any digestible organic material is assimilated as it passes through the alimentary canal. This form of feeding is also called deposit or detrital feeding. The majority of groups using this method have a ventral buccal organ or simple axial pharynx, or in the case of earthworms, a dorsal buccal organ. Most earthworms live on organic remains in the soil, though some feed on organic material such as dead leaves that they drag down from the surface. Of the polychaete groups that are non-selective deposit feeders, some, such as arenicolids and maldanids, live in relatively permanent burrows or tubes and ingest the sediment in such a way that a continuous rain of sand or mud falls in front of them. Others, such as opheliids, do not have permanent burrows or tubes but move about the sediment, eating it as they tunnel. The amount of nutritional value in sediment varies, but shallow-water mud deposits tend to carry the largest numbers of non-selective deposit feeders. The total organic matter in such areas can be as low as 1% and most of the deposit feeders have to eat continuously. The processing time (from ingestion to passage into the rectum) for the sediment can be as little as 15 minutes in taxa such as *Arenicola*.

Selective deposit feeders

Some deposit feeders do not ingest sediment haphazardly, but use their palps to sort organic material from the sediment before ingestion. Selective deposit feeders live in tubes, though the method of sorting varies according to the types of palps that are present. In terebellids that live in vertical tubes the multiple palps are laid out over the sediment surface. Each palp has a longitudinal ciliated groove into which mucus is secreted. The palps are extended by 'creeping' on cilia, and detritus is selected or rejected. The selected particles stick to the mucus and are carried along the groove by ciliary action to the base of the palp. Each palp is wiped periodically on the lower lip, where cilia carry the mucus plus detritus into the mouth. Polychaetes with only a single pair of grooved palps, such as spionids, use a similar method of feeding, sweeping the palps over the sediment surface or waving them in the water column to gather food.

Filter feeders
A number of polychaete groups have the ability to collect particles suspended in the water column. All known filter-feeders live in some sort of tube. The best-known are the Sabellidae and Serpulidae (feather-duster worms) and the Chaetopteridae (parchment-tube worms). In sabellids and serpulids the radiolar crown is expanded out of the tube to form a funnel. Ciliary beating creates a water current through the funnel. Food particles are trapped and carried by cilia towards the base of the funnel, where the material is sorted. Small particles are swallowed while large ones are pushed away from the mouth and drop into the water. Some of these worms sort the particles into three sizes, with mid-sized particles being used in the construction of the tube.

Symbiosis and parasitism
The most famous example of symbiosis in the Annelida occurs in the family Siboglinidae. The mode of nutrition in siboglinids was the subject of debate for many years, since they were said to lack a gut. In fact the gut lumen, through which food passes in other animals, is completely blocked by the expanded gut epithelium. This endodermal tissue is filled with chemoautotrophic bacteria, forming a structure called the trophosome. Siboglinids live in areas where there is a high level of reduced-sulfur (H_2S) or methane in the water or sediment around their tubes.

There is now considerable evidence that most, if not all, of their nutritional requirements are derived from the bacteria in the trophosome, which can make up 15% of the animal's total body weight. Large species, such as those in the genus *Riftia*, live around deep-sea hydrothermal vents from which heated water, hydrogen sulfide and carbon dioxide are ejected. The bacteria in the trophosome appear to be able to oxidise the sulfide and reduce carbon dioxide to organic matter. The palps of siboglinids project from the tube and are richly supplied with blood vessels that serve the respiratory exchanges of the worm, but also transport the sulfide and carbon dioxide that are needed by the symbiotic bacteria. In the genus *Siboglinum* the trophosome contains methanotrophic bacteria, and the worms live in areas where methane seeps up through the sediment. The eggs of siboglinids do not contain bacteria, so how the bacteria enter the juveniles has been an issue of considerable interest. It is now known that a mouth and anus, as well as a lumen in the area of the future trophosome, are present during the early development of siboglinids. It appears that the bacteria probably enter through this transient digestive tract and that the gut wall then thickens, collapsing the lumen.

The best-known parasitic annelids are those leeches that are ectoparasitic blood-suckers of mammals (including humans). Apart from the Hirudinea, there are a number of polychaete groups that can be characterised as being symbiotic or even parasitic. Many of these have been given the taxonomic status of family, though research is still clarifying their relationships. One such family is the Histriobdellidae. These are small worms in the genera *Histriobdella* and *Stratiodrilus* that are found only in association with crustaceans. Several species of *Stratiodrilus* live on the gills of freshwater crayfish in Australian streams and feed on organisms such as diatoms that also live in the gill chambers. Histriobdellids are therefore commensals. Another family, also in the Eunicida, that is definitely parasitic is the Oenonidae, whose members live in the coelomic cavity of other polychaetes or echiurans and appear to derive their nutrition directly from their host.

Most of 150 or more described species of Myzostomida (Fig. 8.5b) are commensals living on crinoids, stealing food with their eversible proboscis from the feeding grooves of their hosts. Many are mobile and will roam around the host, but others remain sessile near a convenient 'feeding' site. Other myzostome species induce the host crinoid to form galls or cysts around them on the arms, the pinnules of the arms, or the oral disc. The cysts can be soft or calcified, stalked or spherical. A small number of myzostome species are endoparasitic, living either in the gut lumen, coelom or gonads of their host.

Nephridia and excretion

In most annelids there are usually two fluid systems, the coelom and the circulatory system, and both (if present) are involved in the excretion of waste. To achieve this there must be ducts that carry the wastes to the exterior. There has been considerable debate over the structure, function and evolution of the excretory system in annelids. Monumental studies by Goodrich on excretory systems in invertebrates, published more than 50 years ago, are still influential. Because of the diversity of their excretory organs, Goodrich focused much of his efforts on annelids. He showed that annelids have connections to the exterior from the coelom via segmental nephridia and gonoducts (also called coelomoducts).

Basically, nephridia are excretory organs and coelomoducts are gonoducts. Some polychaetes have protonephridia (Fig. 8.12a, d) with flame cells at the inner end, as in flatworms and nemerteans. More usually, the inner end of the nephridium is an open funnel or nephrostome. This type of nephridium is called a metanephridium (Fig. 8.12b, c, e, f). Coelomoducts also have an open funnel, the coelomostome, at the inner end. In a few types of polychaete (e.g. capitellids) and in the Clitellata, the two types of duct remain separate, but in many polychaetes they become fused together, in three different ways:

1 As protonephromixia, which combine a protonephridium with a coelomostome grafted onto the nephridial duct (Fig. 8.12a, d).
2 As metanephromixia, which combine a metanephridium with a coelomostome grafted onto the nephrostome (Fig. 8.12b).
3 As mixonephridia, which show complete fusion of a coelomostome with a metanephridial nephrostome (Fig. 8.12c).

The first two combinations occur only in the Phyllodocida, possibly the most ancient lineage of modern polychaetes. Most living polychaetes have mixonephridia.

In many polychaetes, nephridia that have both an excretory and genital function are present in most segments of the body, but sometimes there is a functional differentiation along the body length. Sabellids and serpulids have a single pair of excretory nephridia anteriorly and numerous pairs of genital nephridia (acting as gonoducts) posteriorly. Arenicolids and terebellids have five to seven pairs of nephridia. The more anterior pairs are excretory while the others act as gonoducts. In clitellates, excretory metanephridia occur in most segments of the body, while coelomoduct gonoducts are restricted to a few segments in the mid-body region.

Osmoregulation

Most polychaetes are osmoconformers, and this may explain why very few species live in fresh water. When exposed to low salinities, most polychaetes swell up in an uncontrolled manner. Certain intertidal and estuar-

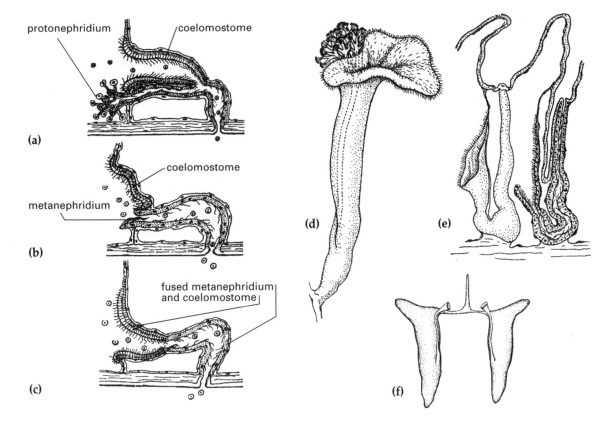

Figure 8.12
(a) Protonephromixium: coelomostome fused with protonephridium.
(b) Metanephromixium: coelomostome fused with metanephridium.
(c) Mixonephridium: coelomostome fused with nephridioduct.
(d) Protonephromixium of phyllodocid, showing coelomoduct fused with protonephridium. (e) Spionid coelomostome grafted onto nephridioduct: whole (left) and in section (right). (f) Anterior pair of nephridia of *Pomatoceros triqueter* (Serpulidae). The nephridia have a single external opening in all sabellid and serpulid polychaetes. (a–e after Goodrich 1945.)

ine forms are able to withstand short periods of stress because they are adapted to changing salinities. The few polychaetes that have been able to penetrate and live in fresh water belong to the families Nereididae and Sabellidae. These are almost certainly osmoregulators, although how they accomplish this is unknown. Because of their soft bodies, polychaetes are virtually incapable of resisting desiccation. There are very few terrestrial polychaetes, and these are found only in very damp environments.

Little is know about osmoregulation in marine clitellates, but earthworms are capable of osmoregulation. When exposed to fresh water they produce copious quantities of dilute urine and there is active resorption of salts by the nephridial walls. Earthworms can also withstand considerable desiccation. Leeches can withstand desiccation more than any other annelids, losing as much as 80% of their body weight. Leeches that live in fresh water are osmoregulators and excrete large amounts of very dilute urine.

Reproduction and development

Asexual reproduction in the Polychaeta

Asexual reproduction in polychaetes takes the form of division of the body and regeneration of the missing parts (schizotomy). Schizotomy has been further divided into two processes, paratomy and architomy. Paratomy is the formation of a recognisable complete individual which then separates from the 'parent' stock; architomy is simple fission or fragmentation of the body with no prior individualisation.

Sexual reproduction in the Polychaeta
Gonochorism and hermaphroditism
Gonochorism (separate sexes) is the most common sexual condition found in polychaetes, but hermaphroditism is widespread and may be simultaneous or sequential. Polychaete groups in which all members are simultaneous hermaphrodites include the myzostomids and the spirorbid serpulids.

Gametogenesis
Most polychaetes that have been studied lack permanent gonads. The origin and proliferation of gametes is poorly understood. Usually, gametogonia or gametocytes are liberated into the coelom from cells lining the peritoneum. Gametes then mature while floating in the coelomic fluid. In a number of polychaete families, however, the females have distinct ovaries and the males have definite testes. Most ovaries lie beneath the gut, on the peritoneum. The peritoneum can have cells in close contact with the developing oocytes, termed follicle cells, and may also be associated with blood vessels and/or nurse cells. When the oocytes are released into the coelom they may be solitary, or continue to be associated with nurse cells.

The testes of polychaetes also lie in the peritoneum, but usually contain only spermatogonia and stem cells. Fully differentiated sperm have not been found in testes. In nearly all polychaetes the sperm develop in the coelom in syncytial masses (rosettes, morulae or platelets). In most polychaetes, gametes are expelled from the body via nephridia, although in some species their release results from rupture of the body wall.

Fertilisation mechanisms
Many polychaetes shed eggs or sperm freely into the water, where fertilisation takes place. There is essentially no interaction between the sexes and no parental care. The rate of successful fertilisation can be very low in these cases. Other polychaetes increase fertilisation success through synchronous spawning. A common reproductive cycle in annelids is one based on an annual system involving the perception of changes in daylength. Other cues for synchronous spawning are external factors such as phases of the moon, tidal cycles and temperature.

Many benthic polychaetes swarm to the surface regions of the water and spawn. This widespread method is accomplished by epitoky (morphological modification that enables the animal to leave the bottom to reproduce). Epitokes can arise by two different processes. Epigamous epitokes are the result of the transformation of a whole individual into the epitoke (Fig. 8.13a). Examples occur in many families, including Glyceridae, Nephtyidae, Nereididae, Phyllodocidae and Syllidae. Schizogamous epitokes arise by modification and separation from the posterior end of the worm. Such epitokes occur in the Eunicidae and Syllidae.

The most famous example of epitoky occurs in the eunicid polychaete *Palola viridis* (the Palolo worm) that lives in burrows in coral rubble on reefs in the South Pacific Ocean. In October and November the worms are cued by the phase of the moon to release the posterior end of the body as epitokes which are filled with sperm or eggs. The rest of the worm lives to spawn again the following year. The epitokes have photoreceptors that allow them to detect moonlight, and they swim upwards to the surface. They then break up, releasing the sperm or eggs. Fertilisation takes place in the water and the resulting larvae swim for a few days before settling to the

Figure 8.13
(a) SEM of a syllid polychaete transformed into an epigamous epitoke. (b) Syllid with early stage embryos attached to the body surface. (c) Anterior end of the syllid shown in b. (d) SEM of a syllid with late-stage larvae attached to the body surface.

bottom and becoming small worms. The timing of spawning is accurate to within one day and people in the area can predict the spawning and have special festivals celebrating the event.

Brooding polychaetes
Some species of polychaetes tend to invest more energy (as yolk) in each egg. They have fewer, larger eggs which provide all the energy needs for the larvae to proceed to full development as a young juvenile. They also tend to care for or brood their eggs and often have a 100% survival rate to the juvenile stage which crawls away from the parent.

Many methods of brooding have evolved. The female may deposit eggs in a jelly mass which contains substances that induce nearby males to release sperm. In other species the presence of sperm in the water induces the female to deposit egg masses. Some of the polychaetes using this method are tube-dwellers that attach their jelly masses to the tube opening. Other tube-dwelling species lay their eggs in the space between the body and the tube, either freely or in capsules. In these animals, the females usually have complex sperm storage mechanisms from which sperm are released at the time of oviposition. This is a common reproductive strategy in sabellids and terebellids. The sabellid genus *Caobangia* shows a more extreme condition of ovoviviparity, in which the eggs are fertilised and brooded inside the body of the female.

Syllids and some other errant polychaete families include species in which the eggs are attached by the female to her body (Fig. 8.13b–d). This is often accompanied by an elaborate courtship ritual in which the male swims around the egg-carrying female and envelops her in sperm. A few species in these families are ovoviviparous, retaining and brooding their embryos internally after fertilisation.

Sperm transfer mechanisms
Many male polychaetes release sperm freely into the water, but others package their sperm, either in spermatophores or as spermatozeugmata. Spermatophores are bundles of sperm surrounded by a sheath or capsule. They may float in the water and be gathered by the female, or may be transferred to the female during pseudocopulation or copulation. Spermatophores are usually stored by the female in spermathecae, as in spionids, but may become attached directly to the epidermis, as in myzostomids. In the latter case the sperm penetrate into the female body through the epidermis

and fertilisation is internal (hypodermic impregnation). Spermatozeugmata (bundles of sperm that lack an external covering) are produced by the males of some species in the families Arenicolidae, Maldanidae, Syllidae and Terebellidae.

The transfer of sperm, spermatophores or spermatozeugmata from male to female takes place in three main ways:

1. Without contact. Sperm or spermatophores released into the water are detected and gathered by the female. Many sabellids and serpulids concentrate the gathered sperm into spermathecae. Several spionid species collect floating spermatophores and transfer these to spermathecae. Males of the syllid *Autolytus prolifera* release sperm in mucus balls into the water during 'nuptial dances'. The sperm balls stick to the body of the female.

2. During pseudocopulation. Sperm, spermatophores or spermatozeugmata are passed from male to female while the two individuals are in close contact. The male of the terebellid *Nicolea zostericola* releases spermatozeugmata through elongate nephridial papillae, to be gathered by the female feeeding tentacles. A jelly mass is then spawned. Dorvilleids of the genus *Ophryotrocha* deposit eggs and sperm into a cocoon during pseudocopulation. Polynoids (scale worms) also pair in a similar way.

3. By means of an intromittent organ. The males of many polychaetes have penis-like structures. In some species the penis may be inserted into a female receptacle in a process of copulation. An extreme case is *Saccocirrus erotica*, in which the male has up to 100 penes arranged segmentally along one side of the body, and the female has a similar number of spermathecae. In other taxa the female has no sperm receptacle and is inseminated directly, by hypodermic impregnation. *Stratiodrilus novaehollandiae*, for example, inserts a chitinised penis at random into the female's body.

Larval development

There is a great range of larval forms and development within the Polychaeta. Polychaetes are usually said to have a basic bentho-pelagic life cycle and a feeding larvae termed a trochophore. This generalisation is being questioned, but until a detailed phylogeny of the Annelida is available the issue cannot really be resolved. Many polychaete families exhibit three or four different ways of reproducing, from free spawning to jelly masses to intratubular brooding. A study of the family Sabellidae has shown that intratubular brooding of a few larvae is the primitive mode of reproduction in this family and that broadcast spawning of eggs and sperm has evolved secondarily.

The early embryonic development of all polychaetes is essentially similar. Though polychaete eggs can range in size from less 50 μm in diameter to more than 1 mm, they all show a sequence of spiral cleavage to the 64-cell stage. The division may be equal or unequal, depending on the amount of yolk in the eggs. The germ layers are formed from a few well-defined cells in the blastula, including mesoderm from the descendants of the 4d cell. Eggs with small amounts of yolk tend to develop into larval forms that have to feed in order to complete development (planktotrophic larvae; Fig. 8.14a). Embryos that have enough yolk to complete a large part of their development without feeding are called lecithotrophic larvae (Fig. 8.14b),

THE ANNELIDA

and those that can develop completely into young worms are called direct developers.

Some of the most obvious features of polychaete larvae are the ciliary rings that are used for locomotion and feeding. The prototroch is the anterior-most transverse ring and lies in front of the mouth; the metatroch is a transverse ring behind the mouth; and the telotroch is a posterior ring encircling the pygidium. A longitudinal band along the ventral part of the larvae is also common and is called a neurotroch. These bands are not always present, and some larvae may only show the prototroch. Direct developers that are brooded by the parent may not have any ciliary bands at all. Many polychaetes also have an apical tuft of cilia at the anterior end of the larva (Fig. 8.14a).

The larval type called a trochophore occurs in annelids and several other marine invertebrate groups, including sipunculids, echiurids and molluscs. This larval form has a prototroch, apical tuft and some of the bands listed above. Some zoologists define the trochophore in a more restricted way, as a larva with all of the ciliary bands listed above, as well as a functional gut and a pair of protonephridia. In the Annelida this form of larva is relatively rare, being found in only a few families such as the Serpulidae (Fig. 8.14a) and Polygordiidae. Many of the families in the Phyllodocida have a larva that closely resembles the strictly defined trochophore, but it lacks a metatroch and appears to feed in a different way. This phyllodocid larval type may have evolved convergently with the true trochophore. Most polychaete families have lecithotrophic or direct-developing larvae, which may be the primitive condition for the group. Further detailed studies on many polychaetes are needed to assess the evolution of larvae in the Annelida.

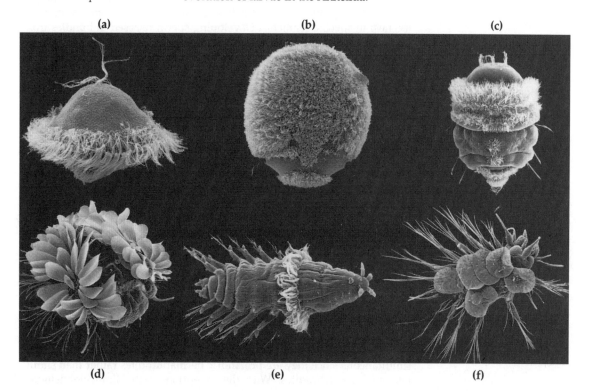

Figure 8.14
(a) Trochophore larvae of *Spirobranchus giganteus* (Serpulidae). (b) Lecithotrophic larvae of *Marphysa* (Eunicidae) with one pair of chaetae. (c) Lecithotrophic larvae of *Filograna* (Serpulidae) with three segments formed. (d) Late-stage larva of a chrysopetalid polychaete, about to settle from the plankton. (e) Late-stage larva of a glycerid polychaete, about to settle from the plankton. (f) Late-stage larva of a polynoid polychaete, about to settle from the plankton.

Life cycles

The traditional view of the polychaete life cycle is that they have external fertilisation and planktotrophic larvae, but of the 306 species for which the life cycle is known only 79 exhibit this form of reproduction. A further 44 species show external fertilisation and lecithotrophic or direct developing larvae. The remaining 183 polychaetes have some form of brooding. Of course, this is by no means an indication of the real proportions of the various reproductive modes among polychaetes; it only reflects our present state of knowledge. The life cycles of most species are still unknown.

It is difficult to develop a system of classification that encompasses the diversity of polychaete life cycles. The more generally used classification splits life cycles into two broad divisions, semelparous or iteroparous. Semelparity refers to breeding only occurring once per lifetime and occurs in many Nereididae. Iteroparity, where breeding occurs several times in a lifetime, can be divided into annual iteroparity (breeding yearly as in the Palolo worm) and continuous iteroparity (breeding taking place over an extended breeding season, as in many brooding polychaetes). The latter condition can in some cases be such that breeding occurs throughout the lifetime after maturity.

Sexual reproduction in the Clitellata

All clitellates are simultaneous hermaphrodites and have complex mechanisms and behaviours for exchanging sperm between individuals. The sperm that are obtained from another worm are usually stored in spermathecae. Unlike most polychaetes, clitellates have distinct reproductive organs restricted to a few anterior segments. The structure of the reproductive system is a feature in the classification of the group. The male component of the reproductive system has one to four pairs of testes (more in leeches) in specific segments. Developing sperm released in bundles from the testes travel to paired sacs called seminal vesicles, complete their development, and are stored. During copulation the sperm travel to the male pores, which may be elaborated to form penes. In all clitellates except the Euhirudinea, the ovaries are posterior to the testes. The female system is simple and consists of paired ovaries and small paired funnels which generally lead into egg sacs. The other main female reproductive structures are the paired spermathecae that open ventro-laterally in the region of the gonads.

Sperm exchange in earthworms occurs when two worms lie head to tail, often held together by special copulatory chaetae. The worms are positioned such that the openings from the seminal vesicles lie over the entrances to the spermathecae of the partner. Where penes are present they can be inserted directly into the spermathecae to deposit sperm. In the Lumbricidae the sperm are deposited on the outer surface of the partner and are then transported to the spermathecae along sperm grooves. After copulation the worms separate and egg-laying begins. The clitellum produces a cocoon that contains albumen and, in the case of terrestrial clitellates, resists desiccation. The cocoon moves forward over the exit ducts from the ovaries and receives a number of eggs (sometimes only one), then continues over the spermathecal openings, where sperm are added. Fertilisation and development occur in the cocoon.

In leeches sperm exchange also takes place, but in some groups it is not simultaneous since they are protandric hermaphrodites rather than simultaneous hermaphrodites. With the exception of the Branchiobdellidae, leeches have lost spermathecae. Instead, in many of the Arhynchobdellida, the opening of the ducts leading from the ovaries acts as a vagina, and a penis is present. In the Rhynchobdellida neither penis nor vagina are present and spermatophores are used. The spermatophores are placed on the external surface of the partner and contain enzymes that dissolve the epidermis, allowing the sperm to enter the body (hypodermic impregnation). The spermatophores are either deposited in a special region on the female that is underlain by vector tissue that transfers the sperm to the ovaries, or anywhere on the body surface. In the latter case the sperm migrate to the ovaries via the coelomic remnants. In all leeches fertilisation is internal, though they place the embryos into cocoons as in other clitellates.

Development in clitellates proceeds from yolky eggs or from secondarily small eggs, with little yolk, that draw on nutrients in the cocoon (as in earthworms), and is always direct. The cleavage patterns and the subsequent development of the embryos are modifications of those found in polychaetes.

Fossil annelids

Annelids lack most of the persistent structures present in organisms with an extensive fossil record. Polychaete jaws (scolecodonts) are common in deposits from certain periods, and various kinds of tube and burrow structures have been referred to polychaetes, but are far less diagnostic than chaetae or jaws. Full-body fossils are rare, but those present in the Middle Cambrian Burgess Shales of Canada and Mazon Creek beds of Pennsylvania have demonstrated that polychaetes were well represented in early Palaeozoic seas.

Wiwaxia has been interpreted as a chrysopetalid-like polychaete; the position of this genus has yet to be fully clarified. *Canadia* and *Burgessochaeta* are probable polychaetes from the Middle Cambrian in Canada. Both have anterior ends with a prostomium-like structure and have many apparently similar segments. They also have obvious tufts of chaetae along the body. Both genera are similar to the Phyllodocida. Representatives of recent families reported from the Devonian period, including Aphroditidae, Phyllodocidae, Hesionidae, Nephtyidae and Goniadidae, are also Phyllodocida. While we know that annelids are an ancient group, there is little evidence in the fossil record to aid in sorting out evolutionary relationships.

Classification of the phylum Annelida

Morphological evidence from living species has been used by Rouse and Fauchald (1997) to construct the cladogram of the extant Annelida shown in Figure 8.15. This tree clearly shows the Siboglinidae (formerly the phylum Pogonophora) well inside the Polychaeta. Further studies involving molecular sequence data may result in taxa such as the Clitellata and Echiura being moved relative to the Polychaeta.

The classification of the major families of the Polychaeta is based on the cladogram of Figure 8.15. Certain families have been added on the basis of other evidence. No Linnean categories above the family, such as superfamilies or orders, were used by Rouse and Fauchald (1997) in their classification, as the meaning of these categories is not clear in relation to their new results.

The classification of the Clitellata sets out some of the major groups. The Hirudinea are included, but their relationship to other clitellates remains unclear and no Linnean catgories are specified for them.

CLASS POLYCHAETA
Annelids with numerous chaetae on the trunk segments.
Polychaeta Scolecida
Parapodia with similar rami; more than two pairs of pygidial cirri. (Families Arenicolidae, Capitellidae, Maldanidae, Opheliidae, Orbiniidae, Paraonidae, Scalibregmatidae.)
Polychaeta Palpata Aciculata
Polychaetes with sensory palps; parapodia supported by acicula.
Eunicida — with muscularised ventral pharynx and usually with jaws. (Families Amphinomidae, Dorvilleidae, Eunicidae, Histriobdellidae, Lumbrinereidae, Onuphidae.)
Phyllodocida — with muscularised axial pharynx, often with jaws. (Families Alciopidae, Aphroditidae, Chrysopetalidae, Glyceridae, Hesionidae, Ichthyotomidae, Myzostomidae (formerly the class Myzostomaria), Nephtyidae, Nereididae, Phyllodocidae, Polynoidae, Sigalionidae, Syllidae, Tomopteridae.)
Polychaeta Palpata Canaliculata
Polychaetes with grooved palps.
Sabellida — palps forming a radiolar crown. (Families Oweniidae, Siboglinidae (formerly the phylum Pogonophora), Sabellidae, Sabellariidae, Serpulidae.)
Spionida — with a single pair of grooved palps. (Families Chaetopteridae, Magelonidae, Spionidae.)
Terebellida — with paired or multiple grooved palps and a gular membrane. (Families Ampharetidae, Cirratulidae, Ctenodrilidae, Pectinariidae, Sternaspidae, Terebellidae, Trichobranchidae.)

CLASS CLITELLATA
Annelids with few or no chaetae on the trunk segments; hermaphrodite, with a clitellum.
Subclass Tubificata
One pair of testes followed by a pair of ovaries; male pore anterior to female pore.
Order Tubificida (Families Tubificidae, Naididae, Phreodrilidae, Enchytraeidae; mostly freshwater and marine, a few enchytraeids terrestrial).
Subclass Lumbriculata
Multigonadial; male pores anterior to female pores; nephridia connected intersegmentally.
Order Lumbriculida (Family Lumbriculidae, freshwater).
Subclass Diplotesticulata
With 8 gonads, basically 2 pairs of testes and 2 pairs of ovaries.
Superorder Haplotaxida — Posterior ovaries reduced (Family Haplotaxidae, freshwater and semiterrestrial).
Superorder Metagynophora — Anterior ovaries lost (Families include Lumbricidae and Megascolecidae, terrestrial earthworms).

HIRUDINEA
Leeches. Clitellate annelids with a fixed number of 34 segments, a posterior sucker and usually an anterior sucker.

Figure 8.15
Cladogram summarising the current classification of the Annelida, with special reference to the Polychaeta. Note that the phylum Pogonophora is shown here as a family, the Siboglinidae, in the polychaete group Sabellida (after Rouse and Fauchald 1997).

Branchiobdellida
Small leeches ectoparasitic on freshwater crayfish; chaetae absent; anterior and posterior suckers; spacious coelom internally (e.g. *Stephanodrilus*).

Acanthobdellida
Small leeches ectoparasitic on salmonid fishes; chaetae on anterior segments; posterior sucker only; coelom partially retained (e.g. *Acathobdella*).

Euhirudinea
Marine, freshwater and terrestrial leeches; chaetae absent; anterior and posterior suckers; coelom reduced. Two subgroups: Rhynchobdellida — marine and freshwater, with an eversible pharynx (e.g. *Glossiphonia*); Arhynchobdellida — freshwater and terrestrial, with a non-eversible pharynx (e.g. *Hirudo*).

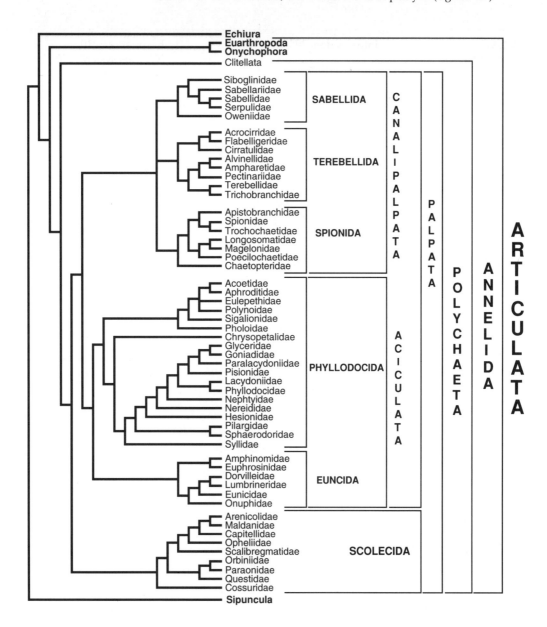

Chapter 9

The Onychophora and Tardigrada

N.N.Tait

PHYLUM ONYCHOPHORA

Introduction 205

External morphology 205
   General body form 205
   Appendages 205
   Body surface and water balance 206

Internal organs and physiology 206
Haemocoel and circulation 206
   Muscle layers 207
   Limb musculature 207
   Locomotion 208
   Slime glands and prey capture 208
   Digestion 208
   Excretion 209

Respiration 210
Nervous system 210
Sense organs 210

Reproduction 211
   Pheromones 212
   Spermatophores and insemination 212
   Female reproductive strategies 213
   Development 213

Evolutionary history 215

Classification of the phylum Onychophora 215

PHYLUM TARDIGRADA

Introduction 216

External morphology 216
   General body form and integument 216

Internal organs and physiology 217

Reproduction 219

Resistant stages 219

Development 220

Phylogenetic relationships 220

Classification of the phylum Tardigrada 221

Phylum Onychophora

Bilaterally symmetrical, metamerically segmented coelomates with reduced coelom; haemocoelic body cavity; soft, weakly sclerotised cuticle, moulted periodically; numerous pairs of lobopod limbs; head with three cephalised segments; brain and paired ventral nerve cords; segmental organs with coelomic end sacs; dorsal coelomic gonads; separate sexes; direct development.

Introduction

Onychophorans are commonly called peripatus (in reference to the first described genus) or velvet worms (because of the texture of their integument). This group of animals is of special significance to evolutionary studies because its members display characteristics also found in annelids and arthropods. The present-day fauna of less than 200 species of onychophorans is disjunctly distributed on land masses derived from the break-up of the Gondwanan supercontinent: Africa, Central and South America, South-East Asia and Australasia.

All extant onychophorans are restricted to the moist terrestrial microhabitats of rotting logs, leaf litter and soil. They are not confined to tropical rain forests, being also found in wet temperate forests, drier open woodlands, and grasslands created by natural or human-induced deforestation. In grasslands, onychophorans survive in crevices in the soil, coming to the surface under the protection of rocks and debris.

Figure 9.1
The peripatopsid *Euperipatoides rowelli*, showing the rhythmic movement of the legs (courtesy J. Norman).

External morphology

General body form

Onychophorans display a uniform body form in which the head merges into an elongate, sub-cylindrical, flexible trunk adapted to life within the narrow crevices of decaying vegetation and soil (Fig. 9.1). They range in length from 15 to 150 mm in different species. Length is difficult to measure, as the body is capable of considerable elongation and contraction, facilitated by the numerous folds which annulate the integument of each segment of the trunk. Males tend to be smaller than females. Some species are pigmented a deep blue-black, shades of brown from fawn to rust, or pure white. Others are overlaid with subtle geometric patterns of colour such as stripes or segmentally arranged diamonds.

Appendages

The head bears three pairs of segmental appendages (Fig. 9.2a). The elongate, cylindrical antennae are annulated and flexible, providing constant

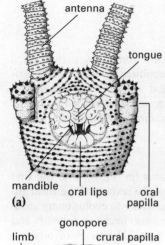

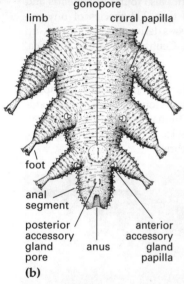

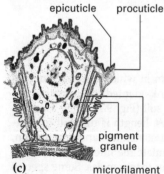

Figure 9.2
Anatomy of onychophorans.
(a) Ventral view of head.
(b) Ventral view of posterior end of male. (c) Schematic view of integument (from Storch and Ruhberg 1993).

surveillance of the environment as the animal moves forward. The tips of the mandibles protrude from the ventral, oval-shaped mouth. Each mandible ends in a pair of sickle-shaped blades, homologous with the paired, curved claws at the tip of each walking limb. There is a short, turret-like oral papilla on each side of the head, and these papillae eject threads of sticky slime for prey capture or defence. The trunk has 13 to 43 pairs of ventro-laterally directed, conical, ambulatory limbs. The number of leg-pairs can vary within a species, and in some species males have fewer legs than females. Each limb terminates in a flexible foot bearing a pair of curved claws (Figs 9.2b, 9.5b). The ventral surface of the limb has a number of crescent-shaped, spinose plantar pads which, together with the claws, provide traction against the substratum during locomotion. The limbs are superficially annulated, like the body, and perform extensive contractions and elongations. The trunk terminates in a conical to spherical anal segment, often erroneously referred to as a pygidium (Fig. 9.2b).

Body surface and water balance

The entire body surface is covered with a thin cuticle, 2 to 3 μm thick (Fig. 9.2c). Biochemical and ultrastructural investigations of the cuticle have distinguished an outer epicuticle and an inner chitin-protein composite procuticle. The procuticle may be sclerotised to produce an outer exocuticle, as in the mandibles and claws. The lack of a surface layer of wax, together with the numerous unclosable spiracular openings to the tracheal respiratory system, predisposes onychophorans to water loss. Onychophorans are negatively phototaxic animals which forage at night. Their nocturnal activity rhythm is maintained in the laboratory even in continuous, total darkness. Body water is replenished by drinking. Some African species are able to absorb droplets of water through eversible sacs at the base of the legs, called coxal organs. Paradoxically, the body surface is highly water-repellent, which may be of importance in preventing drowning in very wet conditions.

The cuticle is secreted by a single layer of cuboidal epithelium, possessing numerous microfilaments which connect the cuticle to a thick underlying dermis of collagenous fibres (Fig. 9.2c). The cuticle is regularly moulted throughout life by splitting down the mid-dorsal line. The exuviae are eaten after release from the posterior end of the body.

Internal organs and physiology

Haemocoel and circulation

The body cavity of onychophorans consists of a haemocoel subdivided longitudinally by partitions into a dorsal pericardial sinus, a central perivisceral sinus and a pair of lateral sinuses confluent with the cavities of the paired limbs (Fig. 9.3). A tubular heart with segmentally arranged ostia extends within the pericardium for much of the length of the body. Communication between the sinuses is provided by other ostia in the partitions. The ostia allow unidirectional flow of haemolymph from the perivisceral to the pericardial sinus, and exchange in both directions between the perivisceral and lateral sinuses. A system of haemal ring canals lie within the plical folds of the body and limbs as spaces between the outer circular and inner oblique muscle. These connect the lateral sinuses with the pericardial sinus via ostia in the dorsal wall of the pericardium. Haemolymph is circulated by contractions of the heart, driving the fluid forward into the perivisceral sinus. The haemolymph lacks respiratory pigments, but contains circulating haemocytes that are presumably responsible for the removal of foreign and waste material. The coelomic spaces of onychophorans are confined to the cavity of the gonads and the end sacs of segmentally arranged excretory organs (see below).

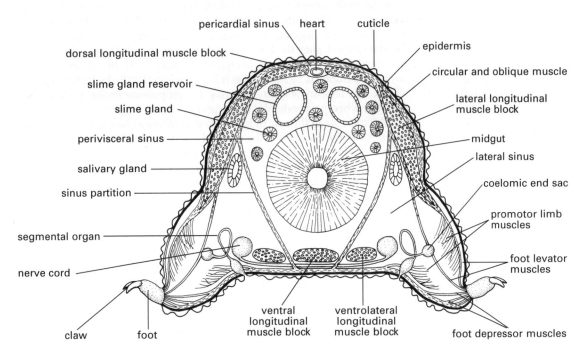

Figure 9.3
Transverse section of an onychophoran.

Muscle layers

The muscle layers of the body wall consist of an outer circular layer, a middle oblique layer, and an inner longitudinal layer arranged as muscle blocks (Fig. 9.3). These muscles cause elongation, contraction and bending of the body and are used to maintain a high hydrostatic pressure within the haemolymph. The haemocoel acts as a hydrostatic skeleton which, together with the dense collagenous dermal layer, provides the rigidity needed to support the limb muscles.

Limb musculature

The limbs have extrinsic muscles which provide the forward (promotor) and backward (remotor) swings of the limbs used in stepping movements. Intrinsic limb muscles extend and contract the limbs (Fig. 9.3). The promotor and remotor muscles are attached to the anterior and posterior internal surfaces of the limbs, and are inserted ventrally and dorso-laterally on the body wall. Other body-wall muscles also pass imperceptibly into the limbs, where their antagonistic action against the haemocoelic fluid assists the elongation and contraction of the limbs. Muscles attached to the lateral and median surfaces of the leg base fan out and cross the lumen to insert on the ventral and dorso-lateral body wall respectively. These muscles regulate the size of the leg base. The movements of the foot and its associated claws are controlled by intrinsic levator and depressor muscles inserted on the upper and lower walls of the foot respectively (Fig. 9.3).

Locomotion

Despite being soft-bodied animals, onychophorans show no undulations, or peristaltic contractions, during locomotion. The head end is cast alternately to the left and right, even when the animal is progressing in a straight line. The antennae are held forward above the ground but occasionally make contact with the substratum and objects, from which they gently recoil. The body is raised above the ground by the ventro-laterally placed legs. The legs move in metachronal rhythm, each leg being placed on the ground after the leg in front has been raised. During the forward recovery stroke, each leg is elongated. Contact with the ground is made with the plantar pads. Sometimes it is also gripped with the terminal paired claws. The legs are shortened during the propulsive stroke, to be at their shortest when perpendicular to the body.

An increase in the rate of locomotion involves a change in gait by altering the relative duration of the forward and backward strokes of the limbs. When moving from a stationary position, there is a rapid forward recovery stroke of the limbs followed by a slower backward propulsive stroke. This is progressively reversed at higher speeds. Hence, in 'low gear' there are more legs on the ground at any one time than when in 'high gear'. An increase in the rate of locomotion is also accompanied by a progressive increase in the length of the body. While there is a strong coordination of the phase difference between the legs down each side of the body, the paired limbs change from in-phase to alternate-phase as the animal accelerates (Fig. 9.1). Onychophorans are also capable of coordinated backward movement when moving in crevices.

Slime glands and prey capture

Onychophorans are carnivorous, feeding on other small invertebrates, particularly arthropods. Their method of prey capture involves the ejection of sticky threads of high elasticity and tensile strength from the oral papillae on either side of the head (Fig. 9.2a). The prey is immobilised on contact with the beads of sticky slime regularly spaced along these threads. Although the glue is ingested along with the prey, this method of capture represents a considerable energy loss if the prey is too small to compensate for the hunting effort or so large that it escapes. Hence, prey assessment is a critical aspect of feeding behaviour. The slime glands occupy much of the perivisceral haemocoel of the posterior two-thirds of the body (Figs 9.3, 9.4). They consist of a pair of tubes from which numerous side branches arise. The tubes and branches are all lined with secretory columnar epithelium. Each gland opens anteriorly into a broad reservoir lined with flattened epithelium, surrounded by muscles responsible for the forceful ejection of the slime from the aperture at the tip of the oral papilla. Chemical analyses of the slime indicate that it is a composite of water, protein, sugar, lipid, and a surfactant (nonylphenol) which may maintain the fluidity of the stored secretion. The sugar component, N-acetyl galactosamine, is associated with high-molecular-weight glycoproteins which may be responsible for the adhesive qualities of the threads.

Digestion

Once immobilised, the prey is torn open by a slashing action of the jaws. Mucus and carbohydrate- and protein-digesting enzymes from the paired salivary glands (Figs 9.3, 9.4) are injected, and the partially digested flesh is sucked back into the mouth; much of the exoskeleton of an arthropod prey

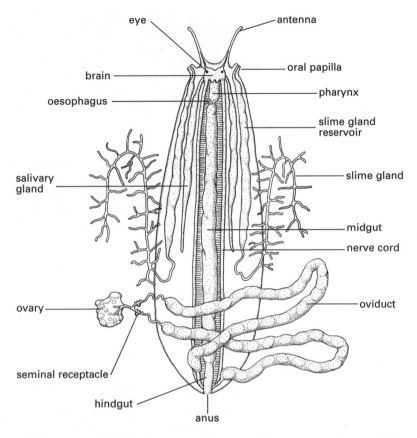

Figure 9.4
Internal anatomy of a female onychophoran.

is left undigested. The salivary glands are modified segmental organs. The alimentary canal, a broad, straight tube extending from the mouth to the terminal anus, is divided into three regions (Fig. 9.4). The foregut, composed of a buccal cavity, pharynx and oesophagus, is lined with cuticle. The midgut extends for much of the length of the animal and is lined with glandular and absorptive columnar epithelium. A cuticle-lined hindgut or rectum completes the gut. A chitinous peritrophic membrane is secreted by the epithelial cells throughout the midgut. The peritrophic membrane is periodically sloughed off and replaced.

Carbohydrate-, lipid- and protein-digesting enzymes, which have been identified in the midgut, enable food to be digested rapidly. The intestinal epithelium is also a site of storage of nutrients. Onychophorans are able to survive long periods without food.

Excretion

A major feature of onychophorans is the segmental excretory organs, one pair of which occur in each leg-bearing segment except for the segment bearing the genital opening (Fig. 9.3). The organs open at indistinct pores at the base of the legs, except for leg pairs 4 and 5 where they open more distally on distinct papillae (Fig. 9.5b). The excretory organs display considerable variation in form and ultrastructure in the different leg-pairs, possibly indicating different functions. They lie in the lateral sinuses, where a coelomic end sac leads into a collecting tubule via a ciliated nephrostome. The nephrostome provides the driving force for the ultra-filtration of haemocoelic fluid between interdigitating cells (podocytes) making up the wall of the end sac. The excretory organs may be primarily responsible for maintaining water and ionic balance. The accumulation of large quantities of uric acid crystals within the peritrophic membrane indicates that the midgut is a major site for the accumulation and release of nitrogenous wastes. Nephrocytes, present in the pericardial and lateral sinuses, accumulate particles injected into the body cavity and hence could also be involved in the removal of waste materials.

Respiration

Onychophorans possess a tracheal respiratory system most probably evolved independently of those of terrestrial arthropods. Microscopic, un-closable spiracles are scattered over the general body surface. Each spiracle leads into a short tracheal trunk that passes through the body wall and branches to form a tuft of tracheal tubules which supply the body organs.

Nervous system

The central nervous system is basically similar to that of annelids and arthropods. A large bilobed cerebral ganglion, forming the brain, lies dorsally above the pharynx and is connected via a pair of circumpharyngeal commissures to the paired ventral nerve cords (Fig. 9.4). Nerves from each side of the brain pass anteriorly to the antennae, latero-dorsally to the optic ganglia, and posteriorly as a pair to each jaw. The nerves to the oral papillae arise from the junction of the circum-pharyngeal commissures with the nerve cords. The widely spaced nerve cords, lying ventrally in the lateral sinuses (Fig. 9.3), are linked by numerous transverse connections and give off lateral nerves to the body wall in each segment. Arising from the indistinctly demarcated segmental ganglia, a pair of large pedal nerves innervates each limb. A pair of giant nerve fibres in each nerve cord mediates the rapid recoil reaction that follows strong stimulation.

While neurosecretory cells have been identified in the brain, nerve cords and pedal nerves, no information is available on the role of neurosecretions or hormones in the regulation of physiological and developmental processes in onychophorans.

Sense organs

The surface of the body is covered with microscopic dermal papillae (Fig. 9.5a). Each papilla is decorated with rows of sculptured cuticular scales, each underlain by an epidermal cell. Many of the papillae support a terminal, sensory bristle. Papillae of similar form are also abundant on the antennae and legs. Ultrastructural studies indicate that these bristles are involved in mechanoreception. Chemosensory sensilla have also been identified on the antennae and the lips surrounding the mouth. The eyes, at the base of the antennae, are each visible externally as a smooth, glistening hemisphere, the cornea, formed by the cuticle. Within each eye, a hemispherical lens is surrounded by a retinal cup composed of pigment cells interspersed with photoreceptive cells, each with a distal rod and proximal axon. The axons converge to form an optic ganglion, from which the optic nerve leads to the brain. Although the eyes may be able to form an image, the dark microhabitats and nocturnal activity patterns, together with the position of the eyes on the head, probably preclude their use for visual perception. Their sole function may be that of light avoidance.

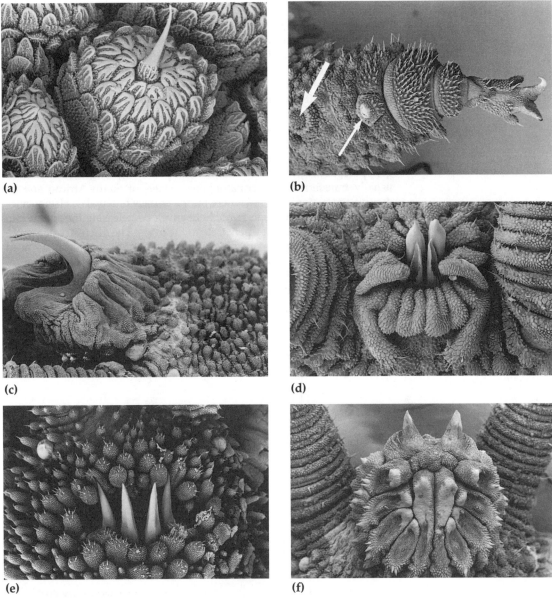

Figure 9.5
Scanning electron microscope (SEM) photograph of external features of onychophorans.
(a) Dermal papilla. **(b)** Ventral view of fourth leg; small arrow indicates position of opening to segmental organ; large arrow indicates crural papilla.
(c–f) Head structures of male onychophorans: (c) *Ruhbergia bifalcata*. (d) *Regimitra quadricaula*. (e) *Planipapillus taylori*. (f) *Ruhbergia brevicorna*.

Reproduction

While body form and feeding strategy are conservative in onychophorans, their reproductive biology is not so constrained. Male, and especially female, reproductive strategies are as diverse as in any other phylum of animals. Most species display a sex ratio biased towards females. This bias may be due to a differential survival of the sexes, since an equal sex ratio occurs amongst embryos before birth. Precocious sexual maturity has been reported in males of some species, which may indicate that males have a shorter life-span than females. One species of onychophoran is parthenogenetic.

Pheromones

Considering the often sparse distribution of onychophorans and the nature of their environment, the question arises as to how the sexes locate each other for mating. In males, crural glands, opening on the ventral surface of the legs as eversible crural papillae, have been shown to produce a female-attracting pheromone (Figs 9.2b, 9.5b). The chemical composition and species-specificity of this pheromone have yet to be determined.

Spermatophores and insemination

The male reproductive system consists of paired testes, seminal vesicles and sperm ducts leading to an unpaired ejaculatory duct that opens at the gonopore, between the last or second last pair of legs (Fig. 9.6). Sperm are usually transferred in spermatophores. Males of South African species deposit spermatophores anywhere on the surface of females. Haemocytes digest the skin and the spermatophore envelope, releasing the sperm into the haemocoel, where they penetrate to the ovary. The discovery of a number of Australian species of onychophorans with distinctive male head structures, involved in sperm transfer, adds to the diversity of male reproductive strategies (Fig. 9.5c–f). A male of one species has been observed delivering a spermatophore from its everted head structure directly onto the posterior female gonopore.

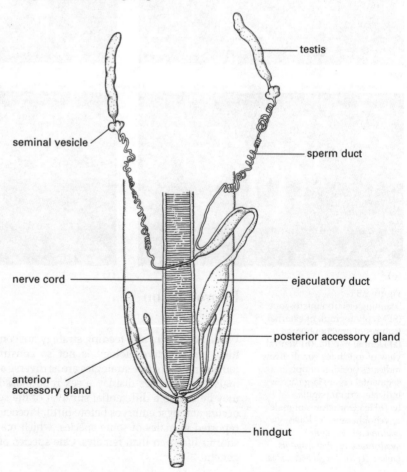

Figure 9.6
Male reproductive system and accessory glands of an onychophoran.

Males of many species possess accessory glands in the posterior region of the body (Figs 9.2b, 9.6). The anterior accessory and posterior accessory (anal) glands open at various sites on the genital and post-genital segments, according to species. They extend into the perivisceral haemocoel as blind sacs and are of variable shape and size in different species. The functions of these glands are unknown.

Female reproductive strategies

The paired ovaries are fused and attached to the dorsal wall of the posterior part of the perivisceral haemocoel (Fig. 9.4). In many species, the paired oviducts each carry a spherical seminal receptacle which functions in sperm storage. The oviducts, sometimes referred to as uteri in ovoviviparous and viviparous forms, loop forward and back before uniting as a short vagina which opens at the gonopore. Australasian species are oviparous, with yolky shelled eggs, or ovoviparous with yolky membrane-enclosed eggs. All others have a more modified gestation, with little or no yolk in the eggs, involving some form of maternal nutritional support for the embryos. A condition of placental viviparity has evolved in Neotropical species, showing extraordinary parallels to that of mammals (Fig. 9.7). Each fertilised egg becomes enclosed in an epithelial sac formed by cells of the oviduct lining. A stalk attaches the embryo to a placental zone, which is a specialised region of the epithelial sac lining the uterus wall. Segment formation and the differentiation of organ systems in the embryo are completed while in the epithelial sac, from which it eventually escapes. The embryo then moves into the distal region of the uterus, vacated by the birth of an older embryo, where growth continues until the juvenile is ready for birth. Embryos attached by placentae are moved by a conveyor belt system maintained by the growth of the proximal ends of the oviducts and the resorption of the distal ends.

Figure 9.7
Reconstruction of the oviduct of *Epiperipatus trinidadensis*; see text for details (from Anderson and Manton 1972).

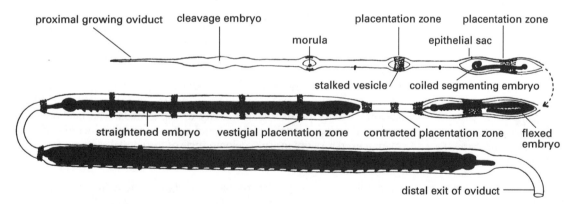

Development

Eggs of onychophorans vary from almost 2 mm to 0.04 mm in length, depending on the amount of yolk. The eggs of ovoviviparous species are contained in an inner vitelline and outer chorionic membrane, which forms a hard, sculptured shell in oviparous species. These membranes are absent in placental species. Despite differences in the volume of yolk, the majority of onychophoran embryos display a uniform pattern of early development. Following cleavage in yolky-egged forms, a blastoderm of small cells is

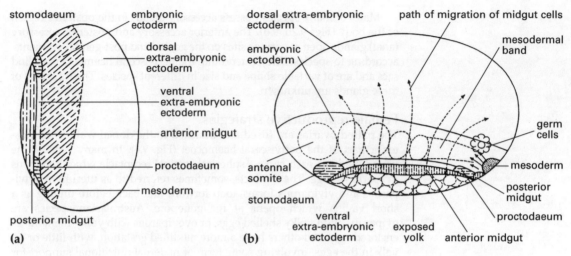

Figure 9.8
Early embryonic development in onychophorans. (a) Fate map of blastoderm. (b) Gastrulation movements in a yolky egg (see text for details). (From Anderson 1973.)

formed, with a characteristic distribution of presumptive areas (Fig. 9.8a). The presumptive anterior midgut splits along the mid-ventral line and invaginates (Fig. 9.8b).

From this band, cells proliferate to form a diffuse epithelium that envelops the yolk. The ventral split closes and the epithelium rounds up to form a continuous tube of midgut. This links up with the posterior midgut, proctodaeum and stomodaeum to form the completed alimentary canal. Cells of the small, posterior patch of presumptive mesoderm sink slightly below the surface and proliferate forward as two mesodermal bands, on either side of the gut. During this proliferation the bands begin to segment into an anteroposterior sequence of paired somites, each with a coelomic cavity. The first three pairs form the somites of the antennae, jaws and oral papillae, and are incorporated into the head. The remaining pairs form the somites of the trunk segments. The cavity within each trunk somite (Fig. 9.9) become subdivided into three compartments, a dorso-lateral coelom, a medio-ventral coelom, and an appendicular coelom (Fig. 9.9). The appendicular cavity pushes out into the limb bud and its mesoderm becomes the musculature of the body wall and limbs. The dorso-lateral coelom extends to the mid-dorsal line, where its mesoderm contributes to the heart, pericardial floor and muscles surrounding the midgut (Fig. 9.9). Part of the coelomic cavity persists as the cavity of the gonads. The medio-ventral coelom develops into the coelomic end sac and coelomoduct of each segmental excretory organ. The coelomoducts of the genital segment forms the reproductive ducts associated with the gonads.

Figure 9.9
Somite differentiation in onychophorans; (see text for details). (From Anderson 1973.)

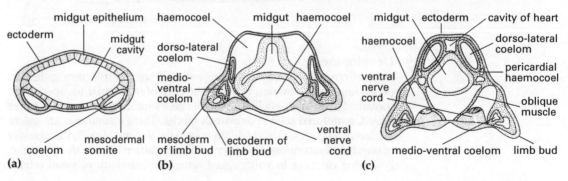

All onychophorans are epimorphic and attain the full complement of segments and limbs before birth or hatching. Development takes a surprisingly long 17 months in an Australian egg-laying species and 12 months in an African ovoviviparous and neotropical placental species. There is no maternal care, and juveniles are able to trap prey immediately after birth. Little detailed information is available on the life history of any species of onychophoran.

Evolutionary history

The first discovered species of an onychophoran — identified in 1826 as *Peripatus juliformis* from the Caribbean — was considered to be an aberrant mollusc, but it was soon recognised that these animals display a curious combination of annelid and arthropod characteristics. The current consensus is that the Onychophora represent a sister group to the arthropods (see Chapter 18). Evidence for this has been derived from a variety of morphological, molecular and palaeontological data. The soft body of onychophorans has resulted in a sparse fossil record. However, recent discoveries of a number of Cambrian fossil assemblages of soft-bodied animals, and the re-evaluation of others, have brought to light an extensive radiation of marine Cambrian lobopodial forms that may be related to Recent onychophorans. Less controversial Carboniferous (300 mya) fossils from the United States of America and France bear a striking resemblance to present-day forms. These, together with recently identified fossil onychophorans in Baltic and Dominican amber (20–40 mya), indicate that terrestrial onychophorans were once much more widespread than at present. This poses the question of why they have become extinct on all land masses of Laurasian origin, but have survived on all Gondwanan fragments with the notable exception of the Indian Plate. Apart from the Dominican amber fossils, no fossil onychophorans have been identified from regions where they presently exist. The identification of further onychophoran fossils in the Northern and Southern Hemispheres may resolve the question of the origin and timing of one of the major evolutionary differences between the Cambrian lobopods and present day onychophorans, the adoption of a terrestrial life style. The unique method of prey capture in modern onychophorans is clearly a terrestrial adaptation and, given its complexity, seems unlikely to have evolved more than once. In their evolution onto land, onychophorans have been constrained by their cylindrical, soft and extensible body, and by limited mobility on stumpy clawed legs, to an environment of moist crevices created by decaying vegetation.

Classification of the phylum Onychophora

The conservative body form of onychophorans is reflected in the taxonomy of the phylum, which lacks defined categories higher than family level. The phylum is divided into two families, the circum-equatorial Peripatidae and the circum-austral Peripatopsidae. It seems likely that the two families had evolved separate latitudinal distributions before the fragmentation of Gondwana. Consequently, the Peripatopsidae and Peripatidae are found, disjunctly distributed in southern and tropical regions respectively, in both Africa and South America. The Wallace Line forms the demarcation between the Peripatidae in South-East Asia (an early fragmentation of

Gondwana as the Shan–Thai–Malay–Kalimantan Plate) and the Peripatopsidae in Australasia.

FAMILY PERIPATOPSIDAE
Circum-austral distribution; 13–29 pairs of legs; 3 spinous pads on legs; opening of excretory organs on fourth and fifth pairs of legs in middle of third spinous pad; genital opening between or behind last pair of legs; oviparous or ovoviviparous with yolky or non-yolky eggs.

FAMILY PERIPATIDAE
Circum-tropical distribution; 19–43 pairs of legs; more than 3 pairs of spinous pads on legs; opening of excretory organs on fourth and fifth pairs of legs above middle of third spinous pad; genital opening between second-last pair of legs; ovoviviparous with non-yolky eggs or viviparous with placenta.

Phylum Tardigrada

Bilaterally symmetrical, metamerically segmented coelomates with a reduced coelom; haemocoelic body cavity; weakly sclerotised cuticle, moulted periodically; four pairs of lobopod limbs; minimal cephalisation, no cephalic limbs; brain and paired ventral nerve cord; dorsal coelomic gonads; separate sexes; direct development.

Introduction

The first observation of the rounded form and stumpy, clawed limbs of these animals inspired the common name water bears, while Tardigrada (slow-stepper) was coined for the resemblance of their slow, lumbering gait to that of a tortoise. Tardigrades are microscopic, with few species exceeding 1 mm in length. The phylum, which contains approximately 700 species, has a world-wide distribution from polar regions to abyssal depths. Marine and fresh-water species are benthic, living amongst sediments or on the surface of vegetation. Terrestrial tardigrades live in the water film surrounding soil particles and on the surface of plants, particularly mosses, liverworts and lichens. Fresh-water and terrestrial species respond to unfavourable environmental conditions by producing resistant stages. These stages, together with the production of resistant eggs, underlie the broad distributions of many species of tardigrades, through dissemination by wind or on the bodies of other animals. Tardigrades feed on detritus, bacteria, algae, protozoans, microscopic animals or the cell contents of plants. A few marine tardigrades are ectoparasitic on invertebrates such as holothurians.

External morphology
General body form and integument
The body varies from cylindrical, to dorso-ventrally flattened, to convex on the dorsal surface and flattened on the ventral surface. An indistinctly demarcated head is followed by four segments, each bearing a pair of stumpy or telescopic lobopod legs. The head carries a terminal or subterminal mouth. The legs terminate in claws or toes with adhesive pads or

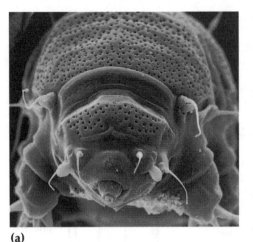
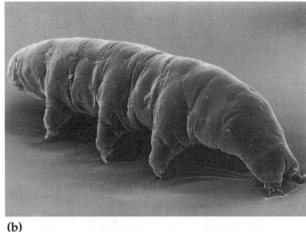

Figure 9.10
SEM photograph of tardigrades. (a) Anterior view of a terrestrial heterotardigrade, *Echiniscus duboisi*. (b) A terrestrial eutardigrade, *Macrobiotus peteri*.

claws. The claws and toes are secreted by glands at the tips of the legs. A separate subterminal anus and preanal gonopore occur in heterotardigrades, while eutardigrades possess a combined, subterminal cloaca.

The body is covered by a cuticle, which is secreted by a single-layered epidermis overlying a thin, collagenous basal lamella. The cuticle is often ornamented and may be thickened as plates on the dorsal and lateral surfaces of the body (Figs 9.10a, 9.11a). The cuticle is composed of an outer epicuticle, middle intracuticle and inner procuticle. The epicuticle is sclerotised, while the intracuticle contains lipid and the procuticle is a composite of chitin and protein. Most species are white or colourless and transparent, but others display a variety of colours — brown through to red, black or green. The colours are due to food in their gut, or granules in the body cavity cells, epidermis or cuticle.

Internal organs and physiology

The body cavity of tardigrades is a haemocoel, as in onychophorans, although a dorsal heart is lacking. The coelom is restricted to the cavity of the reproductive system. The body musculature consists of muscles attached at specific points of procuticular thickening. They extend as dorsal and ventral longitudinal fibres, and transversely from the dorsal body wall to the base of the legs. Each muscle band consists of only a single or a few muscle cells. Locomotion is achieved by a slow, lumbering gait in which the limbs are moved forwards and backwards by antagonistic, extrinsic muscles.

The digestive system consists of an ectodermal fore- and hindgut and endodermal midgut. The foregut and hindgut are lined with cuticle. The complex buccal apparatus in the foregut consists of a buccal cavity, buccal tube, muscular pharynx, oesophagus and stylets (Fig. 9.11c). The stylets are lateral to the buccal tube and at their base may rest on a stylet support, inserted on the buccal tube. Protractor and retractor muscles are attached to the bases of the stylets and are inserted on the buccal tube and pharynx. Stylets penetrate the lumen of the buccal cavity anteriorly and are protruded through the mouth during feeding. After moulting, glands which open into the buccal cavity secrete the buccal tube cuticle, stylets and stylet supports.

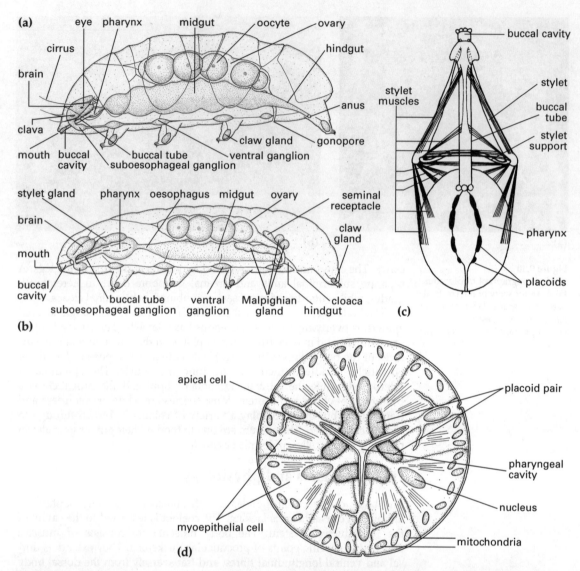

Figure 9.11
Anatomy of tardigrades.
(a) Heterotardigrade, lateral view. (b) Eutardigrade, lateral view. (c) Details of buccal apparatus. (d) Transverse section of pharynx. (a, b after Nelson 1982; c from Greven 1980.)

The pharynx provides suction for drawing food into the gut. The pharyngeal wall is composed of a fixed number of myoepithelial cells radiating from a triradiate lumen lined with cuticle (Fig 9.11d). Each myoepithelial cell is striated and composed of a single sarcomere. The pharynx possesses placoids, which are regions of strengthened cuticle secreted by epithelial (apical) cells between the radiating muscle cells. Placoids consist of three longitudinally arranged rods in heterotardigrades, or six rows of individual plates in eutardigrades. They serve as attachment points for the pharyngeal muscles. A narrow triradiate oesophagus, which secretes a peritrophic membrane, leads into the midgut, where digestion and absorption occurs. The hindgut opens into a cloaca or to a separate anus.

Excretion and osmoregulation are poorly understood in tardigrades. Various organs have been implicated in these activities, including Malpighian glands in eutardigrades, opening into the junction of the midgut and hindgut, and ventral segmental glands in heterotardigrades, occurring at the base of the legs. Excretory wastes may also be released at ecdysis when the buccal apparatus and body cuticle are shed.

As tardigrades are microscopic, diffusion provides sufficient oxygen into the haemocoelic fluids bathing the body organs. Tardigrades are very sensitive to low levels of oxygen and can enter an inactive stage in which they can survive for several days.

The nervous system is segmentally arranged (Fig. 9.11a, b). The dorsal brain is composed of three lobes (the protocerebrum, deutocerebrum and tritocerebrum), although the equivalence between these and similarly named regions of the arthropod brain is obscure. The brain is connected by commissures to the suboesophageal ganglia. From these, a pair of ventral nerve cords connects the four pairs of ganglia in the trunk. There is also a direct link between the brain and the first trunk ganglia.

Sensilla occur on the body, and particularly on the head, as bristle-like cirri functioning as mechanoreceptors, while papillae and clavae with terminal pores are possibly chemosensory (Figs 9.10a, 9.11a). These cephalic sensilla are lacking in many species (Eutardigrada) (Figs 9.10b, 9.11b). Some species also have eyes, each consisting of a single cup-shaped pigment cell enclosing sensory cells.

Reproduction

Tardigrades typically have separate sexes with little sexual dimorphism, although males are often smaller than females. Parthenogenesis is also believed to be common, and hermaphroditism has been reported in some species. Male and female reproductive systems are simple and similar (Fig. 9.11a, b). There is a single gonad, suspended in the body cavity by a ligament attached to the dorsal body wall. A single oviduct in females, and paired sperm ducts in males, open into the hindgut at the cloaca or at a gonopore in front of the anus. In females, single or paired seminal receptacles, opening ventrally into the hindgut, are often present. Mating and fertilisation have been recorded for only a few species. Sperm may be deposited at the cloacal or genital opening, with fertilisation taking place as the eggs are laid. In other species, internal fertilisation occurs in the ovary or female reproductive tract after sperm is stored in the seminal receptacles. Eggs are laid singly or in clusters, attached to the substratum or in the moulted cuticle of the female. Eggs that are laid freely often have thick, variously ornamented shells, while those that are laid in the cuticle have smooth, thin shells. Incubation takes from five to forty days. The emerging juveniles may differ from the adult in the development of reproductive organs, buccal apparatus, the number and form of the claws, and body ornamentation. Growth is accompanied by a number of moults, and the body grows through an increase in cell size as well as cell number in various tissues.

Resistant stages

While the life-span of tardigrades has been reported as a few months to one or two years, this can be considerably increased by the remarkable ability of some tardigrades to develop stages resistant to adverse environmental factors. Several types of resistance have been identified, including encystment and cryptobiosis. In some species of fresh-water tardigrades, low temperatures initiate a moult, after which the animal remains in a contracted state within the new cuticle (a white cyst). In terrestrial and some freshwater species, cyst formation is more dramatic. At moulting, the old cuticle is retained and forms a deep red or black sclerotised cyst wall enclosing a rounded body, which lacks limbs and a digestive system (a red cyst).

Cryptobiosis enables tardigrades to withstand freezing and thawing and is also induced by the onset of low temperatures. The body becomes barrel-shaped and inactive, a condition called a tun. An alternative condition of anhydrobiosis is induced by desiccating conditions and consequently is most commonly encountered in terrestrial species. The body rounds up to form a tun as water is lost. During tun formation, tardigrades undergo a period of transpiration followed by an abrupt arrest of further water loss. The retention of water is essential for the extensive biochemical changes occurring at this time. These include the rapid breakdown of glycogen and lipid reserves and an accompanying increase in the synthesis of dehydration protectants of proteins and membranes, such as glycerol and the disaccharide trehalose. The intracuticular lipids have been implicated in the control of transpiration during tun formation. The anhydrobiotic state is also extremely resistant to other adverse environmental conditions (low oxygen and temperature, radiation and chemicals), as evidenced by the survival of tardigrades on herbarium specimens for up to 120 years.

Development

The yolk of tardigrade eggs is equally distributed throughout the cytoplasm, and cleavage is total. The endoderm delaminates to fill the blastocoel, then forms a single layer of cells surrounding a central cavity, the archenteron. Intuckings of ectoderm to form the stomodaeum (foregut) and later the proctodaeum (hindgut) join the endodermal tube to complete the digestive system. The origin of the mesoderm and coelomic sacs in tardigrades is unclear. Early studies reported that coelom formation is enterocoelic, but recent investigations indicate that the five pairs of coelomic sacs appear after paired mesodermal bands have formed, as in onychophorans. The most posterior pair of coelomic sacs move dorsally and fuse to become the gonad. The cells of the other pouches become dissociated to form muscles and haemocytes.

Phylogenetic relationships

Tardigrades have been traditionally aligned with the pseudocoelomate phyla. Shared features include the form of the buccal apparatus, including stylets and bulbous pharynx, the constant number of cells in certain organs, and the development of resistant stages in the life cycle. More recently, an alternative view that tardigrades are related to onychophorans and arthropods has been proposed. Features shared with pseudocoelomate phyla may be convergent and due to small size, feeding strategies and habitat characteristics. Features shared with onychophorans include the form and chemical composition of the cuticle; claw-bearing, segmentally arranged limbs; sensilla structure; muscle ultrastructure and attachment; and the segmentally based nervous system. Molecular evidence also suggests a close relationship between tardigrades, onychophorans and arthropods (see Chapter 18).

Classification of the phylum Tardigrada

CLASS HETEROTARDIGRADA
Mostly marine but also includes armoured terrestrial species and a few freshwater species; cuticle often thick and forming dorsal segmental plates but secondarily reduced in many species; cephalic sensilla well developed, including cirri, papillae and clavae; separate gonopore and anus.

CLASS EUTARDIGRADA
Freshwater and terrestrial species; cuticle thin; cephalic appendages lacking or consisting only of oral and lateral cephalic papillae; reproductive and digestive systems open into cloaca.

Chapter 10

Introduction to arthropods

D.T. Anderson

PHYLUM ARTHROPODA

Introduction 223

Lobopod features of arthropods 223

Distinctive arthropod features 224
    Arthropod limbs 224
    Jaws 225
    Segmental organs 226

Acron and telson 226

The arthropod subphyla 227
    Subphylum Hexapoda 227
    Subphylum Myriapoda 227
    Subphylum Crustacea 227
    Subphylum Chelicerata 227

Phylum Arthropoda

Bilaterally symmetrical, metamerically segmented coelomates with reduced coelom; haemocoelic body cavity; jointed chitinous exoskeleton with sclerotised plates, moulted periodically; jointed segmental appendages, usually showing regional specialisation along the body; head with several cephalised segments; brain and ganglionated ventral nerve cord; segmental organs in some anterior segments; dorsal coelomic gonads; separate sexes (some hermaphrodite); spiral cleavage (a few examples) or, more usually, modified yolky development, either via larvae or direct; ciliated larvae never formed.

Introduction

Onychophorans and tardigrades, the subject of the preceding chapter, are unfamiliar to most people, but the arthropods include many well-known types of animal such as insects, spiders and crabs. The diversity of arthropods extends over almost a million living species, and many more are being discovered each year. The fossil history of arthropods begins in the early Cambrian, when there were bizarre forms, now long extinct, as well as the early ancestors of the modern arthropod fauna.

Arthropods fall into four living and several fossil groups, all of which are sharply different from one another (see Fig. 1.3). Zoologists have long argued whether these groups are members of a monophyletic clade, phylum Arthropoda, or whether the arthropod grade of organisation has evolved more than once from different pre-arthropod ancestors. Current views on this controversy are assessed in Chapter 18. In Chapters 11–14 we present the four extant groups of arthropods as subphyla of the phylum Arthropoda, but the reader should keep in mind that the evolutionary origin of the arthropods is far from resolved.

Lobopod features of arthropods

The functional organisation on which arthropod diversity is based shares much in common with the lobopod organisation of the Onychophora (see Chapter 9). Arthropods, like onychophorans, are metamerically segmented coelomates with paired ventrolateral segmental limbs, a reduced coelom replaced as a body cavity by an expanded haemocoel, a dorsal ostiate heart and a nervous system based on paired ventral nerve cords. Anteriorly, several segments are cephalised. The surface of the body is covered by a chitin-protein cuticle secreted by an underlying epidermis. The cuticle is moulted periodically during growth.

Internally (Fig. 10.1a), the haemocoel is divided by a transverse pericardial floor into a dorsal pericardial sinus containing the heart and a more general perivisceral cavity extending into the limbs. The coelom persists as the cavities of paired, tubular gonads beneath the pericardial floor, and as cavities in the end sacs of paired segmental organs in certain anterior segments. The gonoducts of arthropods, like those of onychophorans, are a pair of modified coelomoducts (except in some arthropods in which new modifications have evolved). The body wall musculature includes dorsal and ventral longitudinal muscles uniting the segments and muscles operating the limbs. Extrinsic limb muscles, located within the body, move the limbs on the body. Intrinsic limb muscles, located within the limbs, control bending, extension and other limb movements. During locomotion, the segmental limbs work in metachronal rhythm in walking or swimming.

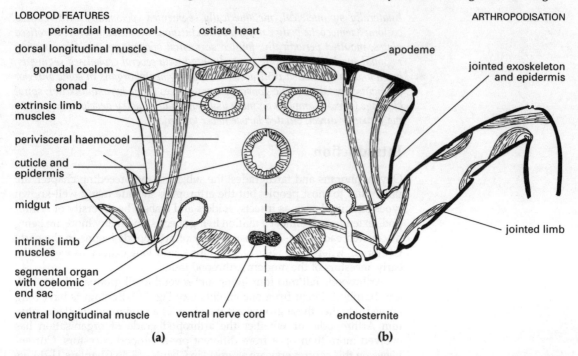

Figure 10.1
Transverse section through a generalised arthropod, showing (a) on the left, the organisational features shared with lobopods and (b) on the right, the additional features of arthropodisation. Segmental organs in arthropods are restricted to a few anterior segments, often to one segment only (see Table 10.1).

Distinctive arthropod features

Despite the many similarities, arthropods differ from onychophorans and other lobopods in a major way (Fig. 10.1b). The external cuticle is thickened and strengthened as an exoskeleton, with flexible joints between the segments, at the limb bases, and at intervals along the limbs. The exoskeleton supports and protects the body and limbs and provides a firm anchorage for arrays of individual muscles, which are able to generate precise, complex movements. The presence of a supportive exoskeleton in arthropods is functionally correlated with the absence of those structures which have a supportive function in onychophorans — the thick dermis, the circular muscle layer and the oblique muscle layer (see Fig. 9.3, page 207).

There are also internal skeletal structures of two kinds in arthropods, adding to the availability of muscle anchorages. These are apodemes (inward projections of the exoskeleton) and endosternites (internal aggregations of dense connective tissue). The cuticle is turned inwards at the mouth and anus, lining the foregut and hindgut. Feeding is facilitated in many arthropods through elaborations of the cuticle of the foregut, serving grinding and filtering functions. The vast diversity of the arthropods is an expression of the potential of the exoskeleton for detailed variation in structure and function.

Arthropod limbs

Arthropod limbs are of two basic types: unbranched (uniramous limbs, Fig. 10.2a); and branched, with either two main branches (biramous limbs,

INTRODUCTION TO ARTHROPODS

Figure 10.2
Arthropod limbs and jaws.
(a) Uniramous limb (walking leg of centipede). **(b)** Biramous limb (walking leg of prawn). **(c)** Polyramous limb (swimming leg of tadpole shrimp, with gnathobase). **(d)** Whole-limb jaw (mandible of centipede). **(e)** Gnathobasic jaw (mandible of prawn). **(f)** Gnathobasic jaw (mandible of tadpole shrimp). (After Anderson 1996.)

Fig. 10.2b) or a number of processes (polyramous limbs, Fig. 10.2c). The function of these limbs usually varies along the length of the body, with anterior limbs specialised as sense organs and mouthparts and those along the body length for locomotory, reproductive and other functions. An associated feature is the division of the body into functional units or tagmata (singular, tagma) comprising groups of segments sharing similar functions. As well as carrying sensory limbs preorally and mouthparts postorally, the several cephalised segments contribute to two major nerve centres, a complex preoral brain and a postoral suboesophageal ganglion. It is well known that arthropods display elaborate patterns of behaviour.

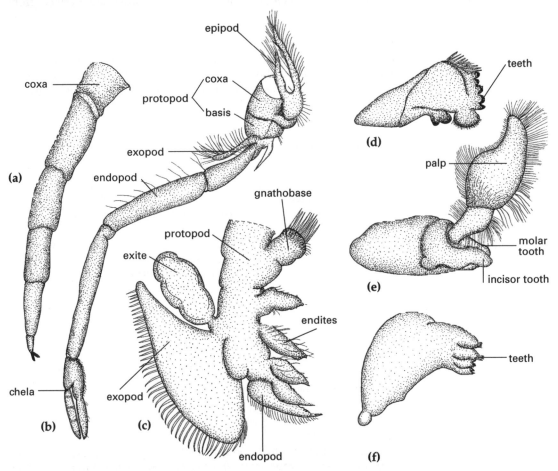

Jaws

There are two main types of jaw in arthropods. In some, the modification of segmental limbs as shortened jaw structures involves the foreshortening of the whole limb, with the tip of the limb being the functional biting surface. Centipedes and insects, for example have this type of jaw (Fig. 10.2d). In other arthropods, jaws are developed as median enlargements of the limb base, called gnathobases (Fig. 10.2c), which provide the biting surface. The more distal part of the limb often persists (Fig. 10.2e) and serves other functions (e.g. sensing, locomotion or food-gathering) or may be entirely lost in more specialised jaws of this type (Fig. 10.2f). Crustaceans have gnathobasic jaws.

Whole-limb jaws, which also occur in onychophorans, are always located anteriorly, close behind the mouth on a distinct head tagma. Gnatho-basic jaws are primitively arrayed serially on the limbs of several to many postoral segments, conveying food forwards to the mouth. In many types of arthropod with gnathobasic jaws, the jaws have become restricted to head segments directly behind the mouth, resulting in a convergent resemblance to arthropods with whole-limb jaws. The jaws of a cockroach and a slater, for example, are similar in form and function, but the former are whole-limb jaws while the latter are derived from gnathobases.

Segmental organs

The regional specialisation of the arthropod body into tagmata is further reflected in the distribution of segmental organs, which are restricted at most to a few anterior segments (Table 10.1). In some arthropod groups only one pair is retained, while in insects segmental organs are lost except for vestigial remnants in a few species.

Table 10.1 Arthropod segmental organs.

Location	Function
Hexapoda	
Premandibular segment	Embryonic; modified as glandular suboesophageal body in some species
Myriapoda	
Premandibular segment	Embryonic; retained as salivary glands in pauropods
Maxillary segment	Salivary glands, in all classes
Crustacea	
Antennal segment	Larval antennal glands (excretory); sometimes retained in adults
Maxillary segment	Maxillary glands (excretory); replacing larval glands in other adults
Chelicerata	
Several prosomal segments (sometimes only one, as in scorpions)	Coxal glands (excretory)

Acron and telson

Two units contributing to the body plan of arthropods are absent from onychophorans. These are an acron, an anterior presegmental unit contributing substantially to the brain, and a telson, a posterior postsegmental unit on which the anus opens. For this and other reasons, it is clear that the onychophorans were not the ancestors of the arthropods, and we need to look to other, extinct lobopods for this role.

It is now possible to see why the evolution of the arthropod body form is such a controversial matter. The functional advances of the arthropod over the lobopod organisation (exoskeleton, jointed limbs and associated functional changes internally) can be interpreted as consequences of the evolutionary modification of the external chitin-protein cuticle into a jointed exoskeleton. It is still not clear whether this evolutionary change, resulting in the development of arthropod characteristics, occurred more than once in evolution. The sharply differing patterns of morphological structure and function in the four subphyla of living arthropods only serve to accentuate the problem, emphasising the need to seek other kinds of evidence, such as that from molecular biology and the early fossil record (see Chapter 18).

The arthropod subphyla

Subphylum Hexapoda

The Hexapoda are primarily terrestrial arthropods with uniramous limbs, although some fossil freshwater insects have multibranched (polyramous) limbs. Hexapods have one pair of antennae, unjointed mandibles formed as whole-limb jaws, three pairs of walking legs, and a posterior (opisthogoneate) gonopore. Hexapods include the winged insects and several other insect-like groups (see Chapter 11).

Subphylum Myriapoda

The Myriapoda are primarily terrestrial arthropods with uniramous limbs, one pair of antennae, jointed mandibles formed as whole-limb jaws, numerous pairs of walking legs, and an opisthogoneate (sometimes secondarily anterior or progoneate) gonopore. Myriapods include the centipedes, the millipedes and two smaller groups (see Chapter 12).

Subphylum Crustacea

The Crustacea are primarily marine arthropods with biramous limbs, two pairs of antennae, mandibles formed as gnathobasic jaws, several to numerous pairs of walking and/or swimming legs, and gonopores in the mid-region of the body. Crustaceans include shrimps, crabs, barnacles and many other groups (see Chapter 13).

Subphylum Chelicerata

The Chelicerata are also primarily marine arthropods, although most living chelicerates are terrestrial. Chelicerates have no antennae, the most anterior limbs being a pair of chelicerae. They have gnathobasic jaws (but lack specialised mandibles) and biramous limbs (often secondarily uniramous). Gonopores are in the mid-region of the body. Horseshoe crabs, scorpions, spiders, ticks and several other groups are members of this subphylum (see Chapter 14).

The following chapters fill out the details of each subphylum and its diversity. Chapter 18 includes an examination of the relationships between the subphyla and discuss current research on the origin (or origins) of the arthropods.

Chapter 11

The Hexapoda

D.F. Hales

SUBPHYLUM HEXAPODA

Introduction 229

Structure and functions of the cuticle 230

Basic functional morphology 232
  Segmental composition of tagmata 232
  Appendages 234
  Water conservation 235
  Diversity 235

Major patterns of form and function 236
  Locomotion 238
  Mouthparts 240

Evolutionary history 241

Physiological processes 242
  Feeding and digestion 242
  Fat body and metabolism 245
  Excretion and osmoregulation 246
  Circulation 247
  Terrestrial respiratory exchange 248
  Aquatic respiratory exchange 250

Control systems 251
  Nervous system 251
  Endocrine system 252
  Sense organs 252

Sense organs and control of flight 255

Reproduction and development 255
  Reproductive systems and processes 255
  Gametes and fertilisation 258
  Embryonic development 259
  Larval development 260
  Metamorphosis 262

Thermoregulation 263

Survival at extreme temperatures 264

Insect societies 264

Classification of the subphylum Hexapoda 265

Subphylum Hexapoda

Primarily terrestrial arthropods with uniramous limbs; one pair of antennae; unjointed whole-limb mandibles; three pairs of walking legs; gonopore posterior.

Introduction

The major group of hexapod arthropods is the class Insecta, ranging from wingless silverfish and their relatives through to dragonflies, cockroaches, moths, bees, and flies. Less familiar to most people are two other groups of primarily wingless hexapods, the classes Ellipura and Diplura. These are small, but not uncommon, dwellers in leaf-litter and similar humid microhabitats.

The adaptations of hexapods to terrestrial life form a major theme of this chapter. Some groups of insects are associated with fresh water, but few insect species inhabit marine or hypersaline environments.

The insects are unique among invertebrates in having evolved wings that enable active, directed flight. Their ability to fly to and colonise isolated habitat patches enhances the survival prospects of individuals, populations and species. The mode of evolution of wings is contentious, but wing development in the life history of an individual is well understood. It occurs gradually throughout early life, with a rapid acceleration just before the adult stage.

The pattern of wing development is linked to the pattern of metamorphosis. Like other arthropods, hexapods go through a series of ecdyses in which the old exoskeleton is shed and the external body form changes to a greater or lesser degree, culminating in a major change as the wings reach their full development at the ecdysis to the adult. Three categories are used in describing hexapod metamorphosis: ametabolous for primitively wingless groups, hemimetabolous (Fig. 11.1a) for those having a gradual change towards the adult form, with wings developing externally (e.g. grass hoppers, cicadas, aphids), and holometabolous (Fig. 11.1b) for those with substantial changes from larva to adult, and an intermediate pupa (e.g. moths, bees). Most holometabolous insects develop their wings (and often other adult structures) internally during the stages preceding the non-feeding pupa. The immature feeding stages of insects are called larvae (or sometimes nymphs in hemimetabolous groups). The term 'instar' is used for the form of the insect between ecdyses. The duration of the period between ecdyses is called the stadium. The adult stage is called the imago (plural, imagines or imagos).

There are many reasons for studying insects, including their sheer numbers and diversity, as well as the aesthetic beauty of many of them. In practical terms, insects are our greatest competitors for food, as well as damaging crops, fibres, building materials and human and livestock health. The course of history has been changed many times by the activities of insects, particularly those that carry major diseases of humans (plague, malaria, sleeping sickness) or of livestock. On the other hand, many insect species provide services to humans as food, crop pollinators, control agents for pests, or producers of honey, silk, etc. Other contributions include recycling of dead organic matter, food web components in all land and

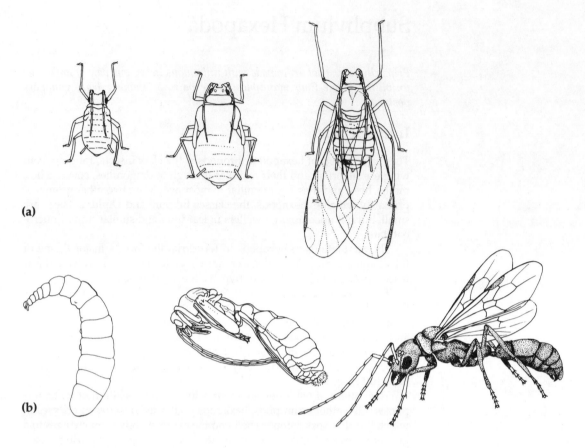

Figure 11.1
(a) The last two nymphal instars and adult of an aphid (*Myzus*, Hemiptera), showing hemimetabolous exopterygote development. (b) Larval, pupal and adult stages of a male bulldog ant (*Myrmecia*, Hymenoptera), showing holometabolous endopterygote development.

freshwater environments, and possible sources of novel chemicals that may be useful in medicine or pest control. Insects are also excellent laboratory or field models in many areas of biological science. The monumental contribution of studies on *Drosophila* to our understanding of genetics and development are the outstanding example, but insects have been used widely in studies of physiology, ecology and evolution. Rapid generation time, small size and the ease of rearing large numbers in the laboratory are some of the reasons for their importance as model systems. Because of these practical advantages, and because of the enormous importance of insects in terrestrial ecosystems, agriculture and health, we have learned far more about their physiology than about that of other invertebrates.

Structure and functions of the cuticle

The cuticle of insects is waterproof. For early insects this probably was important in excluding water (i.e. preventing drowning), as in onychophorans (Chapter 9), but it also allowed the colonisation of new habitats, because it prevented water loss in drier environments. These functions can be understood in terms of the detailed structure and chemistry of the cuticle.

As in other arthropods, the hexapod cuticle (Fig. 11.2a) is made up of an epicuticle and a procuticle. The components of the procuticle are microfibrils of the polysaccharide chitin (a polymer of $\beta(1-4)$N–acetyl glucosamine

(Fig. 11.2c) in a matrix of protein. In hard regions of the exoskeleton, the procuticle is divided into exocuticle and endocuticle. The hard exocuticle is sclerotised (tanned) by cross-linking of the protein chains with various aromatic molecules such as quinones and dopamine derivatives.

Almost every feature of the cuticle, from its chemical composition and its macro- and microsculpturing to the sequence of biochemical events that determine moulting, is determined by the dynamic activities of the single-layered epidermis. For example, the chitin microfibrils that stiffen the cuticle are arranged in a parallel array and are laid down in successive layers (Fig. 11.2b). During the night, each new layer is oriented at a slight angle to the preceding one, but in the day the orientation remains constant. In sections we can see a series of crescentic shapes at high resolution in the night cuticle, and can estimate the age in days since the last moult by the 'growth rings' of the cuticle. The cuticle microfibrils appear to be laid down by microvilli on the apical surface of the epidermal cell. The microvilli are thought to bend in unison when laying down a section of a bundle of microfibrils. Intercellular communication ensures that all cells are coordinated so that the microvilli sweep simultaneously in the same direction.

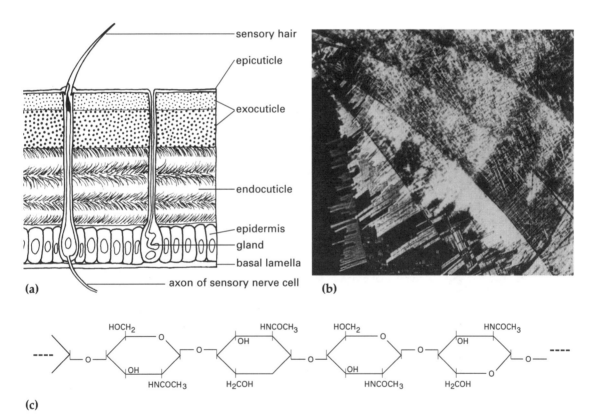

Figure 11.2
(a) Section through the cuticle and epidermis of a typical insect. (b) Micrograph of insect cuticle, broken to show offset layers of chitin microfibrils (from Neville 1970). (c) Part of a chitin molecule, composed of 1,4 β-linked molecules of N-acetyl glucosamine.

The epicuticle is the first part to be secreted at moulting, and the form of the epicuticle determines the appearance of the insect. It also includes the lipid components that protect insects from water loss. The lipids are produced in oenocytes — specialised cells lying beneath the epidermal cells. Sex pheromones and other signalling substances commonly occur as lipids in the superficial wax layer. A cement layer is secreted from dermal glands.

Basic functional morphology

Segmental composition of tagmata

The segments of hexapods are grouped into three tagmata: the head, thorax and abdomen (Fig. 11.3). Each segment is covered dorsally by a cuticular plate, the tergum, which is sclerotised in adult insects, but not always in juvenile stages. Similarly, the ventral part of each segment is covered by the sternum. In the abdomen, successive terga or sterna are joined by unsclerotised regions of cuticle, the intersegmental membranes. Laterally, between the tergum and sternum of any abdominal segment, there is an unsclerotised pleural membrane. The unsclerotised regions of the cuticle can be folded, and unfolding can accommodate increases in volume. Extreme examples are seen in the distended abdomens of honeypot ants, insects that take in large blood meals (mosquitoes, tsetse flies), and female insects whose bodies become distended by the growth of the ovaries, such as some termite queens.

The head is divided into preoral and postoral regions. The preoral region is innervated by three fused ganglionic centres: the protocerebrum, deutocerebrum and tritocerebrum. Together they constitute the brain, lying above the oesophagus. Only the deutocerebrum is associated with appendages visible in the adult (the antennae). Embryological studies have shown that an acron and three segments contribute to the preoral region of the head. In front of the antennal segment is the remnant of a preantennal

Figure 11.3
Lateral and dorsal views of a locust, showing tagmatisation and major morphological features (from *The Insects of Australia*, CSIRO 1991).

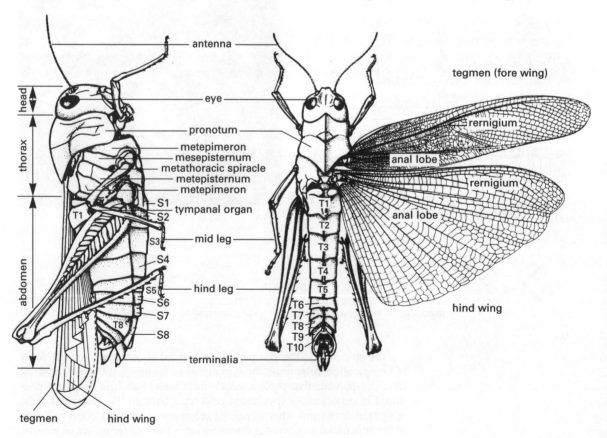

segment. The ganglia of the acron and the preantennal segment together form the protocerebrum. Behind the antennal segment is a premandibular segment with embryonic limbs, which are later resorbed. The premandibular ganglia form the tritocerebrum.

The postoral region of the head also includes three segments but is easier to interpret.

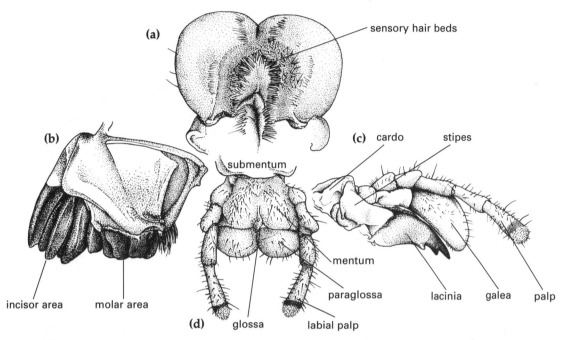

Figure 11.4
Mouthparts of an Australian plague locust, *Chortoicetes terminifera*, in posterior view:
(a) Labrum. (b) Mandible. (c) Maxilla. (d) Labium.

The segments are represented by the three pairs of mouthparts: mandibles, maxillae and labium (= fused second maxillae) (Fig. 11.4). Originally leg-like, these limbs have undergone evolutionary modifications that enable insects to feed on a wide range of food, including free liquids and enclosed liquids such as blood, or the phloem sap of plants. The ganglia corresponding to the paired mouthparts are fused as a single mass beneath the foregut, the suboesophageal ganglion. Apart from the appendages, there are almost no external indications of the segments of the head in postembryonic insects.

The thorax (Fig. 11.3) consists of three distinct segments, the prothorax, mesothorax, and metathorax, each bearing a pair of legs. In winged insects, the meso- and metathorax also have a pair of wings. The sides of the thorax are strengthened by sclerotised plates, which permit the biomechanical activities associated with locomotion. The terga, sterna and pleura of the thorax are all subdivided into separate plates (tergites, sternites, pleurites), each with their own specific functions.

The hexapod abdomen is made up of 11 segments, although the first may be reduced. Segments 1–8 are generally fairly unspecialised, and except in the most primitive groups (Ellipura, Archaeognatha) do not have appendages. Terminal segments in adult insects are specialised for reproduction, and have appendages modified in the male as copulatory structures and in the female as ovipositor valves. The paired appendages of segment 11 are the mechanosensory cerci of many insects. There is great variation among hexapods in the arrangement of the genital structures, and they may often provide the only morphological differences that allow taxonomists to distinguish between closely related species. More importantly, they play a large part in allowing the insects themselves to recognise mates, by species-specific patterns of sensory stimulation.

Appendages

The legs of most present-day insects are uniramous and have five podomeres: the coxa, trochanter, femur, tibia and tarsus (Fig. 11.5a, left). The tarsus has one to five sections, or tarsomeres, and terminates in the pretarsal region, including the claws and sometimes pads between them (the arolium, or the paired pulvilli). Often tarsomeres have pads known as plantulae on their undersurfaces. The correct recognition and counting of tarsomeres is essential when using keys to identify insects. The legs may be modified in various ways, for swimming (water beetles), jumping (grasshoppers), catching prey (praying mantises), etc.

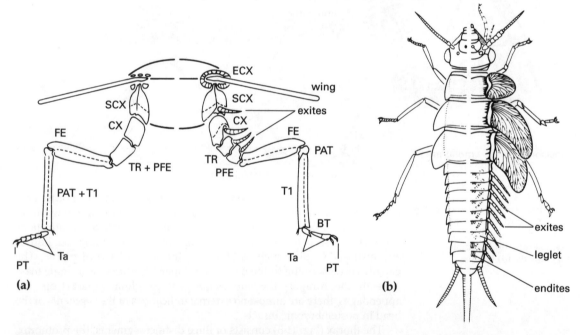

Figure 11.5
(a) Insect thoracic and leg structure. (b) Insect nymphal structure. The left-hand side represents extant insects; the right-hand side represents primitive winged insects from the Palaeozoic (from *The Insects of Australia*, CSIRO 1991).

Some fossil aquatic insects have been interpreted as having as many as 11 podomeres in the legs, with exites (lateral outgrowths) on some or all of the most proximal five (Fig. 11.5a, right). Thus, the limbs of these early hexapods were not uniramous. These fossil insects also had leg-like appendages, with claws, on the abdominal segments (Fig. 11.5b). It is hypothesised that the evolutionary origin of the wings of insects may have been by development of the exites of the most proximal podomeres of the legs, initially on all the legs, including the abdominal ones. These exites probably functioned as articulated gill plates in early aquatic insects. An aquatic origin of wing development is further suggested by the use of wings as sails for locomotion across the water surface by some present-day semiaquatic insect adults. Powered flight may have come about independently in different lines of insects. Flapping aerial flight is a complex activity, requiring the existence not only of wings but also of a complex musculature and sensory systems controlling the orientation of the body with reference to gravity, wind and destination.

Water conservation

The ability of insects to conserve water is largely due to the presence of the lipid layer on and in the epicuticle. This layer is complex, and varies between species. It includes 50–75% hydrocarbons, mainly straight-chain alkanes (up to 34 carbon atoms); the remainder is made up largely of wax esters, free fatty acids and alcohols. Sterols, chiefly cholesterol, are also present. It is suggested that a change in orientation of the lipid molecules results from heating, so that water can escape between them. Water loss increases sharply at a critical temperature, characteristic for each species. Scratching the lipid layer also destroys its impermeability. The composition of the lipid layer is known to vary seasonally in some insects, with more long-chain alkanes in summer. This provides a more impermeable barrier than the shorter-chain molecules predominant in the winter lipids. The non-insect hexapods (Ellipura, Diplura) are less resistant to water loss. They float on water without being wetted, suggesting a hydrophobic wax layer that inhibits wetting and prevents drowning in small drops of water. The surface of the cuticle in peripatus (see Chapter 9) acts in a similar manner.

Insects restrict the loss of water in other ways. Their tracheal respiratory system is a potential site of water loss, but the spiracles are generally kept closed. Water may also be lost with nitrogen excretion or elimination of faeces: insects living in dry conditions excrete their nitrogen metabolites in the form of insoluble uric acid crystals. Insects also resorb water via the wall of the hindgut from faecal material before elimination.

Finally, terrestrial life requires mechanisms of reproduction that include protection of the gametes from water loss. In the insects, sperm, sometimes encased in a spermatophore, are deposited directly in the reproductive tract of the female. In the non-insect hexapods, insemination is similar to that of some myriapods (Chapter 12) and chelicerates (Chapter 14). A spermatophore is deposited on the ground and subsequently taken up by the female. The eggs of insects are protected from desiccation by external membranes which have lipid layers analogous to those on the cuticle, and are often placed in protected microenvironments, such as soil, plant tissue or in other animals.

Diversity

Characteristics that have enabled hexapods (and particularly insects) to diversify include their small size, their ability to utilise specialised resources, and their power of flight. Small body size means that an individual hexapod can use minute cavities for shelter, and needs only small quantities of food. Hence any given environment (say, a single tree) can accommodate many species of insects, feeding within leaves, on leaves, within flowers, fruit or seeds, boring into stems, as predators and prey under bark, or sucking sap from roots. Larger animals, like birds, are excluded from many of these niches. Not all insects are small, of course. Many modern forms are large, and many of the early fossil insects also had large bodies and long wings. The specificity of many insects — to a single species of plant and a single developmental stage of leaves, flowers or fruit, — divides the environment further and reduces competition between species. This degree of specialisation is made possible because flight confers on insects the power of dispersal, so that when one food source is exhausted a new one can be found. A further factor in insect success may be the striking metamorphosis from larva to adult, and the concomitant changes in resource utilisation. Even among the hemimetabolous insects, larvae may inhabit environments that are different from those of adults. Dragonflies, mayflies, and stoneflies have aquatic larvae. In holometabolous insects, the larva is usually different in form from the adult and uses different resources. The leaf-feeding caterpillar and nectar-feeding butterfly are familiar examples of this principle. We can also think of a 'division of labour' within the life cycle, where the main roles of the larvae are feeding and growth and the main roles of the adults are dispersal and reproduction. Adult feeding supports the energy demands of flight and the synthetic demands of gametogenesis.

Major patterns of form and function

For a group that is so diverse in terms of speciation, the hexapods are strikingly uniform in their morphology. Important variations are related to appendages and their modifications for different patterns of locomotion and of feeding. The following paragraphs indicate major phylogenetic subdivisions and the major characters on which they are based. As with all higher-level systematics, this arrangement is in a state of flux as new evidence (particularly molecular evidence) is added.

The non-insect hexapods (classes Ellipura and Diplura) are clearly separated from the insects (class Insecta). All are wingless, and all display entognathy — the downgrowth of the sides of the head to enclose the mouthparts laterally. The Ellipura are small hexapods: the Collembola (Fig. 11.6a) (springtails) and Protura (Fig. 11.6b). All are minute, generally less than 3 mm, although in the Tasmanian temperate rainforests there are springtails as much as 10 mm in length. The Ellipura lack the internal head skeleton (tentorium) typical of the class Insecta. The Protura lack antennae, but both they and the Collembola have appendages on the anterior abdominal segments. In springtails, these are specialised as a ventral tube, a furca (the springing 'tail') and a retinaculum which secures the furca prior to springing. Springtails have an abdomen of only six segments and also differ from other hexapods in having eggs with total cleavage.

The Diplura (Fig.11.6c) are more closely related to the insects, but again lack a tentorium. Anterior abdominal appendages are represented by small styles. The largest diplurans are 50 mm long.

The Insecta are characterised, among other features, by their internal head skeleton or tentorium. Most primitive among present-day insects are the archaeognaths (subclass Archaeognatha), wingless forms differentiated from other insects by the single basal articulation of their mandibles (Fig. 11.6e). A well-known species is the rock-hopper, *Petrobius maritimus*. Archaeognaths have styles on the anterior abdominal segments, and can jump.

The other wingless group, the infraclass Thysanura, or silverfish (Fig. 11.6d), is the sister-group to the winged insects or pterygotes, which are grouped together in the subclass Dicondylia. Thysanurans are superficially similar to the archaeognaths, but have a dicondylar (double) articulation of the mandibles (Fig. 11.6f). In all the groups discussed so far, metamorphosis is slight (ametabolous) and moulting continues after the reproductive stage has been reached. The winged insects (Infraclass Pterygota) also have

THE HEXAPODA

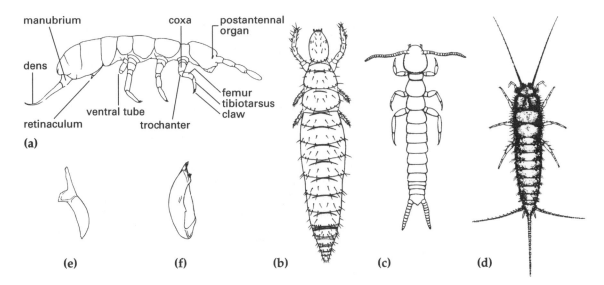

Figure 11.6
(a) Collembolan. (b) Proturan. (c) Dipluran. (d) Thysanuran. (e) Monocondylar mandible of an archaeognath. (f) Dicondylar mandible of a thysanuran. (From *The Insects of Australia*, CSIRO 1991).

dicondylar mandibles. Winged insects are the largest group of arthropods, comprising 27 orders (Box 11.1; also see the classification at the end of the chapter for further details). All have an elaborate metamorphosis during development.

> **Box 11.1**
> **The orders of pterygote insects**
>
> Ephemeroptera........mayflies
> Odonata ....dragonflies and damselflies
> Plecoptera...............stoneflies
> Blattodea.................cockroaches
> Grylloblattodea........ice crawlers
> Isoptera ....termites
> Mantodea.................praying mantises
> Dermaptera.............earwigs
> Orthoptera ....locusts, grasshoppers, etc.
> Phasmatodea..........stick insects
> Embioptera ....web-spinners
> Zoraptera
> Psocoptera..............barklice, booklice
> Phthiraptera.............lice
> Hemiptera ....sucking bugs
> Thysanoptera...........thrips
> Megaloptera ....dobson flies, alderflies
> Raphidioptera..........snake flies
> Neuroptera ....lacewings
> Coleoptera ....beetles
> Strepsiptera
> Mecoptera................scorpion flies
> Siphonaptera...........fleas
> Diptera ....flies
> Trichoptera..............caddis flies
> Lepidoptera.............moths, butterflies
> Hymenoptera..........wasps, bees, ants

The mayflies (Order Ephemeroptera) are thought to be a sister-group to the remaining winged insects, from which they are separated by a suite of characters, including a pre-adult winged stage which gives rise to the free-flying adult after a final moult. The wings of mayflies cannot be folded down by the sides of the body, and are held vertically when not in use. The larvae are aquatic, with lateral gill plates.

Another isolated group is the Order Odonata (dragonflies and damselflies), again with non-folding wings and aquatic larvae. The Odonata are a sister group to the remaining insects, which are collectively called the Neoptera. In these, the wings can be rotated to lie parallel to the body axis when at rest. The holometabolous orders with internally developing wings (the Endopterygota) are generally regarded as the most highly evolved neopterans. One distinctive group of hemimetabolous orders, the Paraneoptera, is characterised by mouthparts modified for sucking and/or scraping.

Larval form varies enormously among and within the endopterygote orders, ranging from essentially adult-like (though wingless) to caterpillar-like (with thoracic legs and a large abdomen supported by prolegs), to legless, to maggot-like (with neither legs nor a visible head capsule).

Locomotion
Legs

Soft-bodied larval insects move by crawling, with movements generated by the segmental longitudinal and dorsoventral muscles, using sculpturing on the cuticle to provide friction. Some lepidopteran and hymenopteran caterpillars have stumpy, soft, abdominal prolegs.

Adult insects (Fig. 11.3) and larvae with well-developed thoracic legs use their legs as levers. The pattern of movement is based on two sets of alternating tripods, with the anterior and posterior legs of one side moving in phase with the middle leg of the other side. The low centre of gravity provides stability. All limb movement in insects results from muscular activity coordinated by the nervous system. (Compare this with the hydrostatic extension of limbs in myriapods, Chapter 12, and chelicerates, Chapter 14.)

Wings

The wing is formed as a double layer of cuticle. Each layer is secreted by epidermal cells, but these break down after the expansion of the wing at the ecdysis to the adult. At an early stage of development, a pattern of 'veins', characteristic for each taxon, becomes apparent in the wings. The veins are defined in the adult wing (Fig. 11.3) by heavier sclerotisation, and contain a blood space, a trachea, and often a nerve. The wing folds in a fan-like pattern at the major veins. The sclerotised bases of the veins articulate with a complex array of thoracic sclerites. Most insects have membranous wings, in which the cuticle between the veins is relatively thin and transparent. In some groups (e.g. cockroaches), the forewings are somewhat sclerotised but still flexible, and are known as tegmina. The forewings of beetles are heavily sclerotised and are known as elytra.

The wing is basically a flat structure and therefore has characteristics different from those of an aeroplane wing or a sail. The leading edge of the wing, supported by veins, is relatively stable, and the trailing edge is flexible. The wing can be rotated by direct flight muscles so that the surface area pushing against the air is maximised during the downstroke (thus maximising lift) and minimised during the upstroke (thus minimising drag).

The characteristics of any wing determine the ratio between the inertial and viscous forces (Reynolds number). The Reynolds number for very small insects is small, meaning that viscous forces are much more significant than inertial forces. This has been likened, on a vertebrate scale, to flying through syrup. (The same calculations have been made for the swimming/filter-feeding activities of small crustaceans.) In small insects the wings are commonly very narrow and trimmed with a fringe of long hairs. This seems to overcome the Reynolds number constraint to some extent. At least some insects (e.g. bees) have aerodynamic flight steering, using the hindwing as a flap that can be raised or lowered to alter direction. The structure of the wing joint is remarkably complex and the insect can change flying 'gears' depending on the position of the wing base relative to the pleural sclerites on the thorax.

Flight can be of two major kinds. Trivial flight occurs within the habitat for seeking food, mates, or shelter. Migratory flight takes insects beyond their initial habitat and allows colonisation of new ones. The migration of the Australian plague locust is an example. Migratory flight is often a precursor to diapause, a physiologically and reproductively inactive state during unfavourable summer or winter conditions. Both migration and diapause require a preliminary phase of laying down nutrient and energy stores. In eastern Australia, Bogong moths migrate during spring to the Australian Alps, where they aestivate in caves. Their fat and protein stores were used as a seasonal food resource by Aborigines.

Flight muscles

Muscles attached to the wing bases (direct flight muscles, Fig. 11.7a) provide the motive force for flight in dragonflies, cockroaches, many grasshoppers, and beetles. The muscles contract once for each arriving nerve impulse and are thus called synchronous. They produce a relatively slow wingbeat. Muscles attached to the walls of the thorax (indirect flight muscles, Fig. 11.7b) provide the main motive force in most other orders. These muscles contract more frequently than one contraction per nerve impulse and are thus asynchronous. As the indirect flight muscles contract, the thoracic wall moves through an unstable position (the so-called click mechanism) and the sudden release of tension stimulates the next contraction in the cycle. Wing beat frequency can be up to 1000 per second in

Figure 11.7
(a, b) Transverse sections through the thorax of (a) an insect with direct flight muscle, and (b) an insect with indirect flight muscle, showing how the contraction of the principal flight muscles changes the shape of the thorax and moves the wings. The intermediate position is unstable and the thorax 'clicks' into the up or down position (from Mordue et al. 1980). (c) Fine structure of part of an asynchronous muscle fibre, showing giant myofibrils and mitochondria.

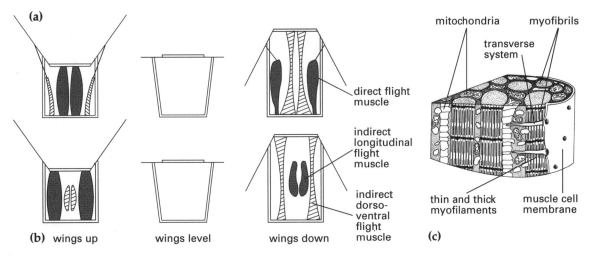

some flies. Contraction of the dorso-ventral (tergo-sternal) muscles stretches the longitudinal muscles attached to inturned cuticle at the ends of the mesothorax and metathorax, and vice versa.

In an insect with indirect flight muscles, the wings operate as a single unit, either coupled or with one pair non-functional or reduced. For example, the forewings and hindwings are linked by minute hook-like hamuli in Hymenoptera, the wings of some moths are linked by a bristle-like frenulum fitting under a retinaculum, and the hindwings in Diptera are greatly reduced.

The flight-muscle cells (Fig.11.7c) are very large and contain large myofibrils. Between the myofibrils are columns of giant mitochondria (sarcosomes), providing for the very high rates of oxidative metabolism supplying the energy requirements of flight. If an insect has indirect flight muscles, the direct flight muscles alter the angle of the wings and fold them against the body.

Muscle attachment to cuticle

Insect muscles are attached to the cuticle (Fig. 11.8), even though there is a continuous layer of epidermal cells between the muscle cells and the cuticle and the cuticle is lost at moulting. Microtubules pass through the epidermal cells, from desmosomes at the muscle–epidermal cell boundary to hemidesmosomes associated with muscle attachment fibres at the epidermal–cuticular boundary. The muscle attachment fibres in the cuticle are maintained during moulting, but they are lost at ecdysis and secreted anew.

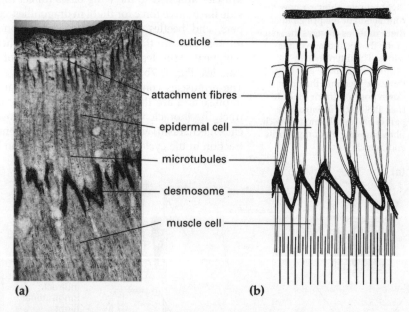

Figure 11.8
Modifications of epidermal cells providing attachment of muscles to the cuticle.
(a) Electron micrograph.
(b) Diagram.

Mouthparts

Hexapods can eat almost any kind of organic material. The basic limb-like mouthparts described above (mandibles, maxillae, labium; Fig. 11.4) constitute the most common arrangement. The mandibles and maxillae have a transverse biting action and are able to manage most kinds of solid or semi-solid food. Some insects have mouthparts modified for feeding on liquid diets: no modification is required for liquids that can be reached directly (drops of water on a surface), but organic liquids are commonly concealed. For example, nectar is often at the base of a tubular corolla. Bees have a greatly elongated labium that allows them to reach the nectar; and moths and butterflies have the galeae (outer lobes) of the maxillae elongated and apposed to form a drinking tube. The phloem or mesophyll cell contents of plants form the food resource for many sucking bugs (Order Hemiptera), which have mouthparts modified as stylets (Fig. 11.9). A pair of stylets derived from the mandibles surrounds an inner pair derived from the maxillae. Together they form a flexible syringe. The serrated tips of the mandibular stylets cut their way through the epidermis, then between cells, and eventually reach their target. Saliva is pumped down one channel between the interlocking maxillary stylets, and food is sucked up a second channel. Stylets can also be used for gaining access to blood. Less extreme modifications occur in lice, thrips, and barklice. Many flies (Diptera) have mouthparts modified for piercing and sucking. Mosquitoes (suborder Nematocera) are the most important example. In these, there are additional stylets derived from other precursors, the labrum and hypopharynx. There are other important modifications of the basic arrangement of mouthparts, particularly in larval forms. The labial 'mask' of dragonfly larvae can be shot out to seize passing prey, and the mandibles and maxillae of antlion larvae form seizing, sucking jaws.

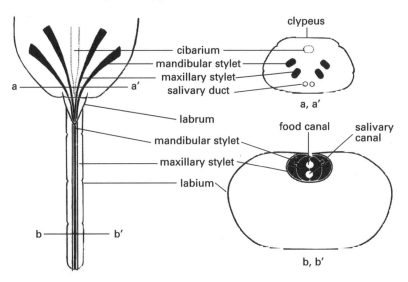

Figure 11.9
Mouthparts of a hemipteran, with transverse section showing interlocking mandibular and maxillary stylets enclosing food canal and salivary canal (from *The Insects of Australia*, CSIRO 1991).

Evolutionary history

The earliest recognisable hexapod fossils are springtails from the early Devonian. Other early fossils include archaeognaths and a now extinct wingless group, the Monura. It is possible that ancestral hexapods first evolved during the Ordovician, when land plants first appeared. The fossil record of many modern orders begins in the Carboniferous or Permian. About fifteen orders recognisable from fossil remains disappeared between the Permian and Cretaceous. Some of these were spectacularly large — Protodonata with 700 mm wingspans, 60 mm silverfish, and cockroach-like palaeodictyopteroids with 30 mm sucking beaks, used for probing into the vascular tissues of ancient treeferns. Often the wings are the only parts of insects that are found as fossils. Their identification depends on understanding and interpretation of the wing venation, which is characteristic for major groups and sometimes even for genera or species. Wing venation is also important in the systematics of extant insects.

Physiological processes

Feeding and digestion

Food finding and recognition

Plant-feeding (phytophagous) insects use successively the senses of sight, smell, touch, and taste to locate plants of the correct species and growth stage. The tarsi and ovipositor also have taste and touch receptors. Laying eggs in the right place ensures there is food for the offspring. Recognition of host plant species may be brought about by stimulants such as sugars and particular plant chemicals; unsuitable hosts are rejected on contact with deterrents.

The feeding process is well-understood in the locust. Food is detected by sight, then by olfactory receptors, and then by tarsal taste receptors. Locusts prefer their food to be oriented vertically: they do not feed readily on cut grass. If the tarsal messages are acceptable, labial and maxillary palps taste the surface and, if the appropriate plant waxes are present, the locust uses its labral taste and touch receptors to select a site for a test bite. Taste receptors within the preoral cavity then signal the presence of feeding stimulants and/or deterrents, and on the basis of this information feeding either continues or ceases. The cessation of feeding is not determined by blood nutrient concentration or by neural adaptation of the taste receptors. It seems to depend on gut stretch receptors, and on the secretion of a peptide hormone which closes the pores in the taste receptors.

Carnivorous insects detect food mainly by sight, but also by touch. The antlion larva is stimulated by sand falling down the walls of its pit. Carnivorous insects may also recognise food by its smell, its temperature, or the emission of metabolites such as carbon dioxide. Mosquitoes can be trapped by using carbon dioxide as an attractant.

Digestive tract

The digestive tract (Fig. 11.10) is divided into three major regions: a foregut, lined with cuticle, a midgut, lined with the sieve-like peritrophic membrane, and a hindgut, again lined with cuticle. The foregut consists of the pharynx, oesophagus, distensible crop, and toothed gizzard. The midgut commonly has caeca (blind-ending tubes) at the anterior end. The hindgut is sometimes considered to be divided into ileum, colon and rectum, but the functions of these parts are not the same as their namesakes in vertebrates. At the anterior end of the hindgut, the Malpighian (excretory) tubules enter.

Some generalisations about gut modifications can be made. Insects that eat solid food usually have a short gut. Insects that eat liquid food usually have a long gut. Insects that eat hard food have a well-developed gizzard. Insects that feed continuously have a short foregut and no storage region, while insects that feed intermittently have a well-developed storage region (the crop) and pass material on slowly to the midgut for digestion. There may also be special modifications; for example, the gizzards of mosquitoes and fleas have fine spines which can rupture blood cells.

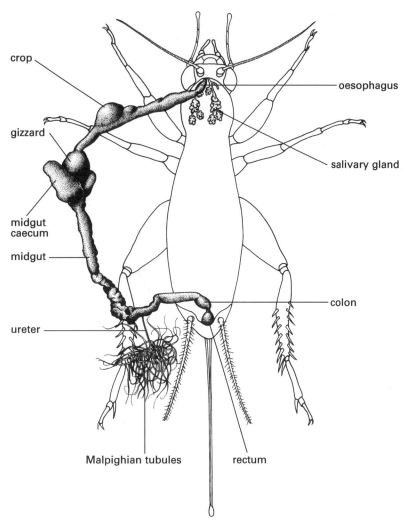

Figure 11.10
Digestive system of a cricket, *Teleogryllus commodus*.

Digestion

The process of digestion in insects is similar in principle to that in other animals. The enzymes secreted by a particular species enable it to digest its normal food; these enzymes may be quite specialised.

The salivary glands produce a watery secretion that lubricates the food and mouthparts and contains enzymes which initiate digestion. The saliva may contain amylase (which breaks down starch), a-glucosidases (which break down sucrose to glucose and fructose), or proteases and lipases (which break down proteins and fats). Special components include pectinases in plant-sucking bugs, venoms in predators, and hyaluronidases in flesh or carrion feeders. The food in the crop continues to undergo digestion, partly through the salivary enzymes and partly through midgut enzymes passed forward into the crop. The foregut does not secrete its own enzymes.

Most digestion occurs in the foregut and midgut, and absorption occurs via the wall of the midgut and its caeca. The types of enzyme secreted are related to the diet. Some of the enzymes in the midgut are summarised in Box 11.2.

> **Box 11.2**
> **Some enzymes of the insect midgut**
>
> **COMMON ENZYMES**
> - exoamylases and endoamylases;
> - glucosidases;
> - trypsin-like proteases, which cut proteins at the carboxyl group of lysine or arginine to give peptides;
> - peptidases;
> - lipases, which hydrolyse glycerol/fatty acid ester bonds.
>
> **SPECIAL ENZYMES IN SOME INSECTS**
> - β-glucosidases, which hydrolyse plant disaccharides such as salicin and cellobiose;
> - collagenases, used by sheep blowfly larvae to break down the connective tissue of skin;
> - keratinases, in clothes-moth larvae feeding on wool;
> - wax-digesting enzymes, in wax-moth larvae feeding on honeycomb.

Absorption

Passive diffusion of hexoses occurs across the midgut and caecal walls into the haemolymph. A concentration gradient is maintained by rapid conversion in the blood of hexoses to trehalose (1,1 α-glucoglucose), a disaccharide of central importance in insect energy metabolism. Plant-sucking bugs, especially aphids, psyllids and some mealybugs, excrete a sugar-rich 'honeydew' which serves as food for a range of other animals, from insects to vertebrates. Lerps, formed from sugary excreta, serve as a covering for psyllids, and were thought to have been the manna that the Israelites ate in the desert.

Amino acids are actively taken up into midgut and caecal cells and then pass passively into the blood. Some are metabolised in the gut cells. Fats are hydrolysed to free fatty acids plus the trihydric alcohol glycerol, then are reformed into diglycerides in the gut cells and passed into the blood in this form. These too are central to energy metabolism.

Nutrition

Insects need the following sorts of molecules in their food:
1. an energy source, generally carbohydrate or fat (or protein in the absence of other energy sources);
2. the amino acids arginine, histidine, isoleucine, leucine, lysine, methionine, phenylalanine, threonine, tryptophan and valine;
3. vitamins: the water-soluble B vitamins (cofactors for enzyme reactions), the fat-soluble vitamins A and E;
4. salts, including trace elements; and
5. unsaturated fatty acids.

They often need additional amino acids, such as glycine, proline, glutamine, aspartic acid, serine, cysteine and tyrosine. Synthesis of these may be possible, but the rate may not meet demand during rapid growth, metamorphosis, silk secretion, etc. All insects need a sterol in their diets. Plant-feeding insects can often modify plant sterols for structural purposes, but carnivorous insects often need cholesterol itself. Some insects with special diets can only use unusual sterols; a cactus-eating species of *Drosophila* can use only the sterol from that plant. There is frequently a requirement for at least a small amount of cholesterol. This may because it is the precursor of the steroid hormone ecdysone. Insects need β-carotene (a vitamin A precursor, required for vision) and plant-feeding insects need vitamin C (ascorbic acid).

Insects require a range of ions similar to those of other animals, and (like mammals) need polyunsaturated fatty acids, which are important to insects in moulting and in reproduction.

Role of microorganisms in nutrition

Many insects with restricted diets (such as those that eat wood, dry cereals, keratin, sap or blood) or varied but poor-quality diets (such as domestic cockroaches) have constant associations with symbiotic microorganisms. These may include bacteria, yeasts or protozoans, and may be intracellular in gut cells or in specialised mycetocytes, or free within the gut. Symbionts may aid nutrition by digesting complex molecules such as cellulose or lignin, synthesising vitamins which are obtained upon digestion of the symbionts, or synthesising and secreting nutrients of limited availability in the diet. Symbionts are transferred between generations via the eggs or by proctodeal feeding, or from secretions from female accessory glands onto the eggs. Sometimes reinfection after each moult is necessary.

Aphids have symbiotic bacteria *Buchnera aphidicola* in their mycetocytes. Molecular studies have shown that the symbionts contain a plasmid which has repeated structural genes for one of the enzymes in the tryptophan synthesis pathway. Other amino acids (especially methionine) may also be contributed by the symbionts. Symbionts in some insects are important in sex determination; in *Stictococcus*, eggs containing bacteria produce females and those without bacteria give males. Some social insects and some beetles and wood-boring wasps use 'external symbionts' to aid their nutrition. For example, some termites and ants have fungal 'gardens' in which the fungi are provided with plant material. The fungi digest the plant material and the insects eat the fungi, or parts of them.

Fat body and metabolism
Structure of the fat body

The fat body of hexapods is a major site of metabolic conversions and storage. It is derived from segmental mesoderm and arranged as sheets of cells. In some insects, modified fat body cells are luminescent; in others they carry haemoglobin. Fat body cells are usually polyploid (4n, 8n). In embryonic and early life the cells are rounded and relatively undifferentiated. Later, their appearance depends on activity. At the start of non-feeding periods (diapause, pupation) they are full of stores of protein, glycogen and fat, but at other times their appearance varies depending on the stage of the moult cycle, reproductive activity, or nutritional status. For example, large amounts of protein are stored in the fat body of queen termites for egg production.

Metabolic functions of the fat body

The fat body has a similar metabolic and storage role to the mammalian liver (Box 11.3).

> **Box 11.3**
> **Some functions of the insect fat body**
>
> Carbohydrate metabolism
> - synthesis of trehalose (the disaccharide which provides a major source of energy for flight);
> - source of glucose for chitin synthesis;
> - synthesis and storage of glycogen;
> - glycolysis and oxidative breakdown of glucose.
>
> Fat metabolism
> - synthesis and storage of triglycerides;
> - synthesis and desaturation of fatty acids;
> - deacylation to form diglycerides, which are used as a source of energy in flight;
> - synthesis of waxes transported by oenocytes to the cuticle.
>
> Amino acid metabolism
> - protein synthesis (structural, storage and yolk proteins);
> - tyrosine storage.
>
> Breakdown of unusable substances
> - including toxins such as plant defence substances and insecticides.

Excretion and osmoregulation

The excretory system (Malpighian tubules plus hindgut) controls both osmoregulation and the elimination of metabolic wastes. The fat body is a major (but not exclusive) site of catabolic activity, and may either pass metabolites back to the blood and thence to the Malpighian tubules or in some cases store them for future recycling.

Structure of Malpighian tubules

The Malpighian tubules (Fig. 11.10) are closed at their distal ends, and open into the gut at the point where the midgut and hindgut join. The number of Malpighian tubules ranges from zero in a few groups of insects, to several hundreds. In some insects the tubules enter one or more common ureters, which then lead into the gut. The wall of the Malpighian tubule is a single cell thick; basal and apical cell membranes are highly infolded, providing a large surface area for taking up or secreting materials. Tracheae and spiralling muscle fibres travel along the outer surface of the tubules.

Urine production

The 'primary urine' is an isosmotic filtrate of haemolymph, driven by the active transport of potassium ions at the apical cell membrane of distal Malpighian tubule cells (Fig. 11.11). Organic compounds usually pass into the tubules intercellularly. There is active resorption of sugars from the lumen, and active transport of some toxins and waste products, especially acid groups, from haemolymph into the tubules. The tubules pass urine into the hindgut, where its composition is modified further. Hence the Malpighian tubules and hindgut are important in controlling homeostasis in insects.

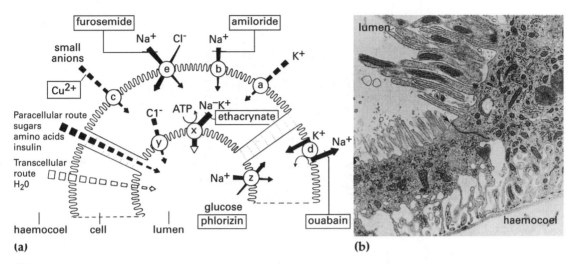

Figure 11.11
(a) The transportation of materials across the wall of a Malpighian tubule. Energy-requiring processes are shown by solid arrows and passive processes by dotted arrows (after Phillips 1981). (b) Electron micrograph of Malpighian tubule cells, showing microvilli on the lumen side and infoldings of the basal cell membrane on the side facing the haemocoel (from Bradley, in Kerkut and Gilbert 1985).

Form of nitrogenous waste

Hexapods can excrete nitrogenous waste in various chemical forms. The final product of nitrogen metabolism depends on the availability of water. Many freshwater invertebrates excrete ammonia, which is toxic to the tissues but diffuses rapidly into the environment. Terrestrial species with moderate water supplies usually excrete urea. If water is limited, or if weight considerations prevent the storage of excretory material in water, uric acid is common as an end-product of amino nitrogen excretion; this is so in many insects. Insects, with few exceptions, also excrete the products of purine breakdown as uric acid.

Storage excretion

Hymenopteran larvae and all pupae store excretory products in the gut and release them at adult emergence. Other insects may store uric acid in special fat body cells called urocytes. The white markings on some bugs and Lepidoptera are the result of stored uric acid. Nitrogen-rich pigments (pteridines) in the wings of some insects can also be regarded as a form of storage excretion.

Role of the hindgut in excretion and osmoregulation

The anterior hindgut (ileum) is active in ion transport, and is permeable to water, sucrose, glucose, serine and proline, which are resorbed. The colon is not active in ion uptake. The posterior hindgut (rectum) regulates the uptake of ions and water to ensure osmotic and ionic homeostasis. Specialised rectal pad cells in the hindgut remove water from the hindgut contents. Their cell membranes (Fig. 11.12) are infolded to form very complex series of parallel compartments. Ions concentrated by active transport in these compartments, in the apical part of the cell, bring about an osmotic flow of water from the hindgut contents to the intercellular areas, while ion resorption in the basal part of the cell allows a hyposmotic (dilute) fluid to be returned to the blood. The material egested from the anus consists partly of nitrogenous excreta and partly of faecal material.

Circulation

Hexapods have the typical open circulation of arthropods (Fig. 10.1). A dorsal heart with ostia lies in the pericardial space and pumps

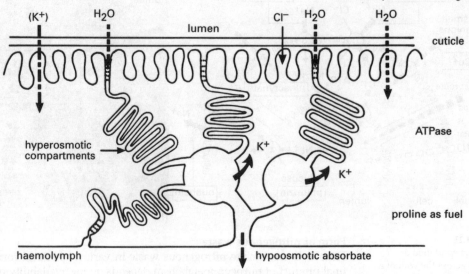

Figure 11.12
Rectal pad cell function. Pathways of water movement are shown by broken arrows, solutes by solid arrows (from Bradley, in Kerkut and Gilbert 1985).

haemolymph forward into the dorsal aorta. The aorta usually opens between the brain and oesophagus, and thereafter the haemolymph circulates throughout the haemocoel. Some groups of insects have short segmental arteries and/or arteries associated with appendages. Pulsatile organs are often associated with the wings and legs. The blood of nearly all hexapods has no respiratory function and contains no respiratory pigment. However, it serves the other general functions of blood, transporting nutrients, hormones, metabolic wastes and blood cells. These act in phagocytosis, encapsulating parasites, and healing wounds.

Terrestrial respiratory exchange
General structure of the tracheal system

The respiratory system of insects is an internal network of branching and interconnecting tracheae. They are lined with cuticle and filled with air, and open at the segmentally arranged spiracles (primitively 12 pairs). The finest tubules of the tracheal system, or tracheoles, develop as cuticular linings of invaginations in terminal cells called tracheoblasts. The tracheal cuticle contains the same layers as the surface cuticle, but lacks the cement layer and the wax layer; hence it is water-permeable. The tracheoles are fluid-filled except when oxygen demand is high. There is a spiral cuticular ridge (= taenidium) including sclerotised cuticle in the tracheae and tracheoles. This prevents collapse.

The tracheal system can occupy nearly 50% of the total body volume. Its cuticle is secreted by epidermal cells similar to those that secrete the cuticle of the body surface. The branching of the tracheal system to individual cells is an efficient method of supplying oxygen, provided the total body size is small. This overrides the need for a respiratory pigment in the blood, except for haemoglobin in a few kinds of insects (e.g. midge larvae) in aquatic environments poor in oxygen.

Passage of oxygen

The movement of oxygen to respiring tissues depends on diffusion. The rate of diffusion depends partly on the molecular weight of the gas, partly

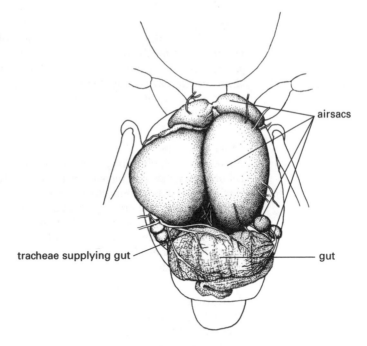

Figure 11.13
Abdominal tracheae and airsacs of a bee.

on the tissue permeability, and partly on the concentration gradient. The diffusion rate of oxygen in air is 100 000 times that in tissue. Diffusion through cuticle is even slower, so that the slowest part of gas exchange is from the tracheole across the cuticle and through the tissue. A muscle fibre with tracheoles on the outside only is limited in diameter according to its energy requirements. For example, a dragonfly flight muscle fibre cannot exceed 20 μm diameter because the diffusion times would be too long. In asynchronous flight muscle the tracheoles are embedded in the fibre, so larger diameters are possible.

A surprising feature of the tracheal system is that cells in conditions of low oxygen concentration may send out cytoplasmic processes, capture tracheoles and draw them towards the oxygen-deprived region.

Spiracles

In most insects the spiracles can be closed by valves, thus limiting water loss. Closure is maintained by nervous stimulation of a closer muscle. Opening is generally in response to high carbon dioxide concentration; sometimes a high potassium ion concentration, which indicates dehydration, inhibits opening. Spiracles also usually have dust filters. In many insects not all spiracles are functional.

Ventilation

In large, active insects, parts of the tracheal system are expanded as airsacs (Fig. 11.13). Diffusion is not fast enough to supply the oxygen needed, and there may be either passive suction (by drop in oxygen pressure on utilisation) or pumping movements (= active ventilation), especially of the abdomen, forcing air from the airsacs to the tissues. Some spiracles are reserved for taking in air and some for expiration, giving a directional airflow. During flight the thoracic airsacs are compressed by the contraction of flight muscle.

Moulting

The cuticular lining of the tracheal system is shed with the exuviae at each moult, except for the lining of the tracheoles. The tracheae are fluid-filled just after moulting, but the fluid is quickly resorbed. The inflation of the tracheal system may contribute to the expansion of body size after moulting, and to wing expansion. There is usually no further growth of the system after adulthood.

Other functions

Airsacs lower the specific gravity of an insect and allow the growth of internal organs without an increase in body size. They also provide insulation. Tracheae serve as a connective tissue. They form a reflecting layer on the back of the eye, and are important adjuncts of tympanic organs. They are used, remarkably, in sound production by some cockroaches, which force air out through the spiracles with a hissing noise. Other cockroaches can squirt out quinones through a pair of abdominal spiracles, thus deterring potential predators.

Aquatic respiratory exchange

While the non-insect hexapods are terrestrial (but limited to humid microenvironments), there are many insect orders in which larvae and/or adults inhabit fresh water, and occasionally even marine environments. These colonisations of water are generally regarded as being secondary, and respiratory exchange is through modifications of the tracheal system, either by collecting gaseous air or using dissolved oxygen. Some aquatic insects make frequent visits to the surface (e.g. the larva and pupa of mosquitoes) or maintain continuous contact with the surface (e.g. the water scorpion and rat-tailed maggot). There are usually spiracular adaptations that prevent the entry of water — a hydrofuge area formed by oily secretions from glands near the spiracle, hydrofuge hairs which cover the spiracle but are spread by surface tension, or hydrofuge hairs within the spiracle.

Other insects have temporary, compressible gas gills, e.g. diving beetles, many hemipterans, and some trichopterans. In these, a bubble of air is held by long hairs, in contact with spiracles, and the utilisation of oxygen causes it to diffuse from the water to the bubble and hence to the spiracles. At the same time the partial pressure of nitrogen in the bubble is raised (as that of oxygen falls) so that nitrogen diffuses back into the water. The size of the bubble decreases and it eventually has to be replaced. Submergence time is shortened by an increase in oxygen consumption and by an increase in depth (which increases the rate at which nitrogen diffuses out). The use of a bubble is combined with haemoglobin as an additional oxygen store in the heteropteran backswimmer *Anisops*. The bubble lasts longer in flowing water, because of the replacement of oxygenated water and physical effects, and at low temperatures or low activity levels.

A third group of insects uses gills that are thin-walled extensions of the body, containing either blood or tracheae into which oxygen diffuses. Blood gills are processes covered with very thin cuticle; tracheal gills are extensions of the body containing tracheal branches. These insects — the larvae of mayflies, dragonflies, damselflies, stoneflies and caddisflies — have no functional spiracles. When the oxygen concentration in the water is low, as it may be at higher temperatures or in stagnant water, only insects with tracheal gills survive.

Some beetles (elmids) and some bugs (belostomatids and naucorids) have permanent gas gills known as plastrons. These also occur on the spiracular gills of some larvae, and on some eggs. Air is held in place by a layer of short epicuticular hairs (2 million per square millimetre). The plastron works in a similar manner to a compressible gill, but because it is incompressible, nitrogen does not diffuse out. Hair plastrons withstand excess pressures of 50–300 kPa (depth of 5–30 metres). Wetting occurs before structural damage, and can also be brought about by low surface tension (e.g. through the action of detergents). Plastron function requires a high partial pressure of oxygen in the water: a low one will extract oxygen from the tissues. Some plastron-using insects leave the water if the partial pressure of oxygen is too low.

Control systems

The integration of insect function and behaviour is carried out, as in other animals, by the nervous system and endocrine system.

Nervous system

The insect central nervous system is made up of segmentally arranged ganglia joined by connectives. The connectives usually appear fused in the abdomen at least, and in advanced insects (flies, bees etc.) all ventral ganglia are fused in a single mass. Segmentally arranged nerves pass to the muscles, and sensory nerves return to the ganglia from the sense organs. A neural lamella around the nervous system is secreted by perineurium cells; it consists of collagen-like fibres in a mucopolysaccharide–mucoprotein matrix. The perineurium cells are involved in nutrition and probably also constitute a blood–brain barrier.

The rate of impulse transmission is directly proportional to axon diameter. Giant fibres are large-diameter (60 μm compared to 5 μm) axons of interneurons. For example, the sensory neurons in the cercal nerve transmit impulses to the cell body of the giant fibre in the last abdominal ganglion, and impulses are then rapidly transmitted to the motor nerves that initiate the escape response. Most of the common insecticides work by affecting the transmission of nerve impulses.

Endocrine system

The neurohumoral system consists of nerve cells which release noradrenalin, dopamine, octopamine or 5-hydroxytryptamine into blood or at a target organ. These substances also act as neurotransmitters in the communication of nerve cells across synapses.

Special nerve cells in the brain and other regions of the central nervous system produce a vast range of peptide hormones that are transported along the cell axons to target organs or to sites of release into the haemolymph. Generally, only one kind of hormone is produced per cell. Other cells that produce peptide hormones occur in the gut and in organs associated with the reproductive tissues. Hundreds of peptide hormones are now known from insects. The corpora cardiaca, a pair of small glands behind the brain (Fig. 11.14), contain nerve endings filled with peptide hormone granules, as well as glandular cells that produce their own hormones. Two other important pairs of glands are the corpora allata, just behind the corpora cardiaca, and the prothoracic glands. These glands have central roles in the control of metamorphosis and moulting. Some examples of different kinds of hormones and their functions are given in Table 11.1.

Figure 11.14
Brain and associated endocrine structures in the cockroach *Periplaneta americana*.

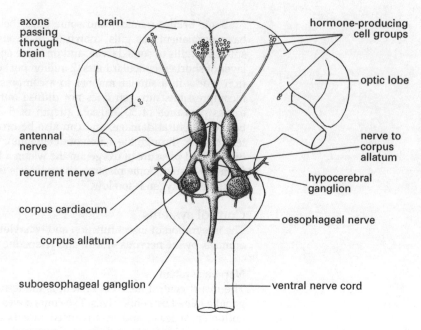

Table 11.1 Some insect hormones and their functions.

Prothoracotropic hormone (PTTH)
Protein hormone, stimulates the prothoracic gland; produced in special cells of the brain and travels along axons to the corpus cardiacum, where it is released to the blood; causes synthesis of ecdysone by the prothoracic glands; PTTH is released by stimuli including proprioception, photoperiod, and temperature.

Juvenile hormone (JH)
Multifunction hormone (there are several closely-related varieties); when present at a moult, JH causes production of another larval cuticle; its absence can induce diapause; required for the synthesis and transport to the oocytes of yolk proteins; JH determines polymorphism in Hymenoptera (queen or worker), termites, locusts, etc.; involved in sex determination in aphids (male eggs are ovulated only if JH is low); many other functions.

Ecdysteroids
Control the process of moulting, much of cell differentiation, and programmed cell death; ecdysone initiates the syntheses required in moulting, by binding to a receptor protein which in turn binds to a particular DNA region and initiates mRNA synthesis; prothoracic gland breaks down at metamorphosis, but flies (and other insects) synthesise ecdysone in the ovary in the adult stage, and this ecdysone then stimulates yolk protein synthesis.

Adipokinetic hormones I and II (AKH I and II)
Short peptides produced by the glandular cells of the corpora cardiaca; AKH I causes lipid release from the fat body as a fuel for flight; AKH II causes hyperglycaemia, perhaps by mobilisation of trehalose from the fat body.

Sense organs

Mechanoreception
Mechanoreception includes the reception of touch, stretch, vibration, gravity and other mechanical stimuli. Touch receptors are articulated bristles (trichoid sensilla, Fig. 11.15b) on the insect's surface. Some trichoid sensilla respond only during movement of the hair; this is a typical response of tactile receptors on the antennae and tarsi. Such hairs may respond to

substratum vibration or to high-intensity sounds. Other trichoid sensilla respond as long as the hair is displaced. This response is typical of proprioceptors at the joints of legs, and of facial hairbeds, which may be stimulated by air moving faster than 2 m/s, allowing the insect to orientate itself during flight with respect to wind and to control yawing. Campaniform sensilla are slow-adapting proprioceptors in groups at wing bases, halteres and leg joints. They respond to compression forces in the cuticle.

Chordotonal organs are attached to the cuticle at one or both ends and are sensitive to vibrations with very small displacements. Johnston's organ, in the base of the antenna, has a phasic response to sound (e.g. courtship dances and calls) or ripples (avoiding collisions by echolocation using ripples, as do freshwater bugs). Tympanal organs also respond to sound.

Some aquatic insects have specialised pressure receptors. The plastron-breathing bug *Aphelocheirus* requires highly oxygenated water, which is usually found close to the water surface. The flattening of hairs surrounding the pressure sensors at greater depths provides information enabling the bug to move to shallower water.

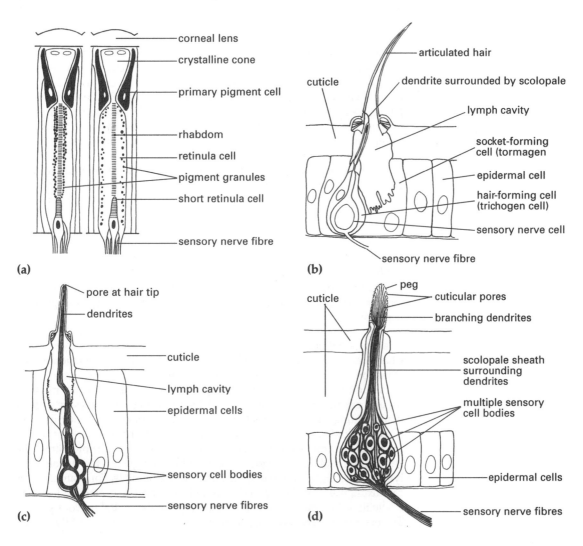

Figure 11.15
Microstructure of insect sense organs. (a) Light-adapted (left) and dark-adapted (right) ommatidia of a compound eye. (b) Hair (trichoid) sensillum, which is responsive to displacement. (c) Contact chemoreceptor (taste). (d) Distance chemoreceptor (smell). (a after *The Insects of Australia: Supplement*, CSIRO 1974; b, c, d after Chapman 1982).

Eyes

The compound eye is made up of ommatidia (Fig. 11.15a) and produces a detailed image called an apposition image in good light. The light falling on any given lens is directed to a specific point on the retina. The aperture is small and the image sharp. The resolving power may be as good as 1° of separation. In dim light, the distribution of pigment surrounding each ommatidium may alter so that light can reach particular points on the retina via several neighbouring ommatidia. Acuity is reduced in these circumstances, but the superposition image makes more efficient use of the available light. Nocturnal insects usually have such eyes.

Each ommatidium contains eight receptor cells whose inner region, the rhabdomere, is folded into close-packed microvilli. The microvilli contain visual pigment consisting of retinene plus the protein rhodopsin. Incoming light causes a molecular change in the pigment, releasing energy which gives rise to a receptor potential.

Insects have up to four different pigments (only one per cell), which are sensitive to different wavelengths, so that colour vision is possible. The pigments have maximum absorption in the ultraviolet, green, blue and sometimes red regions (300–600 nm), corresponding to flower and leaf colours. The wavelengths are filtered in tiers of cells within each ommatidium. Bees can discriminate between wavelengths separated by as little as 8 nm in the ultraviolet/blue–green range. Hymenoptera can detect the pattern of polarised light in the sky when the sun is concealed by clouds, and can use this pattern for navigation. Insects may have almost a 360° field of vision, giving good binocular vision in front and behind. This allows them to judge distance well, and is important in capturing prey and avoiding predators. Compound eyes are well adapted for detecting movement, as the image passes successively across different ommatidia.

Many insects have two or three simple eyes (ocelli) in addition to compound eyes. An ocellus consists of up to about a thousand sensory cells with a single cuticular lens. Ocelli do not form an image, but respond to the intensity and direction of light.

Holometabolous insect larvae have 1–6 simple eyes called stemmata on each side of the head. Each is similar to a single ommatidium, so the image is very poor, resembling a mosaic of up to 12 patches. Visual information can be increased by scanning from side to side.

Photoperiod (daylength), an important environmental influence in insect biology, is not perceived by the eyes but by special neurons, via the head capsule.

Chemoreception

Chemoreception can be divided into olfaction (smell) and gustation (taste). Olfaction occurs via pores in the cuticle, usually on pegs or domes, or on pegs sunk into cavities. Each pore opens into a cavity in which multiple tubules extend to the dendrite (Fig 11.15d). Odorous molecules (from food plant, mating partner, oviposition site, odour trail or colony odour) are captured by lipids on the cuticle and diffuse to the pores. They are carried by the tubules to the dendrite, where they bind to specific acceptor sites. The stimulating molecules are then inactivated by enzymes in the surrounding lymph. Very high degrees of discrimination are possible. For example, bees can distinguish pure benzyl acetate from benzyl acetate containing less than 1% linalool. Discrimination occurs at the level of the central nervous system. The behavioural response depends on the physiological state of the insect.

Insects can taste liquids, and some insects can taste solids such as leaf wax. Taste receptors (Fig. 11.15c) are mainly on tarsi, mouthparts and ovipositor. The physiological mechanisms are similar to those of olfactory receptors. Each taste receptor may have neurons which respond individually to different molecules.

Sense organs and control of flight
Sense organs control all aspects of insect flight, from orientation to the destination (short or long-range), to stability of the body and motor coordination. All proprioceptive organs are influenced by movements in flight, so potentially all of them may be involved in flight control.

In locusts, flight can be initiated by stimulating the wind hairs on the forehead. This causes the pattern of motor output to be initiated, even in experimental preparations of locust head + nervous system. The pattern can persist in decapitated preparations, so the thoracic ganglia seem to be sufficient to keep the pattern of sequential muscle contraction going. Feedback from various mechanosensory organs (trichoid sensilla, campaniform sensilla, chordotonal organs) is used to control body position, and the integration of visual, olfactory and auditory stimuli provides orientation with respect to the target.

For example, a male housefly chases a small, rapidly moving object in its visual field, and keeps the object, as far as possible, frontally on the ommatidia. Deviations are responded to by a change in torque. These responses enable the fly to catch a female for mating. Dragonflies have two types of interneurons in the pro-mesothoracic connectives, one responding to movement of small objects in the visual field, and one to large pattern movements (i.e. one to prey, mate or rival, and the other to landscape movement, equivalent to self-movement).

Responses to auditory stimuli require brain processing and are not local reflex responses. In contrast, navigational responses to olfactory stimuli are simply the result of upwind movement while the stimulus is present.

Flight speed is controlled by aerodynamic sense organs (which monitor speed in relation to air) and vision (which monitors speed in relation to the ground). The aerodynamic sense organs are the Johnston's organs in the bases of the antennae (chordotonal organs); these perceive passive drag on the antennae. In flies a single campaniform sensillum may serve this role.

Altitude is poorly controlled in insect flight. Insects seem to respond to the rate of movement of the landscape against the ommatidia. This is greater at low altitude, so they fly more slowly. Body alignment is controlled by vision, aerodynamic and inertial forces. Gravity is not used.

Reproduction and development

Reproductive systems and processes
The reproductive systems of the hexapods are generally similar in construction in both sexes, with paired gonads each having a duct to carry the gametes to a common median duct leading to a posterior gonopore. There are also accessory glands associated with the gonoducts in both sexes. The female has a spermatheca for storing sperm. The gonads of pterygote insects are each a group of tubular structures along which gametogenesis proceeds from the tip towards the base; these tubes are called follicles in the male and ovarioles in the female. In Collembola, the gonads are sac-like.

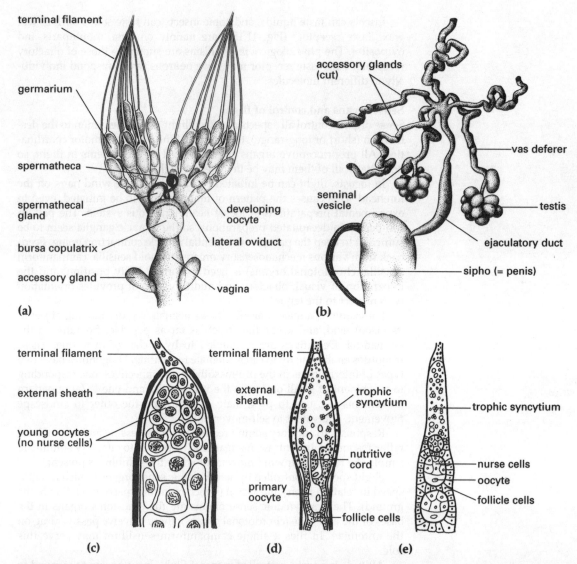

Figure 11.16
(a) Female reproductive system of the fungus-eating ladybird beetle, *Illeis galbula*. (b) Male reproductive system of *I. galbula*. (c) Apex of panoistic ovariole. (d) Apex of telotrophic ovariole. (e) Apex of polytrophic ovariole. (c–e after Chapman 1982.)

The ovaries (Fig. 11.16a) are made up of one to many ovarioles, each with a small germarium at the distal end. In the germarium, immature oocytes begin to grow one at a time and are released sequentially, surrounded by follicle cells, into the ovariole. The germarium (Fig. 11.16c) may contain nurse cells, which provide nutrients to the oocytes and, in some insects, accompany them down the ovariole. The ovarioles connect with the two lateral oviducts. In most insects these join to form a cuticle-lined common oviduct, which in some groups continues as the vagina. At the end of the common oviduct, the spermatheca arises dorsally. In many Lepidoptera there is a separate structure, the bursa copulatrix, with its own external opening, into which sperm are deposited by the male. It has a duct leading to the spermatheca. Accessory reproductive glands also enter the beginning of the vagina. They produce various substances, often used in attaching eggs to a substratum. In cockroaches and praying mantises, the accessory glands are large and produce the materials required to build the egg case (ootheca). In some of the social Hymenoptera, reproductive glands produce venom or trail-marking pheromones.

The paired testes (Fig. 11.16b) each consist of one or more tubular follicles, grouped in some insects in distinct lobes. In the non-insect hexapods, the testes are saccular. Each testis follicle has a short vas efferens leading into the common vas deferens on each side. The vasa deferentia may be expanded proximally to form seminal vesicles before entering the median, cuticle-lined ejaculatory duct. The ejaculatory duct leads to the aedeagus or intromittent organ. Accessory glands enter either the vasa deferentia or the ejaculatory duct. Their products facilitate sperm transfer and contain bioactive substances such as prostaglandins that affect female physiology and behaviour.

Aggregation prior to mating is brought about by environmental and signalling stimuli. Mating swarms of males (especially Diptera) may be triggered by landscape features, including stumps, rocks or smoke plumes. Individual females responding to the same features are quickly mated. Some male butterflies aggregate at the tops of hills or mountains ('hilltopping'), with the same effect.

Mate recognition may include any of an array of sensory stimuli, such as vision (courtship display, or a body pattern including fluorescent and ultraviolet reflectance patterns), sound, pheromones (either recognition or sexual) and touch. Males may present a meal to a female; this may distract her (from moving away, or trying to eat him) and contribute nutritionally to the successful outcome of the mating. Scorpion flies are an example: the male brings a prey insect and the female feeds on it during mating.

Copulation usually occurs by the insertion of the aedeagus into the vagina (or sometimes spermatheca). Variations occur in some insects. Copulatory and oviposition openings may be separate, as in Lepidoptera. There may be secondary sperm storage structures and genitalia, as in male dragonflies. Insemination may be via the female's body wall, as in bedbugs.

Copulatory positions are characteristic for particular insect groups. Primitively, either the male or the female was on top, with the tip of the abdomen folded down to make contact with the genitalia of the mate sex (as in cockroaches and some Orthoptera). In the 'false male above' position the male is on top, but the abdomen is twisted so that it contacts the female from below (e.g. locusts). Twisting of the abdomen is also used by insects that mate side-to-side (e.g. scorpion flies). In insects that mate end-to-end, the genitalia of the male rotate through 180° during metamorphosis (e.g. Heteroptera, some lower Diptera). The male generally uses his legs to hold the female, and there may be sexual dimorphism of the legs. In some Collembola and fleas, the modified antennae of the male are used to hold the female. In most insects the complex male genitalia (usually retracted within the abdomen) are used in a specific way to grip and/or stimulate the female.

While most insects mate, some species can reproduce without mating, that is, they are parthenogenetic. Automictic parthenogenesis involves meiosis, associated with fusion of the two daughter nuclei (in a species of *Drosophila*), or later fusion of nuclei in pairs (in some scale insects), or a preliminary doubling of chromosome number (in some Australian grasshoppers). Some of these options permit segregation of genetic material, so that the offspring are not identical to each other. In apomictic parthenogenesis there is no functional meiosis, the maturation division of the egg being mitotic. The offspring are all female clones, identical to their mother and to each other. Aphids are an example of this. In haplodiploid parthenogenesis, unfertilised eggs give rise to males. This occurs in Hymenoptera and in some scale insects, where fertilised (diploid) eggs give females and unfertilised (haploid) eggs give males. The female hymenopteran can control whether or not sperm are released from the spermatheca to fertilise any given egg. However, it is not the chromosome number as such that determines sex. In at least some hymenopterans, allelic 'sex factors' are required to be heterozygous for female development.

Most insects lay eggs, but viviparity also has a scattered occurrence throughout the Insecta. In ovoviviparity there is no nutritional contribution from the female after the egg is formed, but the young hatches within the female tract. In adenotrophic viviparity, the embryo develops in the egg, hatches in the uterus, and feeds orally on uterine secretions. It is deposited as a fully formed larva, and immediately pupates, as in *Glossina* (tsetse fly) and parasitic hippoboscid flies. Pseudoplacental viviparity, in which a connection develops between maternal and embryonic tissue, is seen in parthenogenetical aphids and in *Hemimerus* (an earwig ectoparasitic on rats). Finally, in haemocoelous viviparity there are no oviducts, and the embryo has no yolk or chorion; the larvae develop in the haemocoel and feed directly on maternal tissues. This occurs in some fungus gnats.

Some insects, such as fungus gnats, exhibit paedogenesis, reproducing while still in the larval stage. Paedogenetic reproduction allows rapid exploitation of an ephemeral environment.

Gametes and fertilisation

Insect sperm are generally filamentous, with head and tail about the same diameter. The nucleus takes up most of the elongate head, while the tail contains the mitochondria and the motor apparatus of microtubule pairs. The sperm are inactive in the vas deferens and seminal vesicle.

As in arachnids and myriapods, non-insect hexapods and apterygote insects produce a spermatophore and place it on the ground. Usually (but not in Diplura) the female is then brought by the male into a position where she can take up the spermatophore. Silverfish males spin silk threads over the female to keep her in the right area. Many pterygote insects also make spermatophores but transfer them directly to the female tract. The female may eat the outer parts of the spermatophore, or she may absorb it via the reproductive tract. Some Orthoptera have a very large spermatophore, most of which protrudes from the female genital opening (Fig. 11.17). It is known as a spermatophylax and is eaten by the female.

Females of many species mate successively with different males, and there are often mechanisms that increase the chances that a particular male's sperm will fertilise the eggs. Dragonfly males use the secondary sexual structures at the anterior end of the abdomen to displace previously deposited sperm, either by scooping them out or by pushing them further in. In Orthoptera, genetic studies have proved that similar activities allow the sperm of a particular male to achieve precedence over the sperm of others.

During oogenesis, the yolk protein vitellogenin is produced by the fat body under hormonal control — by ecdysone in insects that mature their eggs before the adult moult, and by juvenile hormone in insects that mature eggs during adulthood. The yolk protein passes into the haemolymph. Gaps develop between the follicle cells of the ovary under the influence of juvenile hormone, and the protein is taken up pinocytotically into the

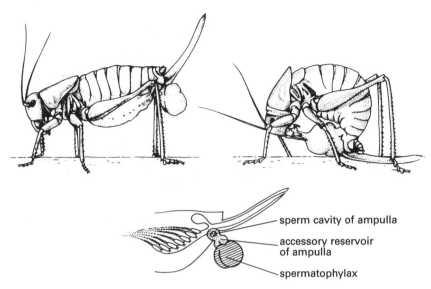

Figure 11.17
A female cricket, *Anabrus simplex*, feeding on spermatophylax (from Gullan and Cranston 1994).

oocyte. The growth rate of the oocyte can be enormous — 100 000¥ in three days in *Drosophila*. Lipid yolk is derived from the fat body and perhaps from the nurse cells.

The egg is enclosed in a vitelline membrane secreted jointly by the follicle cells and the oocyte. Outside the vitelline membrane is the chorion or egg shell, formed as a secretion of the follicle cells. The chorion has layers similar to those of the cuticle, but in the reverse order, i.e. with the wax layer closest to the vitelline membrane. Pores known as aeropyles provide for gas exchange, and there is often a plastron surrounding them. Special pores (the micropyles) provide for sperm entry, one or more sperm being released from the spermatheca as the egg passes its duct. The chorion may have a cap surrounded by a line of weakness which is broken by the insect at hatching.

The female usually deposits the eggs at an appropriate food source for the hatching larva. Often the appendages of the terminal segments are modified to form the valves of an ovipositor, used in digging into soil (as in locusts) or plant tissue (as in sawflies and cicadas) or in penetrating the exoskeleton of other insects. Some parasitic wasps have extremely long ovipositors, and some bore through wood to reach the burrowing larvae of other insects.

Embryonic development

An understanding of insect embryology is basic to our present knowledge of the genetics and molecular biology of development and differentiation, not only in insects but in other animals, including vertebrates. The following gives some indication of the sequence of insect embryonic development.

At the time of laying, the egg is aligned in the same way as the mother, with the posterior end towards her tail. Fusion of male and female pronuclei occurs at about one-third of the distance from the anterior pole of the egg. The egg is filled with granules and droplets of protein and lipid yolk, making up most of its volume. Yolk-free cytoplasm can be found in many insect eggs as a layer around the periphery of the egg and a network of strands between the yolk granules. Cleavage in Collembola is total, but in other hexapods the zygote nucleus undergoes a number of cleavages within the yolk mass (Fig. 11.18) and the nuclei remain in a syncytium: no cell membranes are formed. After several cycles of divisions, some of the nuclei migrate to the surface of the egg and divide further (the 'syncytial blastoderm' stage). Eventually, cell membranes form between and beneath the nuclei, separating them into cells of the 'cellular blastoderm'.

In many holometabolous and some hemimetabolous insects, a group of cells ('pole cells') becomes differentiated at about this time, and they eventually form the germ cells within the gonads. Cell division in the ventral region of the blastoderm then produces a 'germ band', or strip of cells that will form the definitive embryo. The remainder of the blastoderm is extra-embryonic ectoderm covering the yolk mass. The extra-embryonic ectoderm then usually folds over the germ band, forming an inner membrane, the amnion, and an outer one, the serosa. The germ band increases in length, sometimes sinking into the yolk or curling backwards on itself as appendages and segmentation become evident. At the same time, gastrulation is occurring, with the formation of mesoderm and the intucking of stomodaeum and proctodaeum. These two structures carry at their tips the anterior and posterior midgut rudiments, which eventually join, ventrally at first, and finally enclose the left-over yolk by growing up dorsally around it.

The germ band then contracts and straightens. The amnion ruptures and lateral ectoderm and mesoderm grow up to obliterate the serosa: this process is called dorsal closure. The differentiation of the remaining structures proceeds rapidly to produce the mature embryo.

Modern genetic and molecular studies on the embryos of *Drosophila* and other insects have revealed complex and subtle processes during the development of visible form. For example, the segmental units are first established as parasegments, each of which overlaps the front and back of the later-defined body segments. A complex interaction of genes and their products with anterior–posterior gradients of substances in the egg and early embryo also generates the development of regional differences in segmental structure. Such investigations are at the forefront of current analyses of the relationship between genes, development and evolution (see Chapter 18).

Larval development

There are no particular environmental stimuli to hatching. Fully developed embryos increase their volume by swallowing extra-embryonic fluid or air. There may be special hatching muscles, egg bursters, or hatching lines; hatching enzymes may digest parts of the chorion. The embryo is generally clad in an embryonic cuticle which is shed at about the time of hatching.

Growth is more or less continuous, with interruptions during moulting, though the linear dimensions of hard parts can change only at ecdysis. Some insects recycle the exuviae by eating them, some attach the exuviae to the dorsal surface, where they may act as camouflage, but most simply abandon them. Growth is often allometric, occurring to different degrees in different parts of the body, particularly at the adult moult. The number of instars is usually greater in more primitive insects, and the number may vary depending on conditions.

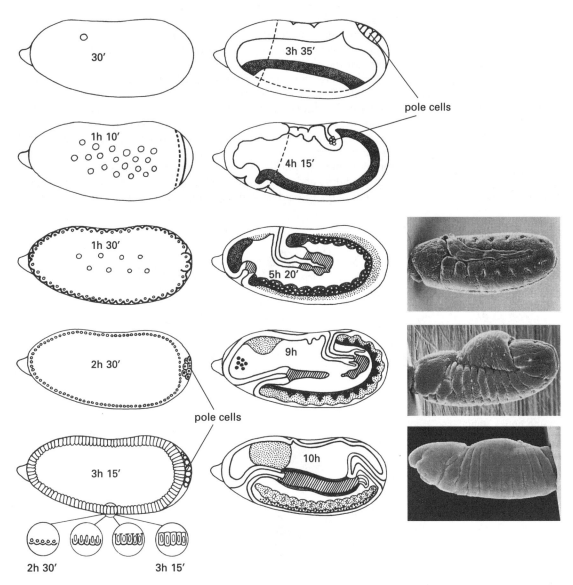

Figure 11.18
Early development of *Drosophila*: dark stipple — mesoderm; light stipple — nervous system; hatch — endoderm. Extra-embryonic membranes are reduced in this embryo, which does not develop an amnion or serosa. Circled enlargements show cellularisation of the blastoderm (after Lawrence 1992). Scanning electron micrographs show surface views of stages similar to drawings immediately to their left. (From Fullilove and Jacobson 1978; see text for further description.)

Pharate ('cloaked') stages exist when the insect has undergone a moult but not yet performed an ecdysis. The new cuticle is formed, with the features of the next instar, but the insect is still 'cloaked' in the old cuticle. For example, the pharate adult still looks like a pupa. There may be a long delay between moulting and ecdysis; some moths overwinter as the pharate adult.

Metamorphosis

In the ametabolous development of non-insect hexapods and apterygote insects, there is no distinction in form between larval and adult stages except for the development of sexual structures. Moulting continues after adulthood is reached. The larvae of hemimetabolous insects are similar to the adults. The wings and genitalia develop externally but are not fully formed until adulthood. In holometabolous insects, the larvae are unlike the adults. Between the last larval instar and the adult there is a pupal stage. The pupa is an essentially immobile stage at the beginning of which the wings evert to the external surface. This kind of development is called 'complete metamorphosis'. The pupa, being immobile, often has protection from predators and parasitoids, such as a cocoon, cryptic colouring, concealment in the ground, jaw-like gin traps on abdominal segments, etc.

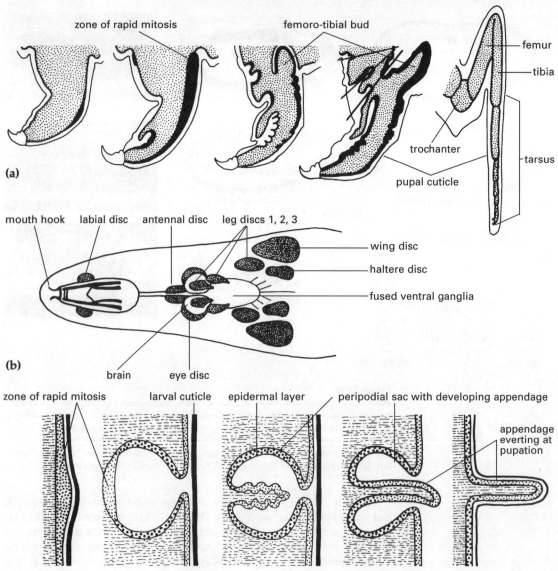

Figure 11.19
(a) Metamorphosis of larval leg to adult leg in a butterfly (zones of major mitotic activity shown in black). (b) Position of imaginal discs contributing to adult head and thoracic structures in *Drosophila* larva. (c) Growth, differentiation and eversion of an appendage-forming imaginal disc. (After Chapman 1982.)

The wings of exopterygote insects grow by mitosis at successive moults as extensions of the pleura and terga of thoracic segments 2 and 3. The accessory sclerites and articulation arise only at the adult moult. Flight muscles are already present in the juvenile stages but undergo rapid growth and differentiation just before the final moult. The external genitalia differentiate gradually at each moult.

In holometabolous insects, some adult structures are set aside as pockets of cells (the imaginal discs, Fig. 11.19b) in the embryo and do not form a functional part of the larval body. They grow by mitosis during larval life (Fig. 11.19c) but are not everted until the larval–pupal moult. This situation reaches its extreme in the higher Diptera, where almost every part of the adult body is derived from imaginal discs, and the larval structures are broken down by enzymes and recycled.

The wings form from imaginal discs in all endopterygote insects, with the wing itself developing within a sac called the peripodial membrane, which forms part of the thoracic wall in the adult. In some endopterygotes the legs also form from imaginal discs (Fig. 11.19b), but in others the larval appendages differentiate further into the adult form (Fig. 11.19a). Individual muscles may persist, or be reconstructed, destroyed, replaced, or developed as new structures. The tracheae send out branches to new tissues and the new tracheoles uncoil as the tissues differentiate. The gut may change dramatically if the adult food is different from that of the larva.

Energy sources for metamorphosis include stored fat and glycogen. Larval structural proteins are broken down to amino acids and adult proteins are synthesised from special larval storage proteins. Waste products accumulate, and are discharged on adult emergence as meconium.

Emergence from the pupa is called eclosion. Eclosion behaviour is initiated by the peptide eclosion hormone. The insect swallows air, increases its thoracic blood pressure, bursts the remnants of the pupal cuticle at the ecdysial line, and wriggles free of the exuviae. It pumps blood into its wings, assisted by gravity. If there is a cocoon, the insect escapes using its mandibles (pupal or adult), or a cocoon cutter. Some silkmoths have an enzyme, cocoonase, in the form of crystals on their galeae. To emerge from the cocoon they secrete buffer to dissolve the enzyme, which then hydrolyses the silk. Higher flies (Cyclorrhapha) escape from the puparium (the persistent last larval cuticle, which is modified to serve a cocoon-like function) by expanding a ptilinum, a balloon-like structure on the front of the head. When the adult has dug its way up through the soil, the ptilinum is retracted and can be detected only by the ptilinal suture, shaped like an inverted U, passing just above the bases of the antennae. Hardening of the cuticle varies between insects and in different parts of the body. Various hormones contribute to its control.

Thermoregulation

Insects may have a body temperature equal to their surroundings, but frequently modify their body temperature by deriving extra heat from the environment or generating heat via their own metabolism (or both). Behaviour patterns may be used to change the body temperature. Examples include perching successively in sunlight or shade (called shuttling), or adjusting posture to maximise or minimise heat uptake by basking. Butterflies may increase the area of wings exposed to sunlight, or locusts may sit so that the sunlight strikes the front of the head only and thus reduce heat uptake. Insects may sit with the body in contact with the ground for heat uptake (called crouching). Alternatively, they may seek an underground cool habitat, or raise the body off ground: this so-called stilting is often seen in long-legged desert arthropods. Butterflies may use the wing surface as a radiant heat exchanger, in this case reducing heat load. Colour may influence temperature: darker colours provide a greater heat uptake. This principle is demonstrated in melanic caterpillars. For example, the wanderer or monarch butterfly has caterpillars with broader black bands in winter, and ladybirds with alleles for black colour survive better in smoke-polluted environments, apparently because of thermal effects rather than differential predation.

Insects generally reduce their flight activity when the air temperature is high, because they are unable to dissipate heat generated by flight. Gliding flight requires less energy and may be favoured by insects at higher ambient temperatures. Social insects such as termites and bees control the nest temperature, either by its architecture or, in the case of bees, by fanning the wings to evaporate water at high ambient temperatures.

Temperature receptors allow insects to respond physiologically to unfavourable temperatures. 'Shivering' enables the flight muscle to warm up prior to flight. The temperature of the body may be maintained by hair layers, or reduced by respiratory heat exchange. The temperature may be regulated by heat loss: the thorax may be 20°C hotter than the abdomen because of flight muscle activity and insulation, and heat is exchanged via the circulation and lost through the surface of the less insulated abdomen. Evaporative cooling occurs from the spiracles or from regurgitated fluid.

Survival at extreme temperatures

The temperature that is lethal to an insect depends on the species, the period of exposure, and the life cycle stage. The upper lethal temperature is around 45–50°C in locusts, but only 20°C in some high altitude insects. The lower lethal temperature is variable; many insects are unable to feed or digest food below a threshold temperature (20°C in locusts). Others die by freezing when the temperature falls below the depressed freezing point of their cellular fluids. Others depress the freezing point of their blood further by secreting glycerol into it, up to a concentration of 5 M. This gives a freezing point around −15°C.

Many insects escape unfavourable periods by diapause, a stage of physiological arrest. Winter diapause or hibernation is common in any life stage but summer diapause (aestivation) occurs especially in countries with hot, dry summers (e.g. Bogong moths and some native aphids in Australia). The life cycle might also be modified in extreme environments. For example, most aphid species have many parthenogenetic generations through spring to autumn, but arctic aphids usually have only a single parthenogenetic generation each summer. Some arctic midges remain in the larval stage for several years, gradually accumulating sufficient 'day-degrees' for development to the adult.

Insect societies

Some insects have evolved social behaviour, in which there is a division of roles between alternative phenotypes (castes). There are usually numerous sterile individuals which contribute to the survival of genes they share with the reproductives through food collection, nest building, defence, care of brood, etc. Termites, some wasps, some bees, and all ants are such eusocial insects; there are arguments for some bark beetles and some aphids to be added to this list. Social insects usually have a limited number of reproductive individuals. In termites and hymenopterans, complex nests are built by the workers. The young and the reproductives are fed within the nest on various foods, depending on the species; food may be gathered and fed to colony members, but is often used indirectly. For example, leafcutter ants feed on fungi farmed on leaf material within the nest, and nectar is processed into honey.

All social insects rely heavily on pheromones for recognising the reproductive caste, trail marking, alarm and aggregation. Complex behaviour patterns also occur, particularly in social hymenopterans. A well-known case is the waggle dance of bees, used to convey to other bees in the hive the position of a food source. The bee communicates the distance to the food and the angle to be travelled relative to the sun. The odour of flowers, carried on the bee's body, gives further information. The initial search is based on visual and olfactory signals, and return navigation is based on the position of the sun. It is not certain how bees measure distance, but they can measure the passage of time and compensate for it in celestial navigation. They also use landmarks: if hives of different colours are stacked together, bees remember the colours of hives next to their own. Studies of bee behaviour have allowed us to investigate their sensory capabilities, such as their ability to distinguish shapes, colours and the threshold concentrations of sugars.

Classification of the subphylum Hexapoda

The classification given here follows that in *The Insects of Australia* (CSIRO 1991). Suborders are mentioned only for large orders.

CLASS ELLIPURA
Small, entognathous hexapods; no abdominal spiracles; appendages on abdominal segments 1–3.
Order Protura — Without eyes or antennae; minute, rare leaf-litter dwellers.
Order Collembola — springtails; 4-segmented legs; springing organ; 6 abdominal segments; common in leaf litter and other humid microclimates; global distribution including subantarctic islands, coral beaches.

CLASS AND ORDER DIPLURA
Small or medium sized, entognathous hexapods; no eyes; abdomen with styles and exsertile vesicles; cerci either thread-like or forceps-like; common leaf-litter dwellers.

CLASS INSECTA
Ectognathous hexapods.
Subclass and Order Archaeognatha
Rockhoppers; mandibles with single basal articulation; large compound eyes; arched thorax, styles on coxae and on abdominal segments; long cerci and median caudal filament.
Subclass Dicondylia
Mandibles with two points of articulation.

Infraclass and Order Thysanura
   Silverfish; similar in superficial appearance to archaeognaths, but with reduced compound eyes and thorax not arched; no styles on thoracic appendages.

Infraclass Pterygota
   Insects with wings in the adult stage (secondary loss in some groups).

Division and Order Ephemeroptera — mayflies; larva and adult with long cerci and usually long median caudal filament; winged subadult and adult stages; wings cannot be folded; larvae aquatic, paired lateral gills on abdomen.

Division and Order Odonata — dragonflies and damselflies; wings cannot be folded; male with secondary sexual structures at front of abdomen; larvae aquatic, gills in rectum (dragonflies) or as three terminal leaflets (damselflies); predacious larvae and adults; larvae with labium modified as prehensile mask.

Division Neoptera — Wings can be folded to the sides of the body.

   Order Plecoptera — stoneflies; aquatic larvae, with gill tufts or plates variously positioned on the body; long, threadlike cerci in larvae and adults; adults with large wings or reduced wings; pollution-sensitive, many species of restricted distribution.

   Order Blattodea — cockroaches; dorso-ventrally flattened, forewings as somewhat hardened tegmina, wings often reduced or absent; eggs in an ootheca, deposited or carried; sometimes viviparous; some cosmopolitan pest species associated with human habitation, many other benign species; omnivorous.

   Order Grylloblattodea — ice crawlers; only at high altitudes of Northern Europe and Asia; wingless; carnivorous.

   Order Isoptera — termites; social insects, small number of reproductives with worker and soldier castes. Fore- and hindwings of reproductives similar in appearance; shed at basal breakpoints; feed on soil, wood, dead vegetation and depend on symbiotic bacteria or protozoans for digestion of cellulose; some species serious timber pests.

   Order Mantodea — praying mantises; mobile head, raptorial front limbs; forewings as tegmina; wings may be reduced; predacious; large, foamy ootheca.

   Order Dermaptera — earwigs; forewings short tegmina, hindwings semicircular; wings often reduced. Cerci as terminal forceps used for prey capture; show maternal behaviour; most omnivorous.

   Order Orthoptera — locusts, crickets, grasshoppers, wetas, Cooloola monsters, etc.; hindlegs usually modified as jumping legs; often communicate by sounds; forewings as tegmina, wings may be reduced; some parthenogenetic, some ant inquilines; mostly plant-feeding.

   Order Phasmatodea — stick or leaf insects; large insects with cryptic body form and coloration, resembling sticks or leaves; forewings as tegmina, wings may be reduced; often parthenogenetic; eggs large, seed-like in appearance; phytophagous, sometimes defoliating eucalypts in eastern Australia.

   Order Embioptera — web-spinners; small insects living in colonies in silken webs, produced from glands on the swollen front tarsi; wings often reduced; feed on mosses, lichens, dead vegetable matter.

   Order Zoraptera (no common name) — small, polymorphic, gregarious, termite-like; often wingless; only 30 species worldwide.

Subdivision Paraneoptera
> Mouthparts tending towards a piercing, sucking structure; postclypeus (at front of head) enlarged, accommodating sucking muscles.
> Order Psocoptera — barklice, booklice; small, winged or wingless; mouthparts asymmetrical; feed on microflora; one genus a pest of stored products.
> Order Phthiraptera — lice; dorso-ventrally flattened, wingless; reduced eyes; parasitic on birds or mammals; highly host-specific, absent from some marsupial groups.
> Order Hemiptera — sucking bugs, with mandibles and maxillae modified as flexible stylets; mainly plant-feeders, some carnivorous or blood-sucking.
>> Suborder Sternorrhyncha — aphids, psyllids, whiteflies, scales, mealybugs; labium originates from prosternum; antennae multi-segmented; forewings and hindwings of similar texture; plant feeders, often vectors of plant diseases.
>> Suborder Auchenorrhyncha — cicadas, leafhoppers, planthoppers, spittle bugs etc.; labium originates from posterior region of head; antenna bristle-like; plant feeders.
>> Suborder Heteroptera — water striders, backswimmers, stink bugs, bedbugs and many others; labium inserted well in front of prosternum; wings lying flat over abdomen, forewing often subdivided into thickened basal and membranous distal region; plant and animal feeders; includes fresh water and marine groups, and blood-sucking disease vectors.
> Order Thysanoptera — thrips; small, flattened, with asymmetrical piercing, sucking mouthparts; wings very narrow, with a fringe of long hairs.

Subdivision Endopterygota
> Order Megaloptera — dobson flies, alderflies; primitive endopterygotes with large membranous wings; pupa with movable mandibles; larva aquatic, predacious, with lateral abdominal gills.
> Order Raphidioptera — snakeflies; elongate membranous wings; long prothorax and ovipositor, terrestrial larvae; predatory.
> Order Neuroptera — lacewings, antlions etc.; small to large, subequal wings with many crossveins; eggs often on stalks; larvae predacious with sucking jaws formed by apposed mandibles and maxillae; one group feeding on freshwater sponges.
> Order Coleoptera — beetles; minute to very large, front wings as hardened elytra, hind wings if present membranous; larvae very variable in form; about 30% of all animals described are beetles, inhabiting environments of all kinds.
> Order Strepsiptera — small, males freeliving, with large hindwings and reduced forewings; first instar larva active, other larvae and adult female legless and grub-like; parasitic in other insects.
> Order Mecoptera — scorpion-flies; subequal membranous wings; head drawn out as a downward pointing rostrum; larvae terrestrial or aquatic, feeding on dead insects; males in some families with enlarged upturned genital segments.
> Order Siphonaptera — fleas; laterally compressed, wingless, with jumping hindlegs and piercing mouthparts; ectoparasitic on birds and mammals; larvae legless, in nests of hosts.
> Order Diptera — flies; forewings well-developed, hindwings reduced to club-shaped halteres; mouthparts piercing and sucking, or sponging;

some vectors of diseases of livestock and humans; larvae legless, often with reduced head.

Suborder Nematocera — mosquitoes, midges, sandflies, craneflies, fungus gnats etc.; antennae threadlike and usually longer than thorax.; larvae often aquatic, with well-formed head capsule.

Suborder Brachycera — house flies, blowflies, robber flies, fruit flies etc.; antennae short and usually with a style or bristle-like arista; larvae with incomplete or no head capsule; terrestrial, aquatic or endoparasitic in invertebrates or vertebrates; include serious livestock pests (screwworm, sheep blowfly).

Order Trichoptera — caddis flies; moth-like but with well-developed maxillary palps; larvae aquatic (one family marine), often in cases; feeding various, occasionally parasitic in other aquatic invertebrates.

Order Lepidoptera — moths, butterflies; galeae of maxillae in more advanced groups elongated and interlocked to form a proboscis; wings with scales; larvae caterpillars, with thoracic legs and usually abdominal prolegs on segments 3–6 and 10; generally phytophagous, sometimes significant crop pests.

Order Hymenoptera — sawflies, wasps, bees, ants; hindwings smaller than forewings, wings coupled by hamuli; first abdominal segment reduced, associated with thorax; larvae caterpillar-like (sawflies) or legless.

Suborder Symphyta — no distinct 'waist' between first and second abdominal segments; ovipositor serrated; larvae phytophagous, caterpillar-like but with prolegs distributed differently from those of Lepidoptera; some economic pests, e.g. wood wasps (*Sirex*) damage pine trees, sawflies (*Perga*) defoliate eucalypts.

Suborder Apocrita — wasps, bees, ants; with distinct 'waist' between first and second abdominal segments; many species with larvae parasitic in other insects, others which collect food for the offspring, and some families social; important pollinators and biological control agents.

Chapter 12

The Myriapoda

N.N. Tait

SUBPHYLUM MYRIAPODA

Introduction 270

Functional morphology and diversity 270
   Class Chilopoda 270
   Class Diplopoda 272
   Class Symphyla 273
   Class Pauropoda 274
   Locomotion 274
   Integument and musculature 275

Feeding 276

Internal organs and physiology 276
   Digestion 276
   Excretion and water balance 277
   Respiration 278
   Circulation 278

Control systems 279
   Nervous system 279
   Sense organs 279
   Neurosecretion and endocrines 279

Defence 280

Reproduction and development 280
   Male reproductive system 281
   Female reproductive system 282
   Courtship and mating 282
   Egg laying and protection 283
   Development 283

Evolutionary history 284

Classification of the subphylum Myriapoda 284

Subphylum Myriapoda

Primarily terrestrial arthropods; uniramous limbs, including one pair of antennae; jointed whole-limb mandibles; numerous pairs of walking legs; gonopore posterior or anterior.

Introduction

While less spectacular in their radiation than other groups of extant arthropods, the myriapods encompass considerable diversity in body form and ways of life. They include four classes: the familiar centipedes (Chilopoda, Fig. 12.2) and millipedes (Diplopoda, Fig. 12.4), the symphylans (Symphyla, Fig. 12.5a) and the pauropods (Pauropoda, Fig. 12.5b). Myriapods have an elongate body divided into a head and a trunk of numerous segments, most of which bear uniramous walking limbs. The head has a single pair of antennae, a pair of mandibles and one or two pairs of maxillae. Myriapods range from 0.5 to 300 mm in length. While most species are darkly pigmented, black through brown, or pale to white, others are more brightly patterned in red, yellow, green or blue.

Present-day myriapods are terrestrial, although a few have secondarily invaded freshwater and intertidal marine habitats. While some species occur in arid regions, most are confined to moist microhabitats in soil, rotting vegetation, the underside of rocks and logs, and under bark on tree trunks. Myriapods make a significant contribution to the invertebrate fauna of caves.

Functional morphology and diversity

Class Chilopoda

The dorsal surface of the head in centipedes is covered by a single, shield-shaped plate, anteriorly bearing a pair of elongate antennae. Eyes are absent in some groups, but others possess one to 40 simple ocelli on each side of the head. In scutigeromorph centipedes, the ocelli are grouped to form a pair of compound eyes. Ventrally on the head, the clypeus is followed by a labrum which forms the anterior rim to the oral cavity. The first pair of postoral appendages, the mandibles, articulate with the sides of the head and curve forwards to the midline, armed with various combinations of teeth, comb-like plates and stout bristles (Fig. 12.1a). The first maxillae consist of a basal coxosternite (fused sternum and coxa) which laterally supports a jointed palp and medially an unjointed coxal process. The second maxillae are more leg-like, with each coxosternite bearing an elongate, jointed palp terminating in a stout claw. When viewed ventrally, the mouthparts are obscured by the fused coxosternites of the first pair of trunk appendages, the maxillipeds, modified as curved poison claws, which deliver venom to the prey via a subterminal pore. Dorsally, the tergum of the maxilliped segment is free, or fused with the tergum of the second trunk segment.

The trunk has a variable number of segments (Fig. 12.2). In each segment, the dorsal tergum and ventral sternum are joined laterally by membranous pleura containing hardened plates (pleurites), including the coxal plates that articulate with the bases of the laterally directed limbs. Except

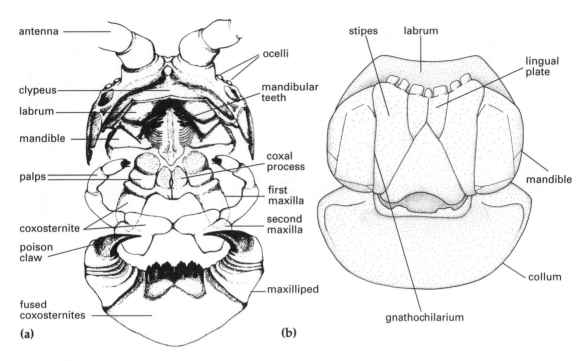

Figure 12.1
Ventral view of mouthparts of (a) Centipede, with mouthparts 'exploded' to show individual details. (b) Millipede.

for the most posterior few, each segment carries a pair of uniramous walking legs terminating in claws. Behind the last leg-bearing segment are three telescoped segments, the second and third of which bear a pair of gonopods and the genital opening respectively. The anus is terminal and surrounded by the telson.

Approximately 3000 species of centipedes have been described. The shape of the trunk and number and form of the limbs distinguish the various orders. Geophilomorphs (Fig. 12.2a) are thin, burrowing centipedes with 31 to 181 pairs of short legs of equal length except for the first pair, which may be reduced, and the last pair, which are directed backwards and often swollen, especially in males. The transverse division of each tergum into a short pretergite and longer metatergite increases trunk flexibility. The sterna are similarly divided into presternites and metasternites.

All other centipedes are running surface dwellers. Scolopendromorphs (Fig. 12.2b) have a broad body with 21 or 23 pairs of legs. The legs tend to increase in length posteriorly. The last pair of legs is long and robust and projects backwards. They are provided with stout spines and are used as prehensile organs for prey capture and in courtship. Lithobiomorphs (Fig. 12.2c) have 15 pairs of legs. Trunk segments 2, 4, 6, 9, 11 and 13 are considerably shorter than the others and the legs increase in length posteriorly. The very fast-running scutigeromorphs (Fig. 12.2d) also have 15 pairs of legs, but the terga are reduced to seven, giving the body greater rigidity. The legs are extremely long, largely because of the annulated tarsi and metatarsi, and increase in length from front to back. The last pair of legs form slender, posteriorly directed sensory organs. The body is not as dorsoventrally flattened as in other centipedes. *Craterostigmus tasmanianus* from Tasmania and New Zealand, at present the sole representative of the order Craterostigmorpha, shares features of the lithobiomorphs and scolopendromorphs (Fig. 12.2e). It possesses 15 pairs of legs, as in lithobiomorphs, but is more scolopendromorph-like in its body form.

INVERTEBRATE ZOOLOGY

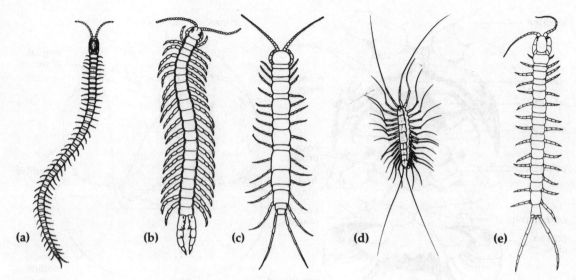

Figure 12.2
Diversity in body form among centipedes. **(a)** Geophilomorpha. **(b)** Scolopendromorpha. **(c)** Lithobiomorpha. **(d)** Scutigeromorpha. **(e)** Craterostigmorpha.
(a, d from Harvey and Yen 1989; b, c from Lawrence 1983; e from Mesibov 1986.)

Class Diplopoda

The head of millipedes is usually convex dorsally, flattened ventrally and tucked under the protection of the legless first trunk segment or collum (Fig. 12.3). The laterally placed antennae are composed of eight podomeres in all millipedes. They are usually held in a flexed position, and tap the substratum as the animal progresses. The eyes, when present, consist of from one to about 90 ocelli on each side, sometimes closely abutting each other and polygonal in shape. The anterior margin of the head is delineated by the labrum. The sides of the head are formed by the well-developed basal regions of the large mandibles (Fig. 12.1b). The distal gnathal lobe of the mandible is provided with grinding and biting surfaces suited to the tough vegetation on which most millipedes feed. Millipedes have only a single pair of maxillae, fused along the mid-line to form a gnathochilarium covering the ventral surface of the head (Fig. 12.1b). The gnathochilarium consists of basal plates supporting a pair of outer stipes and inner lingual plates. Some millipedes have pointed mouthparts for piercing and sucking fluids from living or dead plants and fungi.

Figure 12.3
Segmentation and arrangement of appendages in millipedes (from Hopkin and Read 1992).

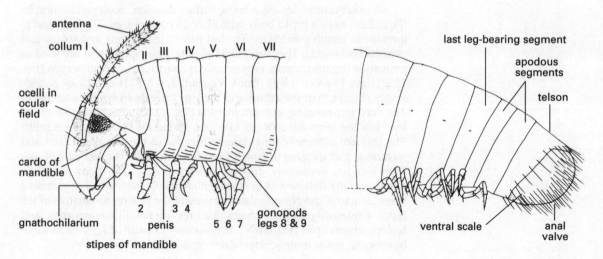

The trunk is often circular in section, but may be hemispherical or (in the flat-backed millipedes) dorsally flattened. Millipedes display various degrees of fusion of the plates covering each segment, from separate tergum, pleuron and sternum to fusion as a complete ring structure. A characteristic feature of millipedes is the fusion of most trunk segments in pairs to form diplosegments, each with two pairs of walking legs (Fig. 12.3) — the source of the name Diplopoda. The first trunk segment, the collum, is legless, and segments 2 to 4 each possess only a single pair of legs. Between the last leg-bearing diplosegment and the telson are one to several legless diplosegments. The legs arise near the ventral midline.

Millipedes have progoneate (anterior) genital openings. The male and female gonopores occur on or posterior to the bases of the second pair of legs. Male gonopores may be associated with a 'penis' and, in some, the paired 'penes' are fused (Fig. 12.3a). Sperm transfer to females is achieved, in most male millipedes, by gonopods formed by modification of the legs further back on the body, on the third diplosegment. These legs initially develop as normal walking legs but regress to buds at a certain instar and then become progressively modified until adulthood.

Millipedes are the most diverse group of myriapods (Fig. 12.4), with some 10 000 described species. They are divided into two subclasses, the Penicillata and Chilognatha. The former contains the single aberrant order Polyxenida or bristly millipedes. These are very small (less than 4 mm), with soft, uncalcified bodies covered in tufts of serrated bristles. They lack specialised gonopods for sperm transfer. The Chilognatha is divided into fourteen orders, including the following that are more commonly encountered. The sphaerotheriids (Southern Hemisphere) and glomerids (Northern Hemisphere) contain species that roll into a ball and are often referred to as pill millipedes. Siphonophorids and polyzoniids are sucking millipedes, with the head and mouthparts drawn out to form a beak. Polydesmids, or flat-backed millipedes, have lateral projections on the hind part of each segment, giving a flattened form to the dorsal surface. Snake millipedes (spirobolids, spirostreptids and julids) have highly calcified, cylindrical bodies.

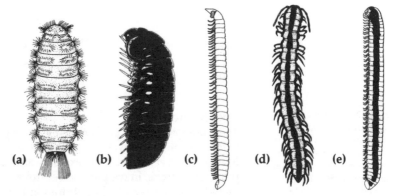

Figure 12.4
Diversity in body form among millipedes. **(a)** Polyxenida. **(b)** Sphaerotheriida. **(c)** Polyzoniida. **(d)** Polydesmida. **(e)** Spirobolidat. (From Harvey and Yen 1989.)

Class Symphyla

Symphylans are a small group of myriapods both in size (less than 10 mm) and number of species (approximately 160). The trunk is composed of 12 leg-bearing segments, covered by 15 to 24 tergites that provide these fleet-footed animals with the flexibility to twist and turn in the narrow crevices of soil and humus (Fig.12.5a). The penultimate segment bears a pair of elongate sensory bristles and a pair of cerci or spinnerets, from which threads of silk are spun to line their chambers. The anus is surrounded by the small terminal telson. Like millipedes, symphylans are progoneate. The genital openings lie between the fourth pair of legs. The head bears a pair of antennae, but eyes are lacking. The mouthparts consist of paired mandibles, partially fused first maxillae and fused second maxillae. The second maxillae of symphylans are convergently similar to the insect labium.

Class Pauropoda

Although frequently abundant in leaf-litter and soil, pauropods are rarely noticed because of their small size (0.5 to 2.0 mm). About 500 species have been described. The trunk of most species has nine pairs of legs but only five terga, due to the lack of terga on even-numbered segments from the fourth onwards (Fig. 12.5b). The first segment (collum) and second last segment are legless, and the trunk terminates in a telson bearing the anus. Each tergum, except that of the collum and second segment, bears a pair of elongate bristles. The head carries a pair of biramous antennae but is eyeless. The mouthparts consist of mandibles and maxillae, the latter fused to form a gnathochilarium. There are no second maxillae. The gonopores are on the third trunk segment. Pauropods thus share a number of features with millipedes.

Figure 12.5 (left)
External morphology of (a) a symphylan, and (b) a pauropod (from Harvey and Yen 1989).

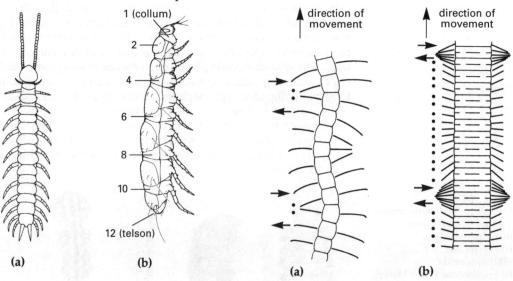

Figure 12.6 (right)
Limb movement during locomotion in (a) a centipede and (b) a millipede (• limb tip on ground; ► limb newly raised from ground; ◄ limb newly placed on ground; see text for details.) (a from Russell-Hunter 1979.)

Locomotion

The locomotory gaits performed by myriapods share much in common with locomotion in onychophorans. The limbs are stretched out during the forward recovery stroke and flexed towards the body during the backward propulsive stroke. In the propulsive stroke, the claws at the tips of the legs provide traction against the substratum to propel the body forward. As a limb is placed on the ground at the beginning of the propulsive stroke, the limbs immediately in front are making their next forward movement while the limbs behind have begun their propulsive stroke, so that the limbs move in metachronal rhythm. Speed can be increased by any combination of (1) increasing the rapidity of the steps, (2) reducing the duration of the propulsive stroke of the step in relation to the recovery stroke, and (3) increasing the angle of the swing of each leg by increasing the length of the legs.

As active carnivores, centipedes move swiftly, using an alternating action of the paired limbs of each segment (Fig. 12.6a). Scutigeromorphs, with a small number of long legs, are the fleetest centipedes. Centipedes use a 'high gear' gait. The propulsive stroke of each leg is of shorter duration than the recovery stroke, giving fast running with more legs off the ground at any one time than are on the ground. The body is held rigidly so that energy-wasteful undulations of the body are prevented. Scutigeromorphs also place much of the jointed distal part of the limb on the ground, providing greater traction. Lithobiomorphs are similar to scutigeromorphs in the arrangement of the limbs and their locomotory gait, except that the limbs are considerably shorter (Fig. 12.2c). Scolopendromorphs have more numerous and relatively short legs on a longer, wider body (Fig. 12.1b). The body is thrown into undulations during fast running, giving increased speed but with some waste of energy. Geophilomorphs, with their numerous short legs and elongate, flexible bodies (Fig. 12.2a), are adapted for crawling and burrowing in loose soil and decaying vegetation. Burrowing is achieved by the propagation of waves of elongation and contraction of the body, achieved by telescoping the short and long divisions of each tergum and sternum. The legs provide traction on the sides of the burrows.

The movement of millipedes is accomplished by quite different locomotory strategies. Millipedes employ a 'low gear' gait for slow, powerful locomotion. The numerous short legs, all of equal length (Fig. 12.4), have short recovery strokes relative to slower propulsive strokes. As a consequence, more legs are on the ground at any one time than off the ground (Fig. 12.6b). The legs of each pair move synchronously in phase, rather than alternating as in centipedes. This, together with the increased rigidity of the body provided by the development of diplosegments, prevents body undulations and increases traction. Dorso-ventral flexibility is retained in many species, which allows them to coil up in a spiral or tight ball. The intucked head, overgrown by the collum in many species of millipedes, presents a smooth surface for pushing aside compacted humus like a bulldozer. Others species use their pointed anterior end as a wedge to increase the width of crevices through which the rest of the body can pass. While lateral undulations of the body are prevented in normal millipede locomotion, members of the spirostrepsid family Odontopygidae, from South Africa, make rapid escape manoeuvres by turning on their backs and vigorously wriggling for about 300 mm. This is then followed by a short period of normal walking and the process is repeated. Even more bizarre is the escape reaction reported in several species of *Diopsiulus* from Sierra Leone, which execute repeated leaps of 20 to 30 mm.

Integument and musculature

The integument of myriapods consists of a collagenous basement membrane, an epidermis, and a thick cuticle that forms the exoskeleton. The cuticle is composed of a thin outer epicuticle and a thicker, inner procuticle. In the hardened sclerites, the procuticle is sclerotised to form an exocuticle of chitin and tanned proteins above the laminated layers of chitin and untanned proteins of an endocuticle. Although myriapods are more susceptible to desiccation than insects and spiders, some species have a thin layer of wax at the surface of the exoskeleton which may impede water loss. The rigidity of the exoskeleton of chilognath millipedes is increased by the incorporation of calcium carbonate, together with magnesium carbonate in some species. In contrast, the exoskeleton of centipedes is less rigid, and there may be an intermediate layer of somewhat hardened but laminated mesocuticle. There is also a gradual transition from the exocuticle and mesocuticle layer in the sclerites to their absence in arthrodial membranes, so that the edges of sclerites are not clearly defined. Centipedes rely to some extent on the hydrostatic pressure of haemolymph to bring about body and limb movements. This is more pronounced in the flexible scolopendromorphs and geophilomorphs than in lithobiomorphs and scutigeromorphs.

Feeding

Centipedes are generally considered to be carnivorous, although some observations indicate that they may also feed on plant material. Depending on the size of the centipede, various invertebrates and even small vertebrates may form their diet. The prey is held by the poison claws, sometimes aided by the anterior legs, while the venom is injected into its body. Scolopendromorphs may initially use the raptorial hind legs to hold prey, while scutigeromorphs have been observed to catch and hold flies using the annulated distal ends of their anterior legs. The poison glands are usually situated within the maxillipeds, but may extend some distance into the trunk. The venom is ejected by contracting the muscles that surround each secretory cell. Little is known of the chemical composition or mode of action of the venom of centipedes. Human fatalities supposedly following bites from centipedes have been largely unconfirmed.

Millipedes generally feed on decaying plant material. Some consume seeds, or graze on fungi or soft plants such as algae, bryophytes and seedlings. Sucking millipedes extract fluids from living or rotting plants. A few are carnivorous or feed on dead animal remains. Most unusual of all is the development of enlarged pectinate lamellae on the mandibles of an aquatic, cave-dwelling species that filters suspended organic material from the water. Many species actively forage for preferred food, such as seeds or fungal fruiting bodies. These patchy and ephemeral resources can attract aggregations of tens, and occasionally hundreds, of individuals.

Symphylans feed on decaying vegetation, although some attack the roots of living plants. Pauropods feed mainly on decaying vegetation or fungi. Some are carnivorous.

Internal organs and physiology

Digestion

The alimentary canal is a straight tube composed of an anterior foregut and posterior hindgut derived from ectoderm and lined with cuticle, and a long midgut derived from endoderm (Fig. 12.7a). The foregut is divided into a muscular pharynx (leading from the mouth) and a narrow oesophagus which usually expands into a crop for storage and a gizzard for further masticating the food. A number of unicellular and multicellular glands have been identified as opening into the oral cavity. Their position and limited analyses of their secretions indicate that some produce a salivary secretion that lubricates the food and may contain digestive enzymes.

Digestion and absorption occur in the midgut, the cavity of which is lined by a peritrophic membrane secreted by midgut cells. It protects the

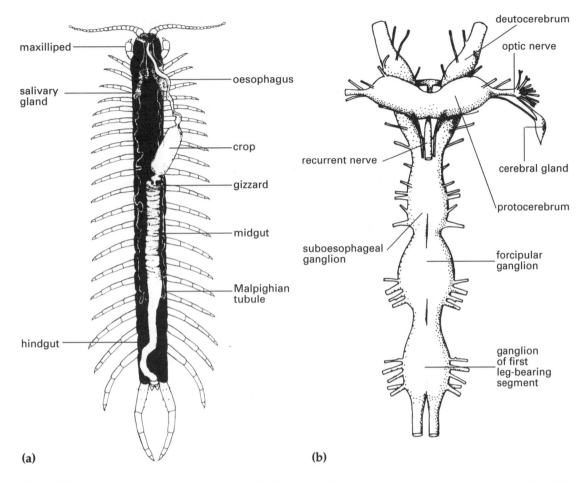

Figure 12.7
Internal anatomy of centipedes.
(a) Digestive system.
(b) Nervous system (after Lewis 1981).

underlying cells from abrasion. As feeders on dead plant material, millipedes rely heavily on microorganisms in their midgut. The gut microflora aids digestion and may also be a source of nutrients. In many habitats, feeding behaviour and associations with microorganisms modulate the rate of plant litter decomposition, which in turn affects where and when nutrients become available to plants. (Terrestrial amphipod crustaceans are more important in this role in Australian forest litter.)

Faeces are formed in the hindgut. In some millipedes the Malpighian tubules are closely associated with the hindgut. The hindgut cells form an osmotic gradient for the resorption of water from the gut contents.

Excretion and water balance

Myriapods usually retain segmental organs in the head, but these are modified to act as salivary glands (see Chapter 10). The segmental organs of lithobiomorphs and scutigeromorphs consist of a pair of coelomic end-sacs in the maxillary segment, associated with two pairs of ducts which open on maxillae 1 and 2. Geophilomorphs and scutigeromorphs lack segmental organs. Millipedes and symphylans have a single pair of maxillary segmental organs and ducts. Pauropods have maxillary and premandibular pairs.

Like insects and many arachnids, myriapods have Malpighian tubules that open, as a pair, at the junction of the midgut and hindgut (Fig. 12.7a).

The tubules appear to be the principal site of nitrogen excretion. Centipedes excrete mainly ammonia, with relatively small amounts of uric acid. Millipedes are more uricotelic.

Respiration

As in many other groups of terrestrial arthropods, myriapods breathe by means of tracheae, lined by cuticle with spiral thickenings (taenidia) and opening to the surface through segmentally arranged spiracles. In centipedes, the spiracles are borne on all leg-bearing segments except the first and last (in geophilomorphs) or only on the longer segments in groups where alternating short and long segments occur (in scolopendromorphs and lithobiomorphs). In scutigeromorphs, the spiracles are merged as an unpaired mid-dorsal slit on each tergite. The spiracles of scutigeromorphs lead into kidney-shaped sacs containing masses of branched but non-anastomosing tracheal tubules, which end blindly in tissue close to the heart. In millipedes there is a spiracle above and anterior to the base of each leg.

Each spiracle leads into an atrium lined with projections of the cuticle, which stop foreign material entering the respiratory system. Some species are capable of closing their spiracular openings by muscles that constrict the atrial cavity, and so restrict water loss.

While respiratory pigments are generally absent in myriapods, haemocyanin is present in the haemolymph of scutigeromorphs in association with the unique arrangement of their tracheal system. A number of circulating haemocytes have been identified, some of which have a phagocytic activity and are presumably involved in eliminating pathogens and other materials.

The tracheal system of symphylans and pauropods is poorly developed or absent. Symphylans have a single pair of spiracles on the head from which tracheal tubules supply only the anterior segments. The very small pauropods generally lack a tracheal system. Oxygen is presumably obtained by diffusion through the thin cuticle.

While myriapods are able to survive considerable submergence in fresh and sea water and are occasionally encountered in aquatic samples, only a few species habitually live in water for at least some part of their life cycle. Amazonian millipedes respire, during seasonal inundation of the forest, by maintaining a layer of air (plastron) over parts of the body. Some species of geophilomorph centipedes inhabit the seashore, but they migrate from this region when immersed at high tide or to lay their eggs and brood their young. Two species of freshwater millipedes are known in Australia. Surprisingly, these species had remained undetected until recently, when they were discovered under stones in a stream in the grounds of Macquarie University in Sydney. Subsequently, both species have been collected from other locations in south-eastern Australia. The more common species leaves the water during winter to mate and presumably lay eggs in cracks on the banks. It is not yet known how these species respire while under water.

Circulation

Myriapods, like other arthropods, have a haemocoel divided into a perivisceral sinus surrounding the body organs and a dorsal pericardial cavity containing the tubular heart. Segmentally arranged, paired ostia open from the pericardial sinus to the heart cavity. In centipedes and symphylans, an extensive system of arteries leads from the heart to the head, including a pair of vessels that arch around the oesophagus to join a longitudinal supraneural artery, from which arise paired segmental arteries. The circulatory system of millipedes lacks this extensive arterial system. Pauropods lack a heart.

Control systems

Nervous system

The central nervous system of myriapods consists of a dorsal brain connected by a pair of circum-oesophageal commissures to a ventral suboesophageal ganglion, which leads to paired ventral nerve cords with segmental ganglia (Fig. 12.7b). The lateral lobes (protocerebrum) of the brain, from which nerves supply the ocelli, include the paired ganglia of a preantennary segment. The anterior lobes (deutocerebrum) are the paired ganglia of the antennary segment, with nerves leading to the antennae. Paired ganglia of the premandibular segment form the tritocerebrum, ventral to the protocerebrum. They give rise to the commissures leading to the suboesophageal ganglion. The latter give off nerves supplying the mandibles and maxillae.

Sense organs

While some species of myriapods are eyeless, others possess a variable number of ocelli on either side of the head. Each ocellus consists of a biconvex lens formed of thickened cuticle (cornea) overlying the epidermis. The underlying cup-shaped retinal layer is composed of photoreceptor cells surrounded by a thin epidermis of sheath cells. Extensions of the distal ends of the retinal cells, called rhabdomeres, possess tightly packed microvilli arranged at right angles to the long axis of the cell. In scutigeromorph centipedes, up to 200 ocelli on each side of the head are aggregated to form a compound eye similar to those of insects and crustaceans, but independently evolved. While the simple ocelli are probably capable only of distinguishing light intensity, prey capture in scutigeromorphs indicates that they can see moving objects. Most centipedes are negatively phototaxic, while millipedes are either negatively or positively phototaxic, depending on the species. Even species lacking eyes are apparently capable of detecting and responding to light.

Myriapods are well supplied with cuticular sensilla, particularly on the antennae, mouthparts and legs. These include mechanoreceptors, such as bristle-like sensilla for receiving tactile or vibratory stimuli, and chemoreceptors in which the cuticular surface of the sensillum is perforated by minute pores for receiving taste or smell stimuli. Many myriapods possess characteristic sensilla called organs of Tömösváry. These are visible as a circular, oval or U-shaped depression between the antenna and the eyes. The rim of the organ is raised and an inner groove is perforated by the pores of unicellular glands. The dendrites of sensory cells pass through a pore on the floor of the depression and spread out in a cavity formed between the epicuticle and procuticle. The function of these sensilla is unknown, but humidity or olfactory reception have been suggested.

Neurosecretion and endocrines

Although neurosecretory cells and accumulations of their axon terminals as neurohaemal organs have been identified in the brain and segmental ganglia of a number of species of myriapods, few studies have been

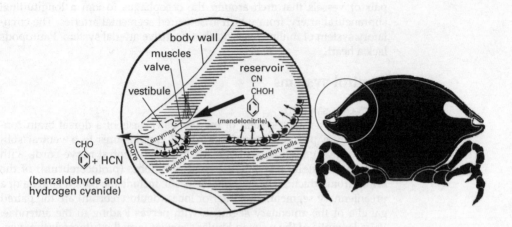

Figure 12.8
Details of the secretion of hydrogen cyanide by the repugnatorial glands of polydesmid millipedes (from Eisner and Eisner 1965).

performed to relate neurosecretory activity to specific control functions. Limited surgical removal and implant experiments indicate that neurosecretion influences such processes as gametogenesis, vitellogenesis and moulting. Injection of ecdysone, the insect moulting hormone, into the centipede *Lithobius* initiates moulting. A diffuse gland in the first leg-bearing segment has been implicated in the endocrine control of moulting.

Defence

While the fast-moving and aggressive behaviour of centipedes is a defence against predation, the slow moving millipedes have evolved a considerable array of anti-predatory devices. These include the hardening of the exoskeleton by calcium impregnation and the ability to roll the body into a spiral or tight ball to protect the more delicate ventral structures. Many millipedes have an association with aggressive animals, particularly species of ants from which the millipedes may themselves be protected by covering their body with ant odours. Most millipedes release noxious secretions from segmental 'repugnatorial' glands, each opening as a pore on the lateral midline of the body or as unpaired, mid-dorsal pores, as in pill millipedes. A large number of compounds have been identified in different species, including hydrogen cyanide, benzoquinones, benzaldehyde, phenol, and aliphatic fatty acids. Hydrogen cyanide is characteristically produced by flat-backed millipedes from glands consisting of an inner reservoir and outer vestibule separated by a muscular valve (Fig. 12.8). The vestibule opens as a pore on each side of the body. The reservoir secretes mandelonitrile, while the vestibule produces an enzyme promoting the dissociation of mandelonitrile into benzaldehyde and hydrogen cyanide on release to the outside. Bioluminescence has been reported in a number of species of centipedes and millipedes, and may act as a warning to predators.

Reproduction and development

Myriapods have separate sexes. Parthenogenesis has been reported in some species or geographic races. The reproductive system is dorsal to the gut in centipedes and ventral in millipedes.

Male reproductive system

The male reproductive system in centipedes is quite variable in the number and arrangement of the testes. In lithobiomorphs there is a single tubular testis leading posteriorly into a sperm duct, while in scutigeromorphs the oval testes are paired, each leading into a sperm duct which join to form the unpaired ejaculatory duct. In geophilomorphs and scolopendromorphs, each testis is joined both anteriorly and posteriorly by sperm ducts to a single median sperm duct. While geophilomorphs have one pair of testes, scolopendromorphs have multiple testes, arranged in pairs or alternating down the length of the sperm duct (Fig. 12.9a). Distally, the ejaculatory duct is enlarged for spermatophore production. It then bifurcates to loop around the gut before opening posteriorly into the genital atrium on the 'penis'. A variable number of paired accessory glands opens into the genital atrium. These glands produce the web on which spermatophores are deposited.

The male reproductive system of millipedes is more uniform and simple. It consists of a single, fused testis or a pair of testes joined by ladder-like connectives, extending for much of the length of the body. The testes lead anteriorly into the paired sperm ducts opening on or near the coxae of the second pair of legs.

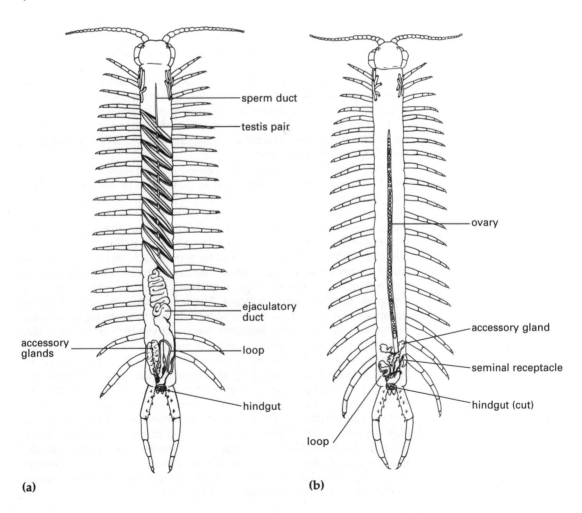

Figure 12.9
Reproductive anatomy of a scolopendromorph centipede. (a) Male system. (b) Female system.

Female reproductive system

The female reproductive system is uniform in the various orders of centipedes. A single tubular ovary leads into a short oviduct, which divides in two and loops around the gut before opening at the genital aperture. Paired accessory glands and seminal receptacles also open into the genital aperture (Fig. 12.9b).

In millipedes, the paired ovaries are fused but lead into a pair of anterior oviducts, each opening near or on the bases of the second pair of legs into an atrium containing a seminal receptacle in its wall.

Courtship and mating

Even though male centipedes possess a penis-like structure, sperm transfer is indirect, either as a naked droplet or invested in a membrane as a spermatophore. As with many groups of aggressively venomous, carnivorous arthropods, mating is engaged upon with caution. Courtship behaviour, although not as elaborate as in many arachnids, is a prelude to sperm transfer. Courtship involves tapping and stroking with the antennae and hind legs, head-to-tail promenading or, in scutigeromorphs, visual cues such as bobbing up and down. The male then spins a web on which a droplet of sperm or a spermatophore is placed. The female, using the web as a guide, picks up the spermatophore in her gonopods and the sperm are released into her reproductive tract for storage. Male scutigeromorphs do not spin a web, but deposit a spermatophore on the ground and guide the female over it.

Courtship behaviour in millipedes can involve robust snatching of females or quite elaborate mating rituals, including touching and drumming with the legs. Males of some species of pill millipedes stridulate by rubbing the last pair of legs against the sides of the telson. Oscillograms of the sounds from closely related species are distinct. The female responds by uncurling, and mating ensues.

Penicillate millipedes do not have gonopods, and sperm transfer is indirect. Males spin a web on which two droplets of sperm are fixed. A guide line directs a passing female to the droplets, which are picked up in her gonopores. In pentazonian millipedes, such as pill millipedes, the male uses his legs as a conveyer belt to transfer sperm from the genital opening to the genital opening of the female. In chilognath millipedes with gonopods, the male crawls up the back of the female and twists around so that anterior ends are ventral-to-ventral. The male flexes his body to transfer sperm from the genital openings, at the bases of the second leg pair, to the gonopods on the seventh segment. The female atria are opened by the gonopods and the sperm are introduced. In the Spirostreptida, complex mating systems have been described that are driven by both female choice of mating partner and intense sperm competition among males. Sperm competition is a form of sexual selection that has resulted in complex male behaviour, such as mate guarding, and elaborate male gonopods for scooping out or mixing up opponents' sperm within the seminal receptacle of the female. Similar behaviour is observed in some dragonflies (see Chapter 11).

Of all the myriapod groups, mating behaviour in symphylans is perhaps the most bizarre. Males deposit a number of spermatophores, each at the end of a stalk. These are eaten by passing females, which store the sperm in buccal pouches. Subsequently, the female uses her mouthparts to remove eggs from her gonopores and glues them to the substratum. Sperm are then smeared from the mouth over the eggs.

Egg laying and protection

The eggs of geophilomorph and scolopendromorph centipedes are laid in batches and brooded by the coiled body of the female. The eggs are coated in a sticky secretion and groomed by the mouth parts, possibly to coat them in an antimicrobial fluid. Lithobiomorph and scutigeromorph centipedes lay single eggs protected by a coating of soil and humus particles. Millipedes lay their eggs, either singly in capsules of earth mixed with saliva, or in clusters in elaborate nests constructed from voided earth mixed with secretions from the rectum.

Development

Myriapods lay yolky eggs (Fig. 12.10a). The embryonic development of centipedes, particularly the large-egged scolopendromorphs, follows a generally similar course to that of yolky onychophoran eggs (Fig. 12.10b; compare with Fig. 9.8b), but includes the development of an acron and telson and the formation of fat body from mesoderm cells, as in insects. The other myriapod groups (symphylans, diplopods and pauropods) share a more specialised pattern of development in which, among other things, fat body is formed from yolk cells. Each class has its own special embryonic features. At the same time, myriapod embryonic development differs sharply from that of hexapods. The labiate mouthparts of the Symphyla are convergent with those of insects.

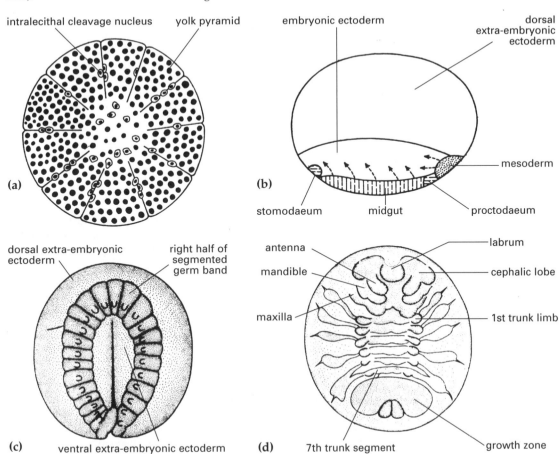

Figure 12.10 Early embryonic development in myriapods. **(a)** Cleavage in scolopendromorph centipedes. The yolk-mass divides into a number of yolk pyramids converging on a central mass of yolk. The cleavage nuclei migrate to the yolk surface by passing along the interfaces between the pyramids. **(b)** Fate map of the chilopod blastoderm. **(c)** Epimorphic germ band formation in scolopendromorph centipede. **(d)** Anamorphic germ band formation in a glomerid pill millipede (from Anderson 1973).

Geophilomorph and scolopendromorph centipedes undergo epimorphic development and hatch with the full complement of legs (Fig. 12.10c). Lithobiomorphs and scutigeromorphs are anamorphic, hatching with only seven and four pairs of legs respectively. Millipedes and pauropods are also anamorphic and hatch with three pairs of legs (Fig. 12.10d), while symphylans hatch with six pairs of legs. After hatching, anamorphic myriapods add segments from a growth zone at the hind end of the body during successive moults, until the full complement of segments and limbs is completed.

Moulting of the exoskeleton follows the digestion of the endocuticle and, in millipedes, the resorption of calcium. In centipedes, the old cuticle splits along suture lines on the dorsal surface of the head or between the head and the first trunk segment. The animal emerges through this split, causing a telescoping of the exuviae. In millipedes, the old exoskeleton initially splits between the head and the collum and subsequently along the midventral line and laterally above the legs. Some species eat their exuviae.

Evolutionary history

The fossil record of myriapods is sparse. The oldest fossil myriapod-like animals come from early Silurian marine deposits, and more contentious myriapod-like fossils have been found in the middle Cambrian. The earliest terrestrial myriapods are reported from an assemblage of land arthropods from the late Silurian. Because of their calcified exoskeletons, millipedes have left a more extensive fossil record than other groups of myriapods. Considerable diversification is seen in the Devonian and Carboniferous, and by the Oligocene, species similar to those of the present can be identified.

Molecular analysis indicates that myriapods are a monophyletic group, although the relationships among the classes remain unresolved, as does the relationship of the myriapods to other arthropods (see Chapter 18).

Classification of the subphylum Myriapoda

CLASS CHILOPODA
Trunk with numerous segments, mostly bearing a single pair of uniramous walking legs; mouthparts consisting of mandibles and 2 pairs of maxillae; first trunk segment with a pair of poison fangs; genital opening on second last segment.

Subclass Epimorpha
Development epimorphic; eggs brooded. (Orders Geophilomorpha, Scolopendromorpha.)

Subclass Anamorpha
Development anamorphic; eggs not brooded. (Orders Lithobiomorpha, Scutigeromorpha, Craterostigmorpha.)

CLASS SYMPHYLA
Colourless, small (less than 10 mm); 12 leg-bearing segments, covered by 15 to 24 terga; Tömösváry organs present but eyes lacking; mouthparts consisting of mandibles, maxillae and labium; last pair of appendages as spinnerets; genital openings between fourth pair of legs; development anamorphic.

CLASS DIPLOPODA

Trunk with a large number of leg-bearing segments. First trunk segment (collum) legless, next three segments each with a single pair of legs; other segments (diplosegments) with two pairs of legs; eyes, when present, as one to 90 ocelli on each side of the head; mouthparts consisting of mandibles and fused first maxillae (gnathochilarium); genital openings on or near base of second pair of legs; development anamorphic.

Subclass Penicillata

Small (less than 4 mm); body soft and covered in tufts of serrated bristles; trunk with 13 to 17 pairs of legs; no gonopods, sperm transfer indirect. (Order Polyxenida.)

Subclass Chilognatha

Body wall impregnated with calcium.

Infraclass Pentazonia — Terga form a hemispherical arch over body; each segment composed of tergal arch, two pleurites and two sterna; last one or two pairs of legs in male modified as claspers for holding female; no gonopods; pill millipedes (see page 273 for details).

Infraclass Helminthomorpha — Elongate cylindrical or flattened body; at least on pair of legs of the third diplosegment modified as gonopods. Contains a large number of orders including sucking millipedes, flat-backed millipedes and snake millipedes (see page 273 for details).

CLASS PAUROPODA

Minute (0.5 to 2.0 mm), barely pigmented; antennae biramous; eyeless; usually nine pairs of legs and five terga; most terga bear elongate sensory bristles; mouthparts consist of mandibles and gnathochilarium; gonopores on third trunk segment; development anamorphic.

Chapter 13

The Crustacea

P. Greenaway

SUBPHYLUM CRUSTACEA

Introduction 287

Basic structure and function 287

Patterns of body form and life style 289
- Class Remipedia 289
- Class Malacostraca 290
- Class Phyllopoda 297
- Class Maxillopoda 298

Functional anatomy and physiology 301
- Locomotion 301
- Feeding and digestion 301
- Gas exchange and transport 303
- Circulation 305
- Osmoregulation and excretion 306

Nervous system and sense organs 307

Hormones 309

Reproduction and development 310

Freshwater and terrestrial radiation 311
- Colonisation of freshwater habitats 311
- Colonisation of land 312

Phylogeny and fossil record 316

Classification of the subphylum Crustacea 316

Subphylum Crustacea

Primarily marine arthropods with biramous limbs; two pairs of antennae; mandibles formed as gnathobasic jaws; several to many pairs of trunk limbs; gonopore in mid-region of body.

Introduction

The Crustacea comprise approximately 75 000 species and include many familiar animals such as crabs, lobsters, prawns, barnacles, woodlice and beach fleas, as well as a host of lesser-known species. Unlike the terrestrial Hexapoda and Myriapoda and the mainly terrestrial Chelicerata, the main radiation of the Crustacea has been aquatic, with the bulk of the species living in marine habitats. There is also a substantial number of freshwater species, but only 2–3% of species live on land. Crustaceans are the dominant arthropods in the oceans, where they occupy benthic, pelagic, planktonic and intertidal niches and lead motile, sedentary, sessile or parasitic life styles. In inland waters they are represented by a more limited range of taxa, but nevertheless have succeeded in virtually all types of water bodies, including fresh water, temporary pools, and even hypersaline lakes. On land the diversity is low, with representatives from only three orders of malacostracans and a poorly studied cryptozoic fauna of microcrustaceans. At the upper end of the size range of crustaceans are large benthic crabs with leg spans of over 3 m (Japanese spider crab) and lobsters which can weigh up to 60 kg, while at the other end are minute planktonic and larval forms. The largest terrestrial species is the robber crab, *Birgus latro*, which can reach 3 kg in weight.

Basic structure and function

The crustacean body is divided into tagmata, typically with recognisable head and trunk regions and with the trunk commonly differentiated into thorax and abdomen. The head generally bears a dorsal shield and consists of a pre-antennal segment (embryonic) followed by five limb-bearing segments; first antennae (antennules), second antennae (antennae), mandibles, first maxillae (maxillules) and second maxillae (maxillae). The mandibles are multiarticulated and act as gnathobasic jaws. The body has a pair of limbs, which are primitively biramous, on each segment. Sexes are generally but not always separate, and development often includes a number of free-living larval stages, characteristically including a naupliar stage, although these may be suppressed or restricted to the egg. Excretory organs are a single pair of segmental organs located in either the antennal or the maxillary head segment. A single naupliar eye and paired compound eyes are usually present. The exoskeleton is calcified in larger species.

The exceedingly flexible body plan of the Crustacea has allowed the evolution of a great diversity of forms. The head is the most conservative tagma, organised in most crustaceans as outlined above. Often the dorsal segmentation of the head is obscured by the outgrowth of the exoskeleton to form a head shield, which frequently extends back over the anterior trunk or thorax as a carapace. The head appendages are generally concerned with sensory reception (the two pairs of antennae) and feeding (the mandibles, maxillules and maxillae), but may be modified for other tasks and may be locomotory in many small forms. The trunk may have a large number of similar segments or be specialised into distinct functional regions, each bearing suitably modified appendages. Thus the anterior trunk appendages commonly help to manipulate food (as maxillipeds) and capture food (as pincers or chelae), while the more posterior limbs are used in locomotion (walking or swimming) and reproduction, as in the eumalacostracans (Fig. 13.1). Other types of body organisation are discussed in the next section. The terminal telson is frequently used in locomotion.

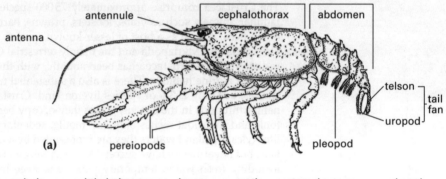

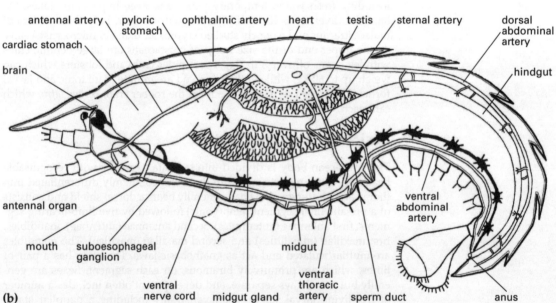

Figure 13.1
(a) External anatomy of a crustacean (crayfish) (after Hale (1927–29). (b) Internal anatomy of a generalised malacostracan.

The cuticle is a multilayered structure composed of chitin and protein complexes, resembling that of insects. It differs from the latter in being impregnated with calcium carbonate, which provides rigidity and mechanical strength in large species, and in usually lacking any waterproofing lipids in the epicuticle. Supplementing the obvious exoskeleton, crustaceans possess an extensive calcified endoskeleton which provides internal support and sites for muscle attachment. As in other arthropods, growth occurs by moulting. The foregut and hindgut are also cuticle-lined (Fig. 13.1b), often resulting in the evolution of grinding and filtering organs in the foregut. The midgut is elaborated into a large midgut gland (the hepatopancreas) made up of numerous fine tubules responsible for enzyme secretion and food absorption.

Crustaceans must exchange respiratory gases with the water. While small species can achieve this across their integument, most of the larger species possess gills. These are outgrowths from limb bases or the body wall and are usually protected under the carapace. The paired excretory organs are anterior to the mouth.

Patterns of body form and life style

The Crustacea show a wide diversity of body form, and this section illustrates the extent of their variability in morphology and life style.

Class Remipedia

The remipedes are a recently discovered group of ten known species which have an anatomy thought to be the closest to the primitive crustacean body form (Fig. 13.2a). The animals are small and elongate (up to 30 mm). The head bears a dorsal shield and is followed by a long vermiform trunk of 20–30 similar segments with biramous limbs. Remipedes swim on their backs by metachronally beating their oar-like thoracic appendages. They are believed to be carnivorous and use the grasping maxillae and maxillipeds to seize their prey, into which they inject digestive fluids from maxillary glands. The prey contents are ingested in semi-fluid form by the sucking action of the foregut, a system similar to that of spiders. Little is

Figure 13.2
(a) A remipede, *Speleonectes*; dorsal view. (b) A stomatopod, *Harpiosquilla harpax*. (c) A syncarid, *Allanaspides*; dorsal view. (a from Yager 1994; b from Caldwell and Dingle 1976; c after Swain *et al.* 1971.)

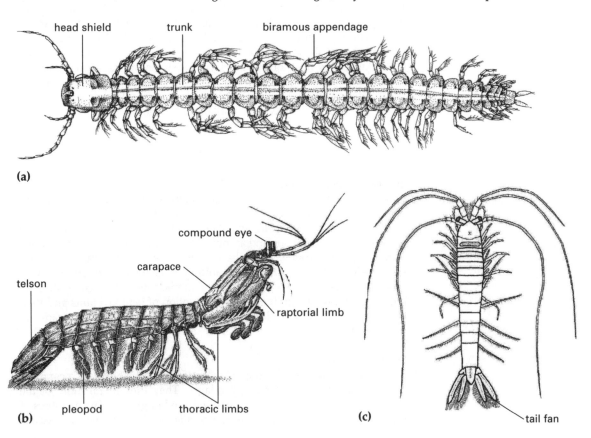

known of the life styles of remipedes. They live in caverns and lava tubes on tropical Atlantic islands where the saline sea water is overlain by fresh water. Remipedes are hermaphrodite.

Class Malacostraca

The malacostracans are the largest class of crustaceans, with more than 21 000 species, including most of the larger and familiar forms such as prawns, crabs and lobsters. The body and appendages are specialised into thoracic and abdominal regions and the head bears sensory and feeding appendages. The anterior thoracic limbs are commonly modified as additional mouthparts and limbs for capturing food, and are often chelate; the more posterior thoracic limbs are locomotory pereiopods. The abdominal appendages or pleopods are less sturdy and typically are used in slow swimming, the production of respiratory currents and, in females, for carrying eggs. The telson and uropods form a tail fan which enables rapid backward swimming when the abdomen is flexed — a behaviour used to escape predators.

Subclass Hoplocarida

The Stomatopoda is the only extant order of hoplocarids with around 350 living species. They are known as mantis shrimps for their convergent similarity in feeding and behaviour to the mantids of the insect world. Stomatopods are marine, generally large (to 340 mm) carnivores which live in shallow seas, particularly in tropical and warm temperate areas. They live in individual burrows from which they emerge to stalk or attack passing organisms. The second thoracic limbs are specialised either to spear and hold swimming prey such as fish and prawns (Fig. 13.2b) or to club and crack the heavy shells or exoskeletons of molluscs and crabs. Locomotion is by walking using the thoracic limbs, but the stomatopods can also swim by metachronally beating the abdominal pleopods, or dart backwards by rapidly flexing the abdomen. The pleopods bear gills. Female stomatopods carry their eggs on the pleopods until they hatch, and the early larval stages commonly remain in the burrow for a while before adopting a planktonic life. After metamorphosis to post-larvae they settle to the bottom and develop the adult mode of life.

Subclass Eumalacostraca

This subclass includes all the other members of the Malacostraca and includes many of the common forms such as crabs, lobsters and crayfishes, spiny lobsters, hermit crabs and their allies, as well as the smaller but equally abundant isopods and amphipods. The carapace is well developed in many of the groups, and gills are often associated with the thoracic limbs. Although the primitive condition is characterised by biramous limbs, there has been a tendency for the thoracic limbs to become strongly built and uniramous in line with the adoption of benthic habits and their development as walking legs. Six main groups of eumalacostracans are recognised.

Order Syncarida

The syncarids are a small group of about 150 species which lack a carapace and have biramous thoracic appendages. They thus retain more of the primitive malacostracan features than most other groups. The members of the suborder Anaspidacea e.g. *Anaspides, Allanaspides,* are small crustaceans (to 50 mm) living in freshwater streams, ponds, caves and crayfish burrows, with a relict distribution in Australasia and South America (Fig. 13.2c). Syncarids belonging to a second suborder, Bathynellacea (e.g. *Bathynella*) are much smaller animals (to 3.4 mm) and are mostly interstitial, living amongst soil particles in dilute ground waters world-wide. Anaspids walk or swim using the thoracic limbs and can dart backwards by flipping the tail fan. Unlike most other malacostracans, the females do not carry the eggs, but attach them to the substratum or vegetation.

Orders Mysida and Lophogastrida

These orders comprise a group of perhaps 800 species (of which 90% are mysids) of generally small and often extremely numerous marine crustaceans. *Gnathophausia* is the largest species, reaching 300 mm. A few species have penetrated fresh water, such as the Great Lakes in North America, but most species are marine and live in or on the substratum or swim in the shallow waters above. Some are benthic during the day but undergo a diurnal migration, moving up into the plankton at night. Locomotion is by walking, tail-flipping and swimming using either the thoracic exopods (mysids) or the pleopods (lophogastrids). Although many mysids are filter feeders (Fig. 13.3a) it is likely that most benthic forms feed on relatively large material scavenged from the substratum. Lophogastrids prey on zooplankton. The carapace is well developed and covers most of the thorax, but is never attached to more than the first three thoracic segments. Stalked compound eyes are present. Both groups have thoracic gills, which in lophogastrids are ventilated by the exopods of the biramous thoracic limbs. Female mysids have a large brood pouch formed by flattened extensions of the thoracic limb bases (oostegites). Developing stages are retained in the pouch and ultimately released as juveniles.

Figure 13.3
(a) A mysid, *Mysis*. (b) Male and female cumaceans, *Manocuma*, in precopulatory behaviour. (c) A tanaid, *Paratanais*. (a from Dales *et al.* 1981; b from Gnewuch and Croker 1973; c after Hale 1927–29.)

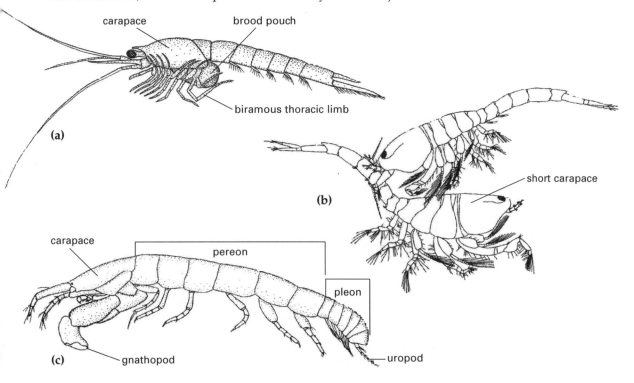

Order Hemicaridea
This order includes the suborders Cumacea, Tanaidacea and Spelaeogriphacea, which are characterised by an inflated carapace attached to the first 3–6 thoracic segments. The carapace forms two branchial chambers ventilated by the epipodite of the first maxillipeds. Behind the carapace the thorax and abdomen are narrow.

The Cumacea (Fig. 13.3b) includes approximately 1000 species of small (to 35 mm) marine crustaceans which occur world-wide in predominantly benthic habitats. They burrow or construct tubes in the sediment, either feeding on organic matter in the sediment or filtering food material from a water current set up by the mouthparts, using the setae on the maxillules. Swimming is by means of the anterior thoracic limbs; the posterior pairs are used in burrowing. Cumaceans may periodically emerge from the sediment and gather in the water column in what are thought to be mating swarms. Females incubate the eggs in a small brood pouch and the stage released is a post-larva which lacks the last thoracic limbs and pleopods.

The 850 species of tanaidaceans are small (0.5–20 mm) marine crustaceans which live in mucous tubes or burrows in muddy sediments beneath shallow waters (Fig. 13.3c). The carapace is smaller than in cumaceans and covers only the first two thoracic segments, to which it is fused. The second thoracic limbs are large and chelate. Tanaidaceans are well adapted morphologically for tubiculous life but are poor swimmers and walkers outside the burrow environment. Some secrete a thread or 'life line' in the manner of spiders, facilitating return to the burrow. Although some species are filter feeders, most are raptorial and feed on pieces of organic material or small organisms. Sex change is common in the group; many species begin life as females, breeding once or more before moulting to become functional males. The female has a marsupium or brood pouch formed by the posterior thoracic limbs, and the young emerge at a similar stage of development to cumaceans i.e. as post-larvae.

Order Edriophthalma
The suborders Isopoda (more than 4000 species) and Amphipoda (about 6000 species) have world wide distributions and have radiated extensively in marine, freshwater and terrestrial habitats. Both groups lack a carapace and have (usually) seven pairs of uniramous thoracic limbs and sessile compound eyes.

The isopods usually have a dorso-ventrally flattened body and flat, biramous pleopods which function both as gills for gas exchange and in swimming (Fig. 13.4). With few exceptions isopods are small animals ranging up to about 20 mm in length, but a few are larger: the deep-water marine isopod *Bathynomus* reaches 440 mm. The main habitat is marine and most species are mobile, either swimming by means of the pleopods or walking on the pereiopods. Some (such as *Corophium*) bore in timber, while others (Epicaridea) are ectoparasites, commonly on fishes. There are numerous freshwater species of isopods (Aselloidea and Phreatoicidea) and over 1000 terrestrial species (Oniscidea). The Phreatoicidea, a mainly Australian family, are bilaterally rather than dorso-ventrally flattened (Fig. 13.4c). In such a large and diverse group it is difficult to generalise about feeding habits. Predators, herbivores, saprovores and parasites are all represented. Most feed on relatively large material and have strong biting and chewing mouthparts, but parasitic isopods are frequently adapted to suck the body fluids of their hosts. Females incubate the eggs in a brood pouch

(Fig. 13.4b) walled by oostegites formed from expansions of the coxal joints of the pereiopods. The young lack only the last thoracic appendages on hatching and release.

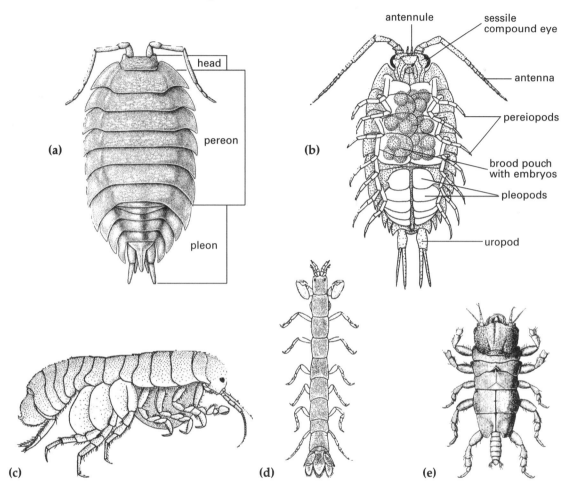

Figure 13.4
Anatomy and range of form in isopods. (a) *Oniscus*, dorsal view. (b) *Ligia*: ventral view. (c) *Phreatoicus*. (d) *Mesanthura*. (e) *Gnathia*. (a from Boolootian and Heyneman 1980; b after Dales *et al.* 1981; c–e after Hale 1927–29.)

The amphipods (Fig. 13.5a) are generally small but range in length from 1 to 250 mm. The body is laterally compressed and at least the first and second pereiopods are modified as chelae to catch and hold food. The gills, in contrast to the abdominal gills of the isopods, are basal outgrowths of the thoracic pereiopods, and there are generally several pairs of uropods. Amphipods are principally marine animals, but there are significant numbers of freshwater species and some semiterrestrial and terrestrial species (Family Talitridae). The benthic mode of life is the most common, walking or jumping being achieved using the pereiopods, but there are a few pelagic forms which swim with the aid of the pleopods and uropods, and also a good representation of tubiculous, burrowing and ectocommensal forms. Only a few amphipods are parasitic. Carnivory and scavenging are the most common feeding patterns, but grazing and filter-feeding are also well represented. As in the isopods, a thoracic brood pouch is formed in which the eggs are incubated, and the young hatch as juveniles.

Order Euphausiacea

The 90 species of krill are marine, pelagic, shrimp-like animals with a shallow carapace, e.g. *Euphausia*, *Meganyctiphanes* (Fig. 13.5b). Species often have wide longitudinal but restricted latitudinal distributions and may be exceedingly numerous, often forming dense feeding swarms. Euphausids range from 40 to 150 mm in length. They are a major food source, especially in the Southern Ocean, for predators such as fish and squid and macro filter-feeders such as baleen whales, and they are of increasing economic importance to humans.

The carapace is fused to the head and thorax but is shallow and does not cover the bases of the biramous thoracic limbs, so the gills are exposed. Swimming is achieved by means of the pleopods and thoracic exopods, which bear oar-like arrays of setae. The thoracic endopods also function in filter-feeding, forming a chamber which filters planktonic organisms from the water as the animal swims forwards. Trapped food is passed to the mandibles.

Eggs are laid several weeks after copulation, and are either shed into the water or carried on the posterior thoracic limbs until they hatch as nauplii. There are 11 or 12 distinct larval stages before the adult body form is reached.

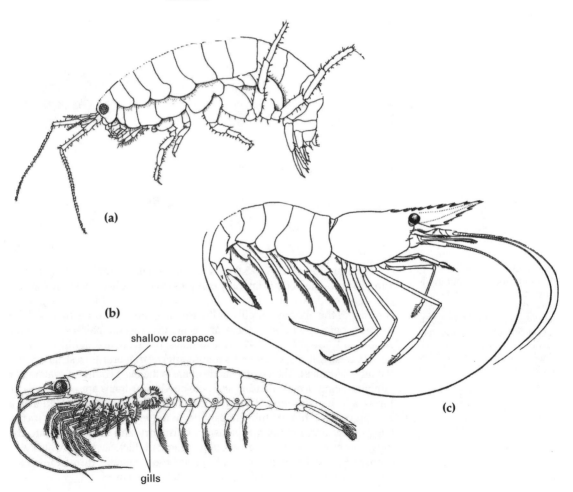

Figure 13.5
(a) *Arcitalitrus*, a terrestrial amphipod. (b) A generalised euphausid. (c) *Palaemon*, a eukyphid prawn. (a from Anderson 1996; b after Mauchline and Fisher 1969; c after Holthuis 1950.)

Order Decapoda

The decapods are a major and conspicuous group of eumalacostracans, with around 10 000 species which include crabs, lobsters, crayfish and hermit crabs (Suborder Reptantia) as well as the pelagic and benthic prawns (Suborders Dendrobranchiata, Eukyphida and Euzygida). They have a large carapace which covers the head and thorax and encloses the gills. The first three pairs of thoracic limbs (maxillipeds) are specialised for food handling; the remaining five pairs are stout, usually uniramous periopods.

The Suborder Dendrobranchiata includes 450 species of peneid and sergestid prawns (such as *Penaeus* and *Sergestes*), which are the basis of many commercial prawn fisheries. They are characterised by the presence of dendrobranchiate gills (Fig. 13.12c). The first three pairs of pereiopods are chelate. Dendrobranch prawns are usually marine, pelagic or epibenthic, although some peneids are benthic and many have estuarine distributions. Females do not brood the eggs, which are instead shed into the water where they hatch as nauplii and proceed through numerous larval stages before reaching the adult form. In all other decapods the females brood the eggs on the pleopods and hatching occurs as a zoea or later-stage larva.

The Suborder Eukyphida (Fig. 13.5c) includes around 2000 species from marine, brackish and freshwater habitats, such as *Caris*, *Pandalus*, *Hippolyte*, *Palaemon* and *Atya*. They are characterised by their phyllobranch gills (Fig. 13.12a), and chelae are present on only two pairs of pereiopods. Suborder Euzygida consists of about 20 species of stenopodid prawns, such as *Stenopus*, which favour reefs and rocky bottoms in tropical seas and are generally scavengers or commensals. The first three pairs of pereiopods in this group are chelate; the third pair are uniquely enlarged.

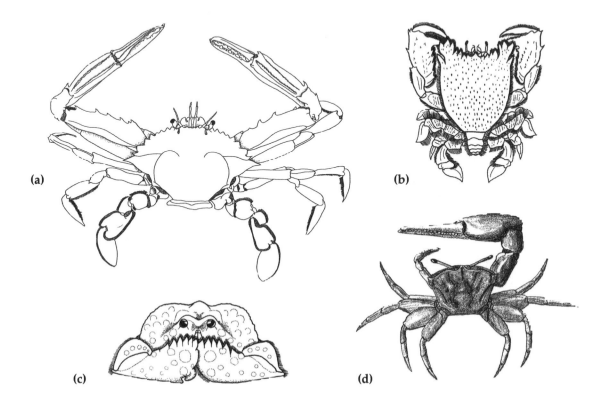

Figure 13.6
Anatomy and range of form in brachyuran decapods.
(a) *Portunus*, a swimming crab.
(b) *Ranina*, a primitive crab.
(c) *Callapa*, a box crab (d) *Uca*, an intertidal fiddler crab. (a after Anderson 1996; b, c after Warner 1977; d from Crane 1975.)

The most conspicuous crustaceans belong to the Suborder Reptantia, a diverse group comprising about 7500 species of predominantly benthic marine crustaceans, including lobsters, spiny lobsters, crabs and hermit crabs, as well as numerous species of freshwater, amphibious and terrestrial crayfish and crabs. All are strongly built and heavily calcified. Locomotion is by walking using the pereiopods, although rapid backward swimming is retained in the lobsters and crayfish. Reptant decapods are generally omnivorous, but there are also specialist deposit feeders, filter-feeders, herbivores and carnivores. Many species have powerful chelae on the first pereiopods for catching and holding food. The basic body plan has been variously modified in the various groups (Figs 13.6, 13.7). Thus the spiny lobsters (infraorder Palinura) lack chelae, while the crabs (infraorder Brachyura) have reduced abdomens tucked under the carapace. The anomalans (infraorder Anomura) include a range of crab and lobster-like body forms, and also include the hermit crabs, which live in abandoned mollusc shells. The shell protects the soft abdomen and serves as a retreat when the animal is threatened.

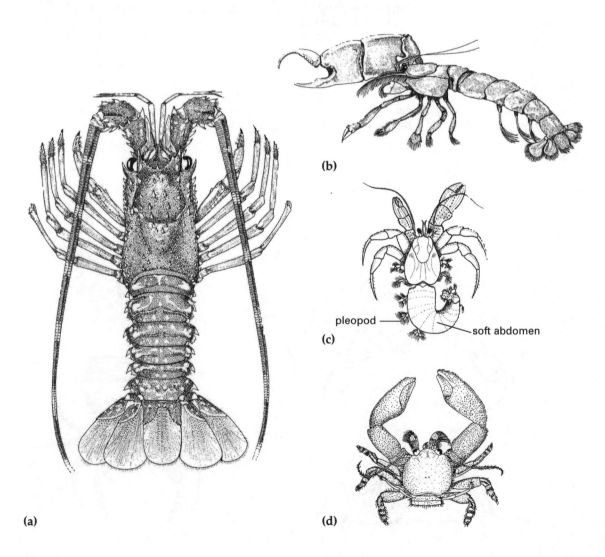

Figure 13.7
(a) *Palinurus*, a spiny lobster.
(b) *Callianassa*, an estuarine burrowing ghost shrimp.
(c) *Pagurus*, a hermit crab.
(d) *Petrolisthes*, an anomalan porcelain crab. (a, b from Holthuis 1991; c after Warner 1977; d from Martin *et al.* 1988.)

Class Phyllopoda

The phyllopods are a large and diverse group of crustaceans which have flattened, polyramous limbs on the thorax, used in swimming and filter feeding. Abdominal limbs are generally reduced in number or lacking and the antennules are generally uniramous.

Subclass Phyllocarida

There is a single order (Leptostraca) in this subclass, containing about 20 species of small mud-dwellers of the marine benthos, such as *Nebalia*. They have a bivalved carapace with the two halves joined by an adductor muscle. Phyllocarids are suspension feeders on bottom sediments.

Subclass Cephalocarida

The subclass Cephalocarida also comprises a single order (Brachypoda), with nine described species of small (2–4 mm) marine, mud-dwelling crustaceans such as *Hutchinsoniella* (Fig 13.8a) which lack a carapace and retain many primitive characters.

Subclass Sarsostraca

The order Anostraca, with 185 species, includes the brine shrimps *Artemia* and *Parartemia* and the fairy shrimps *Chirocephalus* and *Branchinella*. The group is characterised by the lack of a head shield and carapace (Fig. 13.8b). Anostracans are characteristic of inland waters, where they live in ephemeral freshwater pools or salt lakes. They swim using the foliaceous trunk limbs, which beat in metachronal rhythm and propel the animal through the water, usually ventral side uppermost. Feeding results from the swimming movements of the limbs, whose setae separate food material from the water as they swim. Under unfavourable conditions, such as the evaporation of pools or unfavourable temperatures, anostracans produce resistant eggs which can restart the population when suitable conditions return.

Subclass Calmanostraca

The calmanostracans generally have a well-developed carapace which is often bivalved and encloses the body. The group comprises the orders Notostraca (tadpole shrimps, Fig. 13.8c), Conchostraca (clam shrimps, Fig. 13.8d) and Cladocera (water fleas, Fig. 13.8e). Both the tadpole and clam shrimps are typical of ephemeral freshwater pools, where they lead an existence similar to that of the anostracans, which often share their habitat. The notostracans, such as *Triops*, are represented by nine widely distributed species, all of which have a characteristically large flattened carapace. Notostracans move over the bottom of shallow pools, filtering organic material and algae with their setose trunk limbs.

The clam shrimps include about 200 described species, such as *Limnadia* and *Cyzicus* species, and have a bivalved carapace which completely encloses the body. Again they swim and feed using their foliaceous trunk limbs. The cladocerans (about 400 species) inhabit more permanent waters than do notostracans and conchostracans. They are widely distributed in freshwater lakes and ponds; a few marine and terrestrial species have also been described. Benthic cladocerans scrape food material from the substratum, while planktonic species such as *Daphnia* swim by means of the second antennae and use the thoracic limbs for filter-feeding. Parthenogenesis is common among cladocerans, and males may only be present under

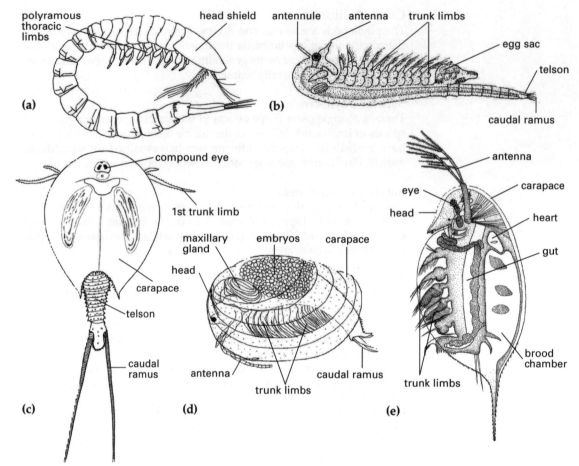

Figure 13.8
Morphology and range of form in freshwater phyllopods.
(a) *Chiltoniella*, a cephalocaridan from New Zealand. (b) *Artemia*, an anostracan brine shrimp from inland salt lakes. (c) *Triops*, a notostracan tadpole shrimp from ephemeral pools.
(d) *Limnadia*, a conchostracan clam shrimp from ephemeral pools. (e) *Daphnia*, a cladoceran water flea from freshwater ponds. (a from Knox 1977; b after Dales *et al.* 1981; c after McLaughlin 1980; d after Anderson 1996; e after van Rensburg *et al.* 1980.)

adverse conditions. The eggs develop in the brood pouch and are released as juveniles. During unfavourable conditions, all three orders of Calmanostraca produce resistant eggs.

Class Maxillopoda

The maxillopods are generally small crustaceans with not more than six thoracic segments and not more than five in the abdomen. The group includes the parasitic fish lice, the extremely numerous copepods and ostracods, and the sessile barnacles.

Subclass Branchiura

There are about 130 described species of branchiurans or fish lice (e.g. *Argulus*, Fig. 13.9a), all of which are highly specialised ectoparasites, commonly of fish. The lice swim between hosts using the thoracic limbs and attach themselves by the two pairs of maxillae, which are modified as suckers and grasping claws. Their flattened bodies and carapace provide streamlining to prevent detachment as the host fish swims.

Subclass Ostracoda

These are the seed shrimps such as *Cypris* — about 8000 species of small crustaceans, mostly around 1 mm (*Gigantocypris* reaches 30 mm). Like the Conchostraca, ostracods have a bivalved, hinged carapace which com-

THE CRUSTACEA

pletely encloses the body but remains open on the ventral side (Fig. 13.9b). The valves are closed by adductor muscles and are opened by the elastic hinge when the adductors relax. The head is well developed and bears two pairs of antennae, the mandibles and the maxillae. The trunk is reduced, and locomotion is by swimming using the antennae or by crawling over the substratum using the caudal ramus. These animals are widely distributed in marine, freshwater and damp terrestrial habitats.

Subclass Copepoda

Copepods are characterised by the presence of a head shield and the absence of a carapace. They have a single median, sessile eye, lack abdominal appendages, and have well-developed caudal rami. The taxon is a large one: about 9000 species have been described, such as *Cyclops*, *Calanus* and *Gladioferens* species (Fig. 13.9c–e). Most are small, in the 1–10 mm range. Copepods occupy a wide range of aquatic and many terrestrial habitats and include a number of parasitic forms. In the sea, calanoid copepods are a major component of the zooplankton, where they swim by means of the antennae and thoracic limbs and feed on phytoplankton or zooplankton. Harpacticoid copepods are benthic. Members of the parasitic orders are ectoparasitic or endoparasitic on a wide range of fishes and invertebrates. Eggs hatch as nauplii (see Fig. 13.14a) and the six naupliar stages are followed by five copepodid larval stages, culminating in the adult body form.

Figure 13.9
Morphology and range of form in the Maxillopoda. (a) *Argulus*, a branchiuran ectoparasitic on fishes. (b) *Eucypris*, a freshwater ostracod. (c) *Cyclops*, a freshwater copepod. (d) *Lernaeocera*, a copepod parasitic on the gills of fishes. (e) *Dermoergasilus*, a copepod ectoparasitic on fishes. (a from Boolootian and Heyneman 1980; b,c from Dales *et al.* 1981; d after Green 1963; e after Ho *et al.* 1992.)

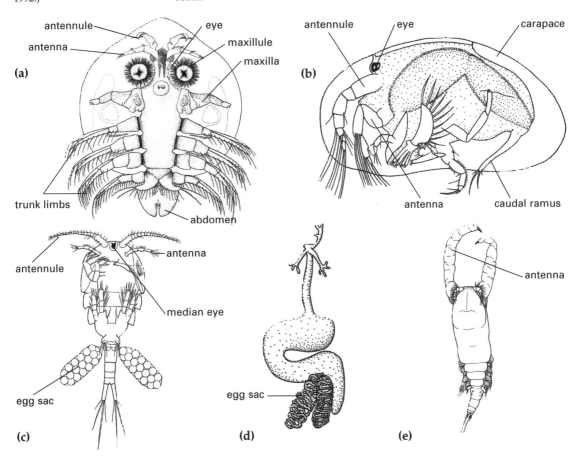

299

Subclass Thecostraca

The thecostracan groups comprise a range of sessile and parasitic marine crustaceans which attach themselves to the substrate or to a host using the antennules. A free-swimming cypris larva acts as the dispersal agent and selects settlement sites or hosts.

The barnacles (Order Cirripedia) comprise most of the 1000+ species in this group. Barnacles are sessile filter feeders which live attached to rocks, floating debris, ships (where they are significant fouling organisms) and animals such as corals, turtles and whales. The pedunculate barnacles, such as *Lepas* (Fig. 13.10a), are characteristic of floating substrata, attaching themselves by means of a tough, fleshy stalk or peduncle cemented to the substratum, the body being protected by a bivalved carapace strengthened by calcareous plates. In other barnacles, such as *Elminius* and *Balanus*, the stalk is reduced to a flattened basis and the calcareous plates completely surround the body. A pair of plated valves at the apex (the operculum) can be opened to allow the filter-feeding thoracic legs to protrude (Fig. 13.10b, c). Unlike most crustaceans, barnacles are hermaphrodite and cross-fertilise their near neighbours by means of a long extensible penis. Eggs brooded in the mantle cavity hatch as nauplii (Fig. 13.14a) and moult into cypris larvae, which are superficially similar in appearance to ostracods. These select a suitable substratum, attach by the antennules, cement themselves in place and then develop into adults.

The Rhizocephala are internal parasites of decapod crustaceans. The infective stage is the free-living cypris larva, which attaches to a decapod host, sheds its trunk region, and gains entry to the host. There it grows long, branching processes which invade all parts of the host's body to extract nutrients. At its next moult the parasite develops an external reproductive mass visible under the abdomen of the crab.

Figure 13.10
Cirripedes. (a) *Lepas*, a stalked barnacle. (b) *Balanus*, an acorn barnacle. (c) Anatomy of an acorn barnacle. (a, b after Boolotian and Heyneman 1980; c after McLaughlin 1983.)

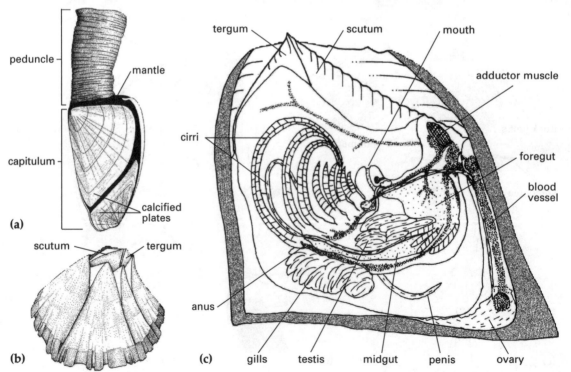

Functional anatomy and physiology

Locomotion

The majority of crustacean species are aquatic. Swimming is the dominant form of locomotion in pelagic species. Epibenthic crustaceans typically have the ability to walk or crawl whilst retaining a strong ability to swim. At the other extreme, heavy-bodied benthic forms and terrestrial species have limited swimming ability but use their strongly developed pereiopods for walking. Swimming limbs need a large surface area to provide thrust, and this is achieved either with flattened appendages (as in phyllopods and pleopods) or by increasing the surface area by means of setae, as in the legs of remipedes (Fig. 13.2a), cladocerans (Fig. 13.8d) and branchiurans (Fig. 13.9a). Many groups combine the two adaptations and have flattened appendages fringed with setae (Fig. 13.11a). The setal fringes of small aquatic crustaceans behave as solid paddles and are therefore very effective in providing thrust. Species with biramous limbs can further increase the contact area by using both rami for swimming, although some increase the versatility of the limb system by using one ramus for a different purpose, such as gas exchange.

Some small planktonic forms may use the first or second antennae for swimming. For example, cladocerans (Fig. 13.8e) use the antennae and copepods (Fig. 13.9c) use the antennules. This often results in a characteristic jerky forward motion. Trunk limbs are used in swimming by representatives of almost all pelagic and epibenthic groups, although in the heavier-bodied eumalacostracans, such as prawns, the anterior trunk limbs are more frequently specialised for walking and the abdominal pleopods assume the swimming role. Many malacostracans escape predators by suddenly darting backwards, using a series of rapid flexures of the abdomen and large tail fan (telson and uropods). The number of limb pairs actually involved in swimming varies from as few as one pair (e.g. in *Daphnia*) to numerous pairs (as in the remipedes and many phyllopods). Where numerous limb pairs are used their movements are usually coordinated in a metachronal sequence and several waves of beating may pass along the body at once, as in remipedes and anostracans. As some appendages are always applying thrust, this generates a much smoother motion than is possible with a single pair of appendages.

Walking limbs are characteristic of the larger malacostracans such as stomatopods, edriophthalmans and decapods, particularly the Reptantia, which have thoracic limb pairs 4–8 specialised as pereiopods (Figs 13.6, 13.7). Typically such limbs are uniramous and stoutly built, with large extrinsic limb muscles. Burrowing and burying forms such as the Scyllaridae and the crab *Corystes* have anterior pereiopods adapted for digging.

Sessile species such as barnacles, and parasites like rhizocephalans and some copepods, are permanently attached as adults to a substratum or host. Their larvae are usually free-living and swim actively to feed and to locate hosts or settlement sites.

Feeding and digestion

Diversity in crustacean body form and size is reflected in feeding mechanisms. The small species are often particulate feeders and have systems based on setae which separate small organic particles or organisms from the water. The feeding current may be generated by the feeding limbs themselves, as in *Daphnia* and barnacles, or by the normal locomotion of the animal through the water. Some animals make use of prevailing water currents in the habitat; for example, the mole crab *Emerita*, which lies buried in sand and extends antennal filters in the backwash of waves. Particulate feeding is characteristic of the phyllopods, barnacles, ostracods, copepods, many of the smaller eumalacostracans, and planktonic larvae of all forms. Epibenthic forms such as some euphausids may kick bottom material into suspension and then filter it (Fig. 13.11b, c). In practice, the 'filtering' limbs in small crustaceans do not actually filter the water. Because of their small size and the high viscosity of water, the setal fringes behave as if they are solid and push water containing food particles to the mouth, instead of straining particles from it. Only in larger animals with larger limbs, where the Reynolds number is higher and setae are more widely spaced, does straining of the water become physically feasible.

Small benthic crustaceans from many groups are grazers or deposit feeders which scrape up bacteria, unicellular algae and organic matter. Many relatively large intertidal crabs, such as the fiddler crab *Uca*, are specialised to separate, minute organic fragments and bacteria from the surface of muddy sand with the aid of setae on the chelae and mouthparts. These setae are fine and flexible in species from muddy habitats, but

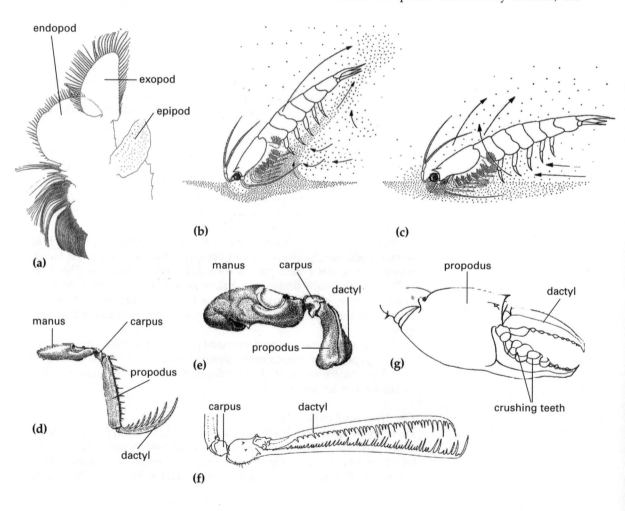

Figure 13.11
(a) A phyllopodous trunk limb of an anostracan.
(b, c) Suspension feeding in euphausiids. Bottom sediment is disturbed by ploughing or kicking, then filtered from the water. Arrows indicate water currents. (d) The second thoracic limb of a stomatopod used to spear prey such as fish and prawns. (e) The second thoracic limb of a stomatopod used as a club to kill heavily armoured prey such as crabs and molluscs. (f) Long, slender chela of the pincer lobster *Thaumastocheles*. (g) Chela of the large swimming crab *Scylla serrata*, modified for crushing. (b, c after Mauchline and Fisher 1969; d, e after Caldwell and Dingle 1976; f after Holthuis 1991.)

species that inhabit sandier substrata, use stiff, spatulate setae to handle the larger, heavier particles. Large species generally require food in larger amounts and thus feed on much larger material as scavengers, predators or herbivores. In the more specialised carnivores such as swimming crabs and stomatopods, thoracic appendages may be extensively modified as chelae, clubs and raptorial limbs for catching prey (Fig. 13.11d–f). External parasites such as *Argulus* show adaptations convergently similar to those of other major parasitic taxa, with mouthparts adapted for piercing and sucking rather than chewing (Fig. 13.9a). The internal rhizocephalan parasites feed in quite a different manner; they develop stolons which ramify through the tissues of the host crustacean, extracting nutrients.

The digestive systems of crustaceans can only process small food particles, and species that feed on large material have developed elaborate systems to sequentially reduce the size of the fragments. The initial phase of processing is undertaken by external appendages. After ingestion the food is commonly subjected to further mechanical processing by internal grinding mechanisms. Many of the paired segmental appendages of the head and thorax have become specialised for particular roles in food processing. Thus the mandibles are responsible for crushing, cutting or grinding the food into particles small enough to be ingested, while the two pairs of maxillae and the maxillipeds are usually involved in holding and manipulating food supplied to the mandibles. In malacostracans the anterior pereiopods are usually chelate and help to manipulate the food while the mandibles work on it. They may also be involved in breaking up food by cutting or crushing (Fig. 13.11g). In the robber crab *Birgus*, the powerful chelae tear material from large food masses and pass it to the more delicate third maxillipeds, which manipulate it for the mandibles.

The food enters the anterior chamber of the foregut (the cardiac stomach), which temporarily stores the food (Fig. 13.1b). The foregut is lined with a flexible cuticle and bears heavily sclerotised teeth or ossicles that grind the food into smaller particles, which are then passed to the pyloric chamber. This consists of a filtration system of chitinous bristles which allows only small particles to pass on to the midgut and shunts large indigestible particles directly to the hindgut.

The midgut is the only part of the alimentary system which lacks a cuticular lining. Although a short tube, it generally has large diverticula (together forming the midgut gland or hepatopancreas), consisting of fine tubules lined with secretory and absorptive cells. This is the site of enzyme secretion. The secretory cells of the midgut tubules produce a digestive juice which flows into the foregut to start digestion. Fluid digesta and small particles from the foregut enter the tubules, where digestion is completed and nutrients are absorbed. The midgut gland also functions as a storage organ for absorbed food, including fat and glycogen.

Undigested material and large particles bypass the midgut gland and move directly to the hindgut (Fig. 13.1b), where they become enclosed in a thin cuticular peritrophic membrane secreted by the posterior region of the midgut and are compacted to form faecal strands. Faeces are periodically evacuated via the anus, which usually opens on the telson.

Gas exchange and transport

Small crustaceans have a high surface area to volume ratio, and gas exchange across the integument is often sufficient for respiration, particularly in groups such as phyllopods that have foliaceous trunk limbs. The

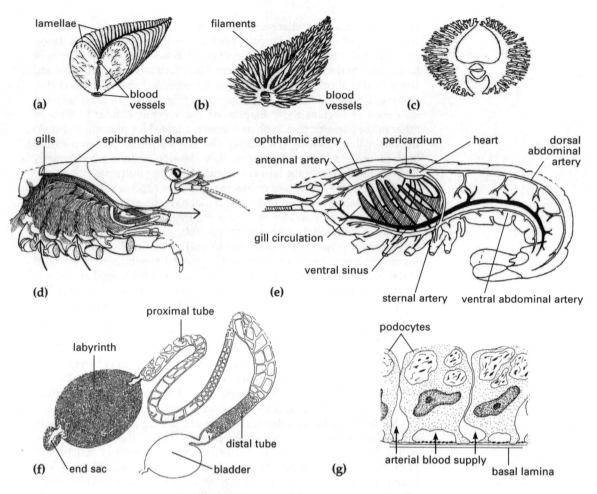

Figure 13.12
(a–c) Principal gill types of decapod Crustacea. (a) Phyllobranchiate gill with flat lamellae. (b) Trichobranchiate gill with tubular filaments. (c) Dendrobranchiate gill with branched filaments. (d) Organisation of the respiratory system of a crayfish. Arrow indicates direction of currents. The scaphognathite draws water over the gills and pushes exhaled water out anteriorly. (e) Circulatory system of a crayfish. Blood is distributed from the heart via several major arteries. Venous blood collects in large ventral sinuses and returns to the pericardial cavity via the gill circulation.
(f) Antennal organ of a crayfish. (g) Section through the wall of the antennal organ (d after Fretter and Graham 1976; e after Meglitsch 1972; f from Felgenhauer 1992; g after Schmidt-Nielsen *et al.* 1968).

carapace, when present, also offers a substantial surface area and is an important site for gas exchange in small crustaceans and larvae. As body size increases, the ratio of surface area to metabolic tissue declines while at the same time the general integument (the exoskeleton) increases in thickness, supporting the larger body mass. In order to maintain adequate levels of gas exchange under these circumstances, the larger crustaceans have evolved gills which increase the area of thin gas-permeable cuticle without compromising general cuticular function. Typically the gills are outgrowths from thoracic limbs, although certain species also utilise the pleopods and some (such as stomatopods and isopods) have developed specialised pleopodal gills. While these outgrowths may be simple in small forms such as amphipods, the need for ever-greater surface area in larger and more active species has encouraged the development of highly organised gill systems. In reptantians, three types of gill have evolved (Fig. 13.12a–c): phyllobranchiate gills with flat lamellae, as in crabs and eukyphid prawns; trichobranchiate gills with tubular filaments as in crayfish and rock lobsters; and dendrobranchiate gills with branched filaments, in the dendrobranch prawns. Several gills may be associated with each thoracic appendage in reptantians. This array of delicate structures is commonly housed under the carapace in branchial chambers on either side of the body (Fig. 13.12d). To ensure that oxygenated water reaches the gills, the branchial chambers are ventilated by specially modified bailers (scaphognathites) on the second maxillae. Lungs have evolved in terrestrial crabs (see later section), while woodlice have pseudotracheae in the pleopods (Fig. 13.15d), a development analogous to the tracheal systems of the insects, myriapods and spiders.

As well as exchanging respiratory gases across the body surface, the animals must transport them between the gas exchange organs and the metabolising tissues. Oxygen is not very soluble in saline fluids such as blood, and the amount that can be carried in simple solution is limited. Where metabolic demands are high, additional transport capacity is needed. This is supplied by respiratory proteins in the blood which bind oxygen and thereby increase the carrying capacity by a factor of 10 or more. The common respiratory pigment in crustaceans is haemocyanin, a high molecular weight (to 9×10^6 daltons), copper-based protein, colourless when deoxygenated and pale blue when oxygenated. Haemocyanin binds oxygen at the high oxygen tensions present at the respiratory surface and releases it at the lower tensions prevailing at the tissues. A few species, particularly phyllopods, use haemoglobin rather than haemocyanin and can increase its concentration in the blood in low oxygen environments. Haemoglobins are also found in some ostracods and copepods. Both haemocyanin and haemoglobin occur in solution in the blood and not in corpuscles.

Circulation

The circulatory system of crustaceans transports respiratory gases, nutrients, wastes, hormones and other substances around the body. Most of these functions do not require a very rapid circulation, but transporting respiratory gases does, and this requirement is commonly the factor which determines the state of development and efficiency of the circulatory system. The importance of the gas transport role varies with the distance over which transport must be effected (i.e. body size) and the speed and magnitude of gas transport needed (metabolic rate) in the particular species. In very small animals, diffusion alone can often satisfy the delivery requirements for respiratory gases, but many small crustaceans have circulatory systems. In larger and more active animals, diffusional delivery becomes limiting and the circulatory system must take over gas transport. In the simplest system, circulation of haemocoelic fluid may be achieved by limb and body movements, but even very small larval forms may have a simple heart and rudimentary arterial system.

The circulatory system of larger crustaceans consists of a dorsal heart and an arterial system, the complexity of which is geared to the size and activity of a particular species (Fig. 13.12e). The blood passes from the terminating branches of the arterial system to haemocoelic sinuses, but usually follows well-defined pathways through the lacunae between tissues and is collected into 'venous' sinuses which drain to large sinuses in the ventral part of the body. From these, blood passes through the gills and returns to the pericardial cavity around the heart via the large gill veins. This open circulation has a relatively large volume, low pressure, and slow circulation time.

The heart, as in other arthropods, is a single contractile chamber suspended in the pericardial cavity by ligaments which run between the exoskeletal body wall and the lateral walls of the heart. As the heart contracts, the ostia close and blood is forced into the arteries. The volume of the heart decreases, which stretches the ligaments and causes a fall in pericardial pressure. This in turn draws blood into the pericardial cavity. When the heart relaxes, the ligaments return it to its original volume, which opens the ostia and allows the heart to be refilled from the pericardial cavity. Back flow from the arteries is prevented by valves.

In primitive crustaceans such as anostracans, as in other arthropods, the heart runs the length of the body, with ostia arranged segmentally. In the large eumalacostracans the heart receives all of its blood from the respiratory system and is more compact, with fewer ostia (Fig. 13.12e). As the gills are generally thoracic the heart is normally located in the thorax, but some groups such as stomatopods and isopods have pleopodal gills and the heart is located abdominally. Small compact species such as copepods and cladocerans have simple bulbous hearts and minimal associated arteries.

The degree of development of the arterial system is related to body size and metabolic activity. The large reptantians have five anterior arteries leaving the front of the heart and one posterior trunk which splits to provide two major arteries supplying the abdomen and the thoracic limbs respectively. The arteries branch progressively, and the finest vessels in the arterial system may be only 1–2 μm in diameter. Some small crustaceans have only a short anterior aorta which discharges into the haemocoel. As gills are often absent in these forms, venous blood returning to the heart enters the pericardial cavity through gaps in the pericardium rather than via the large gill veins seen in the reptantians.

The hearts of crustaceans are under neurogenic control; a cardiac ganglion in the heart tissue acts as a pacemaker which sets the intrinsic rate of contraction. This pacemaker can be speeded up or slowed down by acceleratory and inhibitory nerves from the nerve cord, and its activity is also modulated by hormones and certain metabolites.

Osmoregulation and excretion

Although regulation of the water and salt content of the body is necessary in all animals, the amount of osmotic work to be done varies in different habitats. The body fluids of marine crustaceans generally are similar in concentration and composition to sea water. Osmotic and ionic gradients across their body wall are consequently small and little work is needed to maintain them. In fresh water the body fluids are considerably more concentrated than the external water, and the consequent osmotic entry of water and loss of salts to the medium must be continually corrected. In both environments, there are also inputs of salt and water in the food and losses via the faeces and these too must be compensated. Excess water entering the body has to be discharged and volume regulation is achieved by the activity of the excretory organs, either antennal or maxillary glands

(Fig. 13.12f). Like filtration excretory organs in other animals, those of crustaceans receive arterial blood under pressure. Water and molecules small enough to pass through the filter enter the primary urine in the lumen of the excretory organ (Fig. 13.12g). The molecular weight cut-off of the filter is generally around 10^5 daltons. Filtered fluid passes along the tubular portion of the organ and collects in a bladder, which is periodically emptied.

As a filtrate, the primary urine has a similar concentration and salt composition to the blood from which it is derived. In freshwater crustaceans, most of the filtered salts are reabsorbed along the tubule or in the bladder so that the urine released is dilute. In marine species the urine generally has a similar osmotic concentration to the blood, but magnesium and sulfate are often secreted into the urine along the tubular duct and are elevated in the released urine.

The excretory organ is thus used to eliminate excess salts and water, but salts must also be absorbed from the water to replace losses from the body, particularly in freshwater species. This is achieved by salt-absorbing cells, usually located on the gills. The ultrastructure of these cells is characteristic of ion-transporting tissues and absorption is driven by the activity of ion-transporting enzymes (ATPases) in the cell membranes. The amplification of the cell membranes offers increased surface area for the transporters and may also minimise the distance that ions entering the cell have to diffuse through the cytoplasm before being extruded into the blood. Numerous mitochondria provide the ATP needed to fuel the transport enzymes.

A wide variety of organic waste products is also eliminated by the excretory organs. If the molecules are small they are filtered into the urine as described above, but larger molecules which cannot pass the filter pores may be secreted into the organ. Nitrogenous waste in the form of ammonia is a major excretory product in all crustaceans. In aquatic crustaceans this is eliminated largely by outward diffusion of molecular ammonia or by exchange across the gills of the charged ammonium ion for sodium ions (NH_4^+/Na^+). The excretory organ is not usually a vehicle for nitrogenous excretion in aquatic crustaceans. Terrestrial species, too, excrete waste nitrogen as ammonia; as a gas in the case of isopods, and in the excretory fluid in most crabs (although the robber crab, *Birgus latro,* unusually amongst crustaceans, excretes uric acid via the gut). This is discussed further in a later section on terrestrial adaptations. The midgut gland may also function in excretion of organic compounds.

Nervous system and sense organs

Bodily functions in crustaceans are controlled by a combination of hormonal or endocrine systems and the nervous system.

The gross morphology of the central nervous system varies with that of the body. In the generalised condition, paired ventral nerve cords run the length of the body, with a ganglion in each trunk segment. The anterior ganglia of the head (protocerebrum, deutocerebrum and tritocerebrum) are fused to form a brain (Fig. 13.13a, b) and those of the mouthpart segments are united as a suboesophageal ganglion. In many malacostracans, the anterior thoracic ganglia are also fused. This trend reaches an extreme in crabs, where all the thoracic ganglia are fused into a single massive ganglion and the abdominal ganglia are reduced and fused (Fig. 13.13c), in association with the evolution of a compact thorax and vestigial abdomen. The brain receives major sensory input from the anterior concentration of sensory organs, notably the eyes, antennae and statocysts, and also acts in motor control of the anterior head appendages. The segmental ganglia receive sensory input from the segments they serve and provide motor nerves to the associated appendages.

Crustacean muscles contain varying proportions of fast and slow fibres and may be innervated by fast, slow and inhibitory nerve fibres. Different blends of these components are used to provide the strength, speed and duration of muscle contractions needed to effect such diverse activities as rapid escape reactions (tail flip), fast or slow walking and swimming, gut contractions, maintaining posture, and holding material with chelate limbs. A further layer of response sophistication is added by different types of neuromuscular synapses and neurotransmitters.

Light detection and vision are important senses in many crustaceans. Some small species and many larval forms have only a simple or naupliar eye, consisting of a few ocelli grouped together medially on the head. Most groups have paired compound eyes similar to those of insects (Chapter 11), made up of multiple light-sensitive units (ommatidia) which may be sessile (as in edriophthalmans) or on mobile stalks (as in many other malacostracans). In the latter the ommatidia may wrap around the eyestalk, which together with the mobility of the eye provides a very wide field of vision. In crustaceans that are active nocturnally or in dark habitats, the ommatidia in the compound eye are typically organised to maximise light detection at

Figure 13.13
(a–c) Morphology of the central nervous system in Crustacea: (a) Anostracan. (b) Lobster. (c) Brachyuran crab. (d) Statocysts at the base of the uropods of a mysid. (a after Pennak 1953; b, c from Sandeman 1982; d after Meglitsch 1972.)

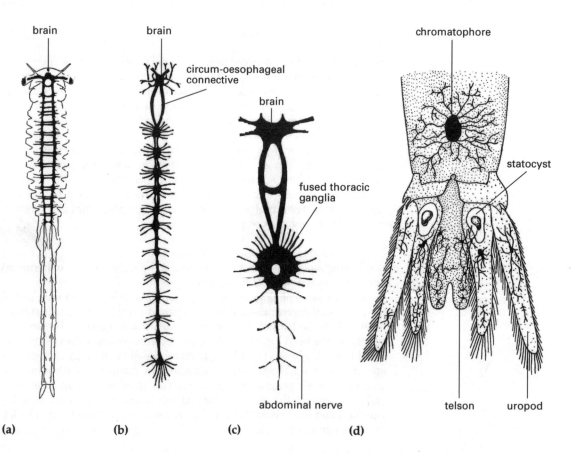

the expense of visual acuity (superposition eyes), and multiple ommatidia combine to produce a single image. Animals which are active in bright conditions generally have apposition eyes which maximise acuity, with sensory input being received separately from each ommatidium. Apposition eyes are commonly found in isopods and crabs, while shrimps and crayfish typically have superposition systems. A number of species can switch from one system to the other as light conditions dictate, by adjusting the position of the screening pigment in the ommatidia (compare Fig. 11.15a). Certain crabs have colour vision, and the overlapping visual fields of the two eyes indicate a capability for binocular vision. Binocular vision is also possible within individual eyes of certain stomatopods. Crayfish have additional photoreceptors in the brain which are thought to be concerned with circadian rhythms.

In aquatic environments, chemoreception is a very important sense for crustaceans; chemoreceptors are present as bristles or hairs known as aesthetascs. While these receptors may be present anywhere on the body, they are most numerous on the antennules, where they are used in the directional location of food sources and sexual partners, and in homing to them. They may also be used to detect predators. Chemoreceptors are also abundant on the tips of the walking legs, where they are used to identify food in the substratum, while the mouthparts and chelae are highly sensitive to dissolved chemicals and are used in sorting and selecting food. Terrestrial crustaceans must detect airborne rather than dissolved chemicals. Other hair receptors are sensitive to touch, vibration and sound (mechanoreceptors) and air or water movements over the body.

Information on body movements, orientation and acceleration is obtained from paired statocysts, usually located at the base of the antennae (as in reptantians) or at the base of the uropods or telson (as in mysids and prawns) (Fig. 13.13d). Each statocyst is a hair-lined pit containing a small statolith which mechanically stimulates the hairs as movements of the animal change its position. The animal also receives information regarding the position of its limbs and flexure of joints from suitably placed mechanoreceptors and internal proprioceptors such as stretch receptors.

Some crustaceans can navigate accurately during migrations, such as the spawning migrations of land crabs and spiny lobsters and it is possible that they, like certain insects, may possess magnetoreceptors for navigation.

Hormones

The major endocrine organs of crustaceans are:
1. neurosecretory cells located in the brain and thoracic ganglia,
2. the pericardial organs in the pericardial cavity,
3. the Y-organ, and
4. the reproductive organs.

Moulting is a complex process controlled by several hormones. Moulting hormone is produced by the Y-organ, an endocrine gland in the head. On release it is converted to the active form, β-ecdysone, which initiates and controls preparations for moulting. Between moults the production of moulting hormone is inhibited by a peptide hormone, which originates from neurosecretory neurons in the X-organ in the brain. This hormone is released into the blood from the sinus gland in the eyestalk. Other hormones may also be involved in initiating the shedding of the old exoskeleton (ecdysis).

Colour changes in crustaceans are effected by the dispersion or concentration of pigment granules within pigment cells (chromatophores, Fig. 13.13d) under the control of a variety of chromatophorins which concentrate or disperse particular pigment types. These are also neurohormones, secreted by the X-organs, post-oesophageal commissures and sinus glands, which act directly on particular types of chromatophores.

The ovaries, and in males the androgenic glands, produce a variety of hormones concerned with reproduction, which control development of the gametes and genitalia. Other hormones regulate activity, salt and water balance and blood glucose levels, but most have been poorly studied.

Reproduction and development

In most crustaceans the sexes are separate, but several groups (such as Remipedia and Cirripedia) are simultaneous hermaphrodites, while others are sequential hermaphrodites. In the latter, only one type of sex organ is active at a time, and a sex reversal occurs some time during the life history. Thus an individual begins its reproductive activity either as a male (protandry) or a female (protogyny) and changes sex after one or more breeding episodes. Many tanaidaceans are protogynous, whilst certain isopods and amphipods are protandric. Several groups of phyllopods (such as cladocerans) and some ostracods are parthenogenetic, the population normally consisting of females which produce further female offspring from unfertilised eggs. Males appear only occasionally, usually under adverse conditions, when fertilised resting eggs are produced.

The reproductive organs consist of dorsal testes with sperm ducts or dorsal ovaries and oviducts. Usually the ducts open ventrally in the mid-region of the body (Fig. 13.1b). The sperm of crustaceans are rarely motile and are commonly delivered in a spermatophore to the female, which may then store and nourish sperm in spermathecae until the eggs are laid, as in brachyuran crabs. Where internal fertilisation is practised the males have appendages specialised to deliver the sperm or spermatophore to the female. Crabs and many prawns use the first pair of pleopods. In barnacles the long penis, which is a modified extension of the abdominal wall, is extended to inseminate neighbours. Mating in motile species may include an elaborate courtship involving visual cues, and pheromones may be used to attract mates.

The marine euphausids and dendrobranch prawns shed their eggs into the water after fertilisation, and syncarids attach them to vegetation; but in many groups the fertilised eggs are brooded by the female, commonly on the pleopods or in a brood pouch formed from modified thoracic or abdominal limbs (Figs 13.3a, 13.4b). In terrestrial isopods the walls of the brood pouch may supply nutrients.

In some crustaceans, of which cirripedes are an example, the fertilised egg begins development with a type of spiral cleavage. Often the hatchling is a larva. At its simplest this is a nauplius, which has only the first three pairs of limbs: the antennules, antennae and mandibles (Fig. 13.14a). Typically the nauplius feeds in the plankton and, with successive moults, adds further segments and associated pairs of limbs. It may undergo periodic metamorphoses which produce a sequence of different larval forms. In decapods this includes the nauplius, protozoea, zoea (Fig. 13.14b, c) and post-larva, although the nauplius and protozoea are often confined to the egg. At the final larval moult the adult body form is attained and remains

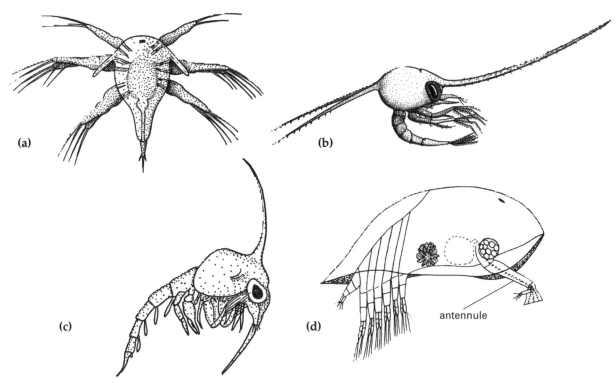

Figure 13.14
Larvae of crustaceans:
(a) Nauplius larva of a barnacle.
(b,c) Zoea larvae of anomalan and brachyuran crabs.
(d) Cypris larva of a barnacle.
(a–c after Green; c from Wehrtman *et al.* 1996; d after Grygier 1983.)

more or less constant during further growth, except for minor changes in species which exhibit marked sexual dimorphism. Other crustacean taxa have different larval types. In the barnacles the nauplius typically metamorphoses into a cypris larva (Fig. 13.14d) and in the copepods a copepodite is characteristic.

There is a continuous spectrum of developmental strategies in the Crustacea, ranging from hatching at the earliest possible stage (nauplius), through various degrees of larval suppression, to hatching as young adults with all larval stages suppressed. For example, the eggs of phyllopods typically hatch as nauplii, those of most marine decapods hatch as zoeae (although dendrobranchiate prawns hatch as nauplii), while in the freshwater crayfish and crabs development is direct and all larval stages are confined within the egg. Larval suppression is usually associated with increased size and yolk content of the eggs.

Sessile and parasitic crustaceans also produce swimming larvae which are able to disperse and locate suitable settling sites or hosts. In freshwater and terrestrial habitats larvae are disadvantageous, so species survival in these habitats has usually been associated with the evolution of direct development.

Freshwater and terrestrial radiation

Colonisation of freshwater habitats

The main barriers to colonisation of fresh water from the sea are not physical obstacles, as the water bodies are directly connected, but rather the osmoregulatory capacity to cope with low salt content, and also the effect of powerful water currents which continually tend to return animals to the ocean. The usual consequence of exposure of permeable marine crustaceans to dilute media is loss of salts and gain of water, which can together cause rapid death, especially in small animals. Freshwater crustaceans have a suite of regulatory mechanisms, including a reduction in the permeability of the body surface, elimination of excess water via the urine, and absorption of salts from fresh water to replace those lost. Commonly, freshwater crustaceans also maintain their body fluids at a lower osmotic concentration than in marine species, minimising the osmotic and ionic gradients across the body surface.

Crustaceans must be able to maintain position and move about in freshwater habitats without being swept downstream, and this has favoured benthic and epibenthic colonists such as crayfish, prawns, crabs, isopods and amphipods over the planktonic and pelagic forms. Planktonic freshwater species such as copepods and cladocerans are largely restricted to lakes, ponds and temporary pools, where downstream flow is not a problem. Larval stages are also at risk in flowing waters, and larvae have almost universally been suppressed in favour of direct development in freshwater malacostracans. The amphipods and isopods all have direct development and were preadapted reproductively for freshwater life. The crabs and crayfish seem to have modified the marine pattern during colonisation of dilute waters. The eggs of freshwater crabs hatch with the adult body form, while those of crayfishes hatch as a non-feeding post-larva which rapidly moults to the final form. Certain freshwater prawns (atyids) still have free-swimming larvae. The anostracans, conchostracans and notostracans, although largely confined to fresh water, retain larval stages and are often planktonic in habit, but they live predominantly in temporary pools and ponds.

Freshwater bodies are usually subject to fluctuations in level, and some dry up altogether either occasionally or seasonally. The extreme is seen in ephemeral desert pools, which may only exist for a few weeks at a time, with long and often irregular dry periods between. The survival mechanisms developed to cope with drying in these groups centre around resistant eggs. Crustaceans of temporary pools have resistant eggs which survive long droughts and provide a dispersal mechanism, as they may readily be transported from pool to pool on the feet of animals or by wind-blown dust. The resistance is conferred by the structure of the egg shell and biochemical adaptations which reduce the metabolism of the embryo. Water and suitable climatic conditions stimulate hatching. Water availability may be brief, and growth to maturity is rapid. *Limnadia*, a conchostracan of temporary freshwater rock pools, and the tadpole shrimp, *Triops*, in desert claypans, both reach maturity in 2–3 weeks. Large animals such as crabs and crayfish often survive dry periods in burrows or buried in mud, and these habits have encouraged the development of amphibious life styles and air-breathing in these groups.

Colonisation of land

The most successful terrestrial crustaceans are the oniscid isopods (1000+ species, distributed world-wide). The decapods (about 200 species, mostly crabs) and amphipods (100+ species, largely restricted to the Southern Hemisphere) have fewer species. The fossil record suggests that both amphipods and isopods may have been terrestrial for around 50 million years (perhaps much longer for isopods), while colonisation by crabs seems to have begun less than 20 million years ago. The Crustacea are relatively recent colonists of land compared with the hexapods, myriapods and arachnids, which became terrestrial more than 400 million years ago. There is an extensive terrestrial fauna of microcrustaceans which mostly live in soil, including representatives of the copepods, ostracods and cladocerans, but little is known of their evolutionary history or biology.

Aquatic animals must undergo major morphological and physiological changes before they can survive and function efficiently on land. The changed conditions particularly affect the respiratory, sensory, osmoregulatory, excretory and musculo-skeletal systems.

The low density of air does not support the delicate gills of aquatic crustaceans. Out of water they collapse and cannot be ventilated. The gills of terrestrial crabs are stiff, with widely spaced lamellae between which air can circulate (Fig. 13.15a). Some terrestrial crustaceans have also developed lungs. In crabs, these are usually formed from the inner lining of the carapace, which may be invaginated or evaginated to increase the surface area available for gas exchange and has an elaborate blood system facilitating gas exchange (Fig. 13.15b). Some intertidal air-breathing crabs have thin areas of cuticle on the leg bases, used in aerial gas exchange. The isopods use pleopods for gas exchange; in many species the more anterior of these appendages have become invaginated and contain airsacs and pseudo-tracheae opening on their lateral margins via one or more spiracles (Fig. 13.15c, d).

Terrestrial animals continually lose water by evaporation. To contain this loss within replaceable levels, the permeability of the body surface ought to be low. But with the notable exception of a few desert isopods such as *Hemilepistus,* terrestrial crustaceans remain quite permeable and must rely heavily on behavioural adaptations to restrict water loss. Isopods are active at night, when humidity is higher, and have moist daytime retreats where they can replenish water stores by absorbing water vapour from the air. Amphipods are restricted to moist habitats such as leaf litter and soil, and many terrestrial crabs utilise the protection of burrows. Ghost crabs (*Ocypode*) use hydrophilic hair tufts to extract water from sandy substrata, then suck water from the hairs and ingest it (Fig. 13.15e).

In most terrestrial animals the lack of a water sink that would enable ammonia to be excreted has resulted in modifications through which ammonia is converted to non-toxic products such as purines. Surprisingly, the only terrestrial crustacean known to follow this pattern is the robber crab, *Birgus latro.* Other terrestrial crustaceans retain ammonia as the main excretory product. In most of these species the gills are still responsible for the outward transport of ammonia, which is periodically released as urine into the branchial chambers. The ammonia is lost when the excretory fluid is finally voided, although isopods and some crabs volatilise ammonia gas from the branchial fluid (Fig. 13.15f). The ghost crab *Ocypode* is unusual in that it excretes ammonia in urine via the antennal organs.

Vision is possibly the most important sense on land, but the different refractive indices of water and air usually make it difficult for focus to be achieved in both media. The flat lenses of the ommatidia of many crustaceans, however, enable focus to be achieved both in and out of water, so that little change to the optics of the eye was necessary during the initial emergence onto land.

The problems of reproduction on land are more severe than those faced by freshwater crustaceans, as mating systems must change in relation to the new environment. In the absence of water, larvae are not usually an option.

Figure 13.15
(a) Gill of a terrestrial brachyuran crab *Geograpsus grayi*, with stiff, widely spaced lamellae separated by small nodules. (b) Dorsal view of the branchiostegal lung of an air-breathing crab *Ocypode*. The gas exchange surface on the inside of the carapace receives a rich supply of venous blood. Oxygenated blood returns directly to the heart. (c) Ventral surface of the abdomen of the terrestrial isopod *Cyclistus*, showing airsacs within the pleopods. (d) Details of a pleopod of *Cyclistus*. (e) Hydrophilic hairs between pereiopods 3 and 4 of the brachyuran *Ocypode ceratophthalma*. The hairs extract water from the sand. The water is then sucked into the branchial chambers via the aperture and passed to the mouth. (f) Patterns of nitrogenous excretion in Crustacea (see text for details). (b after Farrelly and Greenaway 1994; c after Unwin 1931.)

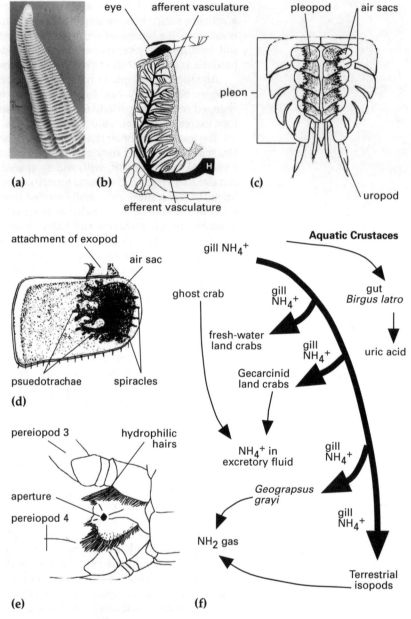

Most aquatic isopods and amphipods have thoracic brood pouches for their eggs, which ultimately hatch as post-larvae with the adult body form. These habits have been retained and refined in terrestrial species, where the brood pouch also protects the eggs from desiccation. In isopods much of the nourishment required by the growing embryo is secreted into the pouch by the female.

Terrestrial crabs which colonised land via fresh water attach their eggs to the pleopods with a secretion from the oviducts and carry them in a pouch formed between the ventral surface of the thorax and the abdominal flap, sealed along the lateral margins by enlarged pleopods. The eggs are large and yolky and hatch as juvenile crabs. Crabs which colonised land directly from the sea, however, have mostly retained planktonic larvae and this has restricted their distribution to coastal regions and islands. The adults mate and produce their eggs on land but must travel to the sea to hatch and release the zoea larvae from their pleopods. In land crabs of the Family Gecarcinidae this results in spectacular breeding migrations. After a period of feeding and growth in the plankton, the megalopa larvae return to the shore, moult into juveniles and take up a terrestrial existence. The retention of marine larvae, although restricting the penetration of land, ensures both a wide distribution and exploitation of island habitats. Importantly, the larvae also utilise a food source and habitat different to those of the adults. The Jamaican crab, *Sesarma jarvisi*, does not migrate to the sea to breed; instead it hatches its eggs as large zoea larvae in water trapped in empty snail shells, where they and the subsequent juveniles are tended and fed by the female. Certain fiddler crabs which live in burrows distant from the sea produce large, non-feeding larvae which complete their development within the burrow.

Body size is an important factor in the determination of the life style in terrestrial crustaceans, and several distinct patterns are recognisable. The very small microcrustaceans have adopted cryptozoic habits and thus follow a similar pattern to their size equivalents in the apterygote insects (e.g. Collembola) and the arachnids (e.g. mites). The isopods are somewhat larger and parallel the ground-dwelling insects in body size, the presence of 'tracheal' respiratory systems and the ability to absorb water vapour from the air. The crabs and crayfish, as in vertebrates, have developed ventilation lungs, a modification which has freed them from the size restrictions imposed on other arthropods by tracheal systems. They are generally much larger than other terrestrial arthropods (commonly in excess of 50 g, ranging up to 3 kg in the robber crab) and represent a novel body plan for medium-sized animals quite different to that of the vertebrates, the only other terrestrial group with representatives in the same size range.

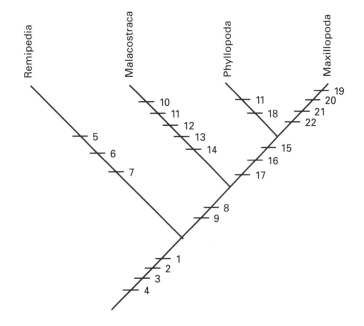

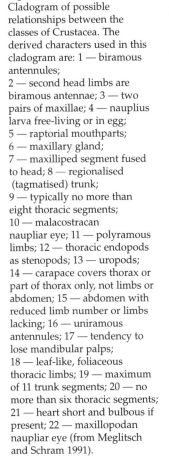

Figure 13.16
Cladogram of possible relationships between the classes of Crustacea. The derived characters used in this cladogram are: 1 — biramous antennules; 2 — second head limbs are biramous antennae; 3 — two pairs of maxillae; 4 — nauplius larva free-living or in egg; 5 — raptorial mouthparts; 6 — maxillary gland; 7 — maxilliped segment fused to head; 8 — regionalised (tagmatised) trunk; 9 — typically no more than eight thoracic segments; 10 — malacostracan naupliar eye; 11 — polyramous limbs; 12 — thoracic endopods as stenopods; 13 — uropods; 14 — carapace covers thorax or part of thorax only, not limbs or abdomen; 15 — abdomen with reduced limb number or limbs lacking; 16 — uniramous antennules; 17 — tendency to lose mandibular palps; 18 — leaf-like, foliaceous thoracic limbs; 19 — maximum of 11 trunk segments; 20 — no more than six thoracic segments; 21 — heart short and bulbous if present; 22 — maxillopodan naupliar eye (from Meglitsch and Schram 1991).

Phylogeny and fossil record

The oldest crustacean fossils date from the early Cambrian, and the group must therefore have originated in the Pre-Cambrian. Whilst the fossil record is quite abundant for many crustacean groups, it provides no clear evidence on the origin of the Crustacea, and there has been considerable debate regarding the primitive or ancestral body form of the first crustaceans. Current opinion favours a long body with many similar trunk segments, two pairs of biramous antennae and a nauplius larva, but is divided on whether the trunk bore biramous or polyramous swimming appendages. If the swimming limbs were biramous, the group closest to the model would be the Remipedia, but if the primitive limb type was polyramous the Cephalocarida are probably the most primitive group.

The possible relationships between the major groups of crustaceans have been analysed in a number of recent studies, and the result of one of these is the cladogram shown in Fig. 13.16. Here the Remipedia are portrayed as the closest group to the ancestral pattern. In the other branch of the clade the animals are distinguished by development of distinct body regions rather than a long trunk of many similar segments. Two of these groups (the classes Malacostraca and Maxillopoda) are generally considered to be monophyletic. The third branch, the Phyllopoda, exhibits such a diversity of body form that a strong case for monophyly is difficult to make, and it is possible that this group may have polyphyletic origins.

Classification of the subphylum Crustacea

CLASS REMIPEDIA
Vermiform animals; head covered by a head shield; trunk of up to 32 similar segments, each bearing a pair of biramous limbs; telson with caudal rami; feeding appendages including a pair of thoracic maxillipeds; gut with segmentally arranged digestive caeca.

CLASS MALACOSTRACA
Typically with a head (six segments); a thorax (eight segments) covered by a carapace; abdomen (six segments) bearing a telson and uropods forming a tail-fan; female and male gonopores on the bases of the sixth and eighth thoracic limbs respectively.

Subclass Hoplocarida

Dendrobranchiate gills on abdominal pleopods.

Order Stomatopoda — first thoracic limbs raptorial.

Subclass Eumalacostraca

Gills associated with thoracic limbs; antennules and antennae biramous, with exopod of antenna usually present as a scale; well-developed carapace in many groups.

Order Syncarida — carapace absent.

Order Mysida — carapace well developed and covering most of the thorax, although attached only to the first three anterior segments; stalked compound eyes; thoracic limbs biramous; large thoracic brood pouch present in female.

Order Lophogastrida — first thoracic limbs as mouthparts.

Order Hemicaridea — carapace reduced, covering only the anterior thoracic segments; suborder Cumacea with an inflated carapace attached to first 3-6 thoracic segments, enclosing two branchial chambers ventilated by the first maxillipeds; suborder Tanaidacea with a small carapace covering only the first two thoracic segments.

Order Edriophthalma — usually seven pairs of uniramous thoracic limbs; sessile compound eyes; lacking a carapace; brood chamber present; suborder Isopoda with dorsoventrally flattened body and plate-like biramous pleopods; suborder Amphipoda usually laterally compressed, with first and second thoracic limbs chelate.

Order Euphausiacea — carapace shallow and fused to head and thoracic segments but not covering limb bases; gills exposed; brood chamber absent; eggs hatch as nauplii.

Order Decapoda — three pairs of maxillipeds and five pairs of usually uniramous periopods; carapace covering the head and thorax and enclosing the gills; four suborders, Dendrobranchiata (prawns), Eukyphida (shrimps), Euzygida (stenopodids) and the Reptantia.

CLASS PHYLLOPODA

Thoracic limbs flattened and polyramous, used for swimming and filter feeding; abdominal limbs generally reduced in number or absent; antennules generally uniramous.

Subclass Phyllocarida
Large bivalved carapace with adductor.

Subclass Cephalocarida
Maxillae polyramous and flattened, resembling thoracic limbs; head with a head shield but lacking eyes; thorax of eight segments, limbless abdomen; telson with large caudal rami.

Subclass Sarsostraca
Head shield and carapace lacking.
Order Anostraca (fairy shrimps)

Subclass Calmanostraca
Carapace present; often bivalved and enclosing body.
Order Notostraca — large flattened dorsal carapace (tadpole shrimps).
Order Conchostraca — bivalved carapace completely enclosing body (clam shrimps).
Order Cladocera — bivalved carapace enclosing trunk but not head (water fleas).

CLASS MAXILLOPODA

Generally small; not more than six thoracic and five abdominal segments; first antennae uniramous.

Subclass Tantulocarida
No limbs on head.

Subclass Branchiura
Fish lice. Head and thorax covered by flat dorsal carapace; thorax with fewer than six segments; attachment organs on head; thoracic limbs modified for swimming.

Subclass Mystacocarida
Less than six thoracic segments; one pair maxillipeds; pincer-like caudal rami.

Subclass Ostracoda
Seed shrimps. Less than six thoracic segments, body reduced and enclosed in calcified, bivalved carapace.

Subclass Copepoda
Head shield present but no carapace; single median, naupliar eye and well-developed caudal rami present; abdominal appendages absent.

Subclass Thecostraca
Sessile or parasitic forms which attach with the antennules; naupliar stages and a cypris larva present.
Order Cirripedia — saccular carapace with calcareous plates; thoracic limbs as feeding cirri; abdomen vestigial.
Order Rhizocephala — parasitic forms with body as a reproductive sac and ramifying feeding stolons; free-living naupliar larvae.

Chapter 14

The Chelicerata

D.T. Anderson

SUBPHYLUM CHELICERATA

Introduction 320

Basic structure and function 321
    The forebody or prosoma 321
    The hindbody or opisthoma 322

Functional diversity in arachnids 323
    The major arachnid orders 323
    Some other arachnid orders 324

Locomotion in arachnids 326
    Stepping 327
    Gaits 328

Feeding in arachnids 328

Internal organs and physiological processes 332
    Digestion 332

Respiratory exchange 332
Circulation 334
Excretion 334
Sense organs 335
Nervous system 335

Reproduction and development 335
Reproductive systems and processes 335
    Development in chelicerates 338

The class Pycnogonida 339
    Form and function 339
    Reproduction and development 340
    Relationship of pycnogonids 341

Classification of the subphylum Chelicerata 341

Subphylum Chelicerata

Primarily marine arthropods, most forms terrestrial; with biramous (often uniramous) limbs including a pair of chelicerae and five (often four) pairs of walking legs; jaws gnathobasic; gonopores in the mid-region of the body.

Introduction

Unlike other arthropods, chelicerates do not have a head and trunk or a head, thorax and abdomen. Instead they have an anterior forebody, the prosoma, and a posterior hindbody, the opisthosoma. The prosoma carries six pairs of uniramous limbs, none of which are antennae or mandibles. A single preoral pair, the chelicerae, are used in feeding. The remaining pairs, all postoral, function variously in walking and feeding. In terrestrial chelicerates the first postoral limbs form a pair of pedipalps.

The opisthosoma of marine chelicerates carries six pairs of biramous limbs, the most anterior used in reproduction and the remainder in respiration. The opisthomal limbs of terrestrial chelicerates are variously reduced and highly modified, but the same functions persist. The gonopores are usually located ventrally on the anterior part of the opisthosoma.

Embryological studies show that the chelicerate prosoma has seven segments behind a terminal acron, and that the opisthosoma has 13 segments in front of a terminal telson. The first prosomal segment and the first opisthosomal segment are always vestigial. Segments 8–13 of the opisthosoma are limbless. These more posterior opisthosomal segments are often suppressed during later development.

Figure 14.1
Merostomatans. (a) *Limulus*, dorsal view. (b) Eurypterid, dorsal view. In eurypterids the opisthosoma is externally segmented, with seven broad mesosomal segments bearing respiratory book gills on segments 2–7, and five narrower, limbless metasomal segments.

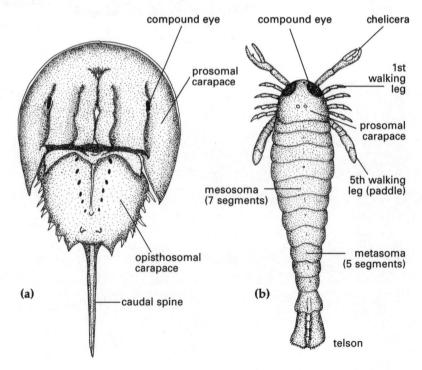

Benthic marine chelicerates, which were already present in the early Cambrian fauna, became diverse in Palaeozoic seas. Beginning in the Silurian, a major diversification of chelicerates also commenced on the land. Three classes of chelicerates are recognised. The class Merostomata are marine arthropods of moderate to large size, with five pairs of opisthosomal respiratory limbs. Merostomatans diverged in the Cambrian and Ordovician as two lines, the subclass Xiphosura (horseshoe crabs), of which a few species still survive (Fig. 14.1a), and the subclass Eurypterida (sea scorpions), which became extinct at the end of the Palaeozoic era (Fig. 14.1b).

A second class of marine chelicerates, the Pycnogonida or sea spiders (Fig. 14.9), are animals of small size and specialised form and habits. In particular, the opisthosoma is vestigial in pycnogonids. The groups is an ancient one, having first appeared in the Devonian. Pycnogonids are numerous in the modern benthic marine fauna.

The terrestrial chelicerates, which emerged in the Silurian and Devonian, have evolved as scorpions, spiders, mites, ticks and some other orders (Fig. 14.3). Although all are grouped in the class Arachnida, fossil and other evidence indicates that the class contains several lines of evolution stemming from different marine chelicerate ancestries. This question is still under investigation.

Basic structure and function

The surviving xiphosurans provide a living example of basis structure and function in merostomatan marine chelicerates. The horseshoe crab, *Limulus polyphemus* (Fig. 14.1a) is distributed along the Atlantic coast of North America. Four species in other genera live on Western Pacific coasts, from Japan to Indonesia. *Limulus* inhabits shallow-water marine habitats, ploughing through the sand surface and feeding mainly as a carnivore. Females grow to 600 mm in length, males somewhat less.

The forebody or prosoma

The prosoma of *Limulus* has a broad, horseshoe-shaped carapace with a pair of compound eyes dorsolaterally and a pair of simple eyes in the midline. The margins of the carapace are reflexed on either side of the narrow ventral surface of the prosoma (Fig. 14.2a), with limb bases set near the ventral midline. The chelicerae are small, chelate limbs flanking a median labrum. Behind the mouth, a midventral food groove extends between the bases of the walking legs. The coxae of the legs carry median, spiny gnathobases acting as jaws. The first four pairs of legs also have chelate tips and are used in food collection. Each leg (Fig. 14.2b) has seven podomeres (the coxa, trochanter, femur, patella, tibia, tarsus and pretarsus), but the patella and tibia are fused. The pretarsus forms the movable finger of the terminal chela.

Limulus walks and burrows by a metachronal stepping action in which the legs, moved by extrinsic leg muscles, swing backwards and forwards on wide transverse joints between the coxae and the prosoma (Fig. 14.2b). These fore and aft swinging movements are accompanied by a raising and lowering of the limbs on pivot joints between the coxa and trochanter and the trochanter and femur, worked by intrinsic leg muscles. The more distal parts of the limbs are flexed and extended during each step. The flexure is due to intrinsic leg muscles, while the extension is due mainly to the hydrostatic pressure of haemocoelic fluid.

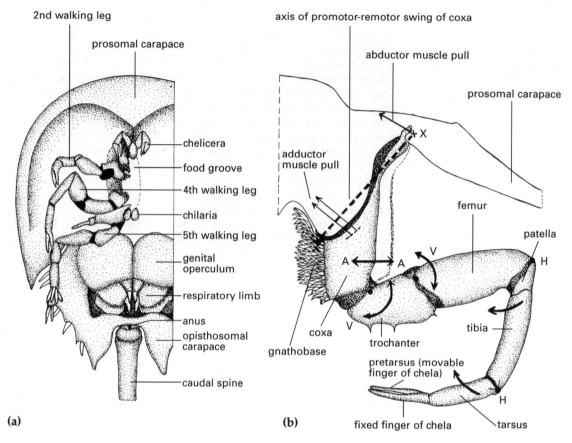

Figure 14.2
Limulus. **(a)** Ventral view.
(b) Walking leg and gnathobase; A–A — movement of gnathobase during chewing; H — hinge joint; V — pivot joint. (After Manton 1978.)

Food gathered by the chelae on the walking legs is placed between the gnathobases. *Limulus* takes worms, molluscs and other prey from the sand, as well as some algal material. The chewing action of the gnathobases is performed by transverse adduction and abduction of the coxae, towards and away from the ventral midline (A–A, Fig. 14.2b). Successive pairs of coxae work alternately, producing a grinding action that moves the food towards the mouth. The extrinsic leg muscles that cause the fore and aft swings of the coxae during walking also produce the adductor–abductor movements of the coxae in chewing. The two patterns of muscular contraction are mutually exclusive, so that *Limulus* cannot walk and chew at the same time.

The fifth pair of walking legs has large gnathobases that are able to generate a powerful bite, used in breaking mollusc shells and other hard foods. Merostomes are the only arthropods in which a strong bite is located at the posterior end of the feeding apparatus. The lateral surface of the coxa carries a short epipod, the flabellum. The tarsus has a ring of four chitinous leaflets, using in sweeping away the sand during ploughing locomotion.

The hindbody or opisthosoma

The opisthosoma is also covered by a carapace, jointed transversely to the prosomal carapace (Fig. 14.1a). Paired lateral spines mark the first six segments. Posteriorly, behind the anus, there is a long caudal spine used as a lever in pushing and balancing during locomotion. Embryological studies show that the caudal spine includes several fused segments, as well as the telson. Xiphosurans retain a vestigial first pair of opisthosomal limbs, the chilaria (Fig. 14.2a), which form a posterior lip to the midventral food groove. Behind the chilaria, the ventral surface of the opisthosoma carries six pairs of flattened, biramous limbs with narrow endopods and broad exopods (Fig. 14.2a). The first pair are fused and form a wide genital operculum, with a pair of gonopores on the posterior surface. The remaining five pairs, covered and protected by the genital operculum, carry close-packed 'books' of gill leaflets posteriorly on the exopods. The metachronal beating of the respiratory limbs ventilates the book gills and can generate a paddling swimming action.

Functional diversity in arachnids

Most arachnids (Fig. 14.3) retain a prosomal carapace and six pairs of prosomal limbs, including preoral chelicerae. The opisthosoma is made up of a maximum of 13 segments, primitively distinct but often fused together and reduced in number. Underpinning their terrestrial lifestyle, the epicuticle of arachnids is waxy, minimising the loss or uptake of water through the body surface. An impervious external cuticle is especially well developed in desert arachnids. The respiratory limbs of the opisthosoma are modified as infolded book lungs, often supplemented or replaced by tracheae. The tracheal systems of arachnids have evolved independently of those in myriapods and hexapods.

Each order of arachnids displays a distinctive pattern of modification of the chelicerate limb pattern. Certain orders—the scorpions, pseudoscorpions, spiders, opilionids and acarines—are of major significance in the modern fauna. The other arachnid orders are restricted in their diversity and numbers of species.

The major arachnid orders
Order Scorpiones
Scorpions are easily recognisable arachnids of moderate size (Fig. 14.3a). The small prosoma carries short chelate chelicerae, large chelate pedipalps and four pairs of flattened walking legs. The opisthosoma is subdivided, as in eurypterids, between a broad mesosoma and a narrow metasoma, ending in scorpions in a telsonic sting. Ventrally, the mesosoma has a small genital operculum, a pair of limbs modified as sensory pectines, and the apertures of four pairs of book lungs (Fig. 14.3c).

Scorpions are mainly nocturnal carnivores, taking arthropods and other small animal prey. The order has a tropical to subtropical distribution from forests to deserts on all continents, including Australia. There are about 70 genera and more than 600 species.

Order Pseudoscorpiones
Pseudoscorpions (Fig. 14.3e) resemble scorpions in having a short prosoma with small chelate chelicerae, large chelate pedipalps and four pairs of slender walking legs, but the segmented opisthosoma is broad and short, with only a genital operculum and two pairs of tracheal openings on the under surface. All pseudoscorpions are small, no more than a few millimetres long. They occur in leaf litter and under bark on all continents, often in large numbers, and are predators on other small leaf litter animals. More than 230 genera and 1000 species have been described.

Figure 14.3 (opposite)
Diversity in arachnids.
(a) Scorpion. **(b)** Spider.
(c) Scorpion, ventral surface.
(d) Spider, ventral surface.
(e) Pseudoscorpion.
(f) Opilionid. **(g)** Acarine.
(h) Solifuge. **(i)** Amblypygid.
(j) Uropygid.

Order Araneae

The Araneae or spiders (Fig. 14.3b) are a major group of terrestrial carnivorous arthropods. Ranging in size from 0.5 mm to over 90 mm, spiders exhibit great diversity of form and habit. The short, broad prosoma carries anterior cheliceral fangs with poison glands, used in paralysing prey. The pedipalps are sensory and serve in males as copulatory organs. The third to sixth leg pairs are walking legs. The opisthosoma, which is connected to the prosoma by the narrow pedicel, is elongate or globular. Primitive spiders retain an externally segmented opisthosoma, but in most species the segments are fused. The under surface has a small genital operculum anteriorly, followed by the openings of two pairs of book lungs, variously replaced in more advanced spiders by tracheae (Fig. 14.3d). The limbs of segments 4 and 5 are uniquely modified as spinnerets, employed in the release of silk. Spiders have a world-wide distribution in all terrestrial habitats. Those with two pairs of book lungs, including funnel-web and trap-door spiders, are mainly restricted to the southern continents, including Australia. Nearly 3000 genera and more than 32 000 species of spiders are classified in numerous families.

Order Opiliones

Opilionids live in tropical to temperate areas of the world, in vegetation, leaf litter and caves. Their body size is normally 5–10 mm. A short prosoma (Fig. 14.3f) is broadly joined to a short, segmented opisthosoma. The centre of the prosomal carapace protrudes as a tubercle, with an eye on either side. The chelicerae and pedipalps are small, but the legs are long, with multi-jointed, flexible tarsi. Opilionids are agile climbers and fast runners, feeding on small prey. There are about 650 genera and 2400 species.

The acarine orders

The acarines (mites and ticks) are a diverse assemblage of small arachnids. They are classified in three orders, the Opilioacariformes, Parasitiformes and Acariformes, but share a number of organisational features (Fig. 14.3g). The opisthosoma is fused with the prosoma as a unitary body, covered by a dorsal shield. The anterior region is differentiated as a capitulum, with a buccal cone which bears the chelicerae dorsally and the pedipalps at the sides. There are four pairs of short walking legs, with a genital plate between the bases of the posterior pair, and one or more pairs of tracheal openings. Acarines have a world-wide distribution and include numerous terrestrial carnivorous and herbivorous species, many pests and parasites, and an important freshwater group. Their diversity is spread over about 1400 genera and more than 6000 species.

Some other arachnid orders

In addition to the seven major orders of arachnids, there are six small orders. Three of them, the Solifugae, Uropygi and Amblypygi, contain large species of great zoological interest. The others, the Palpigradi, Schizomida and Ricinulei, are small and rare and are omitted from this account.

Order Solifugae

Solifuges, or wind scorpions (Fig. 14.3h), are arachnids of tropical to subtropical deserts on all continents except Australia. The prosoma has a small secondary carapace plate behind the main carapace and a pair of massive chelicerae with vertically gripping chelae. The pedipalps are long and

THE CHELICERATA

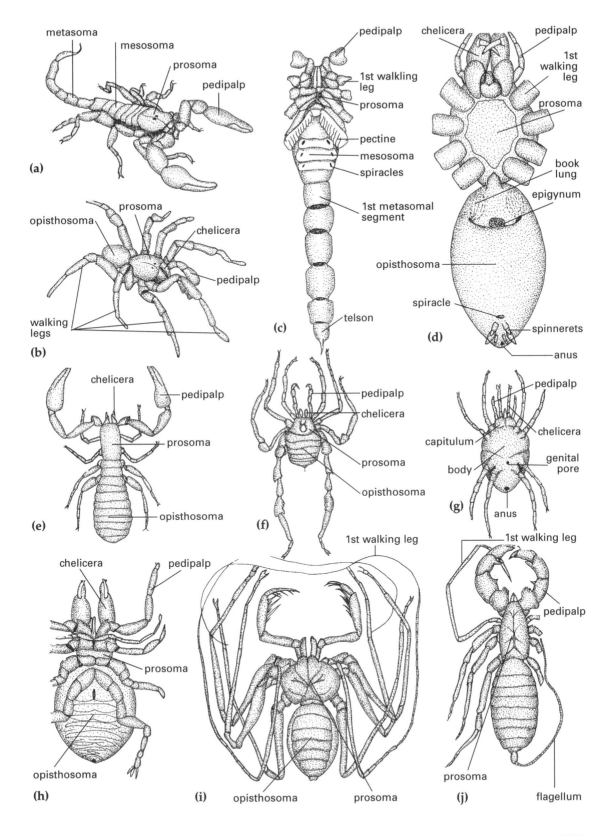

leg-like, but the first walking legs are modified as small, antenniform palps. Hexapod running is performed by the last three pairs of legs. The opisthosoma is long and segmented. Solifuges are fast-running hunters. There are 134 genera and about 600 species.

Order Uropygi

The Uropygi, or whip scorpions (Fig. 14.3j), are found only in the warmer areas of the Americas and East Asia. The prosoma has a single carapace plate. The chelicerae are short fangs, and the pedipalps are stout, curved prehensile limbs used in grasping and crushing prey. The first pair of walking legs is modified as long sensory appendages, so that locomotion is hexapod, as in solifuges, although not as fast. The segmented opisthosoma ends in a long, articulated flagellum. There are two pairs of book lungs, as in generalised spiders. The Uropygi are a small group, with only four genera and about 60 species.

Order Amblypygi

Amblypygids resemble uropygids in having prehensile pedipalps, long sensory first walking legs, hexapod locomotion, and a segmented opisthosoma with two pairs of book lungs, but they have no flagellum. Amblypygids are specialised dwellers under bark and among logs and leaf litter. The body is flattened (Fig. 14.3i), with very long limbs that are reflexed backwards parallel to the ground. Short bursts of fast running can be performed in a crab-like action, either forwards or sideways. Amblypygids are distributed in moist tropical and subtropical habitats on all continents except Australia. Diversity is limited to 18 genera and about 60 species.

Locomotion in arachnids

Most arachnids walk, run, climb or burrow using the third to sixth pairs of prosomal limbs as four pairs of walking legs. The fast-running solifuges, uropygids and amblypygids have evolved hexapod running on the last three pairs of legs. The walking legs retain the seven sections seen in merostomatans (coxa, trochanter, femur, patella, tibia, tarsus and pretarsus), but exhibit a unique feature not found in other arthropods. The coxae are fused to the body and the backwards and forwards (remotor–promotor) swing of the legs during stepping takes place, not at the coxa–body joint as in merostomatans, crustaceans, hexapods and myriapods, but at a more distal joint (R–P, Fig. 14.4). Coxal fusion has the important functional corollary of shifting the power source for walking from extrinsic leg muscles in the body, as in merostomatans, to smaller intrinsic muscles located within the limb bases. The origin of this seemingly retrograde step in chelicerate evolution appears to lie in the dual functions performed by the extrinsic leg muscles of merostomatans—fore and aft swing of the coxae during walking, and adduction–abduction of the coxae during gnathobasal chewing. During the evolution of the benthic marine ancestors of the terrestrial arachnids, the feeding function became concentrated around the mouth and the locomotory function became concentrated on the last four prosomal limb pairs. The loss of gnathobases from these limbs must have been accompanied by a reduction in extrinsic coxal musculature, necessitating a greater role for intrinsic musculature within the limb base in fore and aft stepping by the limbs. Fusion of the coxae to the body, which provides a rigid base for new stepping actions, evolved as a result of these changes. Transient

THE CHELICERATA

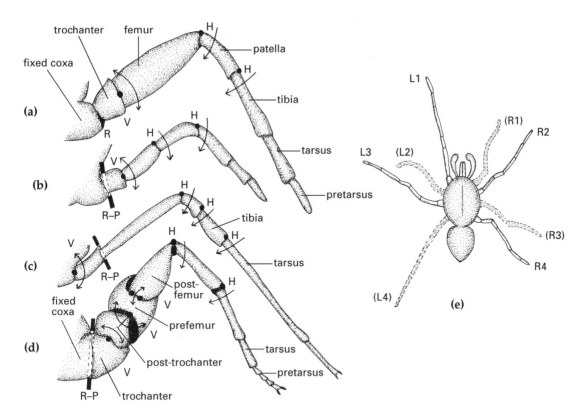

Figure 14.4
(a–d) Arachnid legs and leg joints: (a) Spider. (b) Acarine. (c) Opilionid. (d) Solifuge. **(e)** Basic arachnid gait: legs L1, R2, L3 and R4 are at the end of the forward recovery stroke and about to perform a propulsive stroke; legs R1, L2, R3 and L4 have just completed the propulsive stroke and are about to swing forward in a recovery stroke. (H — hinge joint; R — fore and aft twisting joint; R–P — remotor-promotor joint; V — pivot joint). (After Manton 1978.)

gnathobasal lobes are still formed on the walking legs in the embryos of primitive spiders and solifuges.

Stepping

Different specialisations in limb jointing and walking action are associated with coxal fusion in the different orders. Most arachnids hang down from their legs, with the femur–patella joints projecting as knees (Fig. 14.4a). In acarines, which have short legs, the knee flexure is more distal (Fig. 14.4b).

Scorpions and spiders have not evolved new joints for fore and aft swinging in the walking legs. Instead, these movements are achieved by twisting the legs forwards and backwards on the coxa–trochanter joints (R, Fig. 14.4a). Raising and lowering of each limb during stepping still takes place at a pivot joint between the trochanter and femur, with flexion and extension at a hinge joint between the femur and patella, as in *Limulus*.

Acarines, in contrast, perform remotor–promotor swings of the legs on new vertical pivot joints between the coxa and trochanter (R–P, Fig. 14.4b). Raising and lowering the limbs during walking is still at the trochanter–femur joint, however, with flexion and extension taking place more distally.

Opilionids have followed a different evolutionary route. Their very long legs are raised and lowered at the coxa-trochanter joint (V, Fig. 14.4c), as in *Limulus*, but a strong fore and aft swing takes place at a new distal location, the trochanter–femur joint (R–P, Fig. 14.4c). The legs of amblypygids show the same specialisation, but they are twisted backwards, placing the knees close to the ground and facilitating a fast, crab-like running.

The most specialised leg structure in arachnids has evolved in the Solifugae. The coxa–trochanter joint, as in acarines, has become a vertical hinge joint for fore and aft swinging (R–P, Fig. 14.4d). The proximal part of the limb between the trochanter and the knee is divided into three swollen sections (the post-trochanter, prefemur and postfemur), filled with powerful intrinsic muscles. Complex articulations between these sections combine strength with flexibility, permitting a rapid sustained stepping.

Gaits

Most arachnids move the legs of a pair in opposite phase, with one leg of the pair swinging forward as the other pushes backwards. They also move successive leg pairs in opposite phase. The gait is therefore one in which legs 1 and 3 on one side step in unison with each other and with legs 2 and 4 of the opposite side (Fig. 14.4e). A minimum of four legs, with the weight evenly spread between them, are always on the ground and pushing during walking. This type of gait provides a high level of stability, adjusts well to irregularities in the terrain, and is effective in climbing. In slow-moving arachnids such as scorpions, mites and ticks, the propulsive strokes of the legs are of longer duration than the recovery strokes, maintaining more than four legs on the ground at any one time. But fast-running arachnids, such as some spiders, solifuges and opilionids, do not show the opposite trend. Instead, the stable gait with four legs always on the ground is retained, and speed is achieved by decreasing the pace duration (i.e. performing each step more quickly) and increasing the stride length (i.e. having longer legs). The following examples of pace duration in arachnids were given by Manton (1978):

Acarines	*Ixodes*, a short legged, slow-moving tick	1.3 s
Scorpions	*Buthus*, a crawling, leaf litter scorpion	0.27 s
	Euscorpius, a scuttling rock scorpion	0.12 s
Spiders	*Trochosa*, a long-legged hunting spider	0.07 s
Solifuges	*Galeodes*, a fast-running wind-scorpion	0.06 s

Arachnids have many different patterns of slow or fast running, climbing or burrowing, but all are fundamentally different from the locomotory patterns of insects (Chapter 11) and myriapods (Chapter 12), in which the fore and aft swing of the legs takes place in the usual arthropod manner at the coxa–body joint in all species and is worked by extrinsic leg muscles.

Feeding in arachnids

Arachnids are almost all carnivores, capturing small prey by a variety of means. Three groups—the scorpions, pseudoscorpions and spiders—use venom, secreted by glands in the telson, pedipalps and chelicerae respectively, to subdue the prey. Digestive enzymes are passed out through the mouth, and the food is chewed by an adductor–abductor action of pedipalpal gnathobases. The liquid digest is then sucked in. Any indigestible remnants are ejected directly from the preoral cavity.

Scorpions

Scorpions are wandering, nocturnal predators. Insects and other prey detected by vibration-sensitive hairs on the limbs and pectines are seized by the pedipalps and sometimes stung by the barbed telson. Venom glands in the telson discharge their protein secretion through a sclerotised duct opening on the barb. The venom of most scorpions has little effect on

THE CHELICERATA

vertebrates, but that of a few species is fatal to humans if left untreated, causing respiratory paralysis and cardiac arrest. Species of *Androctonus* of North Africa and *Centruroides* of the southwestern USA and Mexico are the best known examples.

Pseudoscorpions

Pseudoscorpions feed on small arthropods such as mites and collembolans. The prey is caught and paralysed by the chelate pedipalps, which contain the venom glands.

Spiders

Spiders feed primarily on insects, although some species also capture small vertebrates. While many spiders are active hunters, others rely on snaring their prey with silk. The captured prey is pierced by the cheliceral fangs and injected with venom from glands in the chelicerae. Mygalomorph spiders, which have parallel chelicerae (Fig. 14.5a), strike downwards at the prey. Araneomorphs, which have chelicerae with curved, apposed tips (Fig. 14.5b), bite transversely.

Mygalomorph spiders are all lurking predators, usually hunting from a burrow. Araneomorph hunters include fast-moving cursorial groups such as wolf spiders (Family Lycosidae), crab spiders (Family Thomisidae) and jumping spiders (Family Salticidae). Prey are detected by tactile stimuli, often with the use of silken trip wires, and also visually. Wolf spiders and jumping spiders have well-developed eyes (Fig. 14.5c). The Dinopidae or net-casting spiders are highly specialised visual hunters with very large eyes. Dinopids are slow moving and hunt by stealth. The spider weaves silk into a small casting net (Fig. 14.5d) and then waits with the net stretched between the tarsi of the first and second walking legs. The net is thrown over a passing prey, which is then drawn up and consumed.

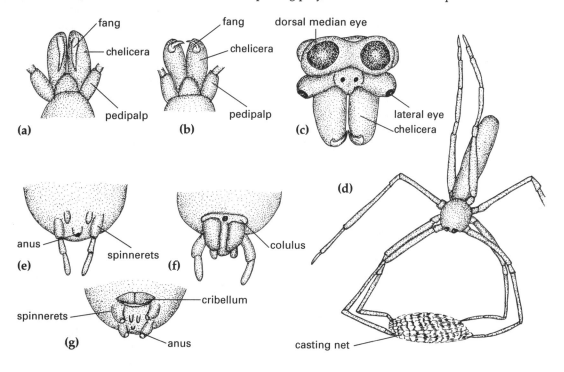

Figure 14.5
(a) Mygalomorph orthognath chelicerae. (b) Araneomorph labidognath chelicerae. (c) Eye pattern in a hunting spider. (d) Dinopid spider with casting net. (e–g) Arrangement of spinnerets in:
(e) mygalomorph,
(f) araneomorph and
(g) cribellate spiders.

Another spider with a specialised prey-catching technique involving silk is the bolas spider, of which *Ordgarius furcatus* is an Australian example. These spiders twirl a thread of silk with a sticky terminal globule. The spider exudes an airborne scent that simulates the pheromone of a female moth. Male moths attracted by the scent become entangled as the spider swings the bolas in their direction. The aim is quite accurate, even though the eyes of the spider are small. It seems likely that vibration sensors are involved. Bolas spiders also occur in Africa and America.

The sedentary, web-building spiders, in contrast, rely on the accidental entanglement of prey in a suspended web. They also have reduced eyes, and detect the presence of captured prey mainly through vibration. The bitten prey is wrapped in silk and stored on the web for later consumption.

Spider venoms

The venom proteins of spiders are neurotoxins of widely varying molecular size and toxic action. Only a few species produce venoms with a significant effect on vertebrates. Genera whose bite causes serious reactions in humans include the mygalomorphs *Atrax* and *Hadronyche* (funnel-web spiders), *Missulena* (mouse-spiders), and *Selenocosmia* (bird-eating spiders) and the araneomorphs *Latrodectus* (redback spiders), *Ixeuticus* (black house spiders), *Lampona* (white-tailed spiders) and *Olios* (huntsman spiders). The bite of *Lampona* causes severe ulceration of the skin as well as a general toxic reaction.

Spider silks

The silk glands of spiders are located in the posterior half of the opisthosoma, with ducts opening through numerous pores on the spinnerets. Mygalomorphs have two pairs of spinnerets (Fig. 14.5e). Araneomorphs have three pairs of spinnerets (Fig. 14.5f) and a median colulus, formed by the fusion of an anterior fourth pair. Silk is emitted as a liquid which hardens when the drawing-out process changes the molecular configuration of the protein. Several kinds of silks are produced. Each silk thread is composed of several fibres released from separate apertures. Mygalomorph spiders use silk mainly to line their burrows, constructing funnel-like extensions and trip lines in some species, and to spin egg cocoons. Araneomorphs, as well as spinning egg cocoons, employ silk in the construction of drag lines, trip lines and complex prey-capturing webs. The web usually contains a mixture of dry framework threads and sticky capture threads produced by different glands. The capture threads have an outer layer of viscous, unpolymerised silk.

Two groups of araneomorphs, the cribellate spiders and the non-cribellate web-spinners, use silken webs in the capture of prey. The cribellate spiders have a plate-like cribellum (Fig. 14.5g) instead of a colulus, and a comb-like calamistrum on the tarsus of each fourth leg. The cribellum has numerous pores through which a gelatinous silk is released. Using the calamistra, this silk is combed, together with normal silk from the spinnerets, to form a broad band of tangled silk called a hackled band. *Hickmania troglodytes*, a primitive, cave-dwelling araneomorph from Tasmania with two pairs of book lungs, incorporates the hackled band into a large sheet-web. The Dictynidae (e.g. the black house spider, *Ixeuticus robustus*) also build sheet webs. Dinopids use a hackled band in their casting nets. The Uloboridae, which are cribellate orb-weavers, lay down a hackled band as the spiral component of their orb-webs.

The non-cribellate araneomorphs have evolved many patterns of web-building. Sheet-webs are constructed by several families, including the Pholcidae (e.g. the daddy-longlegs spider, *Pholcus phalangoides*) and the Lyniphiidae, common in leaf litter and vegetation. One family of araneomorphs, the Agelenidae, builds funnel-webs in vegetation and decaying logs. Tangle-webs are woven by the Theridiidae (e.g. the redback spider, *Latrodectus hasselti*) and orb-webs by the Argiopidae (e.g. *Araneus*, the garden spider, with many species; *Argiope*, the St Andrew's Cross spider; and *Nephila edulis*, the golden orb-weaver).

Feeding in acarines

Mites generally retain the arachnid habit of ingesting liquid food, but have a great variety of diets. Carnivorous mites living in leaf litter and soil eat nematodes and small arthropods. Aquatic mites eat small crustaceans. Spider mites and gall mites have needle-like chelicerae with which they pierce plant cells before sucking out the contents. Other herbivorous mites eat fungi, bryophytes and algae. Organic detritus is exploited by many species. Most soil-dwelling oribatids eat decomposing plant and animal remains. Various storage mites feed on flour, dried fruit, mattress stuffing, hay or cheese. Dust mites (*Dermatophagoides*) browse on skin scales in household dust. Feather mites and fur mites scavenge on the debris on the skin of birds and mammals.

Parasitism is frequent among acarines. Most are ectoparasites, but some inhabit the bronchial passages of terrestrial vertebrates and the tracheal system of insects. In many acarines, parasitism is confined to the larval stage of the life cycle. The larvae of aquatic mites, for example, are parasitic on freshwater insects and bivalves. The larva of trombiculids (harvest mites) parasitise the skin of vertebrates, emerging from eggs in the soil to infest the skin of rodents and other hosts, where they feed on dermal tissue. These larvae, or chiggers, leave the host after a few days, and the subsequent life cycle is free-living. Nymphs and adults in the soil feed mainly on insect eggs. Humans infested with chiggers experience severe itching and dermatitis. In South-East Asia and northern Australia, the chiggers of *Leptotrombidium deliense* transmit the rickettsial bacteria that cause scrub typhus. The bacteria are passed from one mite generation to the next via the eggs.

Figure 14.6
Acarines: (a) The tick *Ixodes*.
(b) The follicle mite *Demodex*.
(c) The scabies mite *Sarcoptes*.

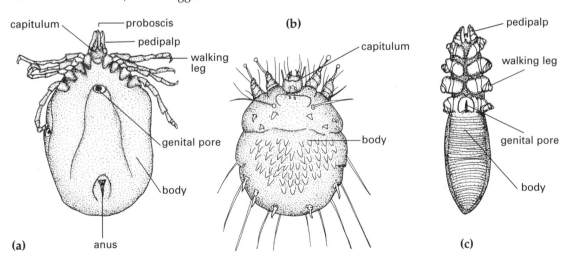

Figure 14.7 (opposite) Internal organs of chelicerates in sagittal section: **(a)** *Limulus*. **(b)** Scorpion. **(c)** Spider.

Other parasitic acarines attach temporarily to a host at each stage of their life cycle. The dermanyssid mites of birds and mammals, such as the red fowl mite, show this pattern, as do the Ixodida, or ticks. Ticks infest all groups of terrestrial vertebrates, penetrating the skin by means of hooked mouthparts and feeding on blood. They are important vectors of mammalian diseases, such as tick fever in cattle caused by the sporozoan *Babesia argentina*. High levels of tick infestation can also be debilitating to the host. The tropical tick *Boophilus microplus*, which was introduced into Australia on water buffalo from Indonesia in the 1870s, is a pest of cattle, horses and sheep in northern Australia. Levels of infestation up to 10 000 ticks per host can develop if left untreated. The bandicoot tick of eastern Australia, *Ixodes holocyclus* (Fig. 14.6a), feeds harmlessly on small marsupials, the natural host of the species, but causes serious allergic reactions when its salivary proteins pass into the blood of dogs or humans.

Some parasitic mites spend their entire life cycle on the host; for example the follicle mites (Demodicidae) and mange mites (Sarcopteridae) of the hair of mammals. Two species of *Demodex* (Fig. 14.6b) are obligatory human parasites: *D. folliculorum* lives in hair follicles, and *D. brevis* in sebaceous glands. The scabies mite, *Sarcoptes scabiei* (Fig. 14.6c), causes scabies by tunnelling in the human epidermis. Infestation by mange mites imported into Australia on domesticated animals and foxes has caused severe debilitation in Australian native marsupials.

Internal organs and physiological processes

Digestion

As in other arthropods, the chelicerate gut begins with a cuticle-lined foregut, continues as a tubular midgut, and ends in a short, cuticle-lined hindgut. In xiphosurans (Fig. 14.7a) the foregut leads forwards from the mouth as a tubular oesophagus, opening into a swollen gizzard with a toothed lining and strong muscles. After the food is macerated in the gizzard, fluids and fine particulate material filter through a valve into the midgut. Any larger and harder residue is regurgitated through the mouth. The anterior part of the midgut receives the ducts of two large, branching digestive glands that ramify through the haemocoel. The digestive glands are the principal site of enzyme secretion and absorption. Any residual waste passes to the hindgut and out through the anus.

In arachnids (Fig. 14.7b, c) the liquid external digest is sucked in by a pumping pharynx and passed along the oesophagus to the long, tubular midgut. Lateral diverticula extend from the midgut in the prosoma and the opisthosoma. The midgut is involved in enzyme secretion and absorption. A short hindgut connects with the posterior anus. Spiders have an accessory pump in the oesophagus, known as the pumping stomach. A second chamber at the posterior end of the midgut in spiders forms a cloacal chamber in which waste is collected before being discharged through the hindgut. Acarines, which take in mainly liquid food, have a morphologically generalised gut but show a wide range of diet and digestive physiology.

Respiratory exchange

The basic organs of respiratory exchange in merostomatans are the book gills on the opisthosomal limbs, but in arachnids these are modified as book lungs, developed as invaginations of the bases of opisthosomal limb buds

THE CHELICERATA

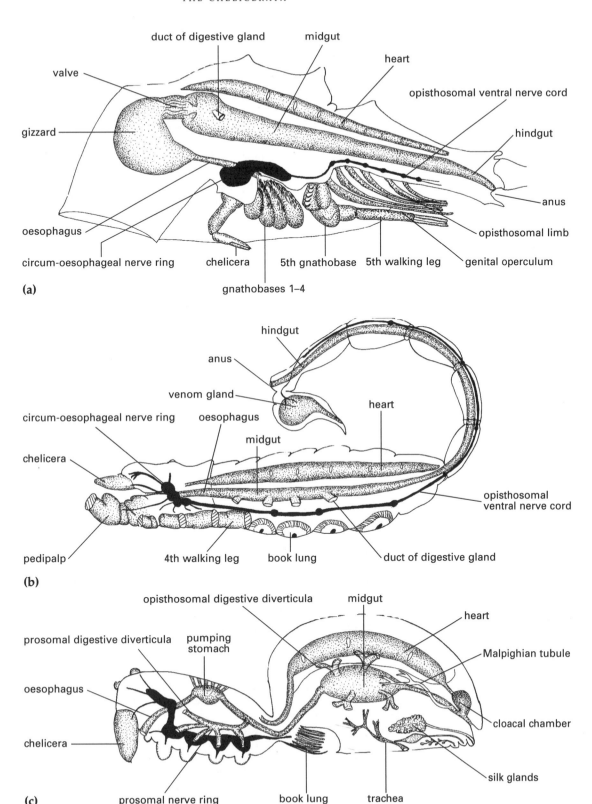

(Fig. 14.7b, c). The wall of one side of the invagination is folded into leaf-like lamellae, providing a large surface for gas exchange. The book lungs open by spiracles on the ventral surface of the opisthosoma. No external trace of the limb buds persists after the book lungs have developed.

Scorpions have four pairs of book lungs, situated on opisthosomal segments three to six. Generalised spiders, uropygids and amblypygids have two pairs of book lungs, on opisthosomal segments two and three. In many arachnids the book lungs are augmented or replaced by tracheae, using the same spiracles. Pseudoscorpions and some spiders have sieve tracheae, arising in bundles from the book lung chambers. Mygalomorph spiders have two pairs of book lungs, but most araneomorphs retain only one pair, the posterior pair being replaced by tracheae that open to a single, ventral spiracle in front of the spinnerets (Fig. 14.7c). Some small spiders have only tracheae.

Opilionids have tracheae opening through spiracles on the first opisthosomal segment. In many active, long-legged opilionids there are secondary spiracles on the tibiae of the legs. Most mites have tracheae, opening by one to four pairs of spiracles on the anterior half of the body. Some small mites have no respiratory organs. Solifuges also have tracheae instead of book lungs.

In contrast to the insect tracheal system, whose fine branches penetrate the internal tissues, the inner ends of the tracheae of arachnids terminate in the haemocoel, and the blood continues to be important in internal gas transport.

Circulation

In xiphosurans the heart (Fig. 14.7a) is a long tube with eight pairs of ostia. Several pairs of arteries carry the blood from the heart to the perivisceral haemocoel, from which the circulation continues into a pair of ventral sinuses supplying the book gills. After oxygenation in the gill circulation, the blood returns to the pericardial haemocoel.

Arachnids have a shortened heart, located mainly in the opisthosoma. Primitively there are seven pairs of ostia (Fig. 14.7b), but in most arachnids the number is reduced (Fig. 14.7c). A large anterior aorta and smaller posterior aorta lead from the heart. The haemocoelic circulation includes a ventral sinus bathing the book lungs. The relatively high blood pressure is important in generating leg extension, opposed by flexor muscles. Mites, in association with their small size, have a reduced circulatory system.

The blood of xiphosurans contains the respiratory pigment haemocyanin, together with amoebocytes which play a part in blood clotting. Haemocyanin is also present in the blood of scorpions and many spiders.

Excretion

The primary excretory system of chelicerates consists of coxal glands, developed from segmental organs in the prosoma. Xiphosurans have four pairs of coxal glands, located in the segments of the second to fifth walking legs. A complex duct drains the four glands of each side and ends in an excretory bladder, opening at the base of the fifth walking leg. The coxal glands contribute to osmoregulation in brackish conditions by producing a dilute urine.

Most arachnids retain coxal glands, but many also have one or two pairs of Malphigian tubules, arising from the midgut near the junction with the hindgut (Fig. 14.7c). Guanine is the major nitrogenous waste in arachnids, together with some uric acid. Scorpions have a single pair of coxal glands, opening at the bases of the third pair of walking legs, and two pairs of Malpighian tubules. Primitive spiders have two pairs of coxal glands, opening onto the coxae of the first and third walking legs, but most spiders retain only the anterior pair, in a reduced form. A pair of Malpighian tubules, opening into the cloacal chamber, acts as the main site of excretion in spiders. Opilionids and acarines also have one pair of coxal glands and a pair of Malpighian tubules.

Sense organs

Among chelicerates, compound eyes occur only in the marine Merostomata. The median eyes of xiphosurans are simple cups of retinal cells. The compound eyes are of an unusual type, with only a few loosely arranged ommatidia. Some fossil marine scorpions from the Silurian had compound eyes similar to those of the eurypterids (sea scorpions), but modern terrestrial scorpions and all other terrestrial arachnids lack compound eyes. Their simple eyes have a characteristic structure. A combined cornea and lens, composed of modified cuticle, covers a layer of epidermal cells enlarged to form a vitreous body. An underlying retinal layer contains photoreceptor cells. Spiders usually have eight of these eyes, arranged in two rows of four on the anterior dorsal margin of the carapace. Jumping spiders (Salticidae) and net-casting spiders (Dinopidae) have exceptionally large median eyes (Fig. 14.5c) of tubular construction, with many photoreceptors, and can register a sharp image.

Sensory hairs on the cuticle of the head, body and limbs are an important component of the arachnid sensory system. Olfactory, tactile and vibration responses are mediated by different hair types. Arachnids also have large numbers of slit sense organs in the cuticle. Each is a pit covered by a thin membrane in contact with an underlying sensory cell. Slit sense organs act as proprioceptors and may detect external stimuli such as vibrations.

Some mites have simple eyes, but most are eyeless. Mites rely mainly on sensory setae, together with innervated pits and slits in the cuticle, as in other arachnids.

Nervous system

The chelicerate central nervous system comprises a preoral brain and postoral ventral nerve cord as in other arthropods, but always shows a large amount of ganglionic fusion. Even in xiphosurans, the ganglia of the prosomal segments are fused with the brain, forming a circum-oesophageal ganglionic ring (Fig. 14.7a). A pair of long connectives leads from this ring to the remainder of the ventral nerve cord, with six pairs of segmental ganglia, in the opisthosoma. Scorpions retain a central nervous system similar to that of xiphosurans (Fig. 14.7b), but in the majority of arachnids the entire ventral nerve cord is fused into a circum-oesophageal ring (Fig. 14.7c).

Reproduction and development

Reproductive systems and processes

Unlike other arthropod groups, the Chelicerata have relatively simple reproductive systems. Sexes are separate. In xiphosurans, the gonads are a pair of branched tubes joined above the gut. A pair of simple ducts (oviducts or sperm ducts) passes to the gonopores under the genital operculum. Mating in *Limulus* begins with the smaller male clinging to the dorsal surface of the female by means of modified, hook-like, first walking legs. There is no copulation or sperm transfer. Eggs are deposited by the female in a shallow sand burrow and fertilised by sperm released by the attached male. The fertilised eggs are covered with sand and left to develop unattended. Each clutch consists of several thousand yolky eggs, 2–3 mm in diameter.

Arachnids have a midventral gonopore on the first opisthosomal segment. In most arachnid groups, sperm transfer is indirect: a spermatophore produced by the male is deposited on the ground and picked up by the female. Usually the male attracts the female to the spermatophore through courtship behaviour, or deposits the spermatophore on or in the female genital opening. The process is hazardous for the male, who is always likely to be mistaken for a prospective meal. Courtship approaches are usually slow and complex.

Scorpions

Scorpions have gonads among the midgut diverticula in the mesosoma, with a pair of gonoducts opening into a midventral genital atrium. Males have spermatophore-forming glands on the genital ducts; females have seminal receptacles. In an elaborate courtship, the male deposits a spermatophore on the ground and manoeuvres the female over it so that a sperm mass is taken up into the female opening. All femals scorpions are viviparous, either retaining yolky eggs which develop in ovarian tubules, or producing small, yolkless eggs whose embryonic development is nourished by secretions passed through a type of placenta. Gestation takes up to a year. After birth, the newborn scorpions are carried on the mother's back for about a week before becoming independent. Juvenile growth is slow—up to six years in some species.

Pseudoscorpions

Pseudoscorpions have a greater variety of reproductive processes. In some species the male deposits a spermatophore on the ground. A wandering female of the same species, encountering the spermatophore, is attracted by a chemical stimulus and takes up sperm into the female aperture. In other species the male spins silk threads leading to the spermatophore, and the female follows the trail of the threads. Some pseudoscorpions use a technique similar to that of scorpions in manoeuvring the female over the spermatophore.

Female pseudoscorpions exhibit complex brood care. A nest is constructed using silk drawn out from glands in the chelicerae. Eggs are deposited into a membranous brood sac on the underside of the female opisthosoma, and develop in this sac. During later development the embryos receive nutrients secreted from the maternal ovaries. Hatching takes place in the brood sac. Growth to maturity takes about a year, and the total life span can be up to five years in some species.

Spiders

Spiders have paired, ventral gonads in the opisthosoma. The female has paired oviducts converging to a median tube that opens through a short, cuticle-lined vagina, midventrally in the epigastric furrow. In some spiders the vaginal opening serves for both insemination and oviposition, with spermathecal ducts opening from the vagina. In other spiders the vaginal

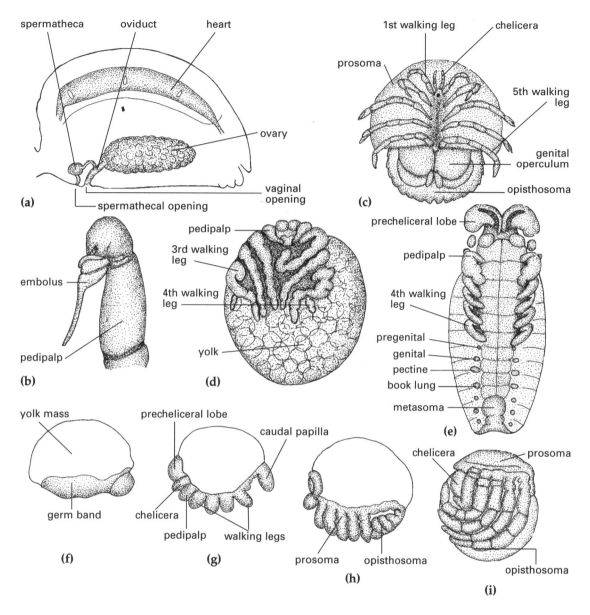

Figure 14.8
(a) Reproductive system of female spider. (b) Copulatory organ (embolus) on pedipalp of male spider. (c) Larval stage of *Limulus*. (d) Late embryo of acarine. (e) Segmented germ band of scorpion embryo. (f–i) Stages in the embryonic development of a spider. (f) Early germ band. (g) Early limb development. (h) Fully segmented germ band. (i) Approaching hatching. (c–i after Anderson 1973.)

opening serves only for oviposition. A pair of spermathecal openings on the epigynal plate in front of the vaginal opening have separate connections to the vagina (Fig. 14.8a). This condition has probably evolved independently several times in spiders.

The testes of male spiders open through two convoluted sperm ducts leading to the male pore, midventrally in the epigastric furrow. During mating, sperm is transferred by copulatory organs developed from the tarsal segments of the pedipalps. Each comprises a bulbous reservoir, ejaculatory duct and projecting embolus (Fig. 14.8b). The transfer of sperm to the reservoir involves external deposition of seminal fluid into a small web of silk, followed by uptake into the pedipalpal reservoirs. During copulation, the emboli are used to pass sperm into the vaginal opening or into the spermathecal openings on the female epigynum. Complex behavioural rituals before, during and after copulation usually ensure that the male does not become a meal. In some species, e.g. *Nephila edulis*, there is a striking sexual dimorphism. The female is much larger than the diminutive male. Such males, by careful stepping, are able to approach the web-bound female without being noticed. They also share the food captured on the female's web.

Once inseminated, a female spider lays her eggs into a silken cup which is then bound up as a cocoon. As in insects, each egg is enclosed in a protective chorion. Mygalomorph spiders brood their cocoons in the burrow for up to a year. Araneomorphs either carry their cocoons, as in wolf spiders, or suspend them on or near the web. Hatching occurs in a few weeks. Many species of araneomorphs have an annual life cycle, in which males die after mating and females die after oviposition is completed. In others, such as the redback spider, *Latrodectus hasselti*, batches of cocoons are produced by the female at intervals during a life span of two to three years.

Acarines
The reproductive system of acarines is relatively simple. In males, a pair of lobate testes is located in the middle of the body. Sperm ducts lead to the ventral median gonopore. In females, a single ovary is connected by a short oviduct to the ventral gonopore. A seminal receptacle and accessory glands are present. The methods used to transmit sperm vary. Many male mites produce a spermatophore which is deposited on the ground. In other mite species the spermatophore is transferred to the female aperture by the male chelicerae or, in some water mites, by the third pair of legs. In most species of acarines a penis effects direct transmission of sperm during copulation. Eggs are deposited generally in soil or leaf litter.

Development in chelicerates

Chelicerates have yolky eggs and retain no trace of spiral cleavage. In xiphosurans, total cleavage precedes the formation of a blastoderm. In arachnids, cleavage is intralecithal and a blastoderm is formed directly. Later embryonic development includes the formation of ventrolateral mesodermal bands and their division into paired, hollow mesodermal somites which develop in the usual lobopod/arthropod manner, but many aspects of chelicerate embryonic development are peculiar to the subphylum. No useful clues to the relationship between chelicerates and other arthropods have emerged from studies on chelicerate embryology.

The yolky embryos of *Limulus* hatch as juveniles (Fig. 14.8c), about 10 mm long, completely formed except for the last three pairs of opisthosomal limbs. They take 9–12 years to reach sexual maturity. The life span of *Limulus* is about 20 years.

Most arachnid embryos hatch as fully formed juveniles (Fig. 14.8e–i). Almost all arachnids have yolky eggs, the major exception being certain families of scorpions in which the eggs are secondarily yolkless and development is intrauterine and placental. Arachnid life spans vary from 20 years or more in some scorpions and spiders to only a few weeks in many mite species. Most arachnid species have an annual life cycle.

Acarines are exceptional among arachnids in hatching as a six-legged juvenile, the so-called 'larva' (Fig. 14.8d), which becomes an eight-legged 'nymph' at the first moult. Only one or two further moults are undergone in becoming an adult.

THE CHELICERATA

The class Pycnogonida

Form and function

The pycnogonids, or sea spiders, are the only extant group of marine chelicerates other than a few species of xiphosurans. Sea spiders are ubiquitous in marine benthic habitats from the littoral to the abyss. About 1000 species have been described. Most are less than 10 mm in body length, but some deep-sea species have a body length of 60 mm and a leg span of 750 mm.

Compared with other chelicerates, a pycnogonid (Fig. 14.9a) is almost all prosoma. The opisthosoma is reduced to a stump with a terminal anus. The prosoma is divided into a head, which incorporates the first walking-leg segment, and a trunk of three (sometimes four or five) walking-leg segments. Each walking-leg segment, including the first, protrudes laterally as a pair of cylindrical processes which carry the legs.

The anterior end of the head is extended as a proboscis (Fig. 14.9b) with a terminal mouth. Paired chelicerae arise at the base of the proboscis, with small pedipalps just behind them. In some species the proboscis is enlarged and the chelicerae and pedipalps have been lost. At the posterior end of the head, dorsally between the bases of the first walking legs, is a tubercle bearing four eyes. A pair of small ovigerous legs in front of the first walking legs are used for grooming and, in the male, for carrying eggs taken up from the female.

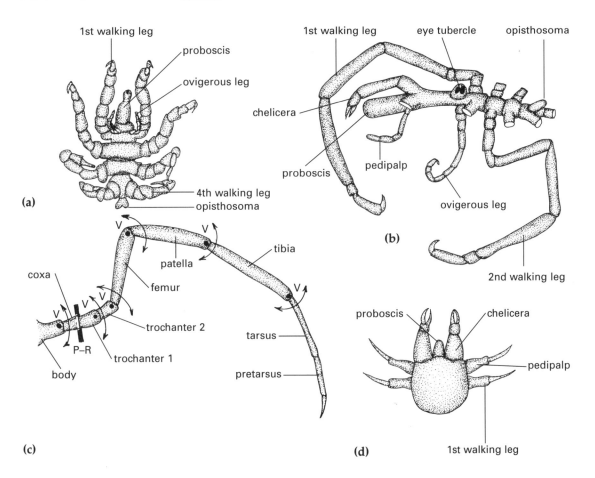

Figure 14.9
Pycnogonids. (a) Ventral view of *Pycnogonum*. (b) Pycnogonid limbs. (c) Walking leg and leg joints; apart from a promotor-remotor (P–R) joint between the coxa and trochanter 1, the remainder are pivot joints (V), facilitating the characteristic swaying movement of pycnogonids. (d) Protonymphon larva. (a after Anderson 1996; b after Nakamura 1987; c after Manton 1978.)

Locomotion

The long, thin walking legs of pycnogonids are fanned out and have terminal hooks. They are used for clinging and crawling on sponges, cnidarians and bryozoans. Pycnogonids can also swim by beating their legs up and down. Movements are slow. The lateral segmental extensions which bear the coxae are sites for the insertion of short extrinsic coxal muscles, which do not extend into the narrow trunk.

In pycnogonids, the principal leg movement takes place at the coxa–body joint, as in *Limulus*, but rather than being a fore and aft swinging action it is a raising and lowering movement about a pivot joint with a horizontal axis (V, Fig. 14.9c). This movement is used by pycnogonids in swaying the body. The next leg joint, between the coxa and trochanter 1, is a promotor–remotor joint. This joint (P–R, Fig. 14.9c) allows a fore and aft swing of the legs used in walking and climbing. No other group of arthropods has the basal joints of the leg organised in this way.

The metachronal rhythm of pycnogonid leg action shows a phase difference of about 0.5 between successive legs, which are therefore in opposite phase, as in arachnids during walking and xiphosurans during gnathobasal chewing. A gentle rise and fall of the body while anchored by the large tarsal claws is displayed by pycnogonids during feeding. Swimming results from a more vigorous performance of the same movement.

Feeding

Most pycnogonids are browsing carnivores, feeding on the tissues of the sessile animals on which they crawl. Some apply the proboscis directly to the surface of the prey. Others use the chelicerae to remove pieces of tissue which are passed to the mouth. Some pycnogonid species browse on algae and microorganisms. The first part of the gut, within the proboscis, is a suctorial and masticatory pharynx. A short oesophagus leads to a long midgut, with lateral caeca extending into the legs. Digestion is intracellular. Waste material is passed through a short hindgut to the anus.

Other organ systems

There are no organs for respiration or excretion in pycnogonids; these exchanges take place through the body surface. Circulation through the haemocoel is due to a dorsal heart. The nervous system begins with a brain beneath the eye-bearing tubercle. Nerves from the brain supply the chelicerae and proboscis. Circum-oesophageal commissures connect with a suboesophageal ganglion, from which nerves pass to the pedipalps and ovigerous legs. Paired ventral nerve cords with segmental ganglia supply the segmental walking legs.

Reproduction and development

Pycnogonids have separate sexes. Males can be identified by the well-developed ovigerous legs (Fig. 14.9b). In each sex a single gonad (either ovary or testis) extends through the trunk above the gut, with paired diverticula branching into the legs. Gonopores are unusual in opening ventrally on the coxae of the prosomal walking legs. The number and location of the gonopores varies in different species.

During oviposition, the male stands over or hangs beneath the female. Eggs are fertilised as they are laid and are gathered by the male onto the ovigerous legs. Glands on the femora of the legs secrete an adhesive by which the eggs are cemented into a mass. The eggs are brooded by the male until they hatch. The hatching stage is usually a protonymphon larva (Fig. 14.9d) with a short proboscis and three pairs of legs, the chelicerae, pedipalps and first walking legs. Development proceeds through a series of moults during which further segments and leg pairs are added.

Relationship of pycnogonids

Pycnogonids have many unique features, including the tagmatisation of the body, the multiple prosomal gonopores, ovigerous legs and male brooding behaviour. Basic chelicerate features include the chelicerae, sense organs, brain structure and type of circulatory system. The relationship of pycnogonids to other marine chelicerates, including the ancestors of the arachnid groups, is not clear. The small simplified body, suctorial proboscis and long, clinging legs with claws are adapted to a swaying and browsing life on sessile marine invertebrates of branching growth habit. Caprellid amphipod crustaceans have a similar shape and action and a similar habit. It seems likely that the pycnogonids diverged at an early stage of chelicerate evolution, before the fixed tagmatisation of the Merostomata and their arachnid descendants had become established.

Classification of the subphylum Chelicerata

CLASS MEROSTOMATA
Marine; prosoma with chelicerae and five pairs of walking legs; opisthosoma with 5–6 pairs of biramous appendages, book gills.
Subclass Xiphosura
  Segmentation indistinct; prosoma with horseshoe-shaped carapace; opisthosoma with six pairs of appendages, the first a genital operculum, 2–6 with book gills; long caudal spine (several fused segments + telson). Limulids. Mainly fossil; five extant species.
Subclass Eurypterida
  Short prosoma with carapace; long, segmented opisthosoma comprising mesosoma of six segments with paired appendages, the first a genital operculum, 2–6 with book gills, metasoma of six limbless segments, telson. Sea scorpions. A wholly fossil group from the Palaeozoic.

CLASS ARACHNIDA
Carnivorous, mostly terrestrial; some secondarily aquatic. Prosoma with chelicerae, pedipalps and four pairs of walking legs; opisthosoma basically of 13 segments + telson, but often reduced; anterior opisthosomal segments with respiratory organs (book lungs or tracheae) except in some small species. Thirteen extant and several fossil orders. Only the major orders are listed below.
Order Scorpiones — Short prosoma with carapace; long, segmented opisthosoma with anterior mesosoma of six segments, pectines on second, book lungs on 3–6; posterior metasoma of six limbless segments; telson with poison spine.
Order Pseudoscorpiones — Small arachnids resembling scorpions, but with opisthosoma not differentiated into mesosoma and metasoma; no terminal sting.
Order Araneae — Prosoma separated from opisthosoma by a pedicel; chelicerae as fangs with poison glands; pedipalps sensory, copulatory in male; four pairs of walking legs; opisthosoma usually unsegmented, with spinnerets.
  Infraorder Mygalomorphae — parallel chelicerae, two pairs of book lungs,

six or fewer spinnerets. Funnel-web spiders, trapdoor spiders and tarantulas. Dark coloured, often large.

Infraorder Araneomorphae — opposed chelicerae, rarely two but usually one pair of book lungs or none, six spinnerets and a median colulus (sometimes a cribellum). Dark or brightly coloured, small to large. Many families.

Order Opiliones — Prosoma and opisthosoma not separated by a pedicel; four pairs of long walking legs; opisthosoma segmented, without book lungs; penis and ovipositor.

Acarine orders

Prosoma and opisthosoma merged; segmentation usually lost; chelicerae and pedipalps on anterior capitulum; up to four pairs of walking legs; book lungs absent.

Order Opilioacariformes — large, brightly coloured mites with opisthosomal segmentation. Omnivorous or predatory, living in leaf litter.

Order Parasitiformes— medium to large mites with an unsegmented body; tracheal system with ventro-lateral spiracles. Free living and parasitic; including ticks.

Order Acariformes — small mites with an unsegmented body; spiracles near mouthparts, or absent. Very diverse in terrestrial, marine and fresh water habitats, stored products and as parasites of plants and animals.

CLASS PYCNOGONIDA

Marine chelicerates with externally segmented body; prosoma divided into head with proboscis, chelicerae, pedipalps and ovigerous legs; trunk of 3–6 segments with long walking legs, short, unsegmented opisthosoma.

Chapter 15

The lophophorates — Phoronida, Brachiopoda and Ectoprocta

P.J. Doherty

	Introduction 344
	Phyletic relationships 345
	Evolutionary histories 345

PHYLUM PHORONIDA	Structure and function 346
	    Digestion 346
	    Circulation and excretion 347
	
	Reproduction and development 348

PHYLUM BRACHIOPODA	Structure and function 349 Feeding and digestion of an articulate brachiopod 352 Circulation and excretion 355 Nervous system 355 Reproduction and development 355

PHYLUM ECTOPROCTA	Structure and function 356
	    The major groups of ectoprocts 356
	    Structure and function of autozooids 358 Digestion 359
	Circulation and excretion 360
	
	Reproduction and development 360

	Tissue regeneration in lophophorates 361 Growth patterns and polymorphism in ectoprocts 362 Induced growth responses 363 Biologically active chemicals in lophophorates 364 Classification of the lophophorates 364

INVERTEBRATE ZOOLOGY

Introduction

This chapter covers three phyla of sedentary and sessile aquatic animals that share common features, most notably a homologous structure called a lophophore that is used for suspension feeding.

The three lophophorate phyla encompass bilaterally symmetrical triploblastic animals with ancestral body plans composed of three parts (protosome, mesosome, metasome), each with a corresponding coelomic cavity (protocoel, mesocoel, metacoel). In modern lophophorate taxa the protosome is assumed to be vestigial or lost as a result of sessility, which makes redundant a head with nervous and sensory tissue. All phyla have a circumoral ridge bearing hollow ciliated tentacles (the lophophore) developed from the mesosome, and a U-shaped gut in the metasome (Fig. 15.1). The anus, when present, exits outside the lophophore. Gonads differentiate from loose peritoneal cells lining the metacoel.

Figure 15.1
Ectoprocts: section through a phylactolaeme zooid, illustrating diagnostic features (after Hyman 1959).

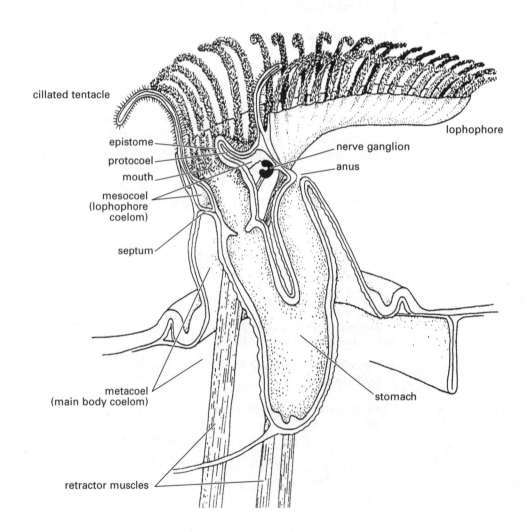

Phyletic relationships

Recent analyses of nucleotide sequences have linked the lophophorates with the protostomes (see Chapter 18), despite their apparent possession of some characteristic deuterostome features. The latter include radial cleavage of the egg, indeterminate development of cells in the blastula, the tripartite body plan and the formation of the coelomic cavities by enterocoely (as paired outgrowths of the embryonic gut) rather than by schizocoely (splitting of the mesoderm). These developmental features are variable in lophophorates, however. In particular, the Ectoprocta are enigmatic because all of the internal tissues of the larvae are autolysed during metamorphosis, so that their body cavities cannot be traced from embryonic tissues. There has been much debate about whether these animals are true coelomates.

Other characters of the lophophorates seem decidedly more protostome, particularly the development of the mouth from the blastopore. Many larval forms appear to be modified trochophores, and the larval protonephridia resemble those of certain polychaetes. Together with the frequent use of chitin, which is very rare in deuterostome phyla, it seems likely that modern lophophorates should be classified as protostomatous coelomates, notwithstanding their indeterminate development (see Chapter 18 for further discussion).

Evolutionary histories

Phoronids are soft-bodied worms with little potential to leave fossil remains. Nonetheless, mud casts from early Palaeozoic sandstones have been attributed to colonies of these animals.

Brachiopods are shelled animals. They are an ancient group, first appearing as fossils in Lower Cambrian strata formed more than 550 million years ago. Some of the lineages leading directly to modern inarticulate brachiopods were present during the Ordovician and have changed little in 400–500 million years. During the first 50–100 million years, brachiopods radiated rapidly. They were then prominent in the marine macrobenthos for a further 250 million years. Palaeontological treatises describe tens of thousands of extinct species classified into thousands of genera, including some huge forms and many bizarre morphologies. Most of this diversity was lost during the Permo-Triassic marine faunal crisis at the end of the Palaeozoic Era. Today a few hundred extant species represent the survivors from this once grand fauna.

Ectoprocts, minute colonial animals with exoskeletons, have also left fossils. As reef-building organisms they have been major contributors to limestone deposits of marine origin. First recognisable in rocks from the Upper Cambrian, these animals also radiated during the early Palaeozoic and suffered mass extinctions at the end of the Permian. One of the surviving classes, the Stenolaemata, rebounded with another burst of radiation during the Mesozoic, but declined at the end of the Cretaceous, when there was a second mass extinction. In the Tertiary, ectoprocts underwent a third wave of radiation, and today a large proportion of the 3000 living species belong to one order, the Cheilostomata. While the evolutionary resilience of Ectoprocta may owe much to their colonial life styles, the recent radiation of cheilostomes has been linked with their unique phenotypic plasticity (see following).

Phylum Phoronida

Bilaterally symmetrical, sedentary coelomate worms of trimeric construction; mesosomal lophophore; U-shaped gut with anus outside lophophore; diffuse nervous system; nephridial excretory system; circulatory system; separate sexes or hermaphrodite, coelomic gonads.

Structure and function

Phoronids are sedentary vermiforms (mostly < 10 cm long) inhabiting secreted chitinous tubes. There are only about 10 species in the modern fauna. Most are infaunal dwellers in soft sediments (Fig. 15.2a), but others bore into calcium, and *Phoronis australis* invades the tube walls of cerianthid anemones. All feed by extending an anterior lophophore into the water.

Ancestral phoronids are assumed to have been free-living worms with bodies divided into three parts; now, in association with a burrowing lifestyle, the phoronid protosome is vestigial. The anterior body is dominated by the mesosome, which bears the lophophore and contains the main nervous system.

The lophophore consists of a ridge on the mesosome that surrounds the mouth and bears a single row of tentacles. As new tentacles form near the mouth, the circular lophophore may invaginate and coil into bilaterally symmetrical spirals. In cross-section each tentacle is oval and hollow, its lumen being continuous with the mesocoel. The tentacles are held erect by fluid pressure and basal connective tissue, but can be moved independently by longitudinal muscles. Lateral cilia along both sides of the tentacles draw water from above the lophophore into the space between opposing rows of tentacles and pass it to the other side (Fig. 15.2b, c, d). Shorter cilia on the inner (frontal) surfaces of the tentacles flick particles from the water towards the base of the tentacles where there is a ciliated food groove, continuous with the crescentic mouth. Mucus may assist the capture and transport of food, but more often seems to be used to bind unwanted material. Particles may be rejected at the angles of the mouth (which can be occluded by the epistome) or rejected in the lophophore. In the latter case, the frontal cilia reverse their beat to move particles to the tips of the tentacles. The anus and nephridiopores discharge downstream into the flow of water leaving the lophophore, and thus their emissions do not foul the mouth or the incurrent flow.

At up to 50 cm long, most of the phoronid body consists of the metasomal region, which is anchored in its tube by a bulbous end. Longitudinal muscle contraction in the trunk allows the anterior to be withdrawn rapidly into the burrow, the lophophore folding as it is retracted. This escape response is coordinated by one or more giant axons extending posteriorly from a nerve ring under the lophophore, which also supplies nerve fibres to the tentacles. Antagonistic circular muscles in body wall compress the fluid-filled metacoel to return the lophophore sluggishly to the top of the tube, where further pressure on the mesocoel extends the tentacles to the feeding position.

Digestion

The metasomal trunk is filled largely by a long recursive gut anchored to four longitudinal mesenteries. Food, mainly diatoms bound in mucus, is

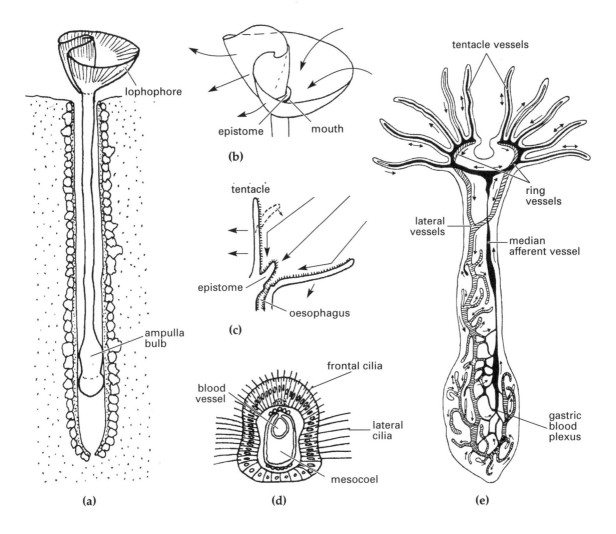

Figure 15.2
Phoronids: **(a)** Adult of *Phoronis psammophila*. **(b, c)** Water currents through lophophore. **(d)** Section through tentacle. **(e)** Circulatory system. (After Emig 1982 and Hyman 1959.)

passed through the gut by cilia, assisted by sluggish peristalsis from muscle fibres in the gut wall. The most important part of the gut is the distended stomach region near the end of the trunk, where food is digested. The upper portion of the gut secretes enzymes which carry out extracellular digestion in the gut lumen, while portions of the stomach wall are places of intracellular digestion. (In brachiopods this function is carried out in special digestive diverticula, which are lacking in the simple guts of phoronids.) From the stomach, the intestine returns anteriorly, terminating in an anus located downstream of the mouth.

Circulation and excretion

The peritoneum surrounding the stomach is permeated by hollow spaces that are part of a semi-closed blood system containing coelomocytes and haemoglobin bound on blood corpuscles. Blood drains from these spaces into a median dorsal afferent vessel and is transported anteriorly to an afferent ring vessel underlying the lophophore, from which a single vessel enters the lumen of each tentacle (Fig. 15.2e). Sluggish bidirectional flows in these vessels circulate the blood and drain near the tentacle bases into an efferent ring vessel. Two major vessels exit from the efferent ring and return posteriorly to the haemal plexus of the gut, although in all but one species they fuse into a single lateral efferent vessel shortly after passing through the septum. There is no heart, but the major vessels are contractile.

Injections of dyed particles show that loose coelomocytes are important agents for moving the products of digestion and removing wastes. Unwanted material is scavenged by these cells as they move through the coelomic spaces and blood system. In a recurrent theme among the lophophorates, loose cellular organisation allows the coelomocytes to migrate through dermal tissues and across basement membranes; they may even transport wastes directly through the body wall. In addition, the metacoel contains a pair of nephridia with elongated, ciliated, funnel-shaped nephrostomes. These structures extract wastes from the coelomic fluid and discharge urine through paired nephridiopores near the anus. During spawning they also act as gonoducts.

Reproduction and development

Most phoronids reproduce sexually, although one species forms large aggregations by asexual fission. Most are hermaphroditic, though some may have separate sexes. Gametes are generated from peritoneal cells surrounding the haemal plexus, and swollen gonads may develop around the lateral vessel. The gametes are shed into the metacoel and escape through the nephridia. Fertilisation may be external in some species but is probably internal in most, with foreign sperm gaining access to the metacoel through the nephridiopores. Embryos are brooded for the majority of their development, either within the tube or more commonly in the centre of the lophophore. In the latter case, embryos emerging through the nephridiopores are attached to the base of the lophophore by adhesives secreted from adjacent glands.

Development is indirect via an actinotroch larva (Fig. 15.3), which appears to be a modified trochophore. Unlike present day adults, larvae

Figure 15.3
Phoronids: actinotroch larva and metamorphic series of *Phoronis muelleri* (after Emig 1982).

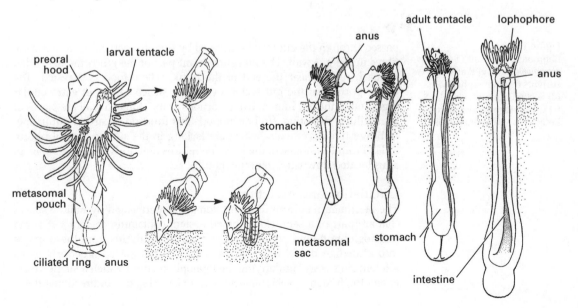

have three body regions and a straight gut. An anterior preoral hood covers the mesosome, which bears 6–24 larval tentacles for feeding. The larva swims near the surface, propelled by a well-developed ring of cilia around the posterior anus.

Beneath the larval tentacles, a metasomal sac develops that will become the definitive trunk of the adult. This pouch everts and drags the gut into its characteristic U-shape. Simultaneously, the larva reorientates 90° from its upright swimming posture and begins burrowing with the pouch. At this time the preoral hood is cast off and resorbed. The epistome arises secondarily as a fold of the body wall near the mouth. The larval tentacles degenerate, while the adult tentacles start to elongate from buds near their bases. Ultimately the lumens of the larval tentacles become the blood vessels inside the definitive tentacles. As the latter grow to form the circumoral lophophore, the distance between mouth and anus is shortened, bringing them close together at the anterior end of the animal. Meanwhile the metasome continues to elongate and secretes the protective tube.

Phylum Brachiopoda

Bilaterally symmetrical, sedentary or sessile coelomates of trimeric construction; mesosomal lophophore, enclosed by bivalved shell with dorsal and ventral valves; U-shaped gut with anus outside lophophore or absent; nervous system with circumoesophageal nerve ring; nephridial excretory system; circulatory system; separate sexes; coelomic gonads.

Structure and function

Brachiopods can be regarded simplistically as phoronids with a hard shell. Superficially, they look like bivalves (Fig. 15.4a), but the shells of brachiopods are dorso-ventral and calcitic or chitinophosphatic, whereas bivalves have lateral shells of aragonite. Furthermore, brachiopods lack the elastic ligament that opens bivalve shells and have antagonistic muscles that both open and close the shell.

Brachiopod shells are lined internally by a thin layer of mantle tissue that is grooved near its margin (Fig. 15.4j). As cells proliferate in this groove, older cells are displaced to the edge, where they roll over to face outwards and progressively occupy more posterior positions inside the shell as growth continues. As they complete this conveyor-belt passage, the cells change their secretions so that the mature shell ultimately consists of an outer chitinous periostracum (deposited first) reinforced underneath by two layers of calcium salts and fibrous protein. The mantle groove is also the site of special cells that secrete long chitinous setae, which protrude beyond the shell margin and have protective and sensory functions. Many of the calcareous shells are punctuated by glandular tissue that communicates with the external periostracum through fine pores (Fig. 15.4j). The purpose of these punctae is not completely known, although oxygen consumption declines when the shells are smeared with petroleum jelly. They may also provide chemical defence against bioerosion and act as glycogen storage reservoirs.

The internal space enclosed by the dorsal and ventral mantles contains the rest of the living animal, which (as in phoronids) has only two regions

Figure 15.4
Brachiopods: **(a)** In situ adult of the inarticulate, *Lingula*. **(b)** Schematic section through the body of *Lingula*. **(c)** Planktotrophic larva of *Lingula*. **(d)** Lecithotrophic larva of the articulate *Calloria inconspicua*. **(e)** Metamorphosis of *Calloria*. **(f)** Adult *Calloria* in situ. **(g)** Section through body. **(h, i)** Pedicle and shell musculature. **(j)** Section through shell margin. **(k, l)** Modern and fossil examples of brachidia supporting lophophores of different form. (After Rudwick 1970 and Hyman 1959.)

in the adult body. The brachiopod metasome, which contains the gut and most of the organs, occupies little more than the posterior third of the shell (in phoronids the metasome is the major body region). The metacoel extends anteriorly, however, as a series of reticulate, blind-ending canals within the mantle tissues. These were once thought to be blood vessels but are now recognised as coelomic spaces which carry nutrients to the growing shell margin. They may also act as accessory sites of gas exchange, and frequently accommodate the gonads. Most of the volume inside the shell contains the large, complex lophophore. Protruding from the rear of the shell is a chitinous pedicle which is important for anchorage and movement.

Brachiopods are sharply divided into those with and those without hinged shells. These lineages, which have been distinct for more than 500 million years, also display other fundamental differences. Inarticulate brachiopods (those lacking a hinge) have more complex musculature than those that are articulate. The inarticulate forms include two orders with very different structure. One is characterised by *Lingula* (Fig. 15.4a) which can be found on any number of muddy intertidal beaches around the world. This burrowing brachiopod has a chitinous hollow pedicle reminiscent of the tube-dwelling Phoronida (Fig. 15.2a), although the pedicle does not contain organs other than intrinsic musculature. The pedicle grows from the ventral body wall soon after larval settlement and contains an extension of the metacoel, which can be deformed to allow burrowing and vertical movements within the burrow. When feeding, *Lingula* extends the shelled portion to the surface of the burrow, where the long anterior setae converge to form three openings (two lateral inhalant and one medial exhalant) to allow water to pass continuously through the gaping shell. This canalisation is the functional equivalent of the fused siphons of infaunal bivalves, separating the flows of water and preventing sediment from falling into the opened shell and clogging the lophophore. The characteristic U-shaped gut with its anterior anus is confined to the metacoel inside the lower part of the shell (Fig. 15.4b).

The other major variant of the inarticulate body plan looks superficially like a limpet and is epibenthic. On a flat surface, its orientation is rotated 90° from that of *Lingula*. Unlike *Lingula*, these hemispherical creatures, the Acrotretida, have no pedicle; instead, the ventral valve is cemented directly to hard substrata. Although both inarticulate orders have similar lophophores, they display differences indicative of a long period of separate evolution.

Articulate brachiopods grow hinged calcareous shells (Fig. 15.4f). Although many variants have evolved, the hinges of modern articulates have interlocking sockets and teeth which restrict valve movements to a small anterior gape. Special anchorage points for the diductor muscles provide leverage to open the shell from a rear position (Fig. 15.4i). Tendons are another innovation in many of these animals. Simple adductor muscles pull the valves together and contain 'quick' and 'catch' filaments with different response rates. The former snap the valves together rapidly in response to disturbance or to create an explosive expulsion of water that cleanses the lophophore when fouled with mucus-bound particles. The 'catch' filaments can keep the shell closed for long periods, resisting predators that would prise the valves apart and allowing some species to survive in intertidal habitats.

Articulate brachiopods have pedicles different in origin, structure and function from those of inarticulates. They arise directly from the lower third

THE LOPHOPHORATES — PHORONIDA, BRACHIOPODA AND ECTOPROCTA

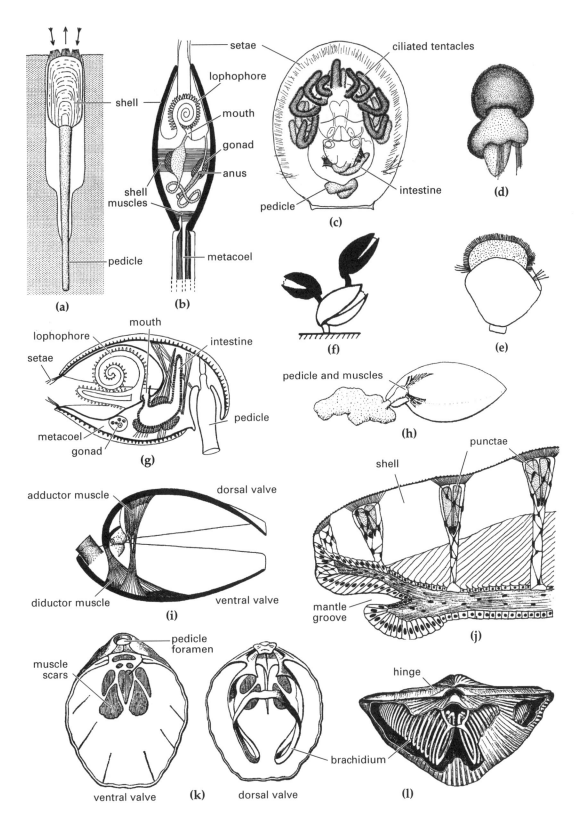

of the larval body, are solid and inert, and have no intrinsic musculature. Typically they are cemented to hard substrates at the distal end. Inside the shell, a set of muscles attached to the proximal end of the pedicle allows the shell to be tilted and swivelled through wide arcs (Fig. 15.4h).

Brachiopods have the largest and most complex lophophores, presumably because of their large unit-body size and the greater rigidity permitted by permanent enclosure within a protective shell. The lophophore is strengthened by thick cartilage-like connective tissue in addition to a substantial hydrostatic skeleton. Some articulates have additional calcareous supports, which were very elaborate in some fossil forms (Fig. 15.4k, l). Larger brachiopods have evolved lophophores with complex, three-dimensional spirals and whorls, which increase the surface area of the filtration apparatus.

Feeding and digestion of an articulate brachiopod

The functioning of the lophophore and food handling are detailed here through a case study of one modern brachiopod. *Calloria inconspicua* is a living brachiopod from the 'long-looped' lineages that dominate the modern fauna. This condition refers to the internal calcified skeleton that suspends the lophophore from the dorsal valve (Fig. 15.4k). The two lateral arms and median spiral of the lophophore support many hundreds of ciliated tentacles. When the lophophore is sectioned transversely, it is clear how it divides the mantle cavity into inhalant and exhalant chambers, allowing continuous filtration (Fig. 15.5).

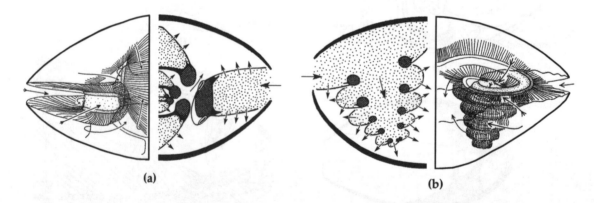

Figure 15.5
Brachiopods: three-dimensional and cut-away views explaining the filtration of food particles (stipple) by (a) plectolophous and (b) spiralolophous taxa. (After Rudwick 1970.)

In profile, each of the three arms reveals its double structure, with opposing rows of tentacles terminating basally in a deep groove that is continuous with the mouth (Fig. 15.6a). The interior of each arm is occupied by a large central canal which, in the adult, does not communicate with any other coelomic space. Together with the calcified brachidia and well-developed connective tissue, this hydrostatic skeleton prevents the elaborate lophophore from collapsing. Close examination of the lophophore reveals tentacles disposed in a double row, in which alternating filaments have different cross-sectional profiles. Long lateral cilia on both types propel water through the feeding apparatus to create the feeding and cleansing currents. Both types are also ciliated densely on their frontal surfaces. Those of the inner row present round profiles to the inhalant flow, while the alternating tentacles of the outer row are distinctly grooved (Fig. 15.6b).

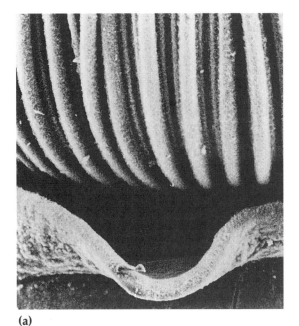 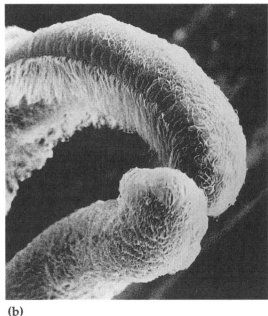

Figure 15.6
Brachiopods: scanning electron micrographs of (a) double row of tentacles and basal food groove in *Calloria inconspicua*, and (b) details of the ciliation on frontal and abfrontal surfaces of the tentacles.

Inhalant flow enters around the lateral margins of the gaping shell and is confronted by the opposing rows of tentacles of the lateral arms. Some water passes straight through the lophophore to the exhalant chamber. The rest flows posteriorly and is filtered through the median spiral. If large particles are introduced into the inhalant flow, two types of rejection are observed. Isolated particles are rejected by distal carriage along the ciliated frontal surfaces of the inner tentacles. Large aggregates are rejected by a different mechanism. In such cases, the inhalant current is halted by nervous inhibition of the lateral cilia. Tentacles in contact with the unwanted material are moved apart by their internal longitudinal muscles to create large gaps in the lophophoral filter. Lateral cilia resume a hesitant beat until the debris is carried through to the other side where it can be disposed of by the ciliated mantle epithelium. If there is heavy clogging, the lophophore is cleansed by a rapid snapping of the adductor muscles, which produces a 'sneeze'.

Smaller particles travel down the tentacles, principally in the grooved surfaces of the outer row, to the basal food groove, where they are moved by cilia to the mouth, often tangled in mucoid strings. Just behind the mouth are a dozen or so short tentacles of uniform profile that are densely ciliated on all surfaces and arranged in a single row. It is not clear whether these are simply precursors of the definitive tentacles in a growing lophophore or are specially modified for particle selection, like the labial palps near the mouths of bivalves. Cilia on these tentacles beat distally, away from the mouth.

Inside the mouth, the gut begins with a short foregut (Fig. 15.7a) lined with a ciliated epithelium and goblet cells secreting copious mucus and digestive enzymes (Fig. 15.7b). The presence of microvilli and micropinocytotic vesicles at the top of most gut cells suggests that extracellular digestion is not localised, but the digestive diverticula opening from the midgut (Fig. 15.7a) are special sites of intracellular digestion. Access to these tubules is through narrow ducts that admit dissolved substances and small

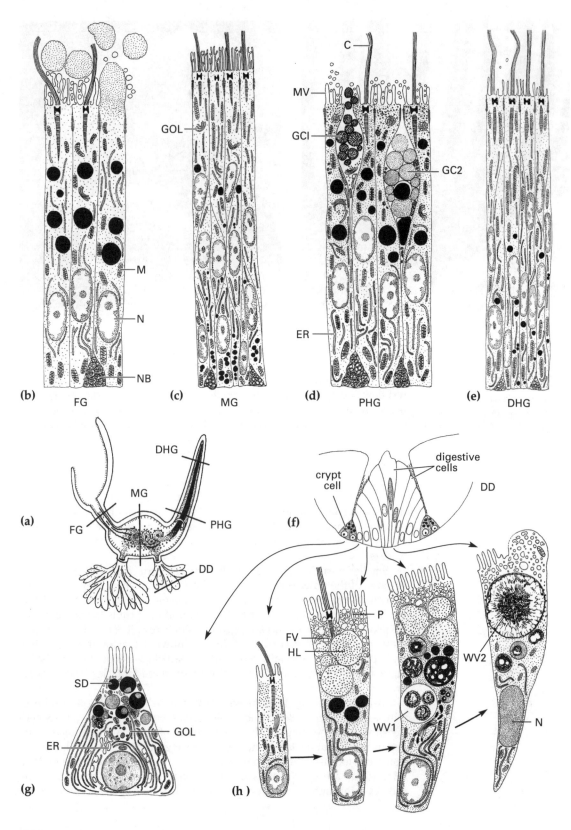

Figure 15.7 (opposite)
Brachiopods: details of the gut epithelium of *Calloria inconspicua*. **(a)** General anatomy and regions of the gut.
(b–e) Epithelial cells of the foregut (FH), midgut (MG), proximal hindgut (PHG) and distal hindgut (DHG).
(f–h) Epithelium of the digestive gland, with crypt cells (g) and digestive cells (h). Labelled organelles include nuclei (N), mitochondria (M), nerve bundle (NB), Golgi apparatus (GOL), cilia (C), microvilli (MV), goblet cells (GC1,2), endoplasmic reticulum (ER), secretory droplets (SD), pinocytotic vesicles (P), food vacuoles (FV), heterolysosomes (HL), and waste vacuoles (WV1, 2).
(Drawn by J.E. Morton.)

particles. Their contents are exchanged with the midgut by rhythmic contractions of circular muscles in the gut wall. In cross-section (Fig. 15.7f), they display lobes of digestive cells separated by crypt cells (Fig. 15.7g) containing large amounts of endoplasmic reticulum. The crypt cells are clearly sites of protein synthesis and secretion, while the digestive cells ingest material through apical vesicle formation. After a bout of heavy feeding, these cells (Fig. 15.7h) become engorged with huge waste vacuoles to the point where they rupture and spill their contents into the lumen. The basal nucleated portions remain *in situ* to restart the cycle.

Glandular goblet cells are scattered throughout the gut wall but are especially abundant in the lower hindgut where food and mucus are bound into a semi-solid bolus, which is rotated by the cilia of the distal hindgut. The purpose of the rotation is unclear. Stirring and mixing of food with digestive enzymes is likely to be enhanced. It also provides a mechanism for periodically purging the gut, as the entire bolus can be expelled through the mouth and ejected from the shell by the muscular 'sneeze'. It is not a direct response to the absence of an anus in articulate brachiopods, however, since similar structures occur in the guts of ectoprocts.

Circulation and excretion

Like phoronids, brachiopods contain two internal fluid-filled systems. The coelomic spaces (Fig. 15.4g), imperfectly separated, are the most extensive and contain free coelomocytes. Their contents are circulated sluggishly by muscular contractions and by peritoneal cilia. In addition, brachiopods have a system of fine blood vessels lying within the coelomic cavities. The contents of this system are circulated by one or more contractile vesicles, which are anchored to the dorsal mesentery over the stomach and connect with a mid-dorsal vessel. Anteriorly, this vessel communicates with haemal spaces around the oesophagus, continues into the mesocoel and terminates in vessels lying in the tentacle lumens. Posteriorly, the dorsal vessel delivers blood to the gut, muscles, gonads and nephridia, with further extensions into the mantle canals. Circulation appears to be completed through tissue spaces, so that the system is not closed and there must be some mixing of the blood and coelomic fluids. Oxygen is transported by the binding pigment haemerythrin, carried on corpuscles. All brachiopods also have one or two pairs of nephridia in the metacoel.

Nervous system

Brachiopods have an extensive nerve net and a circum-oesophageal nerve ring. The latter includes a supra-oesophageal ganglion that supplies nerves to the lophophore and a larger suboesophageal ganglion that supplies nerves to the muscles, pedicle and mantle. The sensory capabilities of brachiopods are poorly known but they respond to many stimuli, including sudden water motion, which may be monitored by the setae protruding around the shell margin.

Reproduction and development

Gametes form from undifferentiated peritoneal cells, and most species are gonochoristic. In inarticulates, testes and ovaries develop in the metacoel near the gut (Fig. 15.4b). In articulates, gonads develop in the mantle canals (Fig. 15.4g). Mature gametes are shed into the metacoel and exit through the nephridia, as in Phoronida. Inarticulate embryos develop

Figure 15.8 (opposite)
Ectoprocts: **(a)** Ctenostome zooids with lophophores extended, partially and fully retracted, and resorbed to provide a brood chamber. **(b, c)** Planktotrophic and lecithotrophic larvae. **(d)** Cyclostome zooids, showing extra coelom. **(e)** Cheilostome zooids, showing frontal membrane, hinged operculum and brood chamber. (After Ryland 1970.)

into planktotrophic larvae that resemble miniature adults, complete with protective shell, rudimentary pedicle (except for *Crania*) and lophophore of three paired tentacles (Fig. 15.4c). Articulate larvae have a very different appearance (Fig. 15.4d). These lecithotrophic trochophore-like larvae are brooded, often in the mantle cavity, sometimes in special pouches. Once liberated, they swim briefly before descending to the substratum and undergoing a complex metamorphosis. The swimming stage has three parts to the body. The posterior lobe becomes the pedicle; the anterior lobe becomes the body and lophophore; and the middle section folds to enclose the body lobe and becomes the mantle that secretes the protective shell (Fig. 15.4e).

Phylum Ectoprocta

Sessile, colonial coelomates with bilaterally symmetrical zooids of trimeric construction; mesosomal lophophore; U-shaped gut with anus outside lophophore; nervous system with circumpharyngeal nerve ring; no excretory or circulatory system; separate sexes; coelomic gonads.

Ectoprocta are individually minute animals (< 0.5 mm long) that clone themselves to form colonies with a wide range of structures (Fig. 15.8). Only one genus, *Monobryozoon*, lives in a solitary state. Ectoprocts gained the common name of moss animals from the macroscopic appearance of the colony surface when the numerous lophophores are extended. Early naturalists attributed plant-like natures to them and they were lumped with other animal polyps in the unnatural group Zoophyta. After being confused with corals and hydroids, it was recognised that these polyp-like creatures have two openings to the gut instead of one as in cnidarians. For a while, this led to them being classified with ascidians. Eventually, they gained phylum status and, like brachiopods, are thought to be derived from phoronid-like ancestors.

Structure and function

All ectoproct colonies start as a single ancestrula zooid, which buds repeatedly to form a colony of interconnected genetic, though not necessarily morphological, replicates. The variety of growth forms within the phylum results in colonies of different size, shape (stoloniferous, arborescent, encrusting) and texture (gelatinous, chitinous, calcareous). A few specialised taxa have limited mobility. One Antarctic species forms spherical floating colonies and is truly pelagic, but most are sessile.

The major groups of ectoprocts

The diversity of life forms encountered in Ectoprocta is assigned to three classes which, based on molecular evidence, may not be monophyletic. Most are marine, but the class Phylactolaemata is exclusively freshwater and considered the most primitive. Phylactolaemes retain traces of a tripartite body plan (Fig. 15.1) and just one type of repeated zooid. The Class Stenolaemata encompasses mostly extinct orders of ectoprocts. These small zooids secrete tubular calcified exoskeletons and develop one type of polymorph. Living stenolaemates have an extra body cavity of pseudo-

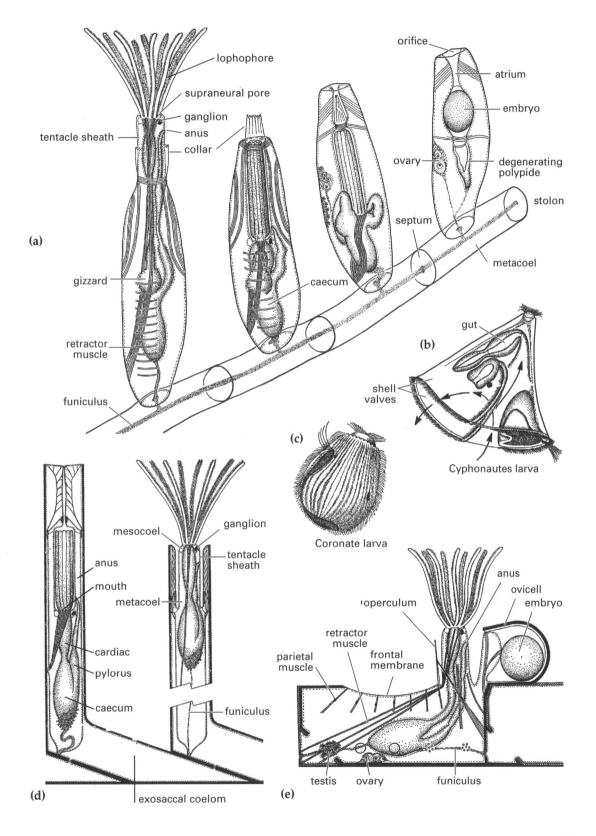

coelomic origin and a unique form of reproduction. The class Gymnolaemata is the most speciose class of living ectoprocts and contains two major subgroups, Ctenostomata and Cheilostomata. The ctenostomes typically have cylindrical zooids, stoloniferous growth forms, and chitinous exoskeletons. The cheilostomes typically have squat adherent zooids, encrusting growth forms, and exoskeletons that have been reinforced with calcium carbonate. This last group contains most of the living diversity in the phylum, and its success seems in part due to highly derived and polymorphic zooids.

Structure and function of autozooids

The basic unit of organisation in all ectoproct colonies is the zooid, which consists of soft tissues within some type of protective exoskeleton (Fig. 15.8a). All colonies contain feeding zooids, termed autozooids, which bear a ciliated lophophore. Although unitary in a functional sense, it is traditional to distinguish two parts in an autozooid: the cystid refers to the protective case and its associated peritoneum; the polypide refers to the lophophore, gut and associated organs. The cystid, once formed, is permanent, whereas the polypide can disintegrate and regenerate more than once during the life cycle of an autozooid. Some colonies contain modified zooids with degenerate or absent polypides, called heterozooids. They perform special functions (discussed below) and can survive only because of cellular connections within the colony that nourish these non-feeding zooids.

The polypide of the autozooid is the 'animal' recognisable as a lophophorate. Except in the Phylactolaemata, a protosome is always absent and the polypide consists of a mesosome bearing the lophophore and a metasome containing the U-shaped gut, gonads and musculature. Each of these body regions contains a fluid-filled cavity, and all these cavities communicate through a coelomopore in an imperfect transverse septum at the base of the lophophore. After the gut, the largest structures in the metacoel are the retractor muscles joining the transverse septum with the posterior wall of the cystid. These muscles pull the lophophore into the cystid.

The ectoproct body plan is easily seen in the cylindrical autozooids of a modern ctenostome (Fig. 15.8a). The everted lophophore in feeding posture protrudes through a terminal orifice in the chitinous cystid, on a tentacle sheath that is an extension of the metasome. When the retractor muscles contract, the tentacle sheath inverts to form a vestibule that accommodates the collapsed lophophore. Ctenostomes are recognised by a pleated collar, outside the everted tentacle sheath, that folds secondarily into a protective cone above the lophophore. As the tentacle sheath inverts, the hydrostatic pressure in the metacoel rises and fluid flows to its distal portion. This action causes the anterior body wall to swell, acting like a diaphragm to the orifice. Circular muscles complete the closure. To return to the feeding position, the circular muscles relax as transverse parietal muscles pull open the diaphragm. This action transfers fluid to the posterior portion of the metacoel and forces the tentacle sheath to evert as the retractor muscles relax.

Autozooids of all classes protect and extend their lophophores by variations on this common theme of muscular retraction and hydrostatic extension. Cyclostomes, which belong to the Stenolaemata, involve an extra body cavity in the process (Fig. 15.8d). An exosaccal coelom surrounds the polypide and closes the 'diaphragm'. Dilator muscles open the diaphragm, raising hydrostatic pressure in the exosaccal coelom, which compresses the metacoel and everts the tentacle sheath. The two models (one versus two fluid compartments) are functionally equivalent, but the latter may be more suited to the larger internal volumes of the elongate stenolaemate cystids. Autozooids in the Phylactolaemata, considered to reflect the primitive condition (Fig. 15.1), often share a common metacoel throughout the colony, which means that each zooid has little capacity to raise the internal hydrostatic pressure. Their lophophores are extended mainly by the contraction of the circular musculature in the local body wall, and they have relatively short introverts. Retraction is accomplished in the usual manner.

Cheilostomes have evolved mechanisms consistent with their squat adherent zooids and loss of near-radial symmetry (Fig. 15.8e). Unlike the cylindrical zooids of ctenostomes, the orifice is not distal but has migrated to the anterior of the frontal surface, where it is covered by a hinged chitinous operculum. The heavy calcification of most walls of the flattened cystid means that hydrostatic pressure changes in the metacoel can only be effected or compensated by a flexible frontal membrane. Transverse parietal muscles pull down on this membrane to extend the tentacle sheath and relax to accommodate the volume change when the lophophore is withdrawn by the retractor muscles. Because a naked membrane would expose the polypide to predators with piercing mouthparts, some ectoprocts have protective spines on and over the frontal membrane, while others have a calcified wall immediately below it. Many ectoprocts have the frontal wall completely calcified, and the hydrodynamic requirements are met by a special sac, the ascus, between the calcified exterior and the flexible membrane covering the metacoel. Pressure changes in the ascus, transmitted as the membrane moves up and down, are accommodated by allowing water to flow in and out through an ascopore in the calcified frontal wall.

Ectoproct lophophores are simpler than those of their larger phoronid and brachiopod relatives. The largest lophophores occur in the phylactolaemes, where they are horseshoe shaped and consist of a double row of tentacles (Fig. 15.1). In the more diminutive zooids of the other classes, the lophophore is circular and composed of a single row of tentacles around the mouth (Fig. 15.8). Ectoprocts are ciliary feeders, taking mainly diatoms. Mucus plays only a minor role in particle capture but may be more important in rejecting unsuitable material. Larger food particles reaching the base of the lophophore may be ingested by suction created by sudden dilation of the pharynx

Digestion

The gut (Fig. 15.8e) has the typical U-shape of all lophophorates. The mouth leads into a muscular pharynx with a ciliated rejection groove, then to a short unciliated oesophagus. The stomach is divided into three sections (cardiac, caecum and pylorus), before the gut ascends to the anus through a slender intestine. In some ctenostomes, the cardiac portion of the stomach is modified into a muscular gizzard in which cells capped with robust denticles rupture the testas of diatoms. The pyloric section is densely ciliated, and these cells are responsible for rotating an ergatula in the manner of brachiopods to mix the gut contents. Extracellular protein digestion takes place in the lumen, while intracellular phagocytosis occurs in all regions of the stomach. Dissolved and particulate nutrients are passed through to coelomocytes in the basal peritoneum. The latter wander throughout the coelomic spaces, circulated by peritoneal cilia, distributing materials and scavenging wastes.

Circulation and excretion

Unlike other lophophorates, ectoprocts lack excretory organs. As an autozooid ages, particulate wastes from intracellular digestion accumulate in the caecal dermis. When these reach critical levels, the whole polypide disintegrates. Some of the tissues of the old individual are phagocytosed and recycled but most of the stomach tissue is compacted into a pigmented brown body. This vacuolated structure may be retained within the metacoel of the new polypide that grows in the same cystid, or it may be engulfed in the new digestive system and expelled through the anus. The presence of multiple brown bodies inside the metacoels of zooids that do not expel them indicates that the cycle of decay and regeneration is a normal part of the life cycle of an autozooid.

Ectoprocts cope without organs for circulation. In all living species, a special cord of tissue, the funiculus (Fig. 15.8d), connects the outside of the stomach caecum to the rear body wall. In stoloniferous species the funiculus passes through pores in the base of the cystid, to join with a communal cord in the stolon. Starch metabolised into glycogen has been followed along funicular cords and may be stored there. Except in the stenolaemates, these cords are continuous throughout the colony and must be critical to the sustenance of non-feeding heterozooids which are a feature of many colonies. The funiculus may also contain nerve cells, although there is little evidence of stimuli being transmitted among zooids. Each polypide, however, has a simple nervous system consisting of a circum-pharyngeal ring with adventitious nerves, to coordinate the operation and protection of its own lophophore. The cystid wall also contains a diffuse nerve net.

Reproduction and development

All mature ectoproct colonies, though not all zooids, are hermaphrodite. Marine species are frequently protandric (male, then female), while the fresh-water phylactolaemes are simultaneous hermaphrodites, with ovaries developing on the body wall and testes on the funiculus. Gametes are derived from peritoneal cells and pass into the metacoel. Since this cavity lacks any permanent communication with the exterior (no nephridia), some species form a temporary supraneural pore between the mouth and the dorsal tentacles. Others fuse the bases of a pair of dorsal tentacles to form an intertentacular organ. Sperm have been observed escaping through fine pores on the tips of specialised tentacles. Since all of these adaptations involve escape from the lophophore, gametes must pass first from the metacoel to the mesocoel, presumably through the coelomopore.

Fertilisation has been described as both internal in the metacoel and external in the water column. A recent survey of a diverse range of gymnolaemates has shown a common mode of fertilisation. Sperm enter the metacoel through one of the temporary openings in the body wall and fuse with primary oocytes in the ovaries, but egg activation is delayed until the eggs are spawned. In species in which larvae develop outside the metacoel (which is the majority of species), the egg is activated only after it has passed through the body wall. Some species release eggs one at a time, which means that the fertilisation membrane may not be raised until three months after the initial fusion of sperm and egg. This unusual mechanism allows gymnolaemates to fertilise a very high proportion of their eggs and to make the most efficient use of batches of sperm captured by the lophophoral currents.

A few taxa activate their eggs and brood the embryos within the metacoel. Presumably, escape of these larvae involves rupture of the maternal coelom and regeneration of the polypide. Phylactolaemes incubate the embryo within a sac formed from an invagination of the body wall but it is not clear whether the egg has to pass externally to be activated. Most ectoprocts brood their embryos externally by various means. Ctenostomes typically extrude a single egg through the supraneural pore and incubate the embryo within the tentacle sheath (Fig. 15.8a). In this case the polypide disintegrates, providing nutrition to the embryo and effectively converting the cystid into a temporary gonozooid. Stenolaemates follow a similar pattern, except that the spacious brood chamber is a specialised heterozooid. This is associated with a rare form of reproduction, polyembryony, in which the zygote divides repeatedly to produce as many as 100 identical embryos. Cheilostomes brood their embryos externally, either in the space under the arch of spines covering the frontal membrane or in temporary ooecia (helmet-shaped brood chambers) developed on the anterior wall (Fig. 15.8e). It appears that the ooecium has a special internal chemistry, because immature embryos cannot survive independently. Maternal nutrition may explain the development of very large embryos.

The large, non-feeding coronate larvae of cheilostomes (Fig. 15.8c) travel only short distances before seeking settlement sites, which are often located by responding to very specific chemical cues. Once attached, all internal tissues in the larva are autolysed and regenerated during metamorphosis. Only the epidermis is retained and becomes the cystid of the founding ancestrula. As in Brachiopoda, some ectoprocts produce planktonic larvae from non-brooded eggs. These larvae are called cyphonautes and are enclosed by triangular shells of chitin, which are lateral as in bivalves (Fig. 15.8b). They possess a mouth and U-shaped gut, allowing them to remain in the plankton for extended periods and may represent a primitive condition of the life cycle among marine ectoprocts.

The phylactolaemes have evolved another variation of asexual reproduction to cope with the uncertainty of freshwater environments. They produce resting cells within spore-like chitinous capsules that are highly resistant to external stress. These statoblasts are formed on the funiculus below the gut. In some species, the statoblasts remain in the cystid after the death of the polypide and regenerate the colony in situ when conditions improve. In others, the statoblasts are dispersed by wind or animal transport to other drainages, which gives them dispersal powers denied to taxa with swimming larvae.

Tissue regeneration in lophophorates

Phoronids have great powers of regeneration, and their reconstitutive powers are not localised. An early experiment showed that an adult cut into six parts grew into six new individuals. Nonetheless, the most common use of this ability is to cast off the lophophore after its partial loss to predators and regenerate a new one. Excision is accomplished by a strong contraction of the circular muscles in the upper trunk, which amputates the lophophore along with anterior blood vessels, nerve ganglia, nephridia, mouth, and anus. In some species the cast-off portion will regenerate the metastome and internal organs, providing another form of asexual multiplication. In at least one species, regeneration has been incorporated into a regular overwintering strategy. In unfavourable conditions the bodies of the animals

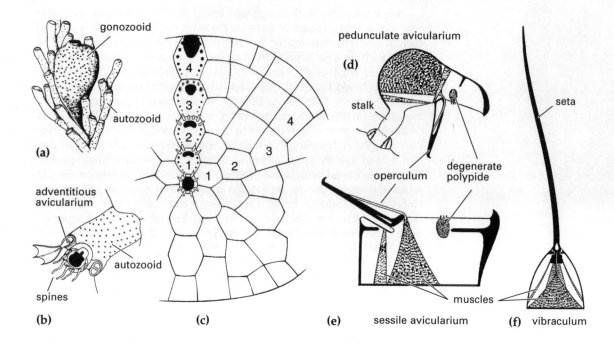

Figure 15.9
Ectoprocts: (a, b) Examples of polymorphic zooids in cyclostomes and cheilostomes. (c) Four cycles of asexual replication from a central ancestrula zooid in a cheilostome colony. (d–f) Detail of cheilostome zooids modified for colony defence. (After Harvell 1984 and Ryland 1970.)

degenerate, leaving just a few cells deep within each phoronid tube, which produce a new generation of adults when conditions become more favourable.

Tissue regeneration is a property shared with the Ectoprocta, which eliminate and regenerate the polypide for a variety of functions. In some taxa the cystid is converted to a temporary brood chamber (Fig. 15.9a). More generally, regeneration in ectoprocts provides a mechanism for dealing with metabolic wastes in a body that lacks excretory organs. It also allows the cloning of zooids to form colonies. Because fission is a multiplicative process, colonies grow geometrically without the physiological constraints that limit the size of solitary animals: i.e. decreasing surface-to-volume ratio. Given no external threats, a colony is theoretically immortal, and there are records of some expanding to millions of zooids. As sessile organisms, ectoprocts compete for space and access to the water column. Rapid growth and early maturation are attributes that have made many of them successful colonists of bare space.

Growth patterns and polymorphism in ectoprocts

All ectoproct colonies expand by adding new zooids at the margin, albeit through various growth forms. Ctenostomes grow and then partition apical buds to produce a segmented stolon. As each new segment is defined, several buds grow laterally on its surface and develop into autozooids (Fig. 15.9b). Encrusting cheilostomes may also have giant dividing buds near the leading edge of the colony, but most consist of close-packed hexagonal zooids (Fig. 15.9c). All sequences result in the oldest zooids being near the founding ancestrula, where they are commonly smaller and less complex.

Ectoproct colonies expand by subdividing the marginal zooids. In phylactolaemes this cloning produces identical autozooids. In other classes it results in polymorphic heterozooids capable of functions other than feeding. Cyclostomes, for example, modify some zooids at branching points in the colony into vase-like gonozooids, which incubate sexually-produced embryos. Gymnolaemates display as many as six polymorphisms from a common genotype, and this plasticity seems to be a large element in their dominance of the modern fauna.

In modern gymnolaemates, the most important trend is the formation of polymorphic heterozooids. These are of five main forms. Kenozooids have very simple contents and no polypide. They function as stolons, rhizomes, and packing cells that allow colonies to change shape. The pore plates that allow cellular connections through the calcified double walls of cheilostomes are highly derived kenozooids. Spinozooids have the frontal surface elongated into one long rigid spine. Gonozooids include the simple brood chambers of the Stenolaemata and the temporary ooecia of cheilostomes (Fig. 15.8e) that are produced cooperatively by contiguous zooids. Avicularia contain an aborted polypide that develops instead into a sensory organ bearing a tuft of setae (Fig. 15.9e). Their main content is enlarged muscles, which operate an operculum modified as a mandible to trap animals moving over the colony surface. Some stalked avicularia resemble grotesque bird heads that swivel and snap like pedunculate brachiopods (Fig. 15.9d). Vibracula have the operculum elongated into a chitinous bristle that is waved about by basal musculature (Fig. 15.9f), discouraging the settlement of other organisms.

This diversity of form and function is produced from a common genotype by differential growth during ontogeny. The greatest number of polymorphisms occurs in the cheilostomes, where the hinged chitinous operculum is the basis for many forms. The development route followed by a young zooid is influenced by a variety of factors, including internal physiological gradients, related to its position within the colony. Zooids that are constricted by space or damage and thus smaller than an average autozooid will develop *in situ* as a sessile heterozooid. Others may grow adventitiously on the frontal surface of autozooids (Fig. 15.9b), especially the pedunculate avicularia and spinozooids.

Induced growth responses

Heterozooids that protect against predators can be induced in ectoprocts by external cues. For example, *Membranipora membranacea* will produce autozooids flanked by four spinozooids and grow spines over the frontal membrane within 48 hours of receiving as little as one hour of exposure to waterborne chemical cues from a nudibranch predator. The spinozooids are permanent but do not develop until induced, and then form only around maturing zooids behind the zone of proliferation. The theory of inducible defences (observed in many plants and lower animals) predicts that large spines should not be programmed into colony growth because they are energetically costly, with defended colonies having reduced growth and earlier senescence than undefended conspecifics. Another circumstance favouring the induction of morphological defences is that single instances of predation rarely destroy the whole colony.

Special heterozooids are also formed at the margin when colonies encounter other encrusting organisms, especially other ectoprocts. For example, *M. membranacea* colonies produce long stolons when they contact smaller conspecifics. These grow across the peripheral zooids of the potential competitor and slow their advance. The response is different when the colonies are of equal size, which implies that the marginal zooids can assess the size of another colony through the chemistry of the contact zone. The vigour of the response also depends on the absolute size of the responding colony, which may reflect the lower cost per individual of producing a response. Other taxa lift their growing edge by extending terminal stolons over the first few rows of zooids of a competing colony. Sometimes this overgrowth is blocked by the other producing stubby stolonal outgrowths that are vertical extensions of the frontal walls of zooids, just beyond the reach of the adventitious stolons. Although some taxa are consistently dominant over others, many encounters between ectoproct colonies end in stalemate, with both redirecting their growth to their free margins. Unlike encounters involving sponges (Chapter 2), there is little evidence for the use of allelochemicals to convert growth hierarchies into competitive networks, and terminal stolons cannot be equated with the stinging mesenterial filaments employed when different scleractinian colonies make contact. The difference might relate to the relatively short life expectancies of ectoprocts.

Biologically active chemicals in lophophorates

The dynamic competitive responses and long evolutionary history of marine Ectoprocta point to them as logical places to find biologically active chemicals. Even though allelochemicals have not been implicated in competitive interactions involving ectoproct colonies, a number of powerful antifouling agents have been isolated. Another group of compounds, known as bryostatins, are particularly cytotoxic and are now in advanced clinical trials as potential human medicines. It is not clear whether these have a natural function in competition or are simply potent antibiotics. Their discovery, however, has focused the attention of bioprospectors on the phylum Ectoprocta, with recent reports of the discovery of active alkaloids and more to be expected.

Modern brachiopods may also make limited use of chemical defences, because bioerosion is rarely observed in their calcareous shells. This may be one of the multiple functions of the punctae which join the mantle tissue and the external periostracum of the shell. Their soft tissues must contain unpalatable chemicals, since they are nearly always rejected by fish and asteroids in feeding trials. For space competition, however, articulate brachiopods rely upon their erect posture and swivelling movements to avoid being overgrown by sessile organisms. In New Zealand waters there are places where *Calloria inconspicua* is the only sessile organism able to resist the spread of large encrusting sponges on subtidal rockfaces. The tiny portion of the almost two-dimensional living space occupied by the brachiopod pedicles may explain why sponges have not evolved specific allelochemicals against brachiopods while showing activity against the more aggressive ectoprocts in the same environment.

Classification of the lophophorates

Details of the distinguishing features of the classes and important orders are given in the relevant sections of the chapter.

PHYLUM PHORONIDA

PHYLUM BRACHIOPODA

Class Inarticulata
    Orders Lingulida, Acrotretida

Class Articulata

PHYLUM ECTOPROCTA

Class Phylactolaemata

Class Stenolaemata
    Order Cyclostomata

Class Gymnolaemata
    Orders Ctenostomata, Cheilostomata

Chapter 16

The Echinodermata

M. Byrne

PHYLUM ECHINODERMATA

Introduction 367

Echinoderm body plan and external characteristics 368

Diversity 369
- Class Asteroidea 369
- Class Ophiuroidea 370
- Class Echinoidea 371
- Class Holothuroidea 373
- Class Crinoidea 374
- Class Concentricycloidea 374

Evolutionary history 375

Form and function 375
- Water vascular system 375
- Body wall and skeleton 377
- Mutable connective tissues 378
- Locomotion 380
- Feeding and digestive system 381

Physiological processes 385
- Digestion 385
- Respiratory exchange and circulation 386
- Excretion 386
- Haemal system 387
- Nervous system 387
- Sensory structures 388

Reproduction and development 388
- Reproductive system 388
- Gametes and fertilisation 389
- Development 390
- Evolution of life histories 392
- Asexual reproduction 393

Classification of the phylum Echinodermata 394

Phylum Echinodermata

Bilaterally symmetrical (as larvae), becoming radially symmetrical pentamerous (as adults), coelomates of trimeric construction (obscured in adult); calcareous endoskeleton of mesodermal ossicles; coelomic water vascular system; nervous system diffuse, radially arranged; no distinct excretory or circulatory system; separate sexes, or hermaphrodites; coelomic gonads, simple gonoducts; deuterostome development, with ciliated larval stages.

Introduction

Echinoderms, with their distinctive shape and brilliant colours, form a conspicuous component of the invertebrate fauna of the world's oceans (Figs 16.1a–f, 16.2a–d). There are six extant classes: Asteroidea (sea stars or starfish), Ophiuroidea (brittle stars), Echinoidea (sea urchins), Holothuroidea (sea cucumbers), Crinoidea (feather stars) and the recently discovered Concentricycloidea (sea daisies). Echinoderms are often the dominant organisms on the sea floor and are ecologically important in many marine communities.

Figure 16.1
(a, b) Class Asteroidea: (a) Two cushion stars, *Patiriella exigua* and *P. gunnii* (left and right), and the multi-armed sun star *P. calcar* (centre). (b) Sea star, *Nardoa novaecaledoniae*.
(c, d) Class Ophiuroidea: (c) Brittle star, *Amphipholis squamata*. (d) Basket star, *Gorgonocephalus caryi*. **(e, f)** Class Echinoidea: (e) A regular echinoid, *Strongylocentrotus franciscanus*. (f) An irregular echinoid, *Clypeaster japonicus*, showing petaloid ambulacra. (a from Byrne and Anderson 1994; b courtesy Dr J. Keesing.)

INVERTEBRATE ZOOLOGY

Figure 16.2
(a, b) Class Holothuroidea:
(a) An aspidochirotid, *Stichopus variegatus*, deposit-feeding.
(b) A dendrochirotid, *Eupentacta quinquesemita*, suspension feeding. (c, d) Class Crinoidea:
(c) A feather star, *Florometra serratissima*, with arms in suspension feeding posture.
(d) A stalked crinoid or sea lily with arms in a parabolic fan for suspension feeding. (a courtesy Dr M.J. Kingsford; c from Byrne and Fontaine 1981; d from Young and Emson 1995.)

Echinoderms have many unique features. These include pentamerous radial symmetry; a complex hydraulic system of water-filled tubes called the water vascular system, which has numerous functions; connective tissues that have mutable (changeable) properties; and a dermal skeleton which gives rise to projecting spines. Despite their unique adult morphology, a link between echinoderms and their deuterostome kin is clearly seen in the structure of the embryos and larvae.

Echinoderm body plan and external characteristics

Although the shape of echinoderms varies a great deal, the body plan of the extant classes is based on pentamerous radial symmetry. Extinct groups had various symmetries. The echinoderm body is organised into five radii or ambulacra with intervening regions called interambulacra (Fig. 16.3a). Each ambulacrum has a branch of the water vascular system which gives rise to tube feet (podia), and a radial nerve cord (Figs 16.3b, c, 16.5b). In contrast to the adults, echinoderm larvae are bilaterally symmetrical (Figs 16.19j, k, 16.20, 16.21). The formation of the adult body is associated with a dramatic metamorphosis.

The external surface is often covered by an array of spines and other calcareous structures. The structure of the skeleton differs greatly between species and is a key character in echinoderm taxonomy. The surface bearing the mouth is called the oral surface and the opposite surface is called the aboral surface. In many echinoderms the anus lies on the aboral surface. Asteroids, ophiuroids and echinoids have the oral surface directed towards the substratum (Fig. 16.1a–f), while crinoids have their mouth directed upwards (Fig. 16.2c, d). Holothuroids can have the mouth directed upwards or downwards (Fig. 16.2a, b).

THE ECHINODERMATA

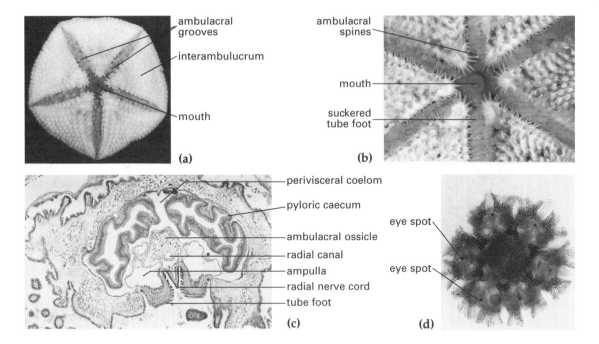

Figure 16.3
(a) Oral side of a sea star, *Patiriella vivipara*, with five ambulacra and suckered tube feet. (b) Oral side of *Patiriella gunnii*, showing mouth, ambulacra, tube feet and protective spines. (c) Cross section of an arm of *Patiriella vivipara*, showing ambulacral system and pyloric caeca in arm coelom. (d) Juvenile *Patiriella gunnii*, showing eye spots on terminal tube feet.

Diversity

Class Asteroidea

Sea stars are among the most conspicuous echinoderms (Figs 16.1a, b, 16.3a). Epibenthic sea stars in the orders Valvatida (e.g. *Patiriella*), Spinulosida (e.g. *Acanthaster*) and Forcipulatida (e.g. *Asterias*) are the most familiar asteroids. Burrowing sea stars in the order Paxillosida (e.g. *Astropecten*) live in and on soft sediments. In many marine communities asteroids are the top predators and their special taste for bivalves often puts them in conflict with aquaculturalists and shellfisheries.

Most asteroids have five arms emerging from the central disc, but multiarmed species having up to 40 arms are also common. The shape of sea stars varies from the star-shaped profile of species that have distinct arms to the cushion shape of species with short, inconspicuous arms (Fig. 16.1a,b). Large species such as the crown-of-thorns starfish, *Acanthaster planci*, have arms up to 350 mm long, while the smallest known asteroid, *Patiriella parvivipara*, has arms 5 mm long.

The central mouth on the oral side of the disc is surrounded by a tough peristomial membrane comprised of connective tissue (Fig. 16.3a, b). A circle of spines is usually present which protects the mouth region. Ambulacral grooves radiate from the peristomium along the oral surface of the arms (Fig. 16.3b). A cross-section of the arm shows a series of ossicles that support the body wall around a spacious arm coelom (Fig. 16.3c). On the oral side of the arm, two rows of ambulacral plates meet in the midline to form a furrow containing the ambulacral groove. The radial water canal runs along the outer surface of the groove. Two or four rows of tube feet are present. A series of movable spines along the arm forms a protective cover over the groove when the tube feet are retracted (Fig. 16.3b). The terminal tube foot at the end of the arm carries a red eyespot (Fig. 16.3d). Epibenthic asteroids have suckered tube feet (Fig. 16.3b, c) and move efficiently across hard substrata, while burrowing species have pointed, suckerless tube feet which are used to excavate sediments.

The aboral surface is usually covered by spines and papulae, and has a central opening for the anus and interambulacral openings for the gonopores. Papulae are short projections of the body wall that emerge between adjacent ossicles, and have a respiratory function. *Acanthaster planci* has particularly long aboral spines which can cause a painful toxic reaction if touched. One or more madreporic plates are also present on the aboral surface (Fig. 16.1a). The madreporite is a distinct, perforated plate that is the fluid inlet for the water vascular system. Forcipulate and valvatid asteroids also have specialised skeletal structures called pedicellariae on the aboral surface. These jaw-like structures can crush zooplankton, and are used to prevent the settlement of epifauna. In some species, pedicellariae are used to catch prey, including small fish. Paxillosid sea stars have distinct borders of large plates along the arms.

Class Ophiuroidea

The Ophiuroidea are stellate echinoderms in which the central flattened disc is sharply set off from the arms (Fig. 16.1c). Most ophiuroids have five simple arms, but six- and seven-armed species also occur. The arms are often long and sinuous, and the ease with which they are spontaneously shed (autotomised) gives ophiuroids their common name, brittle stars. Euryaline species have arms with various degrees of branching; the basket stars have up to 20 highly branched arms (Fig. 16.1d).

Ophiuroids are the most speciose echinoderms and also the most abundant. They commonly dominate the benthos. Where food is readily available, aggregations of thousands of individuals carpet the sea floor. Most ophiuroids are members of the order Ophiurida (e.g. *Ophionereis*) which is characterised by a well-developed skeleton (Fig. 16.5a). The less familiar ophiuroids in the order Phrynophiurida (e.g. *Gorgonocephalus*) which includes the basket stars, are covered by a fleshy skin. The disc ranges in diameter from 2 to 100 mm while the arms range in length from 10 to 650 mm. With their multibranched arms, basket stars are the largest ophiuroids.

Figure 16.4
Oral surface of a brittle star, *Ophiothrix spongicola*, showing the star-shaped mouth, tube feet and bursal slits.

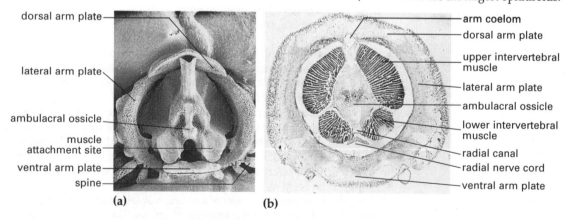

Figure 16.5
(a) Arm skeleton of *Ophiothrix fragilis*, showing the central ambulacral ossicle or vertebra, peripheral arm plates and muscle attachment areas. (b) Cross-section of an arm of *Ophiarachnella ramsayi*, showing the skeleton, muscles, radial canal and nerve cord. (a from Stauber and Markel 1988.)

Ophiuroids lack an anus. The aboral side of the disc has a continuous cover of skeletal plates (Fig. 16.1c). Large skeletal plates called radial shields (Fig. 16.1c) are positioned at the edge of the disc near the base of each arm and are important in ophiuroid taxonomy. The oral surface of the disc has five jaws and a star-shaped mouth (Fig. 16.4). There are also five oral shields, one of which is usually modified to form the madreporite. Two pairs of buccal tube feet extend into the mouth space. Bursal slits, one at either side of the base of each arm, are also usually present on the oral surface (Fig. 16.4). These are the openings of the respiratory bursae, which are sac-like invaginations of the body wall projecting into the coelom.

Ophiuroids of the order Ophiurida have arms with a segmented appearance due to the repeated series of skeletal plates (Fig. 16.1c). In each arm segment the central ambulacral ossicles or vertebrae are surrounded by peripheral plates — the dorsal (aboral), ventral (oral) and lateral arm plates (Fig. 16.5a, b). The central suture in the vertebrae provides evidence that they evolved by fusion and internalisation of a pair of ambulacral ossicles, homologous to those of asteroids. In the soft-skinned phrynophiurids the skeleton is embedded in the integument, while in the basket stars it is reduced to small ossicles. Ophiuroids have a series of spines extending from the sides of the arms which connect with articulating surfaces on the lateral arm plates (Fig. 16.1c). The radial water canal lies internal to the oral arm plate and occupies a groove in the ambulacral ossicles (Fig. 16.5a, b). On the aboral side of these ossicles another groove accommodates the arm coelom (Fig. 16.5a,b). The tube feet emerge on the oral side of the arms through pores between the arm plates. Ophiuroid tube feet do not have suckers. They are used for feeding, respiration and attachment.

Class Echinoidea

The Echinoidea includes the familiar sea urchins (Order Echinoida, e.g. *Heliocidaris*) which inhabit rocky reefs, and the irregular urchins (heart urchins, Order Spatangoida, e.g. *Echinocardium*), and sand dollars (Order Clypeasteroida, e.g. *Clypeaster*) which inhabit soft sediments (Fig. 16.1e, f). Sea urchins are particularly numerous in areas with abundant algae. Their grazing activity is ecologically important and can change the benthos from a lush kelp forest to an urchin-dominated barren ground. Some sea urchins, particularly species of wave-swept habitats, occupy burrows of their own construction and remain fairly sessile, feeding on drift algae. The burrows are excavated by the scraping action of the teeth and the spines. Bioerosion caused by the burrowing activity of sea urchins is a major structuring force in intertidal and subtidal reefs and is important in the dynamics of sedimentation.

The echinoid body is supported by a rigid test covered in spines (Fig. 16.6a–d). The test is an intricate series of closely fitting skeletal plates constructed with an exquisite geometry. Pentamerous radial symmetry is evident in the 10 double rows of alternating ambulacral and interambulacral plates (Fig. 16.6b). The ambulacral plates have minute pores through which connections to the tube feet extend. The radial canals of the water vascular system run along the internal surface of the test (Fig. 16.15). Echinoids range in size from the miniature sand dollars, 10 mm in diameter, to the large commercial species that have test diameters of 120 to 150 mm. Several species of sea urchins are harvested for their gonads, which are considered a delicacy in many countries. Pedicellariae are present on the surface of all echinoids and, like those of asteroids, are used to prevent the settlement of

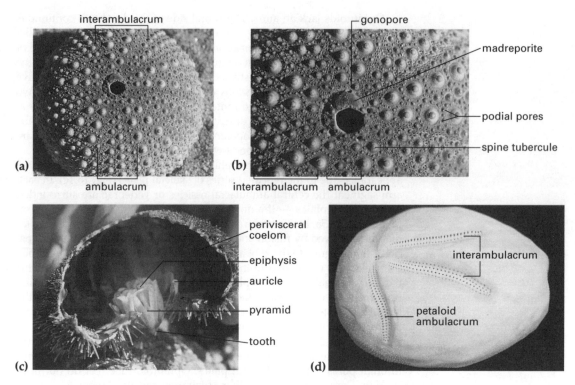

Figure 16.6
(a) Aboral view of the cleaned test of a sea urchin, *Heliocidaris erythrogramma*.
(b) Apical system of *Heliocidaris erythrogramma*, showing the genital plate modified as a madreporite, gonopores and podial pores. (c) Interior of *Strongylocentrotus droebachiensis*, showing Aristotle's lantern complex and coelom. (d) Cleaned test of a heart urchin, *Brissus agassizi*, showing petaloids.

epifauna (Fig. 16.7a, b). Echinoid pedicellariae are intricate structures with a long movable stalk and an apical, tripartite jaw. The large pedicellariae of *Toxopneustes pileolus* have an associated venom gland and cause a painful reaction if touched.

Regular echinoids have a distinct oral–aboral axis with the mouth directed towards the substratum and the anus directed upward (Figs 16.1e, 16.6a, c). The test is globose and is covered by spines which are used in defence and locomotion. Sea urchins also have elongate locomotory tube feet which end in suckers and often extend beyond the spines (Fig. 16.7a). Five pointed teeth project from the mouth opening and form part of a jaw apparatus called the Aristotle's lantern (Fig. 16.6c). The lantern is one of the most distinctive features of echinoids. The mouth is surrounded by a tough peristomial membrane which also has associated buccal tube feet and small bushy structures called gills. At the aboral pole, a periproct supports the anal opening and is surrounded by a series of plates (Fig. 16.6a). These plates include five terminal plates which are aligned with the ambulacra and five genital plates which are aligned with the interambulacra (Fig. 16.6b). Each genital plate has an opening for a gonoduct, and one plate is modified to form the madreporite.

Irregular echinoids include the heart urchins and sand dollars (Figs 16.1f, 16.6d). These echinoids have some degree of bilateral symmetry

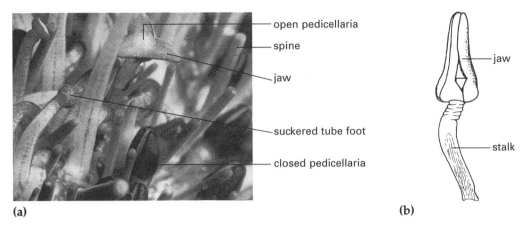

Figure 16.7
(a) Open and closed pedicellaria, spines and suckered tube feet of *Strongylocentrotus purpuratus* (Echinoidea). **(b)** Sea urchin pedicellaria. (a from Anderson 1996.)

superimposed on the radial plan. The formation of an anterior–posterior axis is associated with an infaunal, burrowing lifestyle. In heart urchins the mouth and peristome have moved anteriorly and the anus and periproctal plates have move posteriorly. In sand dollars the mouth is positioned at the centre of the oral surface while the anus has migrated posteriorly. Associated with the change in position of the anus, the ambulacral systems of irregular echinoids form a flower-like arrangement of petaloids (Figs 16.1f, 16.6d). The spines of irregular echinoids are small and function in locomotion like a series of paddles. The spines also form a protective cover over ciliary tracts used for respiration. The buried lifestyle of irregular echinoids is associated with the diversification of podial structure and function. Feeding podia vary in shape from leaf-like to flower-shaped, depending on the mode of feeding, while respiratory podia are flat and thin-walled. The tube feet are also used to build a funnel that enables the animal to maintain a connection with the surface of the sediment.

Class Holothuroidea

Holothuroids differ from other echinoderms in having a soft integument containing small ossicles embedded in connective tissue (Fig. 16.11). These echinoderms are elongated along their oral–aboral axis and are thus commonly know as sea cucumbers (Fig. 16.2a, b). The number of tentacles around the mouth ranges from 10 to 30. Internally the tentacles are associated with a calcareous ring of ossicles, which are the largest skeletal elements in holothuroids (Fig. 16.16a, b). Members of the order Dendrochirotida (e.g. *Cucumaria*) extend an arborescent array of finely branched tentacles into the water column to feed on suspended particles (Fig. 16.2b), while members of the order Aspidochirotida (e.g. *Holothuria*) and the order Apodida (e.g. *Leptosynapta*) use short tentacles to feed on benthic deposits (Fig. 16.2a). Where water motion ensures a constant supply of food, dendrochirotids carpet the sea floor, with thousands of tentacles creating a delicate arborescent cover. The pelagic holothuroids of the order Elasipodida are unique among echinoderms in living in the water column. Some of these have a medusa-like appearance. The smallest sea cucumbers are the worm-like apodids which are just over 10 mm in length, while the largest aspidochirotids are over a metre long. These large sea cucumbers are a spectacular feature of Indo–Pacific reefs. Many tropical and a few temperate sea cucumbers are harvested commercially for the body wall, which is highly prized in Asian countries as beche-de-mer or trepang.

Despite their elongate form, most sea cucumbers have a pentamerous body plan. Many species in the orders Aspidochirotida and Dendrochirotida have five ambulacra with tube feet, running from the oral region to the anus. These are often organised into three ambulacra of locomotory tube feet directed towards the substratum and two upper ambulacra of tube feet which may or may not be reduced. The upper surface of aspidochirotids is often covered by protuberances called papillae, which are modified tube feet. Some species have tube feet scattered all over the body surface, but the Apodida have none at all.

Class Crinoidea

The most striking feature of crinoids, commonly known as feather stars, is their array of feather-like arms (Fig. 16.2c, d). The arms range in number from five to several hundred and have delicate side branches called pinnules. Unlike most echinoderms, crinoids have the mouth directed upwards. The anus is also on the oral surface, elevated on an anal cone. Crinoids are divided into two main groups: the order Comatulida (e.g. *Comanthus*) which includes the feather stars, attached to the substratum by hook-like cirri (Fig. 16.2c), and the order Isocrinida (e.g. *Metacrinus*) which includes the sea lilies, attached to the sea floor by a long stalk (Fig. 16.2d). Feather stars are a conspicuous and colourful component of the fauna of many coral reefs. Sea lilies are found only in deep water. Crinoids are an ancient group, and the species extant today are a small remnant of the diversity seen in the fossil record.

The body of a crinoid, the tegmen, is supported by a cup-like calcareous skeleton called the calyx. Both the cirri of feather stars and the stalk of sea lilies have serial skeletal elements, giving them a segmented appearance. This is also true of the arms. In crinoids the pentaradial symmetry is clearly seen in the five basal regions of the arms as they emerge from the tegmen. Subsequent branching of the arms is based on multiples of five. The radial water canals are internal to the food groove and follow the branches of arms, with terminal branches along the pinnules (Fig. 16.9a). The tube feet of crinoids are minute and are used in suspension feeding (Fig. 16.9b, c).

Class Concentricycloidea

Concentricycloids, or sea daisies, were discovered in 1986 and are still only known from one genus (*Xyloplax*) and two species. They are minute, medusa-like echinoderms with a maximum diameter of 1 cm. The late discovery of the concentricycloids is due to their specialised and inaccessible habitat. Thus far they have only been found in association with wood which has been submerged for some time in deep water. The name Concentricycloidea comes from the presence of two concentric water rings on the outer edge of the disc. The presence of a second oral water ring and the absence of an ambulacral system make these very unusual echinoderms. One species of *Xyloplax* has a simple sheet-like gut which is thought to digest material from decomposing wood, while the second species has a simple blind-ending stomach and can take larger portions of food. Concentricycloids are also unusual in being sexually dimorphic, with the males having a copulatory organ. Fertilisation appears to be internal. The young are brooded in pouches which may be derived from the gonads.

The status of the Concentricycloidea is controversial, with some researchers contending that they are highly derived asteroids. With so many unique features, however, they will probably remain in a separate class until more specimens can be found. The most important information on their relationships with other echinoderms will come with the application of modern molecular phylogenetic techniques. Knowledge of their development would also help in assessing the affinities of these echinoderms.

Evolutionary history

Echinoderms have a long fossil record, extending to the early Cambrian. Because of their calcareous skeleton, fossil echinoderms are often beautifully preserved. At least 15 echinoderm classes are known only from the fossil record, and there are many more fossil species than the species extant today. One feature that helps to designate echinoderm fossils is the ambulacral system of canals. Although modern echinoderms have a body plan based on five ambulacra, many fossil groups had varying arrangements including groups with one, two or three ambulacra. Many fossil echinoderms were attached to the sea floor with their oral surface directed upwards, as seen in modern crinoids. The number of extinct groups, their classification, and how they relate to extant echinoderms and stem chordate groups are major sources of controversy in echinoderm palaeontology. With each discovery of an important new fossil these relationships are revisited.

Of the extant echinoderm classes, the fossil record is most complete for crinoids and echinoids. Crinoids have an impressive species diversity prior to the Permo-Triassic extinction, 250 million years ago, and dominated the fossil record of echinoderms during the Palaeozoic. Most became extinct at the end of the Triassic, 210 million years ago. Crinoids today are a relatively small group, but living feather stars and sea lilies provide important insights into the way of life of the extinct forms. The sedentary, suspension-feeding lifestyle we see today in crinoids with an upwardly directed ambulacral system is considered to reflect the way of life of early echinoderms. Thus, the original functions of the water vascular system were particle collection and respiratory exchange.

The fossil record of echinoids indicates that the benthic grazing lifestyle of these echinoderms is also ancient. Early fossil echinoids had more flexible tests than modern species. Asteroids, ophiuroids and holothuroids have comparatively small skeletal elements which separate soon after death. As a result, the bodies of these echinoderms would not have fossilised well. There is clear evidence, however, that their early evolution took place in the Ordovician.

Form and function

Water vascular system

The water vascular system is derived from the coelom and consists of a series of tubes and channels which form a complex hydraulic system. From their presumed original function as simple respiratory and suspension-feeding structures, tube feet have evolved a broad range of functions in feeding, locomotion, gas exchange, excretion and sensory reception.

The organisation of the water vascular system is most easily illustrated in sea stars (Fig. 16.8). The components of the system are the madreporite, stone canal, oral water ring, radial water canals, ampullae and tube feet. Water enters and leaves the water vascular system through the madreporite, a sieve-like plate usually located on the aboral surface (Figs 16.1a, 16.8). Pore canals in the madreporite open into the stone canal, which in turn connects with the oral water ring. The stone canal is thought to function as a pump that drives the circulation of fluid. In cross-section the stone canal has a scroll-like appearance due to the shape of the calcareous ossicles in the walls. The ring canal often carries a series of sacs called Polian vesicles that serve as reservoirs of fluid (Fig 16.16b). Small pouches called Tiedemann's bodies are also associated with the ring canal. These structures filter water from the stone canal to generate coelomic fluid. Along the arms the radial canal give off lateral canals to the tube feet and ampullae (Figs 16.3c, 16.8). The first podia of the ambulacral series are the buccal podia, which are often used for tasting food.

The tube feet are moved by the antagonistic interaction of hydrostatic pressure and contraction of the podial musculature. Protraction and retraction of the tube feet of asteroids is associated with a movement of fluid between the tube feet and the ampullae. These are muscular bulbs that bulge into the coelom and receive fluid from the tube feet when they contract (Fig. 16.3c). Conversely, tube foot protraction involves contraction of the ampulla, forcing fluid into the tube foot while a valve in the lateral canal is closed off.

The water vascular system of the other echinoderm classes has an organisation similar to that of the Asteroidea, with some differences in each group. In echinoids the oral ring canal encircles the Aristotle's lantern and gives rise to radial canals, which run along the inside of the ambulacral ossicles of the test (Fig. 16.15). The tube feet, which have associated ampullae, emerge through pores in the ambulacral plates (Fig. 16.6b). Sea urchins have one podium per ambulacral plate while sand dollars have numerous podia emerging through each plate. The madreporite is located at the aboral pole (Fig. 16.6b).

In ophiuroids the madreporite is on the oral side of the disc and the stone canal is a short structure leading to the oral ring. Ophiuroid tube feet do not have ampullae, but a basal swelling of the tube feet or small sacs along the radial canal function as reservoirs of fluid. Contraction of the muscles associated with these structures extends the tube feet. Ophiuroid tube feet are often covered with short papillae which provide adhesive secretions for feeding and attachment. These structures may also be sensory.

Holothuroids have several internal madreporites, each of which communicate with the oral ring by a short stone canal. As a result, the fluid in the water vascular system is derived from the perivisceral coelom rather than from sea water. The oral ring encircles the calcareous ring and gives rise to the buccal tentacles and the radial canals, which run inside the body wall in the same manner as in echinoids (Fig. 16.16a, b). The tube feet extend through the body wall (Figs 16.2b, 16.16a). Each tube foot has an ampulla (Fig. 16.16a), and those involved in locomotion generally have an adhesive disc at the tip. The knob-like papillae of aspidochirotid holothurians are modified podia with a sensory function. In the apodids, the water vascular system is limited to the oral ring, buccal tentacles and Polian vesicles.

Crinoids lack a madreporite; instead they have numerous small ciliated canals which traverse the tegmen, connecting the perivisceral coelom with the outside. The stone canals open directly into the perivisceral coelom.

Figure 16.8
Water vascular system of the sea star *Henricia* (from Harrison and Chia 1996).

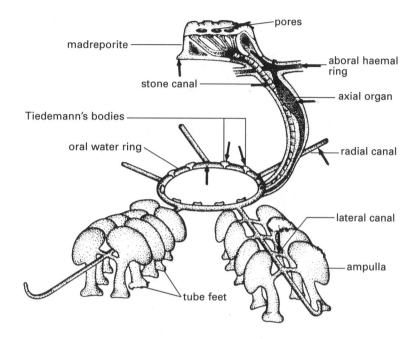

From the oral water-vascular ring, the radial canals of crinoids extend as branches along the arms and pinnules (Fig. 16.9a). The lateral canals give rise to minute podia which are organised in sets of three, with a large, a medium and a small tube foot arising from a common base (Fig. 16.9b, c). These podia produce adhesive secretions which are used to trap suspended food. The podia have short papillae which have a sensory and/or secretory function. Ampullae are not present in crinoids; tube feet are moved by contracting of the radial canals.

Body wall and skeleton

The echinoderm body wall consists of a cuticle, epidermis and dermis. Internally, the wall is lined by the coelomic peritoneum and a layer of muscle tissue. The epidermis contains several cell types, including secretory cells and ciliated cells, some of which are thought to act as surface receptors. Skeletal support is largely a function of the dermis, which consists of two components, the calcite skeleton and collagenous connective tissue.

The echinoderm skeleton is an unusual three-dimensional trabecular network of high-magnesium calcite (Fig. 16.10). It is mesodermal in origin and is therefore an endoskeleton. The porous-trabecular organisation of the skeleton is revealed by removing the associated tissues (Fig. 16.10). The pore spaces of the skeleton are occupied by cells and extracellular matrix. The lattice-like network of plates which forms the skeleton is bound together by bundles of collagen fibres.

Echinoderms vary greatly in their stiffness, depending on the extent of development of the skeleton. Species with closely knit skeletal plates are relatively stiff, while species with less densely packed skeletal plates are less rigid. Sea urchins are the stiffest echinoderms because of their rigid test (Fig. 16.6a), while sea cucumbers have a soft integument (Fig. 16.11) because of the reduction of their skeleton. Despite the stiff consistency of the body wall of many echinoderms, they can be surprisingly dexterous, because the connective tissue between the skeletal elements allows flexibil-

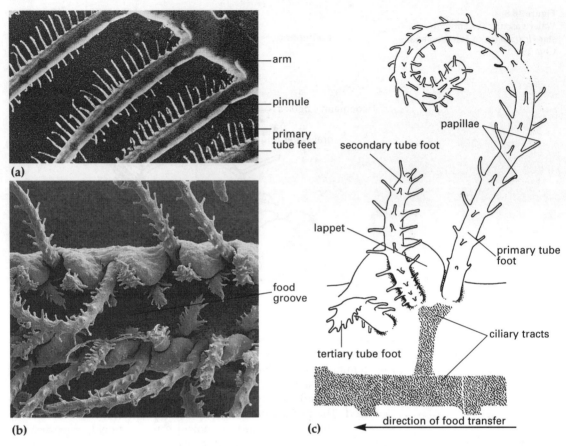

Figure 16.9
(a) Arm pinnules of a feather star, *Florometra serratissima*, showing the primary tube feet extended for suspension feeding. (b) Food groove of *Florometra serratissima* with podial triplets and lappets. (c) Crinoid food groove. (a from Byrne and Fontaine 1981: b, c from Byrne and Fontaine 1983.)

ity and postural change. Many of these changes are facilitated by unusual mutable properties of the connective tissue, which are discussed in the following section.

The development of the body wall musculature varies greatly in the echinoderm classes. Asteroids and holothuroids have the most deformable bodies and have well-developed body wall muscles (Fig. 16.11). The muscles underlie the peritoneum as an external layer of circular muscle and inner bands of longitudinal muscles, accompanied by nervous tissue. Echinoids, with their rigid tests, do not have body-wall muscles. Ophiuroids have a layer of muscles in the disc, but muscular support in the arms is derived largely from the intervertebral muscles associated with the arm skeleton (Fig. 16.5b). Similarly, crinoids have contractile cells associated with the body wall but their major musculature is in the arms.

Mutable connective tissues

It is now widely recognised that one of the characteristic features of the Phylum Echinodermata is the presence of mutable connective tissues that can undergo rapid, nervously mediated changes in their mechanical properties. This is in addition to the conventional roles of echinoderm connective tissues as ligaments and tendons. There are two main expressions of connective tissue mutability: autotomy and 'catch'.

Connective tissue autotomy is exemplified by arm shedding in ophiuroids, asteroids and crinoids, and by evisceration in sea cucumbers.

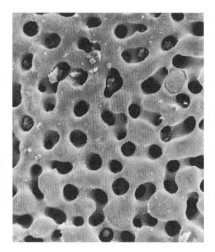

Figure 16.10
Cleaned arm plate from a brittle star, *Ophionereis schayeri*, showing the porous structure of skeleton.

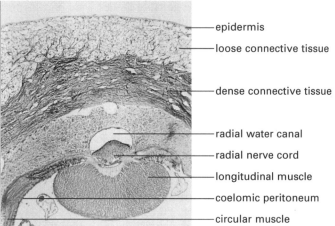

Figure 16.11
Cross-section of the body wall of a sea cucumber, *Eupentacta quinquesemita*, showing the radial nerve cord, water canal and longitudinal muscles.

Figure 16.12
An eviscerating *Eupentacta quinquesemita* discarding its tentacular crown and digestive tract (from Byrne 1985).

Evisceration is a rather dramatic form of autotomy which results in the expulsion of the viscera. This is effected by the rupture of mutable connective tissue structures that anchor the digestive tract to the body wall (Fig. 16.12). The organs are discarded through the breakdown of connective tissue in the body wall. Ophiuroid arm autotomy is accomplished by a rapid (1 s) breakdown of the intervertebral ligament. Asteroids also discard their arms by the rupture of a ring of connective tissue. Similarly, crinoids break their arms at specialised breakage planes equipped with an autotomy ligament. Autotomy in echinoderms is a defensive behaviour that occurs in response to disturbance and to the presence of a predator. The movements of autotomised arms and viscera distract the predator while the main body escapes. With their remarkable powers of regeneration, the discarded body parts of echinoderms are readily replaced. Splitting of the body during asexual reproduction also results from a softening of connective tissue. The two halves of the body walk away from each other, and each subsequently regenerates to make a whole individual.

The second major manifestation of connective tissue mutability is the reversible stiffening-softening changes seen in 'catch' connective tissues. These tissues can switch reversibly from two extreme states: one pliant and extensible and the other stiff and inextensible. This change is used to convert the flexible body into a rigid one and assist in maintaining posture. The best-known example is the circumferential ligament located at the base of sea urchin spines. The spines move by means of muscle contraction and then are held rigidly in place by the stiffening of the catch apparatus. Spine catch helps urchins to maintain a hold on their crevices and burrows. In asteroids, ligamentous stiffening of the body provides a scaffold against which the tube feet work when they open bivalves during feeding. Catch behaviour is characteristic of the holothuroid integument, which is dominated by connective tissue (Fig. 16.11). Ophiuroids and crinoids extend their arms in one posture for prolonged periods during suspension feeding (Fig. 16.2c, d); this is facilitated by the presence of mutable connective tissues which hold the arms in a firm position.

The mutable tissues were once assumed to be specialised muscle tissues, but in fact they are collagenous. The mechanical changes are the result of an alteration in the interfibrillar matrix between the collagen fibres. Nerve-like processes called juxtaligamental cells are though to contain a secretory product which effects these changes.

Connective tissue mutability is without parallel in other phyla. The use of connective tissue for such functions as posture control and autotomy obviates the need to invest in energetically demanding muscle cells for these purposes. Although the physiology of mutability is not understood, it appears to be controlled by an unconventional relationship between the nervous system and the extracellular tissue. This phenomenon remains an active area of echinoderm research.

Locomotion

In most echinoderms the tube feet play a major role in locomotion. This is particularly well illustrated by the sea stars and sea urchins, which have suckered tube feet (Figs 16.3b, 16.7a). The tube feet adhere tenaciously to hard substrata due to the suction-like mechanism of the disc and the adhesive properties of the epidermal secretions. Adhesion and detachment of echinoderm tube feet appear to be based on a two-gland system: the release of one sticky secretory product (responsible for adhesion) is followed by the release of a second product (which effects detachment). These secretions can be displayed by using special dyes to stain the footprints of starfish after they have walked across a glass surface.

Sea urchins also use their spines to move around (Fig. 16.1e). The spines are connected to the test by a ball-and-socket joint which is surrounded by a ring of muscle, enabling the spines to move in any direction. Irregular echinoids that live as infaunal burrowers use their paddle-like locomotory spines in a rowing motion to move through the sediment with their anterior end forward. The small tube feet of irregular echinoids are not used in locomotion. Some heart urchins occupy a semi-permanent burrow and maintain contact with the surface by specialised tube feet and spines.

Ophiuroids move over the substratum by a sinusoidal rowing motion of their arms, propelled by intervertebral muscles working the hinge-like joints between arm segments (Fig. 16.5a, b). Ophiuroid tube feet do not play a major role in locomotion, although some species are capable of walking up vertical surfaces because of the adhesive properties of their podia. Burrowing ophiuroids use their tube feet to displace sediment as they plough into the substratum. Basket stars and other euryalid ophiuroids typically remain with their arms coiled around the substratum for prolonged periods (Fig. 16.1d), although they can crawl about with their arms.

Sea cucumbers range from sedentary suspension-feeding species to mobile deposit feeders that move on top of or through sediments (Fig. 16.2a, b). In the orders Aspidochirotida and Dendrochirotida, locomotion is achieved by the tube feet working with the body wall musculature (Fig. 16.11). Apodids propel themselves through the sediment by undulating contractions of their thin worm-like bodies. Some aspidochirotids can move quite rapidly and swim up into the water column if disturbed by a predator. The deep-sea holothuroids of the Order Elasipodida are remarkable for their spectacular swimming and sailing modes of locomotion. Some elasipods have gelatinous bodies and move through the water by the flapping motion of medusa-like lobes which extend from the anterior body wall. The Elasipodida includes the only truly pelagic echinoderms.

The tube feet of crinoids are not involved in locomotion. Crinoids move with their arms and to a certain extent with either cirri or stalk. The cirri and stalk lack muscles, and the motility of these structures is based entirely on mutable connective tissue. When disturbed from their perch, feather stars exhibit a graceful swimming response, moving alternate sets of arms in ballerina-like fashion. Sea lilies also move with their arms if they become displaced into the water column.

All echinoderms exhibit a rapid righting response when out of position. If asteroids, echinoids or ophiuroids are placed with the oral surface upwards, the tube feet, arms or spines work in concert to return the individual to an oral-side down position. Sea cucumbers twist the body wall and use the tube feet to regain the normal orientation. In crinoids the arms and cirri work to regain the oral-up orientation.

Feeding and digestive system

The digestive system of most echinoderms consists of a mouth, short oesophagus, stomach, intestine, rectum and anus. Echinoids and holothuroids have a pharynx between the mouth and oesophagus. In echinoids the pharynx is contained within the Aristotle's lantern (Fig. 16.6c), while in holothuroids the pharynx traverses the calcareous ring (Fig. 16.16a, b). Echinoids also have a small tube called the siphon which connects the oesophagus directly to the intestine. This acts as a shunt, allowing water ingested with the food to bypass the stomach. Holothuroids have a muscular cloaca which connects with the anus and contracts regularly to propel water into respiratory structures (Fig. 16.16b). An anus is lacking in some asteroids and all ophiuroids (Fig. 16.14b). Indigestible material is egested through the mouth. As in other animal groups, predatory echinoderms have short, fast-acting digestive systems, while herbivorous or deposit-feeding echinoderms have long, slower-acting digestive systems.

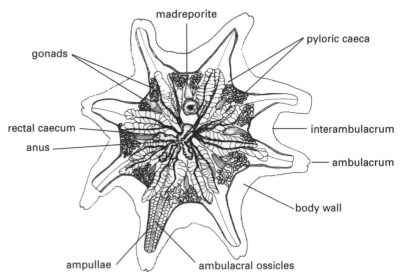

Figure 16.13
Aboral dissection of a sea star, *Patiriella calcar* (from Anderson 1996).

Asteroidea

Sea stars have a capacious stomach with two regions, the oral cardiac stomach and the aboral pyloric stomach (Figs 16.13, 16.14a). Asteroids with extra-oral feeding extrude their cardiac stomach through the mouth to

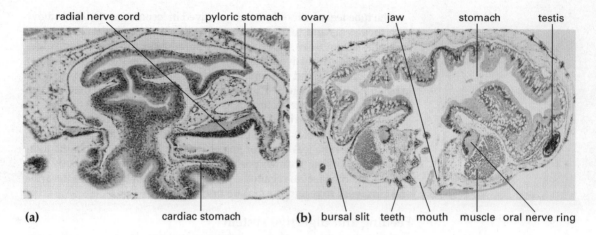

Figure 16.14
(a) Cross-section of a sea star, *Patiriella vivipara*, showing the cardiac stomach extruded through mouth.
(b) Cross-section of a brittle star, *Ophiurochaeta* sp., showing the sac-like gut, mouth, oral nerve ring and gonads. This species is a hermaphrodite, with ovaries and testes. (a from Byrne 1996, b from Byrne 1994.)

surround the food and initiate digestion outside the body (Fig. 16.14a). This is followed by further digestion of prey outside the body, or retraction of the stomach to bring the prey into the body. Extra-oral feeding has a great advantage in that prey too large to be swallowed can be consumed (e.g. *Acathaster planci* feeding on coral tissues). The asteroid digestive system reflects the radial body plan, with two pyloric caeca extending along each arm (Figs 16.3c, 16.13). The pyloric caeca are highly branched digestive diverticula that connect with the pyloric stomach; they are used to digest food and store nutrients. A median duct along the base of the pyloric caeca conveys material from the stomach to the lateral diverticula. On the aboral side the pyloric caeca connect with the intestine, which may give rise to a series of blind sacs called intestinal caeca (Fig. 16.13). The function of these structures is not understood.

Most asteroids are predators and scavengers, taking a broad range of invertebrate prey. Many species focus on bivalves and some species can insert the stomach into gaps as small as 0.1 mm between the valves. The crown-of-thorns starfish, *Acanthaster planci*, is rare among the Asteroidea in being a specialised corallivore. Asteroids such as *Patiriella* species that feed on surface films extrude the stomach over the substratum to initiate feeding. Other asteroids such as *Henricia* are suspension feeders, utilising a muco-ciliary mechanism to obtain food. In this feeding mode fine particles trapped in mucus are propelled to the mouth by ciliary activity. Deposit feeding asteroids such as *Ctenodiscus* engulf organically rich benthic deposits.

Ophiuroidea

Ophiuroids have a simple digestive system consisting of a short oesophagus and a sac-like stomach which occupies most of the disc (Fig. 16.14b). There is no intestine or anus. The function of the jaw apparatus is not understood. Although the structure, with its opposing rows of teeth (Figs 16.4, 16.14b), appears adapted for mastication, there is little evidence for this. The buccal podia appear to have a tasting function (Fig. 16.4).

The feeding habits of ophiuroids range from microphagous suspension feeding and deposit feeding to carnivory. Many species are also scavengers. Most ophiuroids, such as *Ophionereis schayeri*, are opportunistic and use a variety of feeding mechanisms. Suspension-feeding ophiuroids have highly extensible tube feet which project during feeding, presenting a sticky

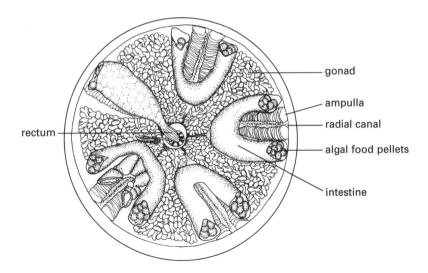

Figure 16.15
Dissection of the aboral half of a sea urchin, *Heliocidaris erythrogramma* (from Anderson 1996).

mucus-covered surface for food capture. The spines also often have a covering of mucus for capturing particles. Particles trapped on the tube feet or spines are compacted by groups of tube feet into small, mucus-bound balls that are passed towards the mouth. Deposit-feeding ophiuroids collect particles from the surface directly with their sticky tube feet. Carnivorous ophiuroids use arm-loop capture, in which an arm coils around the food and brings it directly to the mouth. Basket stars are predatory suspension feeders, and ensnare zooplankton in the terminal tendril-like branches of their arms (Fig. 16.1d). These branches are armed with batteries of hooks that are used to capture prey. When loaded with food, each arm curls under the body to bring the food directly to the mouth.

Echinoidea

Because of their predominantly algal or sedimentary diet, sea urchins have a long digestive tract. From the pharynx the digestive tract coils internally around the test (Fig. 16.15). The jaw apparatus or Aristotle's lantern consists of interacting skeletal elements, muscles and ligaments (Fig. 16.6c). On the test, the lantern is supported by the perignathic girdle, a ring of modified ambulacral and interambulacral plates. These provide attachment surfaces for the lantern muscles and ligaments. Externally the lantern has five pointed teeth that protrude from the mouth. The major skeletal elements of the regular lantern include five pyramids (jaws), teeth, rotulae and compasses. The pyramids have a central calcareous band, with the tooth at the oral end and a soft dental sac at the opposite end where new tooth material is formed. Tooth material is continually synthesised by the dental sacs to renew the teeth as they are abraded by the scraping associated with feeding. On the aboral side of the lantern, the rotulae serve as braces between adjacent pyramids and the small compass ossicles connect with the supporting muscles. The lantern also has muscles which control the protraction and retraction of the teeth during feeding. In irregular echinoids the lantern is reduced or absent. Sand dollars have a minute Aristotle's lantern which is used to crush fine particles. A lantern complex is not present in heart urchins.

Most sea urchins such as *Heliocidaris* and *Centrostephanus* feed on algal material caught by the spines and tube feet. The hardness of the teeth gives urchins the ability to graze on the surface of the substratum, removing encrusting organisms and calcareous algae. This feature also allows regular urchins to burrow into hard substrata. Irregular echinoids are largely deposit feeders, although there are some suspension-feeding species. Sand dollars feed on benthic deposits using a complex particle-picking mechanism. In contrast to regular urchins, they have numerous small podia associated with their ambulacral plates, with some species having hundreds of thousands. The podia on the oral surface probe the sediment and gather food particles. The particles are passed from podium to podium to reach the food grooves, consisting of shallow depressions that converge at the mouth. In the food grooves the particles are gathered into a mucous cord and directed towards the mouth by podia. Suspension-feeding sand dollars also collect particles directly with their tube feet. Some of these echinoids can maintain an erect posture with the edge of the test perpendicular to the direction of water flow. This orientation appears to help capture fine particles.

Heart urchins inhabit very fine sand or mud, either burrowing though the sediment or occupying semipermanent burrows. These echinoids collect organic particles with highly specialised tube feet called pencillate podia. Pencillate podia end in a flower-like array of processes and pick up a large number of particles each time they are applied to the sediment. Some heart urchins maintain a burrow and use their pencillate podia to collect particles from the surface of the sediment and bring them to the mouth. These podia are also used to maintain the burrow.

Holothuroidea

Holothuroids collect food with the buccal tentacles (Fig. 16.2b). These tentacles are highly contractile, and if the animal is disturbed the tentacles are quickly retracted into the body. From the pharynx, the digestive system leads to a small stomach and an elongate intestine which is divided into descending and ascending portions (Fig. 16.16a, b). The intestine connects with a muscular cloaca which opens at the anus.

Holothuroid feeding modes can be broadly divided into suspension feeding, seen in the order Dendrochirotida, and deposit feeding, seen in the orders Aspidochirotida and Apodida. Suspension-feeding sea cucumbers hold their highly branched tentacles in the current (Fig. 16.2b) and collect particles on minute adhesive papillae that cover each tentacle. The numerous fine branches of the tentacles present a large surface area for food capture. Periodically the tentacles are brought into the pharynx, where the food is detached. The diet of suspension-feeding holothuroids includes diatoms, unicellular algae and small invertebrates. Aspidochirotids extend their peltate tentacles over the substratum to push sediment into the mouth (Fig. 16.2a). Adhesive nodes on the surface of the tentacles provide secretions which help in collecting particles. Apodids collect particles with their branched pinnate tentacles. In all deposit-feeding sea cucumbers the tentacles are brought to the pharynx to transfer the sediment to the gut. Although deposit-feeding sea cucumbers are thought to be non-selective, recent research shows that they can focus on organically rich areas. The major food item of the deposit feeders is the bacteria associated with the sediment. Deposit-feeding holothuroids play a major role in reworking sediments and are capable of processing large quantities of sediment. Mucus derived from their feeding and digestive activity is also important in enriching benthic sediments.

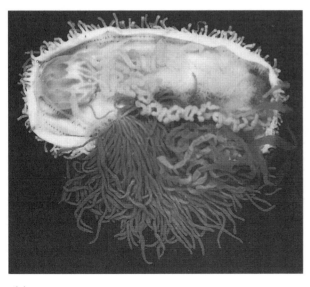
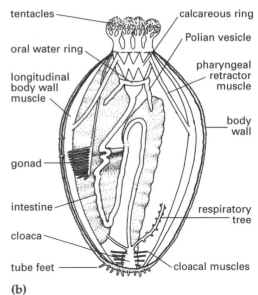

(a) (b)

Figure 16.16
(a) Dissection and (b) diagram of a sea cucumber, *Eupentacta quinquesemita*. The dark dots along the ambulacra are the ampullae (from Byrne 1985).

Crinoidea

In crinoids, which have the mouth and anus directed upwards, the mouth is formed by the confluence of the ambulacral food grooves. The digestive tract has a short oesophagus, an intestine with one major loop and a short rectum.

All crinoids are suspension feeders. They are highly rheophilic and tilt their arms so that the ambulacral groove is facing downstream (Fig. 16.2c, d). They often seek elevated positions above the sea floor. Where aggregations of feather stars inhabit areas with directional currents, it is striking to see all the individuals adopting a similar orientation. Sea lilies are elevated above the sea floor by the stalk, and in a unidirectional flow they orientate their arms in a parabolic fan, with the food groove downstream. During feeding the long primary tube feet are held out of the groove to catch particles on their sticky surface (Fig. 16.9a). The tube feet bring the particles into the food groove with a rapid flicking action. In the food groove the particles are removed by a wiping action of the secondary and tertiary tube feet or by the cilia in the groove (Fig. 16.9b, c). Small skeletal plates along the food groove called lappets are also thought to assist in transferring material from the primary tube feet to the food groove. Captured particles are passed to the mouth along the ciliary tract on the floor of the food groove (Fig. 16.9c).

Physiological processes

Digestion

Digestion in echinoderms is largely extracellular and occurs in the gut lumen. The lining cells of the digestive tract secrete digestive enzymes and mucus which lubricates the food. Most of the cells lining the gut have an extensive brush border of microvilli which increases the surface area for absorption. A haemal sinus in the gut wall appears to store the products of digestion. The role played by the haemal system in the transport of nutrients is not clear. It appears that the coelomic fluid and perhaps coelomocytes play a major role in nutrient transport.

Respiratory exchange and circulation

Distinct circulatory and excretory systems are lacking in echinoderms. Most, if not all, of the circulatory and excretory functions, including gas exchange and the removal of nitrogenous wastes, are accomplished by the coelomic fluid and coelomocytes. The coelom is lined by ciliated cells which assist the circulation of coelomic fluid. The tube feet provide an extensive surface for gas exchange. The coelomic fluid of most echinoderms lacks a respiratory pigment, but in some holothuroids and ophiuroids there are haemocytes which contain haemoglobin. Interestingly, some of these cells lack a nucleus when they are mature, a feature that parallels mammalian red blood cells. Echinoderms that have respiratory pigments often inhabit oxygen-poor habitats such as benthic sediments.

Asteroids, echinoids and holothuroids have a spacious perivisceral coelom (Figs 16.3c, 16.6c, 16.16a). In sea stars the coelom extends into the arms, where it surrounds the gonads and pyloric caeca (Fig. 16.3c). The small, finger-like papulae on the aboral surface of asteroids contain extensions of the perivisceral coelom and provide another surface for respiratory exchange.

In sea urchins, the bushy gills on the peristomial membrane are used in respiratory exchange. They are extensions of the lantern coelom. The thin-walled respiratory podia of burrowing echinoids are important respiratory surfaces. The powerful ciliary currents that traverse the test of burrowing echinoids also play an important role in respiration. These currents ensure that sea water flows continuously across the test.

In ophiuroids the perivisceral coelom is limited to a small space around the gut and a narrow channel on the aboral and lateral sides of the arms (Figs 16.5b, 16.14b). Respiration is an important function of the bursae which protrude into the coelom (Fig. 16.4). Ciliated cells in the bursal epithelium generate currents which circulate sea water through the bursal slits. Gas exchange between the sea water in the bursa and the internal coelomic fluid occurs across the thin bursal wall.

With the exception of the apodids, all holothuroids have specialised structures called respiratory trees or water lungs (Fig. 16.16a, b). A pair of respiratory trees is usually attached to the cloaca and extends anteriorly in the perivisceral coelom. The trees are a system of blind-ending, highly branched tubes which are periodically inflated and emptied of sea water by the rhythmic muscular contraction of the cloaca. The thin walls of the tubes facilitate respiratory exchange between the sea water in the lumen of the tubes and the fluid of the perivisceral coelom.

Excretion

Echinoderms have little or no ion regulatory capabilities, and the coelomic fluid is similar in composition to sea water. Nitrogenous wastes readily diffuse from the body across thin surfaces such as the walls of the tube feet. Coelomocytes also play a role in excretion by phagocytosing waste material. Waste-laden coelomocytes are discharged across thin-walled structures such as the papulae of sea stars, the gills of sea urchins, and the tube feet of echinoderms in general. Sea cucumbers have brown bodies in the perivisceral coelom which are aggregations of necrotic phagocytes. Some sea cucumbers have ciliated pores and channels through the body wall connecting the coelom to the outside. Brown bodies and waste-laden coelomocytes are voided through these conduits.

Sea cucumbers that eviscerate seasonally do so at the end of the summer feeding period. This is thought to be a means of eliminating waste material which accumulates as brown body deposits in the gut wall. Evisceration is followed by regeneration of the gut. Newly regenerated digestive tracts are yellow and develop a brown colour as waste-laden phagocytes accumulate in the gut wall.

Haemal system

Despite its name, the haemal system of echinoderms is unlikely to function as a circulatory system or play a role in respiration. The gelatinous consistency of haemal fluid and its location in ill-defined channels indicates that it is poorly adapted for circulation. In most echinoderms the haemal system parallels the water vascular system, with an oral haemal ring giving off branches that run along the ambulacra beside the water canals. An aboral haemal ring connects to the oral haemal ring by means of the axial gland. The aboral haemal ring gives off branches to the gonads.

The ability of haemal fluid to sequester nutrients indicates that it is involved in storing nutrients. Haemal fluid in histological sections takes up special stains for carbohydrates, providing further evidence for the presence of nutrients. In the gonad, the genital haemal sinus plays an important role in gamete nutrition. One of the major constituents of this sinus in echinoids is vitellogenin, a substance required for egg formation. The presence of vitellogenin in the genital haemal fluid of both male and female urchins indicates that it may serve a general nutritive role in supporting gametogenesis.

Aspidochirotid sea cucumbers have the most highly developed haemal system, with two particularly prominent haemal 'vessels' on either side of the digestive tract. The orange or red pigment of these vessels makes them conspicuous in dissections. A fine network of haemal strands emanating from the vessels was called the *rete mirabile* (wondrous network) by early researchers.

Nervous system

Echinoderms lack a centralised nervous system, and their behaviour appears to be controlled by a series of local reflexes. A nerve plexus is located at the base of the epidermis and is condensed in places to form distinct nerves. The oral nerve ring encircles the mouth and gives rise to radial nerve cords which run along the ambulacra (Figs 16.3c, 16.5b, 16.11, 16.14a, b). Structures around the mouth such as buccal tube feet and tentacles are supplied with nerves that originate from the oral nerve ring. In each radius the radial nerve supplies the tube feet and ampullae. The nervous system includes an extensive ectoneural (sensory and motor) system, mostly associated with the surface, and a more localised hyponeural (motor) nervous system which supplies the muscles. The nerve cords usually contain both ectoneural and hyponeural elements.

The juxtaligamental cells involved in connective tissue mutability are associated with the nerve cords, as individual cells or in distinct ganglia. Juxtaligamental cells contain granules and have processes that extend through connective tissue.

The echinoderm nervous system has proved intractable to study by neurophysiologists, largely due to the small size of the neurons. These are generally less than 1.0 µm in diameter. An exception is the ophiuroid radial nerve cord, which has axons which are 'giant' by echinoderm standards, up to 20 µm in diameter. Ophiuroids have a relatively well-developed nervous system, and their 'giant' axons are thought to be important in the quick reflexes and comparatively rapid locomotion exhibited by some species. Our present knowledge of the physiological basis of neural behaviour in echinoderms is largely derived from research on the nervous system of *Ophiura*, an ophiuroid with particularly large axons and rapid reflexes. The ophiuroid radial nerve cord swells to form serial ganglia that give off nerves to the tube feet, integument and intervertebral muscles. This serially repeated arrangement has facilitated the study of the ophiuroid nervous system.

Light is produced in many echinoderms, often in response to mechanical stimulation. In some ophiuroids this is exhibited by a spectacular green flash along the arms. The bioluminescent cells occur in close association with the nervous system and may be specialised neurons.

Sensory structures

Despite their relatively complex behavioural repertoire, echinoderms generally lack discrete sensory organs. The terminal eyespots of sea star arms and the statocyst-like sphaeridia of echinoids are the only convincing sensory structures seen in the phylum. Asteroid eyespots have pigmented cells that respond to light and are thought to act in photoreception (Fig. 16.3d). The sphaeridia of echinoids, which appear to be involved in orientation, are brightly coloured, knob-shaped appendages. They are present in the ambulacra of most echinoids.

Echinoderms are highly sensitive to light. A visit to a reef at dusk is often rewarded by the appearance of ophiuroid arms emerging from crevices, by crinoids uncoiling their arms, and by echinoids emerging from their holes. Echinoderms often feed at night and return by dawn to their day-time hiding places. This nocturnal activity is thought to be an adaptive feature to avoid predation by fishes that are active during the day. It appears that the subepidermal nerve network and ciliated cells in the epidermis function as general receptors and account for what has long been called the 'dermal light sense' of echinoderms.

The diurnal pattern of light sensitivity in some diademid echinoids (*Diadema*) and ophiocomid ophiuroids (*Ophiocoma*) is associated with a remarkable day–night change in colouration. This colour change is brought about by the migration of pigment cells in the integument. The movement of the pigment cells covers and uncovers underlying neurons which are apparently light receptors.

Many echinoderms exhibit striking chemosensory and rheotactic behaviour in locating food. This is thought to involve general chemosensory cells in the epidermis. In most echinoderms the tube feet or tentacles can taste food, especially those around the mouth.

Reproduction and development

Reproductive system

For the most part, there are no external morphological differences between the sexes in echinoderms. The exceptions are those sea cucumbers that have genital papillae, which differ in the male and female. Larger echinoderm species usually have separate sexes, while the smaller species are mostly hermaphrodites. The reproductive anatomy of asteroids, ophiuroids, echi-

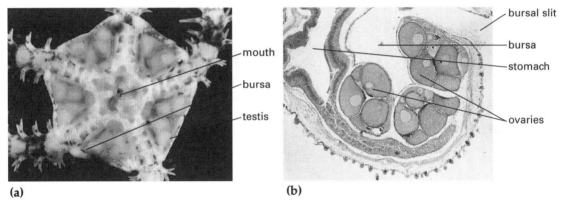

Figure 16.17
(a) Oral dissection of a brittle star, *Ophionereis olivacea*, showing the testes and bursae.
(b) Section through the disc of a brittle star, *Ophiurochaeta* sp. The ovaries contain a few large eggs that are later retained in the bursa, which serves as a brood chamber for the developing young (from Byrne 1985).

noids, and to some extent crinoids, reflects the pentamerous radial organisation of the body (Figs 16.13; 16.15, 16.17a, b).

Echinoderm gonads are simple sac-like structures (Figs 16.17b, 16.18a,b). Asteroids have two gonads in each arm (Fig. 16.13). In mature sea stars the gonads dominate the arm coelom. Each gonad can have one or several gonopores. Echinoids have one elongate gonad attached to the inside of the test in each interambulacrum (Fig. 16.15). The gonopores are located in the genital plates which surround the anus (Fig. 16.6b).

The reproductive anatomy of ophiuroids is associated with the ten bursae, one on either side of the base of each arm (Figs 16.4, 16.17a, b). Depending on the species, each bursa may have one or several gonads associated with it. The gametes are released into the bursae and pass out through the bursal slits at the bases of the arms. Some phrynophiurids are unusual in having gonads in the arm coelom, as in asteroids. In this case there are either one or several gonoducts opening to the outside.

Holothuroids differ from other echinoderms in having a single, highly branched gonad, which is made up of elongate gonadal tubules (Fig. 16.16a, b). The base of the gonad is attached to the anterior region of the body, where the gonoduct is also located. In gravid specimens the gonad dominates the perivisceral coelom (Fig. 16.16a).

Crinoids have a multitude of small gonads on the pinnules. In mature specimens the pinnules are noticeably swollen with gravid gonads. Gonoducts are lacking and spawning occurs by rupture of the gonad wall.

Gametes and fertilisation

The gametes develop in the innermost tissue layer (the germinal layer) and often maintain a close association with the underlying haemal sinus, which provides nutrients for gamete growth. As they mature, the gametes accumulate in the lumen of the gonad prior to spawning (Figs 16.17b, 16.18a, b). Echinoderm eggs range from small (60–100 μm diameter) in species that have feeding larvae to large (200–1000 μm diameter) in species that have non-feeding larvae or brood their young (Figs 16.17b, 16.18a). The sperm develop in spermatocyte columns that extend towards the lumen of the testis (Fig. 16.18b). They are also usually accompanied by somatic cells which send processes along the centre of the columns, providing structural support. Ophiuroids and asteroids have simple, round-headed sperm (Fig. 16.18b) while echinoids have conical sperm.

Free-spawning echinoderms typically release copious numbers of small eggs; *Acanthaster planci* is estimated to produce up to 60 million eggs per

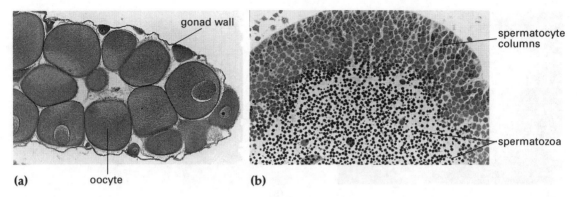

Figure 16.18
(a) Ovary of a sea star, *Patiriella calcar*, containing large eggs.
(b) Testes of brittle star, *Ophiomyxa breverima*, with developing spermatocytes along the wall and spermatozoa in the lumen.

breeding season. Many echinoderms spawn synchronously and in aggregations. Synchronous spawning appears to be cued by environmental factors such as water temperature, photoperiod and the state of the tidal currents, and also by endogenous factors such as hormonal control and the release of pheromones. When spawning is synchronous, fertilisation success would be expected to be high. Many echinoderms, however, do not live close to conspecifics, and in this case it is hard to imagine that fertilisation could be anything more than a haphazard event. Fertilisation in the sea, particularly with echinoderms as model organisms for study, is an active area of research in marine biology.

Development

Echinoderm development follows the deuterostome pattern, beginning with radial cleavage (Fig. 16.19a–c). The blastula is a hollow ball of cells with a fluid-filled blastocoel (Fig. 16.19d). In some echinoderms the blastula develops deep folds in the epithelium, giving a highly contorted appearance (Fig. 16.19e). Hatching occurs at the blastula or gastrula stage (Fig. 16.19f, g). Gastrulation involves invagination of one pole of the embryo to form the archenteron (Fig. 16.19g). The external opening of the archenteron, the blastopore, later forms the anus (Fig. 16.19h, i). Coelom development begins with the formation of a pair of pouches which bud from the anterior end of the archenteron (Fig. 16.19h). Subsequently, the initial pair of enterocoelic pouches divides into three pairs, an anterior axocoel, a middle hydrocoel and a posterior somatocoel. This tripartite body plan is also developed in the other deuterostome groups (see Chapter 17).

The gastrula develops into a bilaterally symmetrical larva, the structure of which differs in each echinoderm class (Figs 16.19j, k, 16.20, 16.21a, b). Larval structure also depends on whether development is of the feeding or non-feeding type (Fig. 16.20). Echinoderms with small eggs (i.e. most of the larger species) have free-swimming planktotrophic larvae with well-developed digestive tracts. These larvae feed on phytoplankton and have an intricate array of ciliated bands for feeding and locomotion (Figs 16.20, 16.21a, b). In contrast, species with large eggs (mainly small species) develop as planktonic or benthic lecithotrophic larvae that lack a functional gut and have a simplified pattern of ciliation (Fig. 16.20). The development of these larvae is supported by the nutritive reserves present in the egg.

Most asteroids have bipinnaria and/or brachiolaria larvae (Figs 16.19j, k, 16.20). The bipinnaria is a feeding larva with two ciliary bands looping around the body (Fig. 16.19j). The archenteron grows forward to form the mouth, and the blastopore becomes the anus (Fig. 16.19i). The digestive

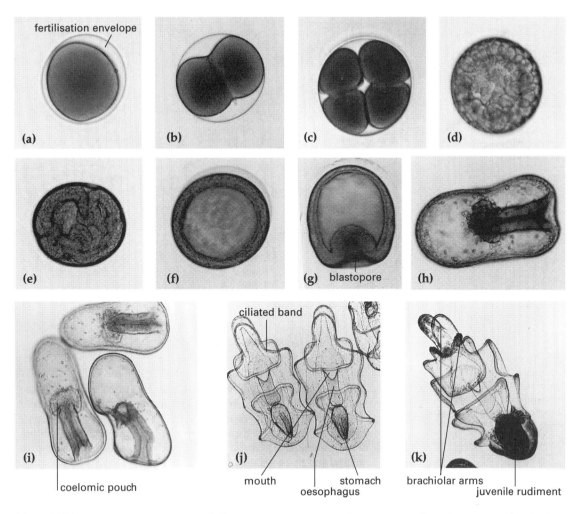

Figure 16.19
Development of the sea star *Paritiella regularis*. **(a)** Newly fertilised egg. **(b)** Two-cell stage. **(c)** Four-cell stage. **(d)** Early blastula. **(e)** Wrinkled blastula. **(f)** Late blastula. **(g)** Hatched gastrula. **(h)** Late gastrula with coelomic pouches beginning to form. **(i)** Early larva with differentiating digestive tract. **(j)** Bipinnaria larva. **(k)** Brachiolaria larva. (From Byrne and Barker 1991.)

tract differentiates into an oesophagus, stomach and intestine (Fig. 16.19j). The bipinnaria usually gives rise to a brachiolaria larva, with the development of the brachiolar apparatus at the anterior end (Figs 16.19k, 16.20). This structure consists of three arms and a central adhesive disc, and is used for benthic attachment by larvae ready to metamorphose. Asteroids with non-feeding development lack a bipinnaria, and the gas-trula gives rise directly to a lecithotrophic brachiolaria which is uniformly ciliated (Fig. 16.20).

Echinoids and ophiuroids have a pluteus larva with a ciliary band that follows the contour of the larval arms (Fig. 16.21a, b). Echinoplutei have up to six pairs of arms, and ophioplutei have up to four pairs. Each arm is supported by a skeletal rod. The similarity of echinoplutei and ophioplutei was at one time taken to suggest a close relationship between echinoids and ophiuroids, but phylogenetic analyses show this is not the case. The presence of a pluteus in these two classes is a striking example of parallel evolution. Echinoids with lecithotrophic development may have a reduced pluteus or have a simple spherical larva with no pluteal structures. Ophiuroids with non-feeding development have vitellaria larvae which are distinguished by their barrel-shape and transverse bands of cilia.

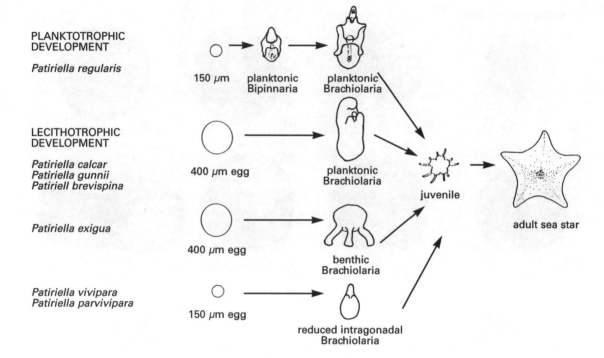

Figure 16.20
Developmental patterns in *Patiriella* (from Byrne and Cerra 1996).

The planktotrophic larva of holothuroids is the auricularia. It has two ciliary bands that loop around the body and looks similar to the asteroid bipinnaria. The auricularia gives rise to a doliolaria larva by a reorganisation of the ciliated bands into transverse rings. Holothuroids with lecithotrophic development have a vitellaria larva. All crinoids have lecithotrophic development through a vitellaria.

Echinoderm metamorphosis involves a dramatic change from the bilateral symmetry of the larva to the pentamerous radial symmetry of the juvenile (Fig. 16.20). The juvenile body develops at the posterior end of the larva (Fig. 16.19k). Metamorphosis involves resorption or loss of the larval body and may occur while the larva is in the plankton or after settlement. In species that care for their young, metamorphosis occurs in the brood chamber and the juveniles walk away from the parent (Fig. 16.17b).

Evolution of life histories

The free-spawning planktotrophic life history of echinoderms is considered to be the ancestral pattern. The presence of vestigial feeding structures in some lecithotrophic larvae, such as a closed gut and reduced pluteal arms, is evidence that lecithotrophy evolved through an ancestor with planktotrophic development. Lecithotrophic development is often associated with a suite of life history traits, including decreased adult size, hermaphroditism, internal fertilisation, and parental care of the young.

The evolutionary changes associated with the switch from planktotrophy to lecithotrophy are exemplified by a genus of sea stars from Australia, *Patiriella* (Fig. 16.20). These asteroids exhibit the full range of life histories, with planktotrophic development at one end of the spectrum and development within the parent at the other end. In contrast to the range of larval forms in *Patiriella*, the adults are quite similar. The close relationship

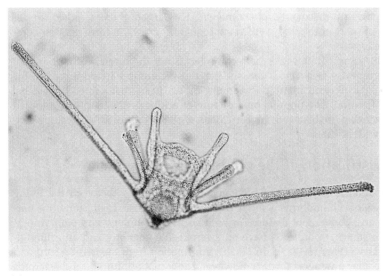

Figure 16.21
(a) Ophiopluteus of *Ophiothrix spongicola*. (b) Echinopluteus of *Heliocidaris tuberculata*. (a courtesy P. Selvakumaraswamy, b courtesy Dr R. Emlet.)

between the *Patiriella* species facilitates a comparison of homologous cells and structures in development, providing a model system to investigate the evolution of development.

P. regularis represents the ancestral state, spawning small eggs which develop as feeding bipinnaria and brachiolaria larvae (Figs 16.19a–k, 16.20). It is also a relatively large species and is dioecious. The intermediate-sized dioecious species *P. calcar*, *P. gunnii* and *P. brevispina* spawn large eggs and have planktonic, lecithotrophic, brachiolaria larvae (Fig. 16.20). The bipinnaria stage has been deleted, but the brachiolar complex is retained due to its importance in benthic settlement. Most asteroids with lecithotrophic development have the type of brachiolaria seen in these species. *P. exigua* is a small hermaphrodite that attaches its large eggs to the substratum. The lecithotrophic brachiolaria of *P. exigua* has a tripod shape, due to the enhanced growth of the brachiolar complex, which serves as a tenacious and permanent attachment device (Fig. 16.20). At the extreme end of the developmental range are the viviparous life histories of *P. vivipara* and *P. parvivipara*. These are the smallest known sea stars. They are hermaphrodites, have internal fertilisation, and have intragonadal development through a minute brachiolaria (Fig. 16.20). Without the need for benthic settlement, the brachiolar complex is reduced. The larvae metamorphose within the gonads into tiny juveniles. These juveniles prey on their intragonadal siblings and are born about a year later as large juveniles which emerge through the gonopore. Viviparity in *Patiriella* is the most derived life history pattern seen in the Echinodermata.

Asexual reproduction

Asexual reproduction is common in echinoderms. In asteroids and ophiuroids asexual reproduction is exhibited by several multiarmed species which reproduce by division or fission of the body. Transverse fission is also common in tropical sea cucumbers. Division of the body is followed by the complete regeneration of each half. The tropical asteroid *Linkia multiflora* has remarkable regenerative capabilities and can form a complete sea star from a single autotomised arm. Asexual propagation by sea star larvae is a recently discovered phenomenon. The larvae of some *Luidia* species bud off part of the body which subsequently grows to form another complete larva.

In some habitats, asexual echinoderms are the numerically dominant invertebrates. *Holothuria atra* is the most common holothuroid in the Indo–Pacific, and many populations appear to be maintained by asexual proliferation. The population structure of echinoderms that have adopted the asexual life history is highly clonal, with most individuals being genetically identical.

Classification of the phylum Echinodermata

The determination of the phylogenetic relationships between the living classes of echinoderms is made difficult by the presence of numerous parallel and convergent features in both adult morphology and embryology. For instance, ophiuroids have an asteroid-like anatomy and an echinoid-like larva (Figs 16.14a, b, 16.21a, b). Information from the fossil record and molecular phylogenetic data, however, show that ophiuroids and asteroids should be grouped together and that the asteroid and echinoid lineages separated early. It appears that the holothuroids are most closely affiliated with the echinoids. Crinoids are quite distant from all the other classes. The concentricycloids are most closely affiliated with the asteroids.

CLASS ASTEROIDEA

The sea stars or starfishes. Stellate echinoderms with radial water canals external to double series of ambulacral ossicles that run down the arms (open ambulacral system); coelom spacious; larval stages bipinnaria and brachiolaria. (Orders Forcipulatida, Paxillosida, Valvatida, Velatida, Spinulosida)

CLASS OPHIUROIDEA

The brittle stars and basket stars. Stellate echinoderms with radial water canals internal to the arm skeleton (closed ambulacral system); ambulacral ossicles in the centre of the arms; coelom largely restricted to the disc; no anus; larval stages ophiopluteus and vitellaria. (Orders Phrynophiurida, Ophiurida)

CLASS CONCENTRICYCLOIDEA

The sea daisies. Discoidal echinoderms with two water rings on the outer edge of the disc external to the skeleton; coelom spacious; males with a copulatory organ; direct development. (Genus *Xyloplax*; only two species known)

CLASS ECHINOIDEA

The sea urchins, sand dollars and heart urchins. Globular echinoderms with test of skeletal plates; ambulacral system internal; coelom spacious; larva an echinopluteus.

Subclass Cidaroidea

Slate-pencil urchins. Primary spines thick, few in number and widely separated. (Order Cidaroida)

Subclass Euechinoidea

The majority of sea urchins, also the sand dollars and heart urchins. Primary spines thin and numerous. (Orders Echinothurioida, Diadematoida, Arbacioida, Temnopleuroida, Echinoida, Cassiduloida, Clypeasteroida, Spatangoida)

CLASS HOLOTHUROIDEA
Sea cucumbers. Elongate echinoderms with skeleton reduced to spicules embedded in leathery body wall; ambulacral system closed; coelom spacious; single gonad; larval stages auricularia and doliolaria, or vitellaria. (Orders Dendrochirotida, Aspidochirotida, Apodida, Molpadida, Elasipodida)

CLASS CRINOIDEA
The feather stars and sea lilies. Central body surround by branching arms with pinnules, supported by ossicles; coelom restricted to central body; ambulacral system closed. Larva a vitellaria. (Orders Isocrinida, Comatulida)

Chapter 17

The invertebrate Chordata, Hemichordata and Chaetognatha

L. Stocker

Introduction 397

PHYLUM CHORDATA

Subphylum Urochordata 398
   Evolution 398

Classes Thaliacea and Larvacea 398
   Diversity 398

Class Ascidiacea 401
   Diversity 401
   Solitary, social and compound ascidians 402
   Ascidian form and function 402

Reproduction and development 405

Fission and fusion in colonial ascidians 406

Subphylum Cephalochordata 409
   Reproduction and development 411

PHYLUM HEMICHORDATA

Introduction 412

Class Pterobranchia 412

Class Enteropneusta 412
   Reproduction and development 413
   Evolution 414

PHYLUM CHAETOGNATHA

Introduction 414

Reproduction 415

Evolution 415

Classification of the phyla Chordata, Hemichordata and Chaetognatha 415

Introduction

There is often a chapter near the end of invertebrate textbooks featuring a collection of seemingly disparate taxa: the urochordates, cephalochordates, hemichordates, and chaetognaths. Many of the species in these groups can be highly abundant and conspicuous in certain seasons or in certain habitats. All are marine. Divers in temperate waters may observe walls and roofs of marine caves completely lined with the delicate sky-blue ascidian *Podoclavella*; and observant sailors are no strangers to the transparent thaliaceans covering the sea surface in great rafts in summer months.

These taxa share with echinoderms the feature of being deuterostomes. Their coelom is formed from the outpocketing of enterocoelic pouches of mesoderm, the basic number being three pairs. The mouth is formed as a secondary opening, with the anus developing from the posterior blastopore. Cleavage in the early development of deuterostome embryos is indeterminate and radial.

Three of these taxa also have chordate connections, visible in later stages of their life history. The urochordates and cephalochordates are subphyla of the phylum Chordata and share their body plan with the vertebrates. The Hemichordata form a separate phylum but share some characteristics with the chordates. The Chaetognatha form a phylum of their own, with no chordate references beyond their deuterostome development. The relationships among the deuterostome phyla are assessed in Chapter 18.

Phylum Chordata

Bilaterally symmetrical deuterostome coelomates; pharyngeal slits, dorsal notochord, dorsal nerve cord; circulatory system with ventral heart; postanal tail; separate sexes or hermaphrodite, coelomic gonads; no ciliated larval stage.

The chordates represent a small fraction of animals on Earth, whether measured by numerical abundance or species diversity, yet are of great interest to humans because of our own chordate status. It may be hard to imagine what characteristics we have in common with certain gelatinous, leathery or worm-like creatures that frequent fairly specialised marine habitats. The answer is threefold. Chordates have in common a notochord, a dorsal hollow nerve cord, and paired lateral slits in the wall of the pharynx.

These characteristics rarely persist during all phases of the life cycle of any given chordate species. The notochord is a slightly flexible shaft which runs the length of the chordate body. In the majority of chordate species, the notochord is obvious only in embryonic or juvenile stages. In adult chordates the notochord is usually lost altogether or is largely superseded, as in the vertebrates, by the vertebral column. The hollow nerve cord lies dorsal to the notochord and develops as the central nervous system. The pharynx is the tube which connects the mouth, or its equivalent, to the oesophagus. During the stage(s) of the life cycle in which the pharynx is perforated, it can have a respiratory or filter-feeding function.

Figure 17.1 (opposite)
(a) *Pyrosoma* colony, showing central cavity and positions of blastozooids. **(b)** *Salpa* zooid with the visceral mass, which is frequently phosphorescent, visible through transparent test. **(c)** *Doliolum* with visceral organs visible through the transparent test. **(d)** *Oikopleura*. **(e)** *Pyura picta*, a solitary ascidian, showing inhalant and exhalant apertures. **(f)** *Perophora namei*, a social ascidian, showing zooids connected to the central stalk. **(g)** *Synoicum otagoensis*, a compound ascidian, showing systems of long parallel rows of zooids which extend along the sides of the colony and discharge through two common cloacal openings. (a from Bullough 1970; b, c, d from Smith 1977; e, g from Millar 1982; f from Kott 1985.)

Subphylum Urochordata

Evolution

'The Ancestor remote of Man,'
Says Darwin, 'is the Ascidian.'
A scanty sort of water beast
That ninety million years at least
Before Gorillas came to be
Went swimming up and down the sea.
— *Andrew Lang, Scottish scholar and poet (1844–1912)*

As the poet suggested, it was once thought that urochordates were ancestral to humans, because of the 'primitive' chordate characteristics exhibited by this group. Other researchers suggested that the urochordates represent some kind of missing link between the vertebrates and the echinoderms. It is more likely, however, that the urochordates, cephalochordates and vertebrate chordates diverged very early in the evolution of the phylum Chordata, and that they represent the only three surviving lines of a number of early chordate prototypes. The lack of hard parts in urochordates makes the task of understanding their evolution a difficult one; they are very poorly represented in the fossil record.

The subphylum Urochordata is composed of three classes: Thaliacea, Larvacea and Ascidiacea. The urochordates are also known as tunicates because of the sack-like tunics surrounding the bodies of these animals.

Classes Thaliacea and Larvacea
Diversity

Members of the Class Thaliacea are free-swimming oceanic animals. They are variously colonial, aggregated or solitary in form, with 'tunics' that are generally transparent and gelatinous in texture. The body and perforated pharynx of the thaliacean zooid is barrel-shaped. The water current passing through the pharynx is used for respiration, feeding and in some cases, locomotion. Water currents created by cilia are enhanced by muscular bands that encircle or partially encircle the body wall of the zooid. Thaliacean individuals which develop from fertilised eggs are called oozooids and have lost the ability to reproduce sexually. They proliferate asexually, producing individuals called blastozooids which are, in turn, sexual protogynous hermaphrodites.

Thaliaceans are composed of three orders: Pyrosomatida, Salpida and Doliolida. *Pyrosoma*, the single genus in the order Pyrosomatida, forms hollow, thimble-shaped colonies of blastozooids (Fig. 17.1a), reaching several metres in length. The rear end of the colony is open. The zooids are embedded in a matrix of transparent tunicin, and are oriented so that the long axis of the pharyngeal cavity spans the colony wall. Water passing through the pharynx of a zooid is exhaled into the central cavity, and eventually out the open end of the colony. Living largely in tropical and subtropical waters, *Pyrosoma* colonies can be highly conspicuous at night because of their phosphorescence; indeed, the name of the genus means 'fire-body'. The light emanating from them may illuminate the water up to 50 cm away. Early naturalists such as Lesueur and Péron gave the first accounts of long trains of fire whose phosphorescing colours varied from intense red to yellow, from gold to orange, to green or to azure blue. The Brazilian navigator

THE INVERTEBRATE CHORDATA, HEMICHORDATA AND CHAETOGNATHA

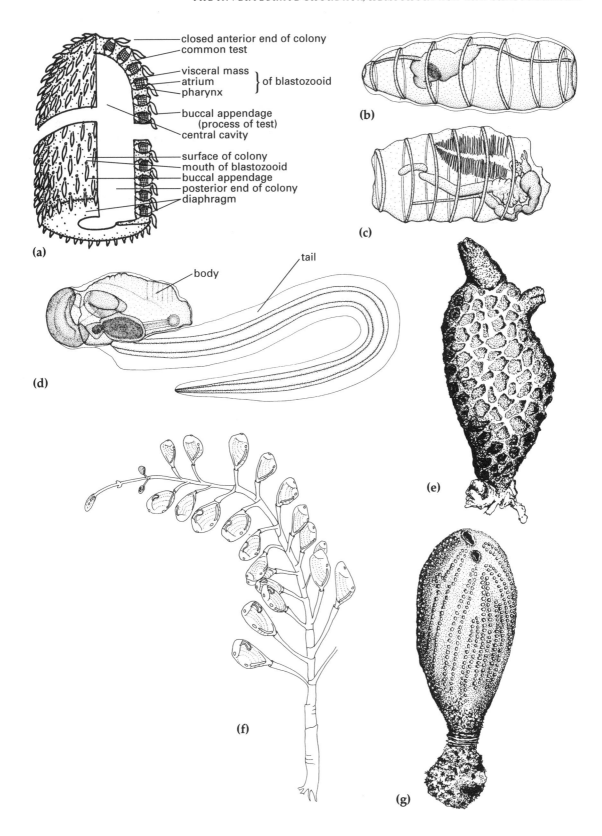

Bibra claimed that he caught six *Pyrosoma* and deployed them to light his cabin. *Pyrosoma atlanticum* has been recorded from Australasian waters but is rare.

Salps inhabit the surface waters of all the world's oceans. They (Fig. 17.1b) are delicately diaphanous, and the functioning of the various organs can be observed through their gelatinous tunic. Salp 'colonies' form chains that are really loose aggregates; the chains fall apart when lifted from the water and will not reaggregate. *Salpa* zooids lie side by side, and each row of zooids is attached to a second row to create a double parallel chain. During the life cycle, generations alternate between the 'colony', formed by the asexual proliferation of a chain of blastozooids, and a solitary oozooid which is produced sexually. As with *Pyrosoma*, there is no free-swimming larval stage. The salp swims with its ventral surface up, by contracting its incomplete circular muscle bands. All zooids in a chain of salps act in concert. Salps, like *Pyrosoma*, are highly phosphorescent and have been called ribbons of fire or even 'sea-serpents' by sailors. In Australian waters the common salp is *Salpa (Thalia) democratica* which occurs in extraordinarily high numbers, especially in spring, even washing up and filling tide pools at times. *Iasis zonaria* is another common salp collected in plankton trawls.

Doliolids (Fig. 17.1c) are primarily tropical and subtropical in distribution but also occur in temperate waters. Like other thaliaceans they are transparent. Doliolids have a very complex life history, featuring three types of blastozooids, an alternation of generations, and a larval form. Unlike the other two orders of thaliaceans, the eggs of doliolids develop through a tadpole larval phase. These larvae result from sexual reproduction by solitary, sexually mature blastozooids, and develop into individual asexual oozooids. Each oozooid buds off blastozooids which, by means of further budding, each produce a colony containing several types of specialist zooids, including gastrozooids (the feeders), phorozooids (the nurses), and gonozooids (the sexual individuals). The gonozooids, which each produce three ova, one at a time, ultimately break away to become the solitary, sexually mature blastozooids. These swim freely by contracting the muscle bands that completely encircle the body wall, which drives water through the pharynx. Among the commoner doliolids in Australasian waters, *Doliolum denticulatum* is frequently recorded in plankton trawls from January to June.

Members of the Class Larvacea, also known as appendicularians, are free-swimming individuals throughout their life history and form a part of the plankton of most of the world's oceans. Species such as *Oikopleura* and *Fritillaria* can occur in very large blooms at certain times of year. Lacking the colonial phase, larvaceans are less spectacular when seen from the deck of a boat than their larger phosphorescing relatives, the thaliaceans. The view of a larvacean down a microscope, however, is impressive. As their name hints, sexually mature adult larvaceans retain the tadpole form (Fig. 17.1d) and associated chordate features displayed in the larval stage of ascidian urochordates (see below). The notochord and dorsal hollow nerve cord are well developed in the tail, which is ventrally attached and is undulated to effect locomotion. The body contains the pharynx, which has only two slits. The larvaceans also exhibit the extraordinary, and unique, habit of constructing a house made of mucus, surrounding the body in *Oikopleura*, and adjacent to it in *Fritillaria*. The house is a filter-feeding device with a two-stage system. The outer filter, with a coarse mesh, is used for screening out inedible particles, while the inner, fine filter is used for catching edible particles which are then ingested. When the outer filter becomes clogged or the house damaged, the whole house is discarded, typically happening every few hours. A fresh one can be constructed in an hour.

The Class Ascidiacea, which forms the third major group of urochordates, is discussed below.

Class Ascidiacea
Diversity
In complete contrast to the pelagic thaliaceans and larvaceans, the Ascidiacea form a major group of benthic, sessile invertebrates. Ascidians (Fig. 17.1e–g) range in form from warty and leathery to dainty and transparent, from extensive sheets to football-shaped sacks, from solitary to colonial, from 10 mm to more than 1 metre long, and they exhibit a multitude of colours. The basic biology is the same among ascidians, except that the colonial forms have connected zooids and associated modifications such as shared exhalant and inhalant siphons. Ascidians are commonly known as sea squirts because of the tendency of the solitary forms to spray water from their siphons when mechanically distorted.

Like other urochordates, ascidians live exclusively in marine habitats. Most families are geographically widespread and have representatives at all latitudes. The abundance of some shallow-water species varies seasonally, and at the peak of their growth period they may overgrow all neighbouring biota. Other species are conspicuous all year round and may be zone-forming in intertidal or shallow subtidal habitats.

Ascidians live on sandy and rocky substrata and often as 'fouling' organisms on boats, jetties and rigs. They feature prominently in the sessile communities of hard substrata, both intertidally and subtidally. In Australia, for example, the best known solitary ascidian is the large *Pyura stolonifera* (or cunjevoi), a brown-green leathery animal forming dense beds on the rock platforms of the east coast. In northern New Zealand *Cnemidocarpa nisiotis* similarly characterises a zone in the shallow rocky subtidal. Immediately below the low-tide mark on rocky shores or pier pilings in New Zealand and Australia, another pyurid, *Pyura pachydermatina*, is found in large numbers. The red colour and stalked habit of this species account for its common name of 'sea tulip'. All three species are highly resilient to the high energy levels of their environments. Colonial ascidians are less identifiable than solitary ascidians to the untrained eye because their appearance can superficially resemble sponges or other colonial organisms. The sandy or brightly coloured gelatinous *Aplidium*, thin crusty *Didemnum*, and *Botrylloides* with its conspicuous colonial systems, are among the commoner taxa of colonial ascidians in Australasian waters. They contribute to the colourful mosaics beneath kelp forests, on subtidal cave walls, overhangs and drop-offs.

The Class Ascidiacea is divided into three orders. In the Order Aplousobranchia, the gut loop is posterior to the pharynx and the body is divided into thorax and abdomen, with (in some species) a post-abdomen. All aplousobranch species are colonial. Examples include *Didemnum, Podoclavella, Cystodytes, Sycozoa* and *Aplidium*. In the Order Phlebobranchia the gut loop is alongside the pharynx, the gonads are on the same side of the body as the gut loop, and the pharyngeal basket is never folded. Some species are colonial. Examples include *Corella* and *Ascidia* (both solitary) and *Perophora* and *Ecteinascidia* (both colonial). In the Order Stolidobranchia the gut loop is alongside the pharynx, but the gonads are distributed on both sides of the body, and the pharyngeal basket is usually folded. Some species are colonial. Examples include *Styela, Cnemidocarpa, Pyura* and *Molgula* (all solitary), and *Botrylloides* and *Alloeocarpa* (both colonial).

Solitary, social and compound ascidians
Ascidians may be categorised as solitary, social or compound. Social and compound ascidians are both colonial. A 'solitary' ascidian (Fig. 17.1e) consists of a discrete individual encased in tunic which is usually either firmly gelatinous or tough and leathery, the latter especially in larger species. Individuals can grow to be many centimetres long. Solitary ascidians are often gregarious, massing together in closely abutting clumps, e.g. *Cnemidocarpa nisiotis* or *Pyura stolonifera*. The key feature of solitary ascidians, however, is that each arises from a separate larva.

A 'social' ascidian (Figs 17.1f, 17.2b) is a colony consisting of a group of zooids which have arisen through asexual budding from a single larva, but are connected only by a stolon or thin basal mat. Zooids thus formed are clones and together form a colony. The tunic is usually thin and delicate but can be encrusted lightly with sand, making the colony difficult to see. The zooids are usually less than 20 mm in the longest dimension, although in some species they can reach several centimetres. Colonies may occupy several square metres on rocky, vertical, subtidal walls. In many parts of Australia's Great Barrier Reef, *Ecteinascidia nexa*, a flimsy green social ascidian, is found in loose clumps, its individual members connected by basal stolons. A more unusual example of a social ascidian is provided by *Clavelina nodula* from South Australia, in which the red zooids arise separately from a firmly gelatinous, translucent branching stalk. New Zealand examples include the pale sand-encrusted dense mats of *Theodorella torus*, and the more sparsely distributed brilliant red domes of *Alloeocarpa minuta*, connected by fine stolons.

A 'compound' ascidian (Figs 17.1g, 17.2b) is a group of zooids which have arisen through asexual budding from a single larva and which are wholly encased in a common matrix or test. Zooids are usually less than 10 mm long, sometimes only 2 mm. In some species with clear gelatinous tests, the bodies of the zooids are visible from the exterior, as in many species of *Aplidium*. In others, systems of zooids can be distinguished by virtue of the inhalant siphons, which have a colour that contrasts with the remainder of the test, as in the purple *Botrylloides magnoecious* with its yellow systems of zooids. In many compound species, however, the tests are so invested or encrusted with sand or with calcareous spicules that zooids are not visible. Compound ascidian species can occur as gelatinous footballs, thin leathery sheets full of sand or spicules, small knobs or cushions, or fleshy sprawling masses.

Two points are important in relation to the description of ascidian form. Firstly, the distinction between social and compound ascidians is an old one which is retained here because of its descriptive power. Social and compound ascidians are usually discussed together as colonial ascidians. Secondly, the above categories cut across taxonomic categories at the level of order and family.

Ascidian form and function
Feeding in ascidians, as in other urochordates, involves the use of a filtering bag called the pharyngeal basket, which occupies the majority of the volume of the zooid (Fig. 17.2a). Water is pumped in through an inhalant

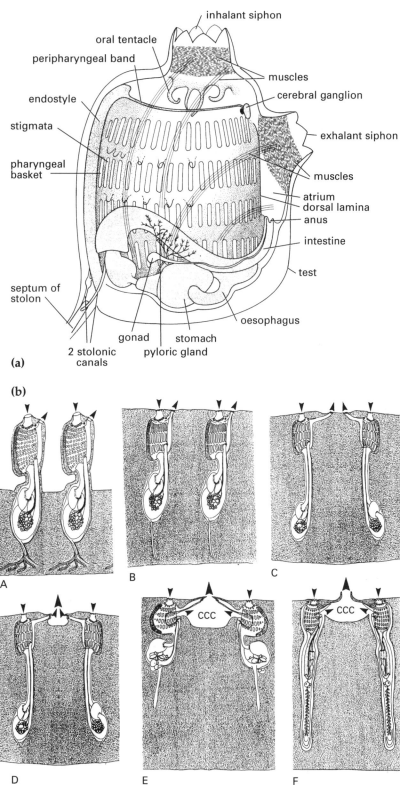

Figure 17.2
Ascidian form. (a) *Perophora*. Individual zooid of social ascidian of the order Phlebobranchia (from Sherman and Sherman 1970). (b) Evolution of cloacal systems in the order Aplousobranchia: A — partially embedded zooids, 'social', no systems (e.g. *Pycnoclavella*); B — completely embedded zooids, 'compound', with exhalant siphons opening separately to the exterior (e.g. *Sigillina*); C — zooids arranged in circles, with exhalant siphons opening separately in the centre to form incipient cloacal systems (e.g. *Cystodytes*); D — zooids with exhalant siphons opening into rudimentary (*Eudistoma*) to extensive (*Hypodistoma*) cloacal cavities; E, F — zooids arranged in well-developed cloacal systems, each of which opens to the exterior by a single cloacal aperture. (ccc, common cloacal cavity; arrows show the direction of current flow). (From Kott 1990.)

siphon to the pharyngeal basket, where food particles and oxygen are extracted, and then pumped out through an exhalant siphon. The external test or tunic which surrounds and protects the body is punctured by the two siphons. A holdfast at the base fixes the ascidian to the substratum. Whether tough or fragile, the test is acellular. It commonly contains a fibrous matrix of tunicin, which is largely cellulose, and also proteins, water, blood cells, and a few migratory mesodermal cells. Blood vessels and calcareous spicules, usually star-shaped, can also occur in the test. Developmentally, the test is secreted by ectoderm tissue.

Inside the test is the body wall, containing longitudinal and circular strands of muscle whose contraction results in the rapid expulsion of water (the squirting of the sea squirt) from both siphons (Fig. 17.2a). Bands of muscles also surround the siphons, allowing them to be closed. The muscles work against the test on one side and against the pharynx on the other. The body wall also contains the nervous system, consisting of a small cerebral ganglion located between the two siphons, from which a plexus of nerves ramifies to the siphons, muscles, organs and pharynx. There are no obvious sense organs in ascidians. The nervous system seems to be used primarily for mechanical reception around the siphons, with some suggestion of a capacity for chemical and light reception, the latter being indicated by the presence of ocelli, or pigment cells. The function of a neural gland lying beneath the central ganglion is unknown.

Feeding and digestion

The most prominent feature of the ascidian body is the pharynx, which is mostly surrounded by the buccal and atrial cavities enclosed by the test. At the base of the zooid lie the oesophagus, stomach, heart and reproductive organs (Fig. 17.2a). In some species there is a distinct postabdomen in which the heart and reproductive organs are located. Water is drawn into the buccal cavity through the inhalant siphon. Oral tentacles at the entrance to the pharynx eliminate any large particles that could not be processed within the pharynx. The sack-shaped pharynx is perforated by slits, known as stigmata. The resulting effect is a basket with a grid-like arrangement of transverse bars separating each of six or more rows of stigmata. The stigmata are about 50 μm wide. Longitudinal bars also exist in stolidobranchs and phlebobranchs. The water passes from the pharyngeal basket, via the stigmata, into the atrial cavity and out through the exhalant siphon. The water current is created by the beating of lateral cilia situated on the bars of the pharyngeal basket. A mucous sheet is produced by gland cells of the endostyle, which lies along the ventral midline of the pharynx. The mucous sheet extruded from the endostyle is moved across the inner wall of the pharynx, where it collects particles of food from the water current. This movement, which is brought about by the beating of frontal cilia on the bars of the pharyngeal basket, carries the mucous sheet to the dorsal side of the pharynx. Experiments have shown that the mucous sheet traps graphite particles of 1–2 μm, but that haemoglobin molecules of 3 nm will pass through. At the dorsal edge of the pharyngeal basket, the mucous sheet with its plankton and other food particles is manipulated by a row of languets (the dorsal lamina) into a rope, in preparation for its passage through the remainder of the digestive system.

The oesophagus, stomach and intestine form a U-shaped tube. Material is passed along this portion of the gut by cilia, as there is no peristaltic action in ascidians. From the oesophagus, food travels to the stomach, where enzyme secretion occurs, and thence to the midgut where absorption occurs. Waste material is passed out through the rectum and anus into the atrial cavity and expelled through the exhalant siphon.

Circulation

Respiratory exchange occurs across the surface of the pharyngeal basket, where oxygen is taken up from the water current. There are no specialised respiratory organs. The heart is located at the base of the pharynx, or in the postabdomen, and is a simple transparent tube. From the heart two blood vessels lead away, one towards the pharyngeal region and one towards the abdominal region. The pharynx is served by a large vessel that runs beneath the endostyle and sends branches into the bars of the pharyngeal basket. On the dorsal side of the pharynx, the vessels converge to form the hyperpharyngeal band vessel. The abdominal organs are served by a ramifying system of vessels that also converge on the hyperpharyngeal band vessel, thus completing the blood-vascular circuit. The vessels are simple lacunae or mesenchymal channels, lacking true walls. Muscular contractions travel along the heart in one direction for a few minutes and are then reversed, so that the blood flows the other way. It is thought about 20 circuits of the blood around the circulatory system are made alternately in each direction. The direction of muscular contractions is controlled partly by internal pacemakers in the heart, but may also be affected by back pressure within the vascular system.

The blood of ascidians is a clear plasma containing several types of cells. One notable type is the vanadocyte, the vacuoles of which in some species accumulate high concentrations of vanadium. The role that vanadium plays is uncertain, although there are some claims that it may be involved in oxygen binding.

Reproduction and development

Ascidians, whether solitary or colonial, are sexually reproducing hermaphrodites, usually exhibiting cross-fertilisation. Many species shed eggs and sperm into the water, leading to external fertilisation. In others, sperm are shed freely but eggs are retained and fertilised within the atrial cavity of the parent. Whichever sequence is followed, the eggs of ascidians are yolky and develop into lecithotrophic larvae called ascidian tadpoles (Fig. 17.3). The tailed larva swims briefly, for a few minutes to a few hours depending on the species, and then settles and metamorphoses to a juvenile zooid, the starting point of a new solitary ascidian or a new ascidian colony. During metamorphosis the tail, with its characteristic chordate notochord, dorsal nerve cord and bilateral muscle bands, is resorbed. Feeding does not begin until the metamorphosed juvenile has developed a functional pharyngeal basket.

The embryonic development of ascidians is a mixture of basic chordate processes and specialised features. Cleavage, gastrulation and organogenesis, in addition to being deuterostome, follow a pattern shared with cephalochordates and vertebrates, especially in the mode of formation of the notochord, dorsal nerve cord and associated dorsolateral mesoderm (Fig. 17.4a–c). In later development the anterior region establishes, more or less directly, the structures carried through metamorphosis into the subsequent zooid. The posterior region, in contrast, develops into the temporary larval tail, in which the notochord, dorsal nerve cord and lateral muscles form an undulatory propulsive unit. Perhaps the most interesting feature of

Figure 17.3 (opposite)
Ascidian larvae and metamorphosis. (**a**) *Cystodytes dellachiajei*, a brooding species.
A — immature larva;
B — maturing larva;
C, D — older maturing larvae before perforation of ampullary fold; E, F — larvae with 5 adhesive organs, before and after perforation of the ampullary fold. (**b**) Ascidian tadpole larva showing pharynx, notochord and dorsal nerve cord. (**c**) Metamorphosis of the free-swimming larva of *Pyura pachydermatina*.
A, B stages in tail resorption;
C, D development of ampullae;
E, development of stalk and siphons. (**d**) Zooid of *Polyclinum glabrum*, with embryos developing in atrial cavity. (a, d from Kott 1992; c from Berrill and Karp 1976; d from Anderson *et al*. 1976.)

the ascidian tail is that, unlike the homologous axial construction in cephalochordates and vertebrates, it shows no trace of metameric segmentation or coelomic cavities. The absence of coelom is due to secondary loss, but the absence of metameric segmentation may reflect an early divergence of urochordates before metameric segmentation evolved within the chordate line. The ascidian tadpole is much too specialised to have provided an evolutionary ancestry for the vertebrates.

Fission and fusion in colonial ascidians

The early naturalist Fredol wrote: 'Here we behold certain animals which eat separately, but which fulfil together as a community very singular functions — a kind of union and communism of which the moral world presents no prototype. With our molluscs [*sic*] we have a score of individuals united. We may consider the entire star as one single animal with many mouths'. The paradox of coloniality in ascidians thus expressed is still exercising biologists and ecologists today as they strive to create models and collect data that make sense of the life styles of these organisms.

Biological models for most solitary organisms describe phenomena such as sexual reproduction, simple age-related growth, and genotype mortality. These, however, may be unimportant or even non-existent in populations of colonial ascidians, for whom reproduction, growth and death can occur by very different means. As a way of providing a conceptual framework for the study of these aspects of colonial and other clonal organisms, the term 'modularity' has become popular; it underscores the importance of a basic unit of construction. In colonial ascidians, zooids, which are the smallest modules, constitute the first level of modularity, but there are other, larger modules, as we shall see. The life histories of colonial ascidians show four features not found in solitary ascidians:
1 complexity and multidimensionality of life-cycles;
2 decoupling of effects of size and age of modules;
3 variations in the relative contributions of asexual and sexual reproduction to total recruitment; and
4 short-range dispersal of sexual and asexual propagules. Each of these will now be examined.

Complexity and multidimensionality of life cycles

In colonial ascidians, 'growth' takes place primarily by the asexual proliferation of joined individuals, rather than by the continued increase in size of a single individual. In many colonial ascidians, most notably in the genus *Didemnum*, there is a second level of modularity, resulting from fission of the whole colony. Fission may follow damage or tissue decay, or may be endogenously controlled. In *Didemnum moseleyi* and *D. molle*, colony fission occurs over a period of one to several days. Asexual propagules, each consisting of a number of zooids, result from whole-colony fission. The propagules may remain *in situ* or move small distances. Each propagule derived from a single genotype (i.e. from a single sexual larva) is called a ramet; ramets sharing a single genotype are collectively called a genet.

The existence in a population of closely related colonies aggregated together also presents the possibility of fusion. Active fusion or rejection in natural populations relies on the ability to recognise self from not-self, as in sponges, cnidarians and ectoprocts. Fusion in colonial ascidians can occur between clone-mates, between parents and offspring, and between siblings. Obviously, the existence of colonies made up of more than one genotype as

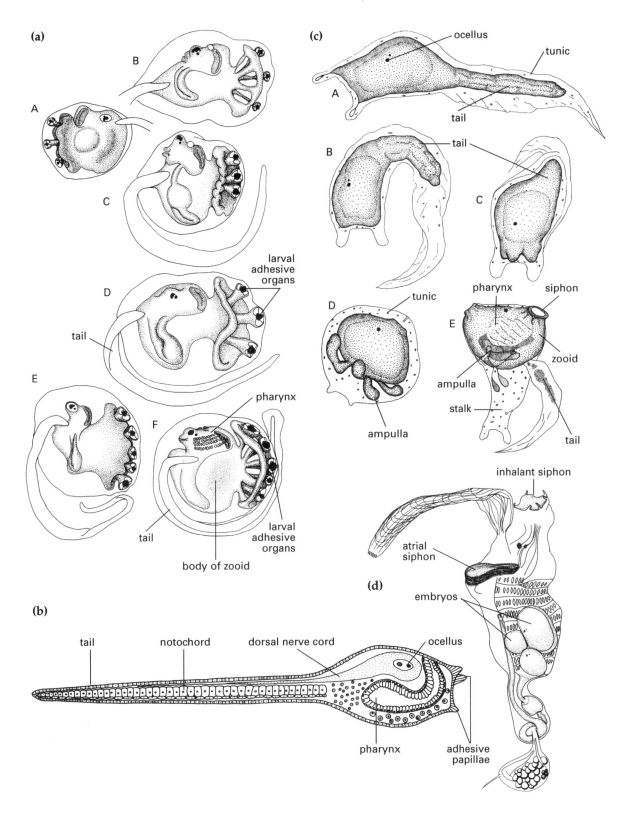

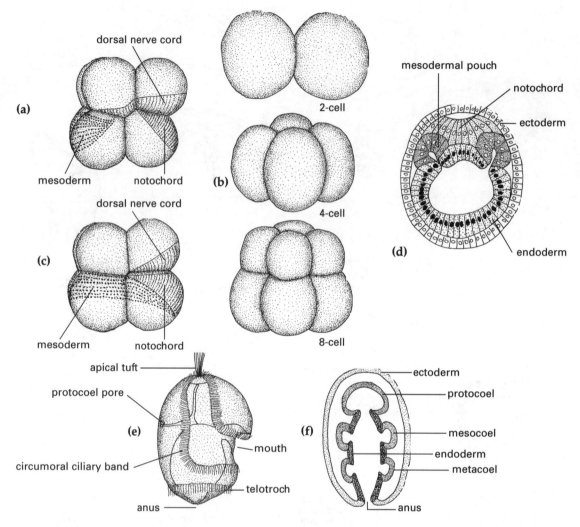

Figure 17.4
(a) Eight-cell stage of ascidian cleavage, showing the ate map in lateral view. (b) Two-cell, four-cell and eight-cell stages of cleavage in *Branchiostoma* (Cephalochordata). (c) Fate map of the eight-cell stage of *Branchiostoma*. (d) Transverse section of *Branchiostoma* embryo, showing the formation of dorso-lateral coelomic pouches.
(e) Tornaria larva of *Saccoglossus* (Hemichordata). (f) Coelomic pouch formation in the larva of *Saccoglossus*. (a–d after Berrill and Karp 1976; e, f after Nielsen 1995).

a result of fusion greatly complicates the concept of a 'colony' in genetic terms.

The death of an individual zooid, or even a whole ramet, does not imply the extinction of the genotype, as the remainder of the genet can continue to grow and reproduce. This curious phenomenon is known as partial mortality. Fission also increases the possibility of genotype immortality, theoretically at least; the presence of a number of ramets makes it less likely that they will all be killed by a single predation or disturbance event. Fission, fusion and partial mortality all influence the spatial distribution, replication and survivorship of the modules, and all contribute in an important way to the fate of the genotype.

Decoupling of effects of size and age of modules

The processes of fission, fusion and partial mortality imply a very poor correlation between size and age in colonial ascidians. Fission and fusion can cause an instantaneous decrease or increase in the size of a ramet. The usual relationship between size and age that is found in other organisms can thus be decoupled in these colonial organisms. The fusion of recruits soon after settlement, and the gregarious settlement of siblings, can also have a dramatic effect on the size of a colony.

Contributions of asexual and sexual reproduction to total recruitment
In many species, like *Didemnum moseleyi* in Sydney Harbour, the ratio of sexual to asexual reproduction varies among sites. Differences in the ratio of sexual to asexual reproduction between sites do not necessarily reflect the existence of adaptive variations in life-history from one place to another. Rather, these microgeographic differences might reflect more immediate physiological demands. Encrusting communities are subject to the vagaries of a variety of physical factors. These vagaries are known to cause lethal or sublethal effects on benthic organisms. They include fresh water, sedimentation, pollution, scouring of various sorts, battering by waves, and boat traffic. High temperatures may also take their toll on certain species. All of these factors could account for differences in life cycles and relative rates of asexual and sexual reproduction in populations of colonial ascidians.

Short-range dispersal of sexual and asexual propagules
The sexual propagules (tadpole larvae) of colonial ascidians are generally in the water column for only a short period of time and do not travel far from their parent colonies. The extent of dispersal after fission of whole colonies may therefore have important implications for the spatial dispersion of the genets. In some species the modules remain close to the parents and are tightly aggregated; in others, modules travel far from the parents. Clearly, there is a continuum between these two extremes. Whether genets will interdigitate physically depends on how far larvae travel from the parent. If larvae have a short range of dispersal, genets may be close enough for modules from different genets to become mingled. The majority of colonial ascidians exhibit a tightly aggregated pattern, so that mingling of genets is likely. Much remains to be learned of these aspects of the dynamics and ecology of colonial ascidians.

Subphylum Cephalochordata

In contrast to the barrel-shaped zooids of the urochordates, the cephalochordates are long, slim, fish-like swimming animals (Fig. 17.5a). Uniquely, the diagnostic features of chordates are all present in adult cephalochordates. The notochord runs along the length of the adult and ends in a point at the anterior end of the animal; the hollow nerve cord runs immediately dorsal to the notochord; and there are numerous pharyngeal slits. The subphylum is represented in the extant fauna only by the genera *Asymmetron* and *Branchiostoma*. The genera are similar in appearance, but differ in that *Asymmetron* bears gonads on the right side of the body only whereas *Branchiostoma* has gonads on both sides. *Branchiostoma* is the better known of the two taxa and is usually referred to by its old name of amphioxus.

Amphioxus occurs in the shallow subtidal regions throughout the world. It inhabits sandy or gravelly bottoms, being often very localised at high densities. In coarse sand or gravel, amphioxus will bury itself completely, in an upright position. Feeding is still possible when fully buried because water can pass between the sand grains. In finer sand the anterior portion of the body remains projecting into the water column. Amphioxus

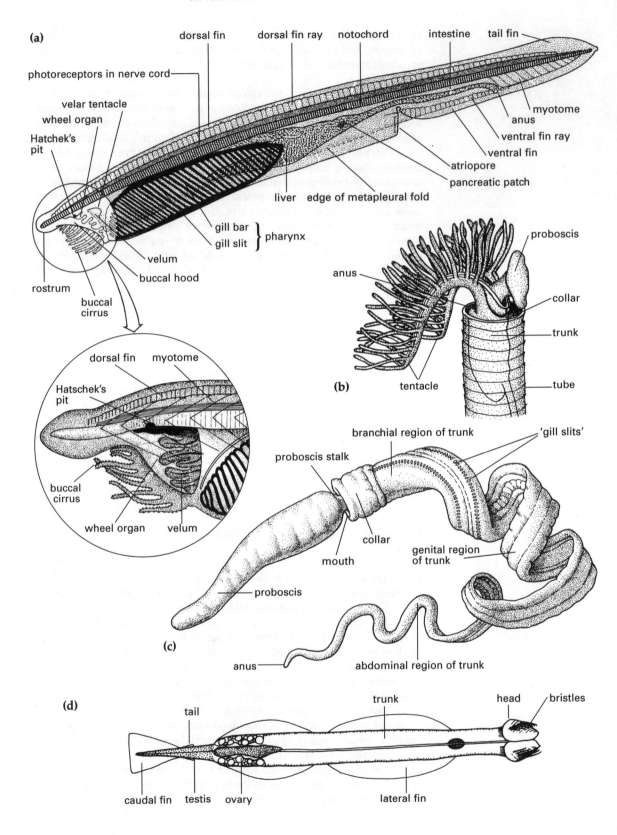

Figure 17.5 (opposite)
(a) *Branchiostoma*: lateral view showing cephalochordate features. **(b)** *Rhabdopleura* (Pterobranchia). **(c)** *Saccoglossus* (Enteropneusta), with distinct proboscis, collar and trunk. **(d)** Chaetognath, showing the division of body into head, trunk and tail. (a from Sherman and Sherman 1970; c from Sherman and Sherman 1970; d from Smith 1977.)

emerges into the water column at night and can move rapidly, swimming like a fish. It can also travel quickly through the sand.

Although amphioxus resembles a fish superficially, it differs in having no specialised head, paired limbs, heart, brain, or kidney, and no skull or vertebrae. The large lateral muscle sheets are divided by septa into metameric blocks, which interact with the notochord to produce the lateral undulations used in swimming and burrowing. There are ventral, dorsal and caudal fins. Amphioxus has a filter feeding system similar to that of ascidians, although the spatial arrangement of the elements is different (Fig. 17.5a). The anterior mouth is covered by an oral hood which is fringed with sensory processes called cirri. Lateral cilia on the long pharyngeal slits create a water current that passes in at the mouth, through a pharyngeal sheet of mucus on which food particles are extracted, out through the pharyngeal slits into the atrium, and then exits at the ventral atriopore, located two-thirds of the way along the body. The mucous sheet is produced by a ventral endostyle, as in urochordates, carried dorsally across the pharyngeal basket by frontal cilia and passed posteriorly into the stomach. Digestion occurs in a blind diverticulum called the hepatic caecum, which protrudes from the stomach along the right side of the pharynx. The posterior portion of the intestine terminates at the anus, located ventrally near the posterior of the animal.

Reproduction and development

Amphioxus has separate sexes. Eggs and sperm are shed into the water and fertilisation is external. The small eggs develop directly into juvenile amphioxus, which do not begin to feed until the pharyngeal basket becomes functional. There is no larval phase in cephalochordate development.

The embryonic development of the amphioxus egg is a model for embryonic development in metamerically segmented chordates. It combines deuterostome development with the typical chordate formation of notochord, dorsal nerve cord and dorsolateral mesodermal pouches (Fig. 17.4c, d). A trimeric structure is established at first through the formation of three pairs of mesodermal pouches (protocoel, mesocoel and metacoel), as in echinoderms and hemichordates. With further development of the mesoderm, the metacoels proliferate the paired metameric somites that develop into the paired muscle blocks of the body, and the initial trimeric construction is obscured. One persistent feature in adult amphioxus, however, is Hatschek's pit (Fig. 17.5a), an opening from the exterior into the coelom of the left side under the oral hood. This opening first develops in the embryo as an outgrowth from the left protocoel and is homologous with the hydropore of echinoderms (Chapter 16) and the proboscis pore of hemichordates (see below).

Phylum Hemichordata

Bilaterally symmetrical deuterostome coelomates of trimeric construction; pharyngeal slits; dorsal nerve cord; circulatory system; separate sexes, extracoelomic gonads; ciliated tornaria larva in some.

Introduction

The hemichordates possess only two of the three diagnostic features of the chordates: pharyngeal slits and a dorsal hollow nerve cord. They also possess a structure which superficially resembles a notochord, but which is considered to have evolved independently of the chordate notochord. Hemichordates have a tripartite body structure (protosome, mesosome, and metasome), each containing a pair of coelomic cavities (protocoel, mesocoel and metacoel, as in echinoderms). There are two classes of hemichordates: the Pterobranchia, which are very obscure, and the Enteropneusta, which are very well known.

Class Pterobranchia

The relatively tiny pterobranchs (Fig. 17.5b) comprise the genera *Cephalodiscus*, *Rhabdopleura* and *Atubaria*. Pterobranchs are sessile and dwell in deep water. *Cephalodiscus* and *Rhabdopleura* secrete tubular houses from a shield-shaped prosomal proboscis. Both genera reproduce by asexual budding. In *Rhabdopleura*, dredged from depths of 100–300 m in the Atlantic, the clone mates form genuine colonies. The individuals are connected by stolons within the branching translucent tubes that are attached to stones or shells. In *Cephalodiscus*, clone mates establish unconnected, but often dense, aggregates.

The pterobranch body is divided into protosomal proboscis, mesosomal collar and metasomal trunk. The trunk curves back on itself, the gut is U-shaped, and the mouth and anus both open to the top of the tube. Each individual has branched tentacles arising from the mesosomal collar: two in *Rhabdopleura* and up to 12 in *Cephalodiscus*. These function in catching plankton from the water column. The processes are out-stretched for feeding but can be retracted when disturbed. *Cephalodiscus* has one pair of pharyngeal slits, while *Rhabdopleura* has a pair of dorsolateral grooves in the pharynx. Both genera have a single pair of gonads which lie in the anterior part of the metasomal trunk. Pterobranchs do not have a free-swimming larval stage; they are direct developing. The developmental stages bear a marked resemblance to young enteropneusts (see below), with a tripartite worm-shaped body and terminal anus.

Class Enteropneusta

The enteropneusts (Fig. 17.5c) are called acorn worms because of their prominent acorn-shaped proboscis and worm-shaped body. There are less than 100 species known. Major genera include *Balanoglossus*, *Saccoglossus*, *Dolichoglossus*, *Glossobalanus*, *Ptychodera* and *Schizocardium*. In Australia and New Zealand *Ptychodera flava* is common on sheltered sandy shores.

The most common enteropneust habitat is in the sandy intertidal zone in temperate waters. Species such as *Ptychodera flava* live in U-shaped, long-term burrows. The animal moves through the sand by peristalsis, using its muscular proboscis for burrowing, and lines the burrow with mucus secreted from the collar. Acorn worms are often detected by their piles of coiled castings at the surface at one end of the burrow, or by their distinctive iodine smell. Most hemichordates have a halogen odour of some kind. Typically 200–250 mm in length, these worms can only be removed from their burrows with the most careful of digging, as their fragile bodies easily rupture.

The enteropneust body is divided, as in pterobranchs, into the proboscis, collar and trunk (Fig. 17.5c). The proboscis is the key to nutrition for the deposit-feeding enteropneusts. The proboscis secretes mucus which collects food particles. Cilia carry strands of food-laden mucus to the ventral, posterior portion of the proboscis, where they are bundled together into a more substantial rope. This rope is then passed into the mouth, located ventrally between the proboscis and the collar. Material which is unacceptably large or distasteful is rejected. The mouth is connected to the pharynx by a short, simple tube that passes through the collar. The pharynx, located at the anterior end of the trunk, is perforated dorsally by slits that vary in number according to the species. The bars of the pharynx bear lateral cilia that drive a water current through the mouth into the pharynx. Water then passes out through the pharyngeal slits to the gill sacs, which surround the slits, and thence to the exterior of the animal through two lines of gill pores. A minor amount of food may be collected on the pharynx in the manner of the filter-feeding urochordates and cephalochordates. A third means of collecting food is by the direct ingestion of the sandy substratum with its content of organic particles. In whatever way the food is collected, it passes from the pharyngeal region into the oesophagus, and then into the long intestine where digestion and absorption occur. Digestive enzymes are secreted by the wall of the midgut and by specialised cells of hepatic sacs, often visible externally on the dorsal surface as dark green sacs. Faeces are passed out through the terminal anus.

Gas exchange is thought to take place across the pharyngeal slits as well as the general body epithelium. The blood system is simple. Blood flows anteriorly through a dorsal longitudinal channel and posteriorly through a ventral longitudinal channel, both of which are contractile.

The nervous system of enteropneusts is also simple. A nerve plexus ramifies between the ciliated epidermis and its basement membrane. In places the nerve fibres align longitudinally to form nerve tracts. One dorsal and one ventral tract run along the trunk region; they are connected by a circular nerve tract surrounding the posterior end of the collar. The dorsal hollow nerve cord, a hemichordate feature shared with chordates, runs anteriorly from the circular nerve along the collar. The proboscis is innervated by a dorsal nerve, a second circular nerve, and fringe of associated nerves.

Reproduction and development

Sexes are separate in enteropneusts. The gonads are in the coelom of the anterior portion of the trunk, occasionally creating swellings visible on the exterior. Each gonad opens separately to the exterior via a simple gonopore. Fertilisation is external and results in the formation of a ciliated tornaria larva (Fig. 17.4e) in many species, although in others development is direct. The embryonic and larval stages bear a notable similarity to those of echinoderms, although the tornaria larva has an additional band of cilia, the telotroch, at the posterior end. Internally, the larva develops paired mesodermal pouches (Fig. 17.4f) forming the protocoel, mesocoel and metacoel, and the protocoel gains a pore to the exterior homologous with the echinoderm hydropore and Hatschek's pit in cephalochordates.

Asexual reproduction by fragmentation of the body is also common in enteropneusts. In some species, the tendency of the animal to rupture when disturbed in its burrow is related to a natural ability to reproduce asexually by breaking into vegetative fragments. These fragments subsequently redifferentiate within the parent's burrow and eventually grow into adult acorn worms.

Evolution

Fossil graptolites, common among Ordovician and Silurian beds but then not in the lower Carboniferous, are believed to be related to the pterobranchs. The graptolites appear to have been colonial animals about 100 mm in diameter, with numerous long rows of serrated cups each of which probably held a zooid. Remarkably, and in contradistinction to contemporary pterobranchs, the graptolites were not sessile but pelagic free-floaters. A key to the graptolite relationship to pterobranchs is the proteinaceous exoskeleton of the graptolites, which contains some amino acids similar to those of the pterobranchs.

Phylum Chaetognatha

Bilaterally symmetrical deuterostome coelomates(?) of trimeric construction; fish-shaped, planktonic; brain, circumenteric nerve ring, ventral ganglion; no circulatory system; postanal tail; hermaphrodite, with coelomic gonads; direct development.

Introduction

This phylum consists of less than 100 species, commonly known as arrow worms (Fig. 17.5d). They look like tiny transparent arrows as they sprint after their planktonic prey. The species are all marine, living largely in the plankton from the surface to several hundred metres. Most species are 20–100 mm in length. Chaetognaths occur in high concentrations at certain times of the year, when they play a significant role in the planktonic community. They are predacious, feeding on any small plankters from diatoms to juvenile fish. Chaetognaths, in turn, form a major part of the diet of many fish. They are highly sensitive to changes in salinity and temperature, and for this reason are useful indicators of the movements of particular water bodies in the ocean.

The body of a chaetognath is bilaterally symmetrical and tripartite. The coelom is divided into paired head, trunk and tail compartments. The trunk and tail bear distinctive lateral and caudal fins, which are a diagnostic feature of the phylum. The fins are used to stabilise the animal and to maintain its position in the water column. On the ventral side of the head a large chamber leads to the mouth. Along the edges of the head hang bristles or spines used for catching and holding prey. It is this feature that gives the chaetognaths their name, which means 'bristle-jawed'.

The digestive system is relatively undifferentiated. The mouth leads to a rounded pharynx and thence to a narrow oesophagus within the head region. Entering the trunk region, the canal continues as a simple straight intestine, leading to the anus located immediately anterior to the end of the trunk. The breakdown of food occurs at the posterior region of the intestine and is partially mechanical. Absorption involves specialised secretory and absorptive cells. There appear to be no organs or tissues for respiration or excretion. Nor is there a circulatory system; coelomic fluid may play the role of a circulatory agent.

The nervous system consists of two ganglia, one dorsal to the mouth and one ventral to the intestine nearly half way along the trunk. The two ganglia are connected by commissures. The dorsal cerebral ganglion also gives rise to a ring of nerves that surrounds the pharynx and innervates the head, including the eyes. The ventral subenteric ganglion innervates the trunk and tail.

Reproduction

Chaetognaths are hermaphrodites. Paired ovaries occur in the trunk and paired testes in the tail region. It is thought that self-fertilisation can occur, and in some species reciprocal sperm transfer occurs between individuals. The small fertilised eggs are released into the water and develop directly into juvenile chaetognaths, without larval intervention.

Evolution

It is difficult to be certain of the evolutionary origins of the chaetognaths. Although tripartite, their body form is radically different from that of the hemichordates, cephalochordates or urochordates except during their early deuterostome development. The chaetognaths have been interpreted as coelomate, but unlike other coelomates they do not have a peritoneum lining the coelom. This characteristic, together with their undifferentiated muscles, is similar to that of the pseudocoelomates (see Chapter 18 for further discussion).

Classification of the phyla Chordata, Hemichordata and Chaetognatha

The following is a basic summary of the classification of the three phyla dealt with in this chapter. See the text for further details of classes and orders.

PHYLUM CHORDATA
SUBPHYLUM UROCHORDATA
Class Thaliacea
　　Orders Pyrosomatida, Salpida, Doliolida
Class Larvacea
Class Ascidiacea
　　Orders Aplousobranchia, Phlebobranchia, Stolidobranchia

SUBPHYLUM CEPHALOCHORDATA

PHYLUM HEMICHORDATA
Class Pterobranchia
Class Enteropneusta

PHYLUM CHAETOGNATHA

Chapter 18

Metazoan phylogeny

R.A. Raff

The problem 417

Isn't there a consensus yet? 422

Methods in phylogeny 422
    Why do phylogeny? 422
    How phylogenies are inferred 423
    Adding genes to the phylogenetic tool kit 425
    How to do gene sequence phylogeny 426

Some major questions 428
    Who were the ancestors of the Metazoa? 428
    Are the diploblasts primitive? 428
    Which are the most primitive Bilateria 428

Do the pseudocoelomates constitute a clade? 430

Did the coelom originate only once? 431

The protostomes 431
    Coelomate protostomes 431

Arthropod phylogeny 432
    Did arthropods originate only once? 432
    Molecular approaches to arthropod phylogeny 434

The deuterostomes 435

The lophophorates 437

Summing up: molecules and metazoans 437

The problem

With the publication of Darwin's *Origin of Species* in 1859, it became obvious to zoologists that the animal phyla must share a phylogenetic history: that is, the presently distinct phyla must have had common ancestors — a remarkable conclusion. Working out the evolutionary relationships among the phyla became a major enterprise. In the last quarter of the 19th century, German zoologist Ernst Haeckel drew the first phylogenetic trees, and most subsequent metazoan phylogeneticists have used his approach — a research program based on finding similarities among embryos of various phyla. For example, although adult molluscs and adult annelids are dissimilar, their embryonic development (spiral cleavage development) and larval form (a trochophore) give them away as more closely related to each other than either is to echinoderms or chordates. There is an ongoing revival of interest in metazoan phylogeny. New tools, including microanatomy, gene sequences, and cladistic methods for inferring phylogenetic relationships, have made this a lively and active area of research.

The animal phyla appear in the fossil record during the Cambrian metazoan explosion, over 530 million years ago. This visible radiation is of course the culmination of an earlier evolutionary radiation that took place in the late Precambrian. Just how rapidly the Cambrian radiation unfolded is still not known, although there are suggestions that it required only a few tens of millions of years. How are the phyla related in evolution to each other? There are three possibilities. The first (Fig. 18.1a) is that the phyla arose independently from unicellular (protist) ancestors. That idea requires that multicellularity and a whole suite of metazoan features (including nervous tissue, muscle, connective tissue, and extracellular materials such as collagen) were acquired independently. However, a common ancestor eliminates the problem of independent origins of animal features, and requires only a single metazoan phylogenetic tree. A common ancestor is consistent with the other two possibilities for phylum-level relationships. One (Fig. 18.1b) is that phyla have a linear evolution, with simpler ones giving rise to the more complex. This requires that the living phyla are ancestral to one another. This is an idea that had a strong hold on early 20th century zoologists, particularly with regard to the origin of vertebrates. Figure 18.2 shows one such scheme, in which early armoured vertebrates were thought to have derived from a primitive chelicerate by closing the old chelicerate mouth and losing the old arthropod appendages. In their place a new mouth was imagined to open on the other side of the animal, and new vertebrate appendages were evolved. In effect, the animal has been turned upside down to convert a ventral central nervous system into a dorsal one.

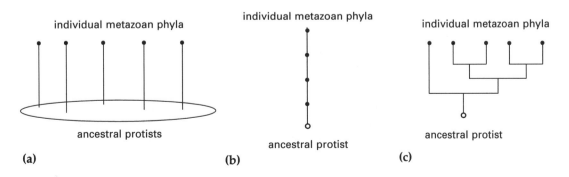

Figure 18.1
Three hypotheses for the origins of metazoan phyla:
(a) Independent origins of each phylum from unicellular (protist) ancestors. **(b)** Linear scheme of phylum evolution, in which all phyla stand in a single line of ancestors and descendants. **(c)** Branching phylogenetic tree, in which metazoans have a single ancestor and evolve by progressive branching.

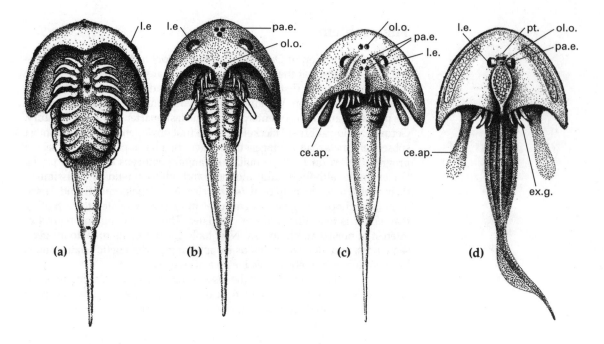

Figure 18.2
Hypothetical transformation of a chelicerate arthropod into an early vertebrate: **(a)** and **(d)** are based on fossils; **(b)** and **(c)** are hypothetical intermediates. (From Patten 1912.)

The final possibility (Fig. 18.1c) is that phylogeny resembles a bush, with a single ancestor and progressively splitting branches. The existing evidence best supports this branching pattern of evolution.

However, there are a lot of ways phyla might be related within a branching tree. If only three organisms are to be placed in a tree, there are only three possible arrangements (Fig. 18.3). With four organisms the number grows to 15 trees, 105 for five organisms, 34 million for ten organisms, and so forth. For the 30-plus animal phyla the number of possible tree topologies is stupendous. We need to consider how phylogenetic trees are inferred, and how we might judge the 'correctness' of the result.

The basis for much currently accepted metazoan phylogeny comes from the writings of Libbie Hyman in the 1950s. In her great compendium, *The Invertebrates*, she evaluated both anatomical and embryological traits of the phyla in terms of constructing a coherent phylogenetic tree. The tree shown in Figure 18.4 reflects her approach, and the corresponding tree of Figure 18.5 shows some of the anatomical features used. In this scheme, the most primitive metazoans (the mesozoan phyla and the sponges) are multicellular and have differentiated cell types, but their cells are not organised into tissues. The first branch in the metazoan tree is thus between these cellular-grade animals and phyla with tissues and organs. Some phyla are radially symmetrical, but the majority of more complex animal phyla are bilateral. The radial cnidarians and ctenophores have a tissue level of construction, whereas the bilateral animals have organs. There is a second major distinction. Cnidarians and ctenophores have two body layers — an outer ectoderm (skin, including a diffuse nerve net and muscle fibres), and an inner endoderm (gut) — and are thus 'diploblastic'. They possess a non-cellular mesoglea (a hydrated supportive connective tissue), which lies between the ectoderm and endoderm. All other phyla have a 'triploblastic' body organisation, with an outer ectoderm and an inner endoderm, and a novel intermediate cellular layer (the mesoderm) that lies between the

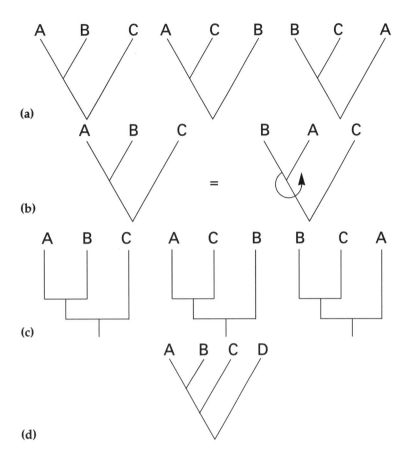

Figure 18.3
How phylogenetic trees work:
(a) The three phylogenetic trees, showing possible relationships among three organisms.
(b) Equivalent topologies; i.e. rotation around a node does not produce a different tree.
(c) A different display format. These three trees are identical in topology to those shown in (a).
(d) One of the 15 possible trees for four organisms.

inner and outer cellular layers. This layer has made possible the evolution of complex tissues, discrete organs, and powerful body-wall muscles. The coelom, developed within mesoderm, is also a major feature of complex metazoan phyla.

The majority of bilaterian animals have a mesoderm-lined body cavity or 'eucoelom'. The eucoelomate phyla include some of the most familiar living animals, such as chordates, arthropods, molluscs, annelids, echinoderms, brachiopods and ectoprocts. It is still a matter of dispute whether the eucoelom was acquired only once, as shown in Figures 18.4 and 18.5, or more than once. Several phyla of small animals (the rotifers, gastrotrichs, nematodes, nematomorphs, acanthocephalans, kinorhynchs, priapulans and loriciferans) possess a 'pseudocoelom'. This kind of body cavity lacks the mesodermal cell lining characteristic of the eucoelom, and supposedly arises differently in development as a persistent blastocoel. The pseudocoelomates have, in the past, been classed as a distinct and more primitive superphylum (Aschelminthes) than the eucoelomates. However, it is probable that they are simplified animals that are not related to each other, and molecular evidence supports this view (see Chapter 5).

Some features of adult anatomy offer potentially uniting features. For instance, in metamerically segmented animals the body is divided into repetitive elements that contain a portion of the coelomic cavity as well as serially repeated organs, musculature and appendages. In some schemes of

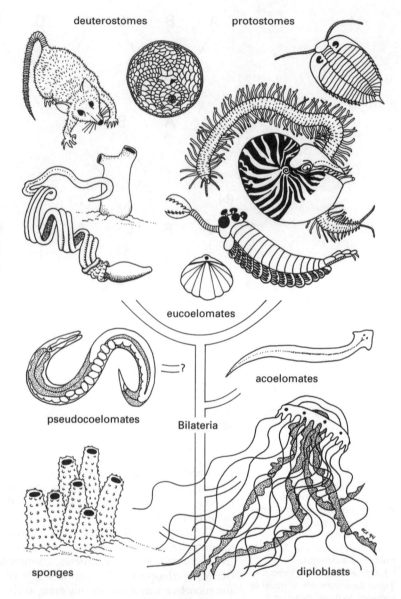

Figure 18.4
Examples of the diversity of animal body-plans, depicted in approximate phylogenetic order. Sponges represent the most primitive grade of metazoan body organisation. Cnidarians represent the diploblastic grade of organisation. The most primitive bilaterally symmetrical, triploblastic animals are platyhelminth flatworms. The pseudocoelomates, such as the nematode worms, are traditionally considered to represent a primitive clade, but that interpretation is being overturned. The major coelomate phyla are divided into two major superphyla: protostomes and deuterostomes. The protostomes here include *Nautilus*, a polychaete and a trilobite (extinct arthropod). The Cambrian animal *Opabinia* has been not assigned to a living phylum, but its anatomy suggests that it is a protostome. A brachiopod is also shown. Molecular sequence data places brachiopods within the protostomes. The deuterostomes include an acorn worm (hemichordate), an early Paleozoic edrioasteroid (extinct echinoderm), an ascidian (urochordate) and a mammal (chordate). (From Raff 1996.)

phylogeny, the annelid worms have been united with the arthropods on the basis of such segmentation, into a group called the 'Articulata' (see Chapter 8). Some phyla, such as echinoderms, share only the barest minimum of adult features with other phyla. From adult anatomy, about all we can say is that echinoderms are eucoelomate animals.

The traditional remedy has been to compare larval forms. Fortunately, a number of phyla share similar larvae, and this approach has been very powerful in some cases. For example, until the mid 19th century, adult barnacles were thought to be molluscs. However, when their larvae were observed to be a nauplius similar to the those of shrimp and other crustaceans, it became clear that barnacles are arthropods — and what's more, crustaceans. In some cases, adult anatomy offers no clues, and relationships have to be inferred from embryonic and larval features. Thus, echinoderms

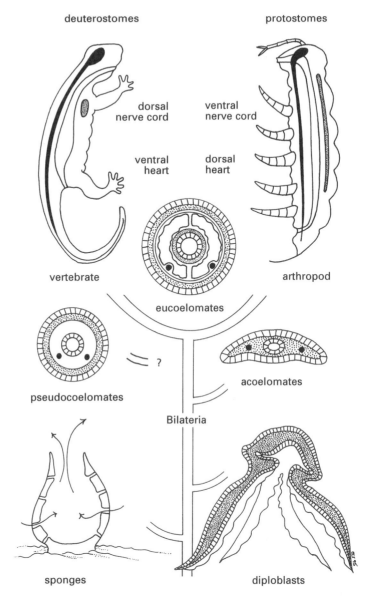

Figure 18.5
Principal elements of metazoan body-plan organisation. Shared anatomical features have been used to infer phylogenetic relationships among metazoan phyla. Sponges have multiple cell types but are not organised into discrete tissues. Diploblasts have tissues and two cellular layers (ectoderm and endoderm), separated by an acellular mesoglea. The majority of phyla have bilateral symmetry combined with a triploblastic arrangement of ectodermal, endodermal and mesodermal cell layers. The most primitive of these, the flatworms, have no coelomic cavity (acoelomates). The space between their body wall and gut is occupied by mesodermal cells. The eucoelomate phyla have a coelom lined by a mesoderm sheet (the peritoneum) that supports the internal organs. Pseudocoelomates are traditionally defined by a body cavity that lacks a peritoneum. This condition may represent a primitively retained blastocoel, or a secondary simplification from a eucoelomic ancestry. A deuterostome and a protostome are contrasted to illustrate the profound topological differences that separate them. The vertebrate has a dorsal nerve cord and a ventral heart. The arthropod has a ventral nerve cord and dorsal heart. (From Raff 1996.)

can be related to the chordates, our own phylum, but only on the basis of embryology. The molluscs have posed a continuing problem as to how we should weigh conflicting features in building a phylogeny. Molluscs share a lot of embryological features with annelids, but they lack a distinct coelom and segmentation. The pericardial cavity of molluscs may correspond to a reduced eucoelom, but that is an interpretation, not necessarily a fact. The primitive monoplacophoran molluscs have repeated muscle elements, which may represent a trace of primitive segmentation, but this interpretation is not convincing to most zoologists (see Chapter 7). The lack of a coelom in molluscs may represent a loss of a feature their ancestors once possessed. Molluscs might have split from the annelid and arthropod lineages before segmentation arose, or they might have had segmentation early on and lost it.

Isn't there a consensus yet?

As yet we do not have a certain knowledge of relationships among metazoan phyla, for three reasons. Firstly, the phyla are distinct. There really are not that many features shared among them. That reduces the obviousness of their evolutionary relationships. Secondly, the methods used to infer evolutionary relationships are themselves rapidly evolving, and new approaches have reopened century-old questions. Thirdly, we have no reliable way of testing for the correctness of phylogenies. The lack of consensus is all too evident. For example, a group of researchers has recently collected the major metazoan phylogenies (based on morphological features) produced since the 1950s. The result was 13 different phylogenies for the animal phyla. Two of the more prominent ones are shown in Figure 18.6. We cannot tell by inspection which of these conflicting hypotheses (if any) is correct. As we shall see below, molecular data offer a possible source of new data, and renewed hope of resolving phylum-level relationships.

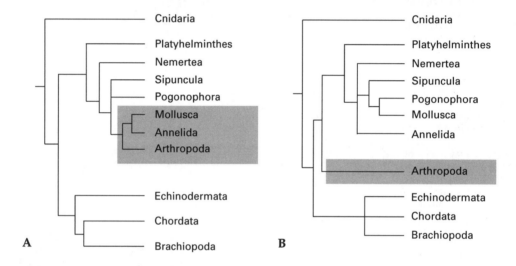

Figure 18.6
Two alternative metazoan phylogenies derived by cladistic analysis of morphological features of animal phyla. In these trees the platyhelminths are place into the protostomes, but in some other trees they are the sister groups of all eucoelomate phyla. In tree A, arthropods are the sister group of annelids and molluscs. In tree B, the arthropods are no more closely related to annelids than to other protostomes. Both morphological trees place brachiopods in the deuterostomes, a result that is inconsistent with results from current gene sequence data. (From Raff 1996.)

Methods in phylogeny

Why do phylogeny?

Reliable phylogenies are crucial in three ways for understanding the evolution of the phyla. The first is that we should like to be able to infer the structure of the Cambrian radiation itself. The second is that we need to understand the origins of new features in evolution. If metazoans had a single origin, the basic structures of animals must be homologous (that is derived from a common ancestor). However, if the phyla arose independently from protistan ancestors, complex structures characteristic of the various phyla originated independently. This would mean that many important features (for example, bilateral symmetry, a coelom, segmentation, a central nervous system) have had independent origins in various phyla. We would have very different expectations about the genetic controls for similar-looking developmental processes, depending on whether they had a shared phylogenetic history or different histories.

Thirdly, it is only with a known phylogeny that we can determine the direction of change in the evolution of developmental features and mechanisms. For example, a knowledge of phylogeny allows us to infer that echinoderms had a bilaterally symmetrical ancestor. Their five-sided symmetry thus evolved later than the features they share with other deuterostomes.

How phylogenies are inferred

A phylogenetic tree represents an evolutionary history. Such trees can be inferred in a variety of ways from the information available. In the past, phylogenetic relationships were largely determined on the basis of overall similarity. This approach seems like a common-sense one, because similar organisms should be more related in evolution than dissimilar ones. However, overall similarity does not discriminate between features that were acquired from a common ancestry (homologous features) from those that arose convergently (analogous features). Convergent similarity (homoplasy) is common, and the ability to distinguish this false homology is an absolute necessity.

We want to build a tree that presents the actual evolutionary history. To do that it is necessary to identify informative homologous features that allow us to unite organisms in the tree. For example, we can unite chelicerates, crustaceans, insects and myriapods into one phylum, Arthropoda, because they share a complex of features, such as a hard cuticle and jointed appendages. The evidence supports these features as a shared homology, derived from a single common ancestor. However, it has been also suggested that the hard cuticle and jointed appendages of arthropods might have been acquired independently, by convergence.

The traditional approach tried to defuse this and similar problems by using all morphological and embryological features to define relationships. This approach has no defined procedure to recognise convergences except through the expert investigator's application of an extensive knowledge of the group. A more precise and disciplined approach, called cladistics, has been devised and has now become the dominant school of systematics. Its principles are relatively simple, and they can be applied to both morphological and molecular data.

Two aspects of evolution are apparent in phylogenetic trees: the branching (cladistic) pattern as taxa separate from each other, and the amount of evolutionary change that accumulates with time along any branch. The result is a tree in which a single ancestral form gives rise to a set of branching and diversifying descendants. Within such trees we seek monophyletic lineages, or clades. A monophyletic lineage is one which arose from a single ancestor and which includes all descendants of that ancestor. Finding monophyly is crucial. If features are used that appear to be homologous, but are in fact false homologies (homoplasies), a group of animals considered to be a true clade might actually represent a polyphyletic assemblage. For example, suppose that we decided that jellyfish and starfish constituted a hypothetical group of radially symmetrical animals, the 'Radiata'. (This combination of cnidarians and echinoderms based on shared radial symmetry was actually used in the 19th century.) These 'radiates' would really be a polyphyletic group whose members independently acquired the features we used to unite them. Cnidarians are diploblastic in organisation, whereas echinoderms are triploblastic coelomates that had a bilaterian ancestry. The anatomical features that echinoderms share with other coelomate animals overwhelm the small number of similar features used to create the radiate group. Such polyphyletic groups are not acceptable if true evolutionary histories are sought. Nevertheless, confusing false homologies might fool us into inferring untrue phylum relationships.

There are three kinds of homologous similarity available to us for inferring a metazoan phylogeny:
1 primitive features shared by all phyla;
2 derived features shared by some phyla but not by others; and
3 features confined to a single phylum.

These distinctions are crucial to a cladistic approach to organising phylogenetic data.

To use cladistics, we have to remember that not all shared similarities are used in the analysis. As primitive traits are shared by all metazoans, they provide no useful information for distinguishing phyla. Primitive features, shared by all members of a clade of organisms, are called symplesiomorphies. Although they represent true homologies, they cannot inform us about relationships among members within the clade.

How should lineages such as phyla be defined? Obviously there are advanced features that occur only in a single group. Thus, butterflies are very different from other arthropods and even from other insects; butterflies have features, such as unique wings and mouth parts, that set them apart. Such features are called autapomorphies. In traditional phylogenies, autapomorphies were given important status in defining groups. However, as such features are unique to a single group, they do not help us in recognising relationships nor the pattern of branching among sister groups. Cladistic methods do that by using a third kind of homologous similarity: shared derived features, called synapomorphies. These are features that are shared by two or more groups. They represent states derived from the primitive features shared by all members of the clade. For example, although butterflies, beetles and wasps are recognisably different, they share a suite of anatomical features in their body plan that unite them as endopterygote insects. The result of this approach is a nested set of lineages: arthropods contain (among other clades) the insects, which contain further nested branches, such as beetles, flies, etc.

Let us for a moment reconsider the hypothetical 'Radiata' (the cnidarian plus echinoderm clade). Some features supported that clade, even though overwhelmed by a majority of features that disputes radiate monophyly. This pattern turns out to be common. That is, not all features support the most likely tree. Because of convergent evolution, some features that look like perfectly good homologies, and perfectly good synapomorphies, are seen to be in fact false homologies. The decision as to what constitutes the best phylogenetic solution has to be supplied by an outside criterion, which is based on how we think evolution works. The most commonly applied concept is called 'parsimony'. The concept of parsimony comes from the medieval English philosopher William of Occam. His principle, best known as 'Occam's razor', states that in seeking an explanation for some phenomenon, the hypothesis requiring the fewest assumptions is preferred over those requiring additional assumptions.

How does this apply to phylogeny? The concept of parsimony can be employed in two different ways. The first is methodological, and it provides a good approach to organising human reasoning. To explain our observations, we prefer straightforward parsimonious hypotheses over ornate and elaborate ones. The second way is the assumption that natural processes — for our purposes, evolution — actually operate parsimoniously. Of course, a law of parsimony is really a statistical statement. We understand that it means evolution generally operates in a parsimonious way, and we therefore assume that evolutionary history follows the line of fewest changes. Evolution does not necessarily do that in every instance. However, since evolution is descent with modification, more similar forms are in general likely to be more closely related than more distinct ones.

The aim of cladistics is to produce a diagram of the branching order of relationships of evolving lineages; that is, a cladogram. The operation is simple in principle; and if not too many lineages are involved, it can be simple in practice as well. Shared derived features, the synapomorphies, have to be scored to yield nested sets of branches. In order to do that, the features being used must be scored quantitatively. Generally, the primitive state of a feature is assigned a 0, and the derived state a 1. If a second derived state is present, it can be assigned a 2. Suppose we wish to derive a cladogram for three organisms: species A, B and C. A and B share synapomorphy state 1, and C has state 0. We would thus construct a cladogram, as in Figure 18.7, that links species A and B together in an A–B branch, with C branching off earlier. If several shared characters are present, the cladogram requiring the fewest steps is selected as the best one. If more than a very few lineages are included, computer methods of inference must be used.

To order the features being scored, some criterion of polarity is required: for the example shown in Figure 18.7, we show that state 0 is primitive. This is usually done by using an outgroup — a related lineage that can be reasonably inferred to lie outside the lineages being ordered in the cladogram. Its features are considered to mark the primitive states of shared derived features inside the cladogram; that is, state 0 for each one.

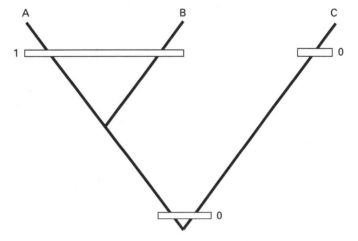

Figure 18.7
A simple cladogram for three taxa, based on a character with two states. The primitive state (0) is present in species C, and is inferred to have been present in the ancestor of A and B. Species A and B share the advanced state (1) and are therefore inferred to share a more recent common ancestry.

Adding genes to the phylogenetic tool kit

Despite a century of remarkable effort by zoologists, it has not been possible to erect a metazoan phylogeny that satisfies all investigators. Cladistic methods have helped in defining the problems more clearly, in rationally ordering the data, and in producing more clearly defined trees. With all this, the cladistic analysis of morphological features has not solved the problem of animal phylum relationships. Phyla by definition are disjunct in their features, and we may have exhausted most of the usable morphological features. Not only are features few, but morphologists often differ in the interpretation of anatomical features. Two anatomists may well interpret the same feature in diametrically different ways. Because phylogenetically informative features are few at the phylum level, mistakes in interpretation will distort the inferences we draw.

Molecules offer a new source of data. An average animal genome contains one to a few billion base pairs. There is thus potentially a gigantic amount of phylogenetic information in DNA. This new source of data does not supersede morphology, but adds to it and allows the possibility of new interpretations. One reason for this optimism is that gene sequences allow comparisons even when there is little or no morphological similarity. Thus, for example, molecular phylogeneticists can study relationships among such dissimilar creatures as fungi, plants and animals, because they share numerous homologous genes. In the remainder of this chapter I will discuss metazoan evolution from a perspective that blends molecules and morphology, as this appears to be the inevitable direction of phylogenetic studies.

Molecular phylogeny is not without its difficulties. Computer algorithms are available and easily used, and thus it is simple to generate gene-based trees. However, these are not literal read-outs of the truth. They are hypotheses. Phylogenies should be drawn from more than a single gene, and those that are used must yield concordant results. It is still so early in this enterprise that multiple gene comparisons are now only becoming possible, and molecular biology and morphology will continue to complement one another in understanding evolutionary histories.

How to do gene sequence phylogeny

In molecular phylogenetic analyses, homologous gene sequences from the species to be included in the analysis are aligned to give an optimum match of the bases in the sequence. Bases in an aligned set of sequences are considered to be homologous in terms of position in the sequence. A sample alignment of a segment of DNA from a gene as represented in four organisms is presented in Figure 18.8. This alignment represents part of a hypothetical gene sequence. Typically, hundreds to a few thousand base pairs are aligned for a gene. To achieve the best match among these sequences, a 'gap' has been inserted into the species 2 sequence, which represents a loss of a base in that sequence with respect to the others. Gaps are valid because insertions and deletions of bases are known to be reasonably frequent mutational events.

Phylogenies can be inferred from molecular data by cladistic methods, as well as by methods that measure the number of bases that differ between sequences. In a cladistic analysis of the sequences in Figure 18.8, position 5 of the sample alignment given above is informative, because its alternative

Figure 18.8
A DNA sequence alignment.

Base Position	1	2	3	4	5	6	7	8	9	10
Species 1	G	A	T	C	A	G	A	T	T	T
Species 2	G	A	T	C	A	G	–	T	T	G
Species 3	G	A	T	C	G	G	A	T	T	A
Species 4	G	A	T	C	G	C	A	T	T	C

METAZOAN PHYLOGENY

derived states are shared by two pairs of species. Species 1 is united with 2 and species 3 with 4. Only one mutational step is needed here; other tree topologies would require two mutational steps. Position 5 thus provides a molecular synapomorphy for cladistic analysis in this small dataset. The robustness of results can be tested by several approaches.

Most gene phylogenies have used the small subunit RNA found in all ribosomes. This molecule is suitable for deep-ranging phylogenetic studies for several reasons:

1. the gene is present in all organisms;
2. it contains a significant number of bases (about 1800);
3. it evolves slowly; and
4. it appears not to be subject to horizontal gene transfer from species to species.

A consensus metazoan phylogenetic tree based on ribosomal gene sequences is presented in Figure 18.9. This tree shows a number of important features that reflect on the metazoan radiation. Sponges branch off earliest, which is consistent with their primitive morphological organisation. Cnidarians branch next, which is consistent with the early origin of diploblastic animals in metazoan history. The primitive bilaterian platy-

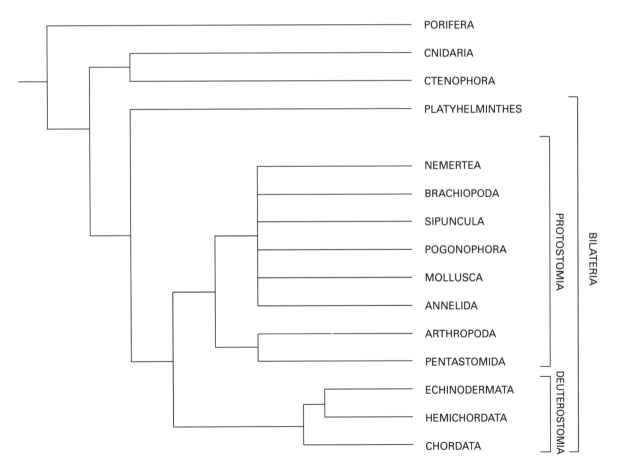

Figure 18.9
A summary molecular phylogenetic tree of relationships among major metazoan phyla, inferred from ribosomal DNA sequences. The tree represents a consensus based on the results of several laboratories. (From Raff 1996.)

427

helminths form the next major branch. The gene sequence data indicate that eucoelomate phyla are monophyletic, with distinct deuterostome and protostome clades. These results are consistent with Hyman's view of metazoan phylogeny, and with a massive radiation of coelomate animals in a relatively short interval of time prior to the great Cambrian radiation of the animal phyla recorded in the fossil record.

Some major questions

Who were the ancestors of the Metazoa?

Metazoans have traditionally been shown as arising from one or more unicellular ancestors among the 'protists', a collective name for a highly diverse range of unicellular eukaryotes. Through the application of gene sequence data, the protists themselves are now seen not as one group, but as a large number of unicellular kingdoms. This approach has made the determination of the sister group of the animals possible as well. The first problem has been to determine the relationship of animals to the other two great multicellular eukaryotic kingdoms, the fungi and plants. The most recent reassessments are consistent with fungi and animals as sister groups.

The sponges (Chapter 2) have a cellular grade of organisation, and only a few cell types. It has long been debated whether they are really metazoans or possibly 'parazoans', with a separate ancestry within the protists from true metazoans. Sponges circulate water by the action of specialised, collared, flagellated cells, the choanocytes. Zoologists have noted that there is a group of colonial protists known as choanoflagellates that possess a similar collared flagellated cell. Gene sequences now confirm that choanoflagellates are the sister group of the metazoans, including the sponges.

Are the diploblasts primitive?

Diploblastic animals (cnidarians and ctenophores, Chapter 3), and possibly a few others, such as the simply organised animal *Trichoplax adherens*, which has been placed in its own phylum, the Placozoa, have traditionally been seen as primitive animals on morphological grounds. Ribosomal gene sequences confirm this view, and place the diploblastic phyla as the sister group to the triploblastic bilaterians.

Which are the most primitive Bilateria

Most animals are bilaterians; that is, they are bilaterally symmetrical. The origin of the bilaterian, triploblastic level of organisation was an evolutionary innovation that allowed the evolution of complex animal body plans. This radical departure from the tissue grade of organisation of diploblastic animals occurred prior to the Cambrian radiation. A few living phyla have been traditionally considered to lack coelomic cavities, and to be basal bilaterians. The placement of these phyla — the platyhelminths (flatworms), nemertines and molluscs — has been controversial, and even monophyly of the acoelomate condition has been disputed.

The platyhelminths (Chapter 4) have been the most crucial group for phylogenetic speculations on bilaterian origins. Cladistic analysis of morphological features of the flatworms suggests that they are monophyletic. Analyses of ribosomal RNA sequence data also supports the monophyly of the platyhelminths. The diversity of free living and parasitic flatworms thus apparently represents a coherent platyhelminth radiation, and not a polyphyletic assemblage of superficially similar, morphologically simple, acoelomate bilaterian lineages. They are the sister group of the coelomate animals. We do not know if the flatworms are the last remnant of a radiation of acoelomate phyla early in metazoan history. It seems likely that it is the evolution of a coelom that made possible the exploitation of most metazoan niches, and thus triggered the evolution of the body plans of the coelomate phyla. The flatworms may offer a view of the primitive bilaterian body plan.

What of nemertines and molluscs? Nemertines (Chapter 4) are traditionally placed with platyhelminths in text books, including this one. Molluscs (Chapter 7) have been argued by some zoologists to lack a coelom, and to have evolved independently from a flatworm ancestry. These ideas have been tested by gene sequence phylogenies. Nemertine embryos show spiral cleavage and form a larval mouth in the protostomous manner. Nemertines, unlike flatworms, which have no body cavities other than a gut, have two kinds of body cavities: a rhynchocoel, which houses the long eversible proboscis with which they capture prey, and a system of cell-lined cavities. These latter cavities have been traditionally considered to be a blood vascular system. If they are circulatory, the cavities may be homologous to invertebrate blood spaces. If however, they are coelomic cavities, then the phylogenetic position of the nemertines must be interpreted very differently. Recent ultrastructural study of the nemertine spaces indicates that they share features with coelomic cavities: their position (lateral), their properties (cell lined as in coeloms versus lined with extracellular matrix as in invertebrate blood vessels), and their origin (from mesoderm). The molecular data also falls on the side of nemertines being eucoelomate animals, and not a sister group of the platyhelminths.

Molluscs have been similarly controversial. They develop in a very similar way to annelids. However, that is not enough, because flatworms also have spiral cleavage. It has not been possible on anatomical grounds to decide whether molluscs posses a coelom. The mollusc pericardial cavity might be a coelomic homologue, as it is mesodermal and is associated with the gonads. This is a controversial issue. It also has been suggested that molluscs are not coelomates but are part of a platyhelminth–nemertine–mollusc clade. It is interesting that ribosomal rRNA sequences and a protein sequence conform to the former of the two favoured phylogenetic hypotheses drawn from morphology. Annelids, pogonophorans and molluscs appear to form a eucoelomate protostome sister group. This means that the molluscan pericardial cavity is probably derived from a coelom. The characteristic segmentation of annelids may have been acquired after the divergence of the annelids from the molluscan lineage, or (less likely, see Chapter 7) primitive molluscs may have had body segmentation but subsequently lost it. Segmentation might also have been lost in other unsegmented, protostome coelomate phyla, although there is no direct evidence that their ancestors were segmented.

As we have removed nemertines and molluscs from the ranks of the truly acoelomate animals, we are left with only the platyhelminths as a major living group with that grade of body organisation. Perhaps these animals are the last remnant of a great unrecorded radiation of acoelomate phyla early in metazoan history. Alternately, they may have always been limited by lack of a coelom, which opened most metazoan niches and triggered the Cambrian radiation. If this scenario is correct, the flatworms are a living fossil group that has continued the primitive bilaterian body plan to the present.

Do the pseudocoelomates constitute a clade?

About 10 phyla — almost a third of all living body plans (rotifers, priapulids, gastrotrichs, kinorhynchs, nematodes, nematomorphs, acanthocephalans and loriciferans, Chapters 5 and 6; and questionably the entoprocts and gnathostomulidans, Chapter 4) — make up a bizarre clade, the pseudocoelomates, also called aschelminths. They pose a real phylogenetic quandary. Inclusion in this group traditionally has been based on the presence of a body cavity lacking a peritoneal lining. The pseudocoelom has been thought to be an embryonic blastocoel that persists in the adult. Most pseudocoelomates are very small and are often parasitic animals. The largest free-living pseudocoelomates are found among the priapulids, some of which have lengths of up to 200 mm.

The nematodes are the most diverse pseudocoelomates, and make up a sizable portion of the world's animal biomass and diversity. As discussed in Chapter 5, many are parasitic in plants and animals. Some, including hookworms, pinworms, ascarids, and the filarids that cause elephantiasis, have decidedly unpleasant associations with humans. One nematode, *Caenorhabditis elegans*, in contrast, is an important model system in the study of animal development, providing what is widely regarded as a primitive animal.

Despite the clear importance of understanding their phylogenetic relationships, the anatomy of the various pseudocoelomates gives few solid features that allow them to be comfortably linked with each other as a monophyletic group. The body cavities of pseudocoelomates do not all arise in the same way. Thus the pseudocoelom is not an informative feature uniting these phyla. Pseudocoelomates have been thought be primitive because they are simple in anatomy. However, it may well be that the properties often seen in pseudocoelomates are the result of extreme size reduction from larger eucoelomic ancestors. For example, many pseudocoelomates exhibit eutely; that is, their bodies have a relatively small number of cells that are constant in number and highly consistent in their positions within the animal. Eutely appears to be a consequence of small size. Copulation likewise is required of very small animals that cannot produce sufficient gametes to broadcast spawn, as do most larger invertebrates.

The phylogeny of pseudocoelomate animals is highly controversial. Clearly, some pseudocoelomate groups can be united on morphological grounds; for example, the rather similar nematodes and nematomorphs. Nevertheless, the most likely conclusion seems to be that the pseudocoelomate condition was attained independently by a number of lineages, and that the pseudocoelomates are a polyphyletic assemblage. The pseudocoelom is probably a convergently acquired developmental modification resulting from abbreviated coelomic development, leaving a blastocoel-derived cavity in small adults. Most pseudocoelomate phyla have entire guts, an anus, and other complex anatomical features indicating that the ancestors of pseudocoelomates were complex animals, not primitive ones. These ancestors were probably eucoelomates that independently underwent extreme size reduction. The analysis of ribosomal DNA sequences obtained from a few pseudocoelomate phyla fails to support monophyly, and suggests that at least some pseudocoelomates arose from ancestors among the coelomate protostomes and arthropod relatives. This issue is further discussed in the section on molecular approaches to arthropod phylogeny (page 434).

Did the coelom originate only once?

The great Cambrian radiation of animals probably resulted from the innovations in body plans, burrowing, and the increased size made possible by possession of a coelom. The coelom can arise in more than one way in development. Did it arise only once in evolution?

Various cladistic analyses of morphological features do not consistently support coelomate monophyly. The ribosomal DNA trees have been more consistent with coelomate monophyly, but they have not proved the point. The consequence is that we are left with a major problem in discovering the origin(s) of one of the defining features of advanced animal body plans. Coelomate animals have traditionally been divided into two great superphyla, the protostomes and the deuterostomes, on the basis of embryological features. Gene sequence trees have been generally (but not universally) consistent with this division.

In the remainder of this chapter we will assume that the coelom arose only once, but this is something that we do not know absolutely. Along with an assumption of the monophyly of the coelom, the protostome and deuterostome superphyla form the basis for the following discussion.

The protostomes

With the exception of chordates and echinoderms, the protostomes contain most of the major phyla of large animals, including annelids, molluscs and arthropods, as well as the sipunculans, pogonophorans and echiurans. These phyla have been united by a few common features of adult anatomy, and by similar modes of development among non-arthropods. However, there are substantial phylogenetic problems. These are sketched in Figure 18.6, which presents two distinct views of the relationship of arthropods to the other protostome phyla, based on recent cladistic analyses of morphological features. Figure 18.6a presents a result in which the protostomes consist of two deep clades. In this tree, the arthropods are equally distant from all coelomate protostomes. Figure 18.6b shows only a single protostome clade. The arthropods are a part of that clade, and have a close relationship to annelids and molluscs. These results are different because their authors evaluated the available morphological traits differently.

We will conditionally consider the relationships of arthropods shown in Figure 18.6a to be correct in the discussion below, because it is more consistent with molecular results and because it allows us to discuss the two great protostome groups as clearly separate evolutionary lineages.

Coelomate protostomes

The coelomate protostomes represent a distinct and major radiation of coelomate phyla that had occurred by the beginning of the Cambrian. Thus this branch of the metazoan phylogenetic tree arose in the late Pre-Cambrian, along with other major branches. The coelomate protostomes originated long enough after the origin of the eucoelom to have left a strong signal in ribosomal DNA. The coelomate protostomes produced an astonishing radiation in animal body plans. Many of these phyla share common features in their development, most notably a trochophore larva.

A major consequence of the discovery that there was a distinct protostome coelomate radiation is that we have to realise that arthropods and annelids are not each other's sister group to the exclusion of other phyla. Thus, gene sequence data cast doubt on the Articulata hypothesis. If arthropods are the sister group of all coelomate protostomes, that either requires that segmentation is primitive in the ancestor of the arthropod plus protostome coelomate clade, or that annelids and arthropod evolved segmentation independently. At the moment we do not know which is correct.

Anthropod phylogeny

Did arthropods originate only once?

Arthropods (Chapters 10–14) constitute the most diverse animal phylum, and not surprisingly they display some of the most complex and interesting phylogenetic problems of any animals. Perhaps surprisingly, not the least of the problems with arthropods has been whether they belong to a single phylum at all (see Chapter 10). Arthropods are characterised by their 'arthropodisation'. That is, unlike annelids, they are covered with a hard cuticle. But they are not merely worms dressed in armour. The evolution of a hard covering has resulted in a profound reorganisation of body plan so that it is very different from that of annelids or any other protostome phylum. Annelid use their hydrostatic skeleton for a kind of locomotion in which the body wall musculature works on the fluid-filled coelom. Arthropods, with their stiff exoskeleton, cannot alter segment sizes; thus they have lost their presumptive ancestral hydrostatic locomotory system, and have evolved jointed appendages and the associated muscles.

Arthropods share a suite of major features, including an external and internal body segmentation. Except for some insects, they add segments in a manner similar to that of annelids, via growth and patterning of a posterior growth zone. The layout of the arthropod nervous system is similar to that of other protostomes. The coelom has been drastically reduced and replaced by a haemocoel (blood) cavity. The dorsal heart characteristic of protostomes has been retained. However, instead of circulating in a system of closed blood vessels, the blood of arthropods enters the haemocoel, which forms the body cavities in each segment and constitutes an open circulatory system.

A number of shared anatomical features of arthropods (alpha-chitin cuticle, haemocoel, dorsal blood vessel with paired ostia, pericardial sinus, complex brain, segmental organs, appendages with extrinsic and intrinsic muscles and terminal claws, and centrolecithal eggs) form the basis for the view that arthropods are monophyletic. However, there has been a very long debate over whether arthropods are monophyletic or polyphyletic. Much of this debate centred on how to account for the Onychophora, and for arthropodisation. Zoologists recognised that onychophorans share features of both arthropods and annelids, and it had long been suggested that onychophorans formed a link between the two phyla. On the basis of such shared features as a uniramous leg and whole-limb jaw, some envisioned the Onychophora as direct ancestors of insects and their relatives (the uniramous arthropods). The crustaceans were hypothesised as arising from a separate ancestor. We are left with the conclusion that no matter how arthropods arose, some convergent evolution had to have occurred. Some interesting arguments have been made for both polyphyly and monophyly.

The distinct locomotory structures and embryonic development of the major living arthropod groups, the chelicerates, crustaceans and unirames, have been interpreted as support for separate origins for these groups, possibly among annelid ancestors.

It has been argued that the highly divergent embryology of onychophorans and the major arthropod groups supports arthropod polyphyly. For example, the onychophorans, myriapods, insects and other minor groups that make up the uniramians all exhibit a syncytial mode of cleavage with subsequent formation of a cellular blastoderm. Crustacean development differs from unirame development, with some crustaceans showing a type of spiral cleavage and all having a nauplius stage in their development. Finally, chelicerate development differs from both of these. Here too, a counter-interpretation — divergent evolution of developmental mechanisms subsequent to a common arthropod ancestor — is also possible.

The simplest level of arthropod organisation is one in which all body segments are similar. Some primitive crustaceans have this kind of organisation, as do the more familiar millipedes and centipedes. The divergence in structure and arrangement of segments (tagmosis) is a major feature of arthropod evolution. The tagmosis patterns of onychophorans and of the major arthropod groups are distinct. The onychophorans have a much simpler tagmosis pattern, with only three head segments; the true arthropods have five or six head segments behind the acron, the anterior end; and onychophorans have no acron. Each arthropod subphylum has its own tagmosis pattern. Thus, in chelicerates there is no distinct head. The most anterior appendages (the chelicerae of the adult) arise behind the mouth in the embryo and migrate forward. Crustaceans have two pre-oral head appendages (one migrates to a pre-oral position during development) and three that are post-oral; and unirames have one pre-oral and two or three post-oral appendages. The nature of the appendages is also characteristic of the group. These patterns exhibit an immensely long evolutionary conservation through hundreds of millions of years, yet the tagmosis patterns of the earliest known arthropods are in most cases quite different from those of the long-persisting major lineages. One interpretation sees these highly divergent tagmosis patterns so early in the history of the arthropods as indicating a hidden and independent evolution of convergent 'arthropods' from non-arthropod ancestors. The counter-interpretation of monophyly envisages divergent evolution in tagmosis patterns early in arthropod history.

Thus, the fully evolved gnathobasic mandible of crustaceans and the whole-limb mandible of uniramians might not be derivable from one another. However, both may have descended divergently from a primitive arthropod head appendage that had both a gnathobase used in feeding and a more distal endite used in walking. In the lineage leading to crustaceans, evolution of a gnathobasic mandible was selectively favoured. In the line leading to insects, a whole-leg mandible evolved.

Other complex organ systems have been interpreted as being consistent with arthropod monophyly. For example, the compound eye is present in living and fossil arthropod groups, but not in any non-arthropod phylum. Compound eyes are thus part of the suite of features that define arthropodisation. This homology is also supported by the long-conserved mechanisms of eye development.

Arguments based on differences in embryology depend on estimates of how much evolutionary change is possible. It has been traditionally held that early development is highly constrained and resistant to evolutionary change. Until recently, there has been little reason to suppose otherwise, but recent studies show that quite dramatic alterations can occur in early development over short evolutionary times. Furthermore, the new molecular genetic descriptions of development show that genes whose expression controls segmentation in the fruit fly *Drosophila* operate similarly in other insects and in crustaceans. These studies show that despite differences in development among arthropods, a considerable commonality exists. Arthropods share an extensive common genetic underpinning upon which the visually large variations in their development have been played out during evolution.

Molecular approaches to arthropod phylogeny

The question of arthropod monophyly, as well as that of how the major arthropod groups are related to each other, has come under molecular scrutiny. The results, although not answering all questions, are revealing. The first major conclusion of these molecular studies is that arthropods are monophyletic. The second is that they do not emerge as any closer to annelids than they do to molluscs, and thus provide no support for the classic annelid–arthropod clade, the Articulata. It has been less easy to define the relationships among arthropods, although some tantalising information has emerged.

All of the gene trees inferred to date indicate that the chelicerates stand distinct from the crustaceans and the uniramians. However, the gene sequence trees do support a mandibulate clade composed of insects (as representative unirames) and crustaceans. The myriapods are found close to insects in some trees, but are found deep in the other trees, breaking up the unirames.

Some of the molecular studies of arthropod phylogeny have included small phyla thought on anatomical or embryological grounds to be related to arthropods. The pentastomids, a small phylum of parasitic 'worms' that live in the nasal passages of tetrapod vertebrates, have been related to crustaceans on the basis of sperm morphology and nauplius-like larva. The ribosomal RNA sequences support a link to crustaceans. Larvae (less than a millimetre long) of these animals have, remarkably, been found in the late Cambrian fossil record. Some primitive morphological features that connect pentastomids to arthropods are present in these fossils, but none that tie them to any specific arthropod group. Intriguingly, they seem to have evolved a parasitic life-style before their current hosts, vertebrates, had evolved. The Onychophora (Chapter 9), which play such a central role in arguments about arthropod origins, are also part of the clade which includes the arthropods but not the annelids. A third phylum of tiny arthropod-like animals, the tardigrades (Chapter 9), have only very recently been considered by molecular phylogeny. Not too surprisingly, considering their morphological features, they are a close sister group of the arthropods. (These tiny animals have been recently discovered to be exquisitely preserved as fossils in Middle Cambrian rocks.) Finally, and more surprisingly, some pseudocoelomates, namely the priapulids, kinorhynchs, nematodes and nematomorphs, are also seen by ribosomal DNA sequences to be part of the clade that contains the arthropods, onychophorans, tardigrades, and pentastomids. The other great protostome clade contains the molluscs, nemerteans, sipunculans, annelids, pogonophorans and echiurans, and also the lophophorates (see page 345). This clade, on ribosomal DNA sequence grounds, also contains the rotifers, traditionally grouped with the aschelminths. Figure 18.10 summarises the molecular view of the expanded arthropod phylogeny, identifying a major clade, Ecdysozoa.

Figure 18.10
A molecular view of arthropod relationships. Analyses of ribosomal DNA sequences supports a clade of animals, the 'Ecdysozoa', that shed their cuticle during growth. (Based on the results of Aginaldo *et al.* 1997.)

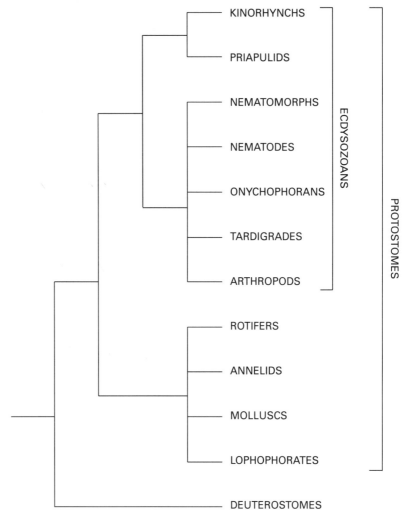

The deuterostomes

The deuterostomes make up the second major superphylum of eucoelomate animals. This superphylum has fewer members than the protostomes, but it has a certain charm for zoologists in that it contains our own phylum, the vertebrates. The generally listed membership of the deuterostomes, based on embryological considerations, is the chordates (urochordates, cephalochordates and vertebrates), hemichordates, echinoderms and chaetognaths (Chapters 16 and 17), and sometimes the lophophorate phyla (Chapter 15). The inclusion of chaetognaths and lophophorates as deuterostomes is in considerable doubt. Chaetognaths are just downright weird, and the lophophorate phyla share enough unique features that they have sometimes been considered to constitute a third superphylum. Not surprisingly, the application of molecular data may be necessary to answer the most basic of questions of membership and relationship. They might also illuminate a classic problem: Which of the invertebrate deuterostome taxa is the sister group of the chordates?

Ribosomal DNA sequences supported a monophyletic origin of the deuterostomes. However, when their 18S rDNA sequences were examined, not all phyla proposed to be deuterostomes were confirmed as members. The results placed lophophorates elsewhere in the metazoan tree altogether, in fact, among the protostomes (see Figure 18.10). We will return to the lophophorates in a moment. The placement of chaetognaths among the deuterostomes was traditionally justified on the basis of several basic, seemingly solid deuterostome traits. The mesoderm of chaetognaths arises from the archenteron, their coelom is tripartite, and their subepidermal muscles arise from mesoderm derived from the archenteron. However, it is not so unambiguous as we might hope. Chaetognath nervous systems are predominantly ventral, and protostome-like. In addition, chaetognaths possess chitin, a probable protostome trait, and they share some pseudocoelomic features, such as lack of a peritoneum. Ribosomal DNA phylogenies confirm that chaetognaths are not deuterostomes. They may be protostomes, but we are still not certain what the relationships of chaetognaths really are.

Let us first put chaetognaths and lophophorates aside and consider the deuterostome core phyla, the echinoderms, hemichordates, and chordates. Cladistic analysis of morphological features of these three phyla indicates that three possible deuterostome phylogenetic trees are supported. The first is the orthodox hypothesis found in many textbooks. It places hemichordates as the sister group of the chordates, with the major shared features being the possession of pharyngeal clefts and a dorsal hollow nerve cord. The second tree places hemichordates as the sister group of the echinoderms; the major shared feature supporting this pairing is the heart–glomerular complex. The third tree places the echinoderms as the sister group of the chordates, a hypothesis based on the structure of some early extinct echinoderms. All three hypotheses can be tested. Results from ribosomal DNA sequences weakly supported the hemichordates as the sister group of the echinoderms. The most robust phylogeny resulted from combining morphological and molecular datasets. When that was done, a solid case could be made for considering the second alternative tree (hemichordates as the sister group of echinoderms) as correct.

If the echinoderm–hemichordate clade is confirmed by further research, it will be important in understanding the evolution of deuterostome features. The precise ordering of the relationships of echinoderms to other deuterostomes will also provide a strong set of clues as to the transformation of a bilaterally symmetric ancestral deuterostome body plan into the highly modified, pentamerally symmetrical echinoderm body plan that was derived from it. We can infer that the ancestral deuterostome was bilaterally symmetrical and would have possessed a dorsal hollow nerve cord and pharyngeal clefts. The chordate lineage that ultimately produced vertebrates added a notochord and somites to produce a mobility system that allowed swimming by side to side motion, and a high degree of cephalisation (an elaborate anterior brain). In their own odd way, echinoderms underwent a remarkable transformation into uncephalised yet successful creatures. They added a unique water vascular motility system, a unique endoskeleton, and pentameral symmetry. In losing their heads, they also acquired a nervous system with a novel topology, consisting of a ring-shaped central nervous system that surrounds the mouth and sends radial nerve projections into each of the arms of the animal. There is no anterior end, and any side can take the lead in locomotion.

The lophophorates

Lophophorates (phoronids, brachiopods, and ectoprocts; see Chapter 15) are often interpreted as constituting a third coelomate superphylum. These phyla are united by a major shared structure, the lophophore, a hollow set of tentacles that surrounds the mouth. The anus is located outside the tentacle ring. Some zoologists have suggested that the lophophorates are related to the protostomes, but others have argued that lophophorates are deuterostomes or a sister group to the deuterostomes. Lophophorates have morphological features that might indicate membership in the deuterostomes: a tripartite coelom, a radial pattern of embryonic cleavage, and the origin of coelomic pouches in development from the sides of the archenteron. Some lophophorates also form the larval mouth in a deuterostome manner. Others do so in the protostome manner from the blastopore. But the overall embryological features of lophophorates are consistent with a deuterostome relationship for the group.

In view of the developmental features of brachiopods, the results of analyses of brachiopod ribosomal RNA sequences were surprising. The molecular data placed the brachiopods squarely into the coelomate protostomes. The brachiopod sequence showed a very high similarity to those of coelomate protostomes, but not to deuterostomes. Sequences from the ribosomal DNAs of other lophophorate phyla, the ectoprocts and the phoronids, have shown that they, as well as the brachiopods, are related to annelids and molluscs.

When morphological and gene sequence results seem to be in such wide disagreement, as in the case of lophophorate ribosomal gene sequences, we have to consider the possibilities. One is that lophophorates really are deuterostomes, but that during their history a horizontal transfer of ribosomal genes from a protostome has occurred, and the original deuterostome ribosomal gene sequence was lost. Horizontal gene transfers have been observed in a few cases in nature, but this is an unlikely scenario. However unlikely, it does have the virtue that it can be tested by determining the relationships of other lophophorate gene sequences. The prediction is that even if a horizontal transfer of ribosomal genes has occurred, other lophophorate genes would reveal the deuterostome genetic background. It seems more probable that lophophorates are protostomes by ancestry, as indicated by the ribosomal gene sequences, and that their deuterostome-like developmental features are the result of convergent evolution. This tells us that although the characteristics of early development and larvae can be highly informative about phylogeny, these features also evolve. Caution about the phylogenetic value of developmental features is more than a little desirable, as we have seen in the discussion above about arthropods.

Summing up: molecules and metazoans

Science thrives on the unanswered questions and the prospects for more research. We do not have the final answers about metazoan evolution and phylogeny, but we do have some challenging mysteries, as well as some intriguing new tools that can be applied to understanding the evolutionary history of the Metazoa.

The action in much of biology has moved to the sphere of genes, and that is true of much of evolutionary biology, including phylogeny. At this time the bulk of gene-sequence data used for phylum-level metazoan phylogenies come from small, subunit ribosomal genes. The results have been striking, but phylogenies based on a single gene cannot be the final story. The extraction of robust phylogenies from gene sequence data will require sequencing additional genes to see if they give the same phylogenies obtained with ribosomal genes. Ribosomal RNA genes are only the beginning. It is also important to note that molecular phylogeny is very dependent on mathematical methods used to infer phylogenetic trees. New inference methods may improve tree construction, and should provide better statistical tests of inferred phylogenies. As the sequences of other suitable genes enter the database, they will allow us to test the initial molecular results, and to resolve branch points that are unresolved in the present gene trees.

Gene phylogenies have resolved some controversies in phylum-level relationships, but not all molecular results have been robust or consistent. What is notable is that most phylogenetic inferences drawn from molecular data show a reasonable concordance with morphological data. A full understanding of metazoan evolution will require an open-minded approach that uses information from all available sources. The most interesting results may well be those yet to come.

Bibliography

Chapter 1 & general
Anderson, D.T., *Atlas of Invertebrate Anatomy*, University of New South Wales Press, Sydney, 1996. (Note: The classification of arthropods included in this book has now been superseded.)
Boardman, R.S., A.H. Cheetham, and A.J. Rowell (eds), *Fossil Invertebrates*, Blackwell, London, 1987.
Brusca, R.C. and G.J. Brusca, *Invertebrates*, Sinauer Associates, Sunderland, Massachusetts, 1990.
Meglitsch, P.A., and F.R. Schram, *Invertebrate Zoology*, 3rd Edition, Oxford University Press, Oxford, 1991.
Nielsen, C., *Animal Evolution*, Oxford University Press, Oxford, 1995.
Pearse, V., J. Pearse, M. Buchsbaum, and R. Buchsbaum, *Living Invertebrates*, Blackwell, Palo Alto, and Boxwood Press, Pacific Grove, 1987.
Ruppert, E.E., and Barnes, R.D., *Invertebrate Zoology*, 6th Edition, Saunders College Publishing, New York, 1994.

Chapter 2
Porifera
Bergquist, P.R., *Sponges*, Hutchinson, London, 1978.
Garrone, R., *Phylogenesis of Connective Tissue. Morphological Aspects and Biosynthesis of the Sponge Intercellular Matrix*, S. Karger, 1978.
Simpson, T.L., *The Cell Biology of Sponges*, Springer-Verlag, Berlin, 1984.

Chapter 3
Cnidaria and Ctenophora
Douglas, A.E., *Symbiotic Interactions*, Oxford University Press, Oxford, 1994.
Dubinsky, Z., (ed.), *Coral Reefs*, Elsevier, Amsterdam, 1990.
Endean, R., *et al.*, Envenomation of the box jellyfish *Chironex fleckeri*: how nematocysts discharge. *Hydrobiologia*, 216/217: 641–648, 1991.
Harrison, F.W., and J. Westfall (eds), *Microscopic Anatomy of Invertebrates. Volume 2, Placozoa, Porifera, Cnidaria and Ctenophora*, Wiley-Liss, New York, 1991.
Hopley, D., and P.J. Davies, The Evolution of the Great Barrier Reef, *Oceanus*, 29, 7–12, 1986.
Kirkpatrick, P.A. and P.R. Pugh, *Siphonophores and Velellids: Keys and Notes for the Identification of the Species*, E.J. Brill, London, 1991.
Schick, J.M., *A Functional Biology of Sea Anemones*, Chapman and Hall, 1991.
Wallace, C.C., R.C. Babcock. P.L. Harrison, J.K. Oliver and B.L. Willis, Sex on the reef: mass spawning of corals, *Oceanus*, 29, 38–42, 1986.

Chapter 4
Platyhelminthes, Nemertea, Entoprocta and Gnathostomulida
Cannon, L.G., *Turbellaria of the World. A Guide to Families and Genera*. Queensland Museum, Brisbane, 1986.
Gibson, R., *Nemerteans*, Hutchinson, London, 1972.

Gibson, R., 'Nemertean genera and species of the world: an annotated checklist of original names and description citations, synonyms, current taxonomic status, habitats and recorded zoogeographical distribution', *Journal of Natural History*, 29: 271–561, 1995.

Harrison, F.W., and B.J. Bogitsh (eds), *Microscopic Anatomy of Invertebrates: Volume 3, Platyhelminthes and Nemertea*, Wiley-Liss, New York, 1991.

Hyman, L.H., *The Invertebrates, Volume II: Platyhelminthes and Rhynchocoela*, McGraw-Hill, New York, 1951.

Smyth, J.D., and D.P. McManus, *The Physiology and Biochemistry of Cestodes*, Cambridge University Press, Cambridge, 1989.

Sterrer, W., M. Mainitz, and R.M. Rieger, 'Gnatho-stomulida: enigmatic as ever', pp. 181–199 in *The Origins and Relationships of Lower Invertebrates*, ed. C. Morris *et al.*, Systematics Association Special Volume 28, Oxford University Press, Oxford, 1985.

Chapter 5
Aschelminthes

Anderson, R.C., *Nematode Parasites of Vertebrates*, Commomwealth Agricultural Bureau, U.K., 1992.

Bird, A.F., and J. Bird, *The Structure of Nematodes*, 2nd edition, Academic Press, San Diego, 1991.

Crompton, D.W.T., and B.B. Nickol (eds), *Biology of Acanthocephala*, Cambridge University Press, Cambridge, 1985.

Harrison, F.W., and E.E. Ruppert (eds), *Microscopic Anatomy of Invertebrates, Volume 4: Aschelminthes*, Wiley-Liss, New York, 1991.

Nicholas, W.L., *The Biology of Free-living Nematodes*, 2nd Edition, Clarendon Press, Oxford, 1984.

Shiel, R.J., and W. Kosta, descriptions and keys to Australian rotifers; papers in the Transactions of the Royal Society of South Australia, 1979–1993.

Siddiqui, M.R., *Tylenchida, Parasites of Plants and Insects*, Commonwealth Institute of Parasitology, U.K., 1986.

Wood, W.B. (ed.), *The Nematode Caenorhabditis elegans*, Cold Spring Harbor, New York, 1988.

Chapter 6
Sipuncula and Priapula

Edmonds, S.J., 'Sipunculans (Phylum Sipuncula)', pp. 299–311 in *Marine Invertebrates of South Australia, Part 1*, ed. S.A. Shepherd & I.M. Thomas, South Australian Government Printer, Adelaide, 1982.

Rice, M.E., 'Sipuncula', pp. 237–325 in *Microscopic Anatomy of Invertebrates, Volume 12: Onychophora, Chilopoda and Lesser Protostomata*, ed. F.W. Harrison, Wiley-Liss, New York, 1993.

Storch, V. 'Priapulida', pp. 333–350 in *Microscopic Anatomy of Invertebrates, Volume 4: Aschelminthes*, ed. F.W. Harrison and E.E. Ruppert, Wiley-Liss, New York, 1991.

Chapter 7
Mollusca

Boss, K.J., 'Mollusca', pp. 945–1166 in *Synopsis and Classification of Living Organisms, Volume 2*, ed. S.P. Parker, McGraw-Hill, New York, 1982.

Hazprunar, G., On the origin and evolution of major gastropod groups, with special reference to the Streptoneura, *Journal of Molluscan Studies*, 54: 367–441, 1988.

Hyman, L.H., *The Invertebrates, Volume VI: The Mollusca*, McGraw-Hill, New York, 1967.

Lamprell, K. and T. Whitehead, *Bivalves of Australia, Volume 1*, Crawford House Press, Bathurst, Australia, 1992.

Ponder, W.F., and D. Lindberg, Towards a phylogeny of gastropod molluscs —

an analysis using morphological characters. *Zoological Journal of the Linnean Society*, 119: 83–265, 1997.

Ross, G., and P. Beesley (eds), *Fauna of Australia, Volume 5, Mollusca*, Bureau of Flora and Fauna, Canberra, 1997.

Taylor, J.T., *Origins and Radiation of the Mollusca*, Oxford University Press, Oxford, 1996.

Wilson, B.R., *Australian Marine Shells. Prosobranchia, Volumes 1 and 2*, Odyssey Publishing, Perth, 1994–95.

Wilbur, K., and C.M. Yonge (eds), *The Mollusca* (12 Volumes), Academic Press, New York, 1984–1988.

Chapter 8
Annelida

Anderson, D.T., *Embryology and Phylogeny in Annelids and Arthropods*, Pergamon, Oxford, 1973.

Brinkhurst, R.O., 'Phylogeny and Classification Part 1', pp. 165–177 in *Aquatic Oligochaeta of the World*, ed. R.O. Brinkhurst and B.G.M. Jamieson, Oliver and Boyd, Edinburgh, 1971.

Dales, R.P., The polychaete stomodaeum and the interrelationships of the families of the Polychaeta, *Proceedings of the Zoological Society of London*, 139: 289–328, 1962.

Fauchald, K., The polychaete worms. Definitions and keys to the orders, families and genera. *Natural History Museum, Los Angeles County, Science Series*, 28: 1–188, 1977.

Fauchald, K., and G.W. Rouse, Polychaete systematics: past and present, *Zoologica Scripta*, 26, (in press), 1997.

Jamieson, B.G.M., On the phylogeny and classification of the Oligochaeta, *Cladistics*, 4: 367–410, 1988.

Rouse, G.W., and K. Fauchald, Cladistics and the Polychaeta, *Zoologica Scripta*, 26 (in press), 1997.

Chapter 9
Onychophora and Tardigrada

Dewel, R.A., D.R. Nelson, and W.C. Dewel, 'Tardigrada', pp. 143–183 in *Microscopic Anatomy of Invertebrates, Volume 12: Onychophora, Chilopoda and Lesser Protostomata*, ed. F.W. Harrison, Wiley-Liss, New York, 1993.

Kinchin, I.M., *The Biology of the Tardigrada*, Portland Press, 1994.

Reid, A., Review of the Peripatopsidae (Onycho-phora) in Australia, with comments on peripatopsid relationships, *Invertebrate Taxonomy*, 10, 663–936, 1996.

Storch, V., and H. Ruhberg, 'Onychophora', pp. 11–56 in *Microscopic Anatomy of Invertebrates, Volume 12: Onychophora, Chilopoda and Lesser Protostomata*, ed. F.W. Harrison, Wiley–Liss, New York, 1993.

Tait, N.N., and D.A. Briscoe, Peripatus: up there for thinking, *Australian Natural History*, 22: 573–578, 1989.

Tait, N.N., and D.A. Briscoe, Sexual head structures in the Onychophora: unique modifications for sperm transfer, *Journal of Natural History*, 24: 1517–1527, 1990.

Chapter 10
Arthropods

Gupta, A.P. (ed.), *Arthropod Phylogeny*, Van Nostrand Reinhold Co., New York, 1979.

Manton, S.M., *The Arthropoda: Habits, Functional Morphology and Evolution*. Clarendon Press, Oxford, 1977.

Raff, R.A., *The Shape of Life*, University of Chicago Press, Chicago, 1996.

Willmer, P., *Invertebrate Relationships. Patterns in Animal Evolution*. Cambridge University Press, Cambridge, 1990.

Chapter 11
Hexapoda

Anderson, D.T., 'The development of hemi-metabolous insects', pp. 95–163 in *Developmental Systems: Insects, Volume 1*, ed. S.J. Counce and C.H. Waddington, Academic Press, New York, 1972.

Anderson, D.T., 'The development of holometabolous insects', pp. 165–242 in *Developmental Systems: Insects, Volume 1*, ed. S.J. Counce and C.H. Waddington, Academic Press, New York, 1972.

Chapman, R.F., *The Insects: Structure and Function*, Hodder and Stoughton, London, 1982.

CSIRO Division of Entomology, *The Insects of Australia, Volumes 1 and 2*, Melbourne University Press, Melbourne, 1991.

Groombridge, B. (ed.), *Global Biodiversity: Status of the Earth's Living Resources*, Chapman and Hall, London, 1992.

Gullan, P.J., and P.S. Cranston, *The Insects: an Outline of Entomology*, Chapman and Hall, London, 1994.

Horne, P.A., and D.J. Crawford, *Backyard Insects*, Melbourne University Press, Melbourne, 1996.

Kerkut, G.A., and L.I. Gilbert (eds), *Comprehensive Insect Physiology, Biochemistry and Pharmacology*, Pergamon, Oxford, 1985.

Lawrence, P.A., *The Making of a Fly*, Blackwell, Oxford, 1992.

Mordue, W., G.J. Goldsworthy, J. Brady, and W.M. Blaney, *Insect Physiology*, Blackwell, Oxford, 1980.

Wilson, E.O., *The Insect Societies*, Harvard University Press, Cambridge, Massachusetts, 1971.

Chapter 12
Myriapoda

Hopkins, S.P., and H.J. Read, *The Biology of Millipedes*, Oxford University Press, Oxford, 1992.

Lewis, J.G.E., *The Biology of Centipedes*, Cambridge University Press, Cambridge, 1981.

Chapter 13
Crustacea

Anderson, D.T., *Barnacles: Structure, Function, Development and Evolution*, Chapman and Hall, London, 1994.

Harrison, F.W., and A.G. Humes (eds), *Microscopic Anatomy of Invertebrates: Volume 9, Crustacea; Volume 10, Decapoda*, Wiley-Liss, New York, 1992.

Schram, F.R., *Crustacea*, Oxford University Press, Oxford, 1986.

Chapter 14
Chelicerata

Bonaventura, J., C. Bonaventura and S. Tesh (eds), *Physiology and Biology of Horseshoe Crabs*, A.R. Liss, New York, 1982.

Doggett, S., Ticks, *Nature, Australia*, 25(8): 40–47, 1997.

Foelix, R.F., *Biology of Spiders*, Harvard University Press, Cambridge, Massachusetts, 1982.

Gray, M., Spider silk: picking up the threads, *Australian Natural History*, 23(2): 112–121, 1989.

Gray, M., Funnel-webs: separating fact from fiction, *Australian Natural History*, 24(3): 32–39, 1993.

King, P.E., *Pycnogonids*, Hutchinson, London, 1973.

Locket, A., Night stalkers (scorpions), *Australian Natural History*, 24(9): 54–59, 1994.

Main, B.Y., *Spiders*, Collins, Sydney, 1976.

Polis, G.A. (ed.), *The Biology of Scorpions*, Stanford University Press, Stanford, 1990.

Savory, T.H., *Arachnida*, Academic Press, New York, 1977.
Reader's Digest Services (eds), *Australia's Dangerous Creatures*, Reader's Digest, Sydney, 1993.
Weygoldt, P., *The Biology of Pseudoscorpions*, Harvard University Press, Cambridge, Massachusetts, 1969.

Chapter 15
Lophophorates
Emig, C.C., The biology of the Phoronida, *Advances in Marine Biology*, 19: 1–89, 1982.
Harvell, C.D., The evolution of polymorphism in colonial invertebrates and social insects, *Quarterly Review of Biology*, 69: 155–185, 1994.
Hyman, L.H., *The Invertebrates. V. Smaller Coelomate Groups*, McGraw-Hill, New York, 1959.
Rudwick, M.J.S., *Living and Fossil Brachiopods*, Hutchinson, London, 1970.
Ryland, J.S., *Bryozoans*, Hutchinson, London, 1970.

Chapter 16
Echinoderms
David, B.A., J-P. Feral, and M. Roux, *Echinoderms Through Time*, Balkema, Rotterdam, 1994.
Dubois, P., 'Biological control of skeletal properties in echinoderms', pp. 17–22 in *Echinoderm Research*, ed. C. De Ridder, P. Dubois, M.C. Lahaye and M. Jangoux, Balkema, Rotterdam, 1990.
Flamming, P., 'Adhesion in echinoderms', pp. 1–60 in *Echinoderm Studies, Volume 5*, ed. M. Jangoux and J.M. Lawrence, Balkema, Rotterdam, 1996.
Harrison, F.W., and F-S. Chia (eds), *Microscopic Anatomy of Invertebrates: Volume 14, Echinoderms*, Wiley-Liss, New York, 1994.
Lawrence, J.M., *A Functional Biology of Echinoderms*, Johns Hopkins University Press, Baltimore, 1987.
Lawrence, J.M., and M. Jangoux, *Echinoderm Nutrition*, Balkema, Rotterdam, 1981.
Schick, J.M., 'Respiratory gas exchange in echinoderms', pp. 67–110 in *Echinoderm Studies, Volume 1*, ed. M. Jangoux and J.M. Lawrence, Balkema, Rotterdam, 1983.

Chapter 17
Chordata, Hemichordata and Chaetognatha
Jackson, J.B.C., and A.G. Coates, Life cycles and evolution of clonal (modular) animals, *Philosophical Transactions of the Royal Society*, B313: 7–22, 1986.
Mackie, G.O., From aggregates to integrates: physiological aspects of modularity in colonial animals, *Philosophical Transactions of the Royal Society*, B313: 175–196, 1986.
Millar, R.H., The biology of ascidians, *Advances in Marine Biology*, 9: 1–100, 1971.
Stocker, L.J., *Ascidian Identification Guide*, Leigh Marine Laboratory, University of Auckland, 1985.

Chapter 18
Metazoan phylogeny
Aginaldo, A.M., J.M. Turbeville, L.S. Linford, M.C. Rivera, J.R. Garey, R.A. Raff, and J.A. Lake, Evidence for a clade of nematodes, arthropods, and other moulting animals, *Nature*, in press.
Raff, R.A., *The Shape of Life. Genes, Development, and the Evolution of Animal Form*, University of Chicago Press, Chicago, 1996.

The following texts also provide discussions of aspects of metazoan phylogeny, offering a wide range of interpretations:

Brusca, R.C., and G.J. Brusca, *Invertebrates*, Sinauer Associates, Sunderland, Massachusetts, 1990.

Nielsen, C., *Animal Evolution. Interrelationships of the Living Phyla*, Oxford University Press, Oxford, 1995.

For readers interested in current views on the invertebrate origins of the vertebrates:

Gee, H., 1997, *Before the Backbone: Views on the Origin of Vertebrates*, Chapman & Hall, London, 1997.

Glossary

Abductor Muscle acting to move a structure away from the midline in a process of abduction. (See also *adductor*.)

Aboral surface In radially symmetrical animals, the body surface opposite the surface on which the mouth is located. (See also *oral surface*.)

Aciliary (Of spermatozoa) lacking a propulsive flagellum.

Acinus Small, terminal, blind-ending sac on a branched gland.

Acron Anterior presegmental part of an arthropod body, in front of the first definable segment. (See also *telson*.)

Acrosome Region at the anterior tip of a spermatozoon, containing enzymes that facilitate the penetration of an ovum.

Actinula Polypoid larval stage following the planula in the life cycle of some hydrozoan cnidarians.

Adductor Muscle acting to move a structure towards the midline in a process of adduction. (See also *abductor*.)

Aesthetes Small, numerous light-sensing organs terminating on the surface of the shell plates of chitons.

Ambulacrum Radial zone or groove on the surface of an echinoderm, containing podia or tube feet.

Ametabolous (Of hexapod arthropods) those in which the body form does not undergo major morphological change during postembryonic development. (See also *hemimetabolous, holometabolous*.)

Amoebocyte Cell capable of moving by pseudopodia.

Ampulla Muscular vesicle connected to a tube foot in the water vascular system of echinoderms.

Analogy Similarity of structure and function, but not of evolutionary origin. An alternative term is **homoplasy**. (See also *homology*.)

Ancestrula. Founder zooid of an ectoproct colony.

Anhydrobiosis. Maintenance of life in the absence of water.

Aragonite Crystalline form of calcium carbonate, less stable than calcite.

Archenteron Internal cavity, lined at least in part by endoderm, developed in an embryo at the gastrula stage; later giving rise to the gut cavity.

Aristotle's lantern Jaw apparatus of echinoid echinoderms, incorporating five calcareous teeth.

Arolium Adhesive pad between the pretarsal claws of a thoracic leg in insects. (See also *pulvillus*.)

Autapomorphy A distinctive morphological feature unique to a single group of animals. (See also *synapomorphy*.)

Autozooid Basic feeding zooid of an ectoproct colony.

Benthic Living on the seabed, a lake bed or a stream bottom. (See also *nektonic*.)

Biramous (Of arthropod limbs) having two branches or rami. (See also *uniramous*.)

Bioerosion Erosion caused by living organisms.

Biogenic Formed through the activities of living organisms.

Bipectinate Comb-like, with a row of projections on either side of the axis. (See also *monopectinate*.)

Blastomere Cell formed by the cleavage divisions of an egg.

Blastozooid Zooid formed by asexual budding.

Blastula Hollow ball of cells formed during early embryonic development. (See also *morula*.)

Bolus Large rounded mass, usually referring to food.

Brachiolaria Second larval stage (settling stage) in asteroid echinoderms.

Bursa copulatrix Reproductive sac receiving sperm during copulation.

Byssus Bundle of tough threads of sclerotised protein, secreted by the foot in many bivalve molluscs and used in anchoring the animal to the substratum.

Calcite Crystalline form of calcium carbonate. (See also *aragonite*.)

Calyx Skeletal cup housing an individual polyp in corals; cup-shaped aboral wall of crinoids; cup-shaped body of the zooid in entoprocts.

Captacula Thread-like tentacles arising from either side of the head in scaphopod molluscs, used in food capture.

Cardinal teeth Large, centrally located teeth on the hinge line of the shell in heterodont bivalve molluscs.

Centrolecithal egg Egg uniformly and densely filled with yolk granules, as in many arthropods.

Cephalisation Formation of a head.

Cerata Horn-like projections of the dorsal body wall in nudibranch molluscs.
Cercus One of a pair of tail-like appendages at the end of the abdomen in insects.
Chaetae Chitinous bristles of polychaete worms (sometimes called **setae**).
Chitinophosphatic Shell composition in inarticulate brachiopods, combining chitin and protein with calcium phosphate.
Chlorocruorin Respiratory pigment in blood, greenish-red, containing iron bound to porphyrin.
Chordotonal organ Mechanoreceptor sense organ of insects; an elongate structure attached at both ends to the body wall.
Chorion Outer egg shell in insects and other arthropods.
Cingulum Postoral ciliary band encircling the apex of a rotifer. (See also *trochus*.)
Cirrate Applying to octopod molluscs with numerous small, coiled tentacles (cirri) on the arms. (See also *incirrate*.)
Cirri Tentacle-like appendages found in various animal groups.
Clade Monophyletic group arising from a common ancestor. (See also *grade*.)
Cladistics Method of phylogenetic analysis aiming to identify monophyletic groups.
Cladogram Diagram of the pattern of branching of an evolutionary lineage, derived from cladistic analysis.
Cnida Intracellular explosive capsule of cnidarians, secreted within a cnidocyte, used in prey capture and adhesion.
Cnidocil Trigger-like sensory spine (a modified flagellum) on the cnidocyte in hydrozoans and scyphozoans.
Cnidocyte Cell in cnidarians which secretes a cnida.
Coeloblastula Hollow blastula in which the external wall of cells encloses a fluid-filled cavity, the blastocoel. (See also *stereoblastula*.)
Coelom Body cavity between the body wall and gut wall, lined by an epithelium derived from mesoderm.
Coelomoduct Duct, of mesodermal origin, opening from the coelom to the exterior.
Coelomopore Pore in a partition between two coelomic compartments.
Coenenchyme Tissue connecting the polyps in octocorallian (alcyonarian) corals.
Collagen A tough, fibrous protein.
Colloblast Glue-secreting cell on the tentacles of ctenophores.
Columellar muscle Muscle attached to the columella (central column) of the shell in gastropod molluscs, acting to withdraw the head and foot into the shell.
Commensalism Close association between individuals of two different species in which one gains benefit, for example by taking a share of the food collected by the other, while the other is neither helped nor harmed.
Conchiolin A quinone-tanned protein material forming the periostracum (outer layer) of the shell in gastropod molluscs.
Cryptobiosis Hidden life: continuing at a low metabolic level, without obvious signs of activity.
Ctene Comb of fused cilia, forming part of a comb row in ctenophores.
Ctenidium Comb-like gill of molluscs.
Cypris Second larval stage (settling stage) of barnacles and other thecostracan crustaceans, with a bivalved carapace.
Cystid Exoskeleton and body wall of an ectoproct zooid. (See also *polypide*.)
Dactylozooid Tentacle-like zooid of a colonial hydrozoan cnidarian, used in food capture and defence
Day-degrees Cumulative time (in days) during which the ambient temperature is higher than a certain minimum.
Demibranch (In bivalve molluscs) one half of a lamellibranch gill, comprising two lamellae. (See also *lamellibranch*.)
Dendrobranchiate Branching gills (as in dendrobranchiate prawns).
Dermanyssids A family of mites.
Desmosome Intercellular junction that bonds together the plasma membranes of adjacent cells.
Deuterostome Animal in which the anus is formed from the blastopore during development and the mouth arises as a secondary invagination.
Diapause Period of arrested development or suspended animation.
Diductor Muscle which causes the valves of the shell in brachiopods to separate.
Digenean Parasitic life cycle passing through at least two hosts; usually applied to trematode flatworms. (See also *monogenean*.)
Dioecious Having male and female reproductive systems in separate individuals. An alternative term is **gonochoristic**. (See also *monoecious, hermaphrodite*.)
Diploblastic Describing an animal in which the body is developed from two cell layers, ectoderm and endoderm. (See also *triploblastic*.)
Diverticulum Blind-ending outpouching.
Doliolaria larva Barrel-shaped second-stage feeding larva of holothuroid echinoderms.
Ecdysis Moulting; the process of shedding the skin.
Eclosion Hatching, or escape from the egg shell, by a larval insect.
Ectocommensal Living as a commensal on the external surface of a host. (See also commensalism, *endocommensal*.)
Ectoparasitic Living as a parasite on the external surface of a host. (See also *endoparasitic*.)
Elytron Thickened, sclerotised forewing in insects of the order Coleoptera. (See also *tegmen*.)
Emergence Escape by an adult insect from either the last nymphal cuticle, the pupal cuticle, or (in some Diptera) the puparium.
Endocommensal Living as a commensal within the

body of a host. (See also *commensalism, ectocommensal*.)

Endocytosis Active process that carries large molecules or particles into a cell via invaginated vesicles of plasma membrane; includes phagocytosis and pinocytosis.

Endoparasitic Living as a parasite within the body of a host. (See also *ectoparasitic*.)

Endopod Median branch of a biramous arthropod limb. (See also *exopod*.)

Enterocoel Coelomic cavity formed by the formation of an outward pouch in the wall of the embryonic archenteron. (See also *schizocoel*.)

Enteron Digestive cavity.

Entognathy Enclosure of arthropod mouthparts within folds of the head capsule.

Ephyra Juvenile medusa released from the polypoid stage in the life cycle of a scyphozoan cnidarian. (See also *scyphistoma*.)

Epibenthic Living on the surface of the seabed, a lake bed or a stream bottom.

Epicuticle External layer of the cuticle of onychophorans, tardigrades and arthropods.

Epifauna Fauna of the surface of the seabed, a lake bed or a stream bottom. (See also *infauna*.)

Epipelagic Living in the upper waters of the sea, down to about 200 metres.

Epipodite Lateral process on the protopod of a crustacean limb, often modified as a gill.

Errant Wandering (usually applied to polychaete worms.)

Eulamellibranch A type of lamellibranch gill in bivalve molluscs in which the gill filaments are joined by tissue junctions. (See also *filibranch, pseudolamellibranch*.)

Euryaline Referring to the Euryalina, a group of ophiuroid echinoderms with vertically coiling, grasping arms, including the basket stars.

Eusocial (Of insects) having the highest level of social development.

Eutely Condition in which the body is composed of a constant number of cells, or contains a constant number of nuclei, in all adults of a species; common in aschelminths.

Exhalant Current of water or air passing out of the body of an animal. (See also *inhalant*.)

Exocytosis Process through which materials are expelled from a cell, often via a vacuole or vesicle. (See also *endocytosis*.)

Exopod Lateral branch of a biramous arthropod limb. (See also *endopod*.)

Exumbrella Upper (aboral) surface of a cnidarian medusa. (See also *subumbrella*.)

Exuviae (plural form only) Moulted cuticle of an arthropod.

Filibranch A type of lamellibranch gill in bivalve molluscs in which the gill filaments are joined by ciliary junctions. (See also *eulamellibranch, pseudolamellibranch*.)

Flame cell Hollow excretory cell containing a tuft of cilia, located at the inner end of a protonephridium. (See also *solenocyte*.)

Foliaceous Leaf-like.

Follicle Small sac or vesicle, especially describing parts of ovaries or testes.

Furca Fork; usually applied to the forked process on the telson of many crustaceans.

Furcula Springing organ of Collembola, formed by fused limbs of the fourth abdominal segment. (See also *retina-culum*.)

Fusiform Spindle-shaped, tapering at both ends.

Galea Terminal lobe of the maxilla of an insect.

Gastrodermis Lining of the digestive cavity in cnidarians and ctenophores.

Gastrovascular cavity Internal cavity of cnidarians and ctenophores, opening to the exterior by the mouth and having digestive and circulatory functions.

Germarium Region containing the primordial germ cells (spermatogonia or oogonia) in an insect gonad.

Gonochoristic Having male and female reproductive systems in separate individuals. An alternative term is dio-ecious. (See also *monoecious, hermaphrodite*.)

Gonoduct Duct leading from the gonad (ovary or testis) to the exterior.

Gonophore Medusa bud of hydrozoan cnidarian colony which remains attached and acts as a reproductive zooid, producing gametes. (See also *sporosac*.)

Gonozooid Reproductive polypoid zooid of hydrozoan cnidarian colony, producing medusa buds.

Grade Level of animal organisation. Without phylogenetic significance, since similar grades of organisation may exist in unrelated groups. (See also *clade*.)

Haemal Referring to the blood system.

Haemerythrin Respiratory pigment in blood, violet in colour, containing iron not bound to porphyrin. (See also *haemoglobin*.)

Haemocoel Body cavity formed as an expanded blood system, associated with reduction of the coelom.

Haemocyanin Respiratory pigment in blood, blue when oxygenated, colourless when deoxygenated, containing copper.

Haemoglobin Respiratory pigment in blood and other tissues, red in colour, containing iron bound to porphyrin.

Hemidesmosome One half of a desmosome. (See *desmosome*.)

Hemimetabolous (Of hexapod arthropods) having a body form which gradually undergoes major morphological change during postembryonic development. (See also *ametabolous, holometabolous*.)

Hermaphrodite Having male and female reproductive systems in the same individual. An alternative term is monoecious. (See also *dioecious, gonochoristic*.)

Heterozooids Specialised zooids of ectoprocts, with a variety of functions.

Hexoses Simple (monosaccharide) sugars with a carbon chain containing six carbon atoms.

447

Histocompatibility system System that identifies cells as self or non-self.

Homeostasis Maintenance of an internal steady state by self-regulation, relative to a fluctuating environment.

Holometabolous Of hexapod arthropods; those in which the body form undergoes rapid major morphological change during a stage of postembryonic development called the pupal stage. (See also *ametabolous, hemimetabolous*.)

Homeobox genes Genes controlling basic pattern-forming processes in development.

Homology Similarity of evolutionary and developmental origin of a stucture in a monophyletic lineage. Homologous structures often perform different functions in different species of the lineage. (See also *analogy, homoplasy*.)

Homoplasy False homology, resulting from the convergent evolution of functionally similar structures in different line-ages. An alternative term is **analogy**. (See also *homology*.)

Hyaluronidase Enzymes found in the head of a sperm, which dissolve the egg coat and allow sperm penetration.

Hypertonic Solute concentration higher than that of some other solution. (See also *hypotonic*.)

Hypobranchial gland Mucous gland on the roof of the mantle cavity in prosobranch gastropod and primitive bivalve molluscs.

Hypopharynx Fleshy lobe located mid-ventrally behind the mouth in insects, carrying the openings of the paired salivary glands.

Hypostome Projecting oral surface of the polyp in hydrozoan cnidarians, with the mouth at the apex.

Hypotonic (Of solutions) having a solute concentration that is lower than that of some other solution. (See also *hypertonic*.)

Imago Adult stage of an insect.

Incirrate Lacking cirri. (See also *cirri, cirrate*.)

Infauna Fauna living in sediment on the seabed, a lake bed or a stream bottom. (See also *epifauna*.)

Inhalant Current of water or air drawn into the body of an animal. (See also *exhalant*.)

Iridocytes Light-reflecting cells in the skin of cephalopod molluscs.

Interambulacrum Interradial zone on the body surface of echinoderms, between two ambulacra. (See also *ambulacrum*.)

Interstitial Located in the small spaces (interstices) between larger structures.

Isosmotic (Of solutions) having equal osmotic concentration. An alternative term is **isotonic**.

Juxtaligamental cells Nerve-like processes in the connective tissue of echinoderms, thought to control the unique connective tissue mutability of these animals

Labrum Upper lip in arthropods, overhanging the mouth.

Lacuna Enclosed space or cavity.

Lamella A layer or sheet.

Lamellibranch Sheet-like, modified form of ctenidium or gill, found in most bivalve molluscs.

Lamellibranchiate Bivalve mollusc with lamellibranch gills.

Lecithotrophic Feeding on yolk reserves. (See also *planktotrophic*.)

Linalool An open-chain terpene alcohol, a component of some flowers.

Lobopodium Segmental appendage of an onychophoran or other lobopod animal, filled with haemocoelic fluid. (See also *parapodium*.)

Macrophagous Feeding on food items that are large relative to the size of the animal. (See also *microphagous*.)

Macromere Large cell formed during the cleavage divisions of an egg. (See also *micromere*.)

Madreporite Porous plate in the skeleton of echinoderms, leading into the water vascular system.

Malpighian gland Excretory gland attached to the gut in some tardigrades.

Malpighian tubule Excretory tubule arising from the gut in many terrestrial arthropods.

Mantle cavity External cavity, overhung by fold(s) of the body wall called mantle fold(s) and open to the surrounding water; in molluscs, brachiopods and cirripede crustaceans.

Manubrium Central projection, with mouth at apex, on the subumbrellar surface of the medusa in hydrozoan cnidarians.

Medusa Umbrella-shaped, sexually reproducing, adult stage of scyphozoan, cubozooan and many hydrozoan cnidarians; absent in anthozoans.

Meiobenthos Fauna of small invertebrates living in the interstices between particles of benthic sediments.

Mesentery Sheet of tissue suspending an internal organ in an animal.

Mesogloea Gelatinous layer between the epidermis and gastrodermis in cnidarians.

Mesosoma Anterior part of the opisthosoma in scorpions and eurypterids. (See also *metasoma*.)

Mesosome Middle section of the body in a trimeric animal. (See also *protosome, metasome*.)

Metachronal rhythm Pattern of rhythmic activity that passes sequentially along the body.

Metameric segmentation Division of the body into a series of similar subunits or segments, each based on a pair of mesodermal somites.

Metanephridium Nephridium in which the inner end terminates in an open nephrostome. (See also *protonephridium*.)

Metasoma Posterior part of the opisthosoma in scorpions and eurypterids.

Metasome Posterior section of the body in a trimeric animal. (See also *protosome, mesosome*.)

Micromere Small cell formed during the cleavage divisions of an egg. (See also *macromere*.)

Microphagous Feeding on food particles that are

small relative to the size of the animal. (See also *macrophagous*.)

Microvillous A condition in which the surface of an epithelial cell is extended as microscopic, finger-like projections (microvilli), greatly increasing the surface area.

Monoecious Having male and female reproductive systems in the same individual. **Hermaphrodite** is an alternative term. (See also *dioecious, gonochoristic*.)

Monogenean Parasitic life cycle passed on a single host. (See also *digenean*.)

Monopectinate Comb-like, with a row of projections on one side only of the axis. (See also *bipectinate*.)

Monophyletic Arising from a common ancestor. (See also *paraphyletic, polyphyletic*.)

Morula Solid ball of cells formed as a result of cleavage of an egg. (See also *blastula*.)

Motile Moving actively from place to place. (See also *sedentary, sessile*.)

Mucopolysaccharide Complex, large carbohydrate composed of monosaccharide units, sometimes associated with protein.

Mucoprotein High-molecular-weight protein found in mucus; also known as mucin.

Müller's larva Ciliated larva of some polyclad turbellarians.

Myoepithelial cell Epithelial cell containing contractile fibres; also called **musculo-epithelial cell**.

Myofibril Contractile fibril, composed of actin and myosin, within a muscle fibre.

Nauplius First larval stage of Crustacea, with three pairs of limbs (antennules, antennae and mandibles.)

Nektonic Swimming in the water column, either in the sea or in fresh waters. (See also *planktonic, benthic*.)

Nephridiopore External opening of a nephridium. (See also *nephrostome*.)

Nephrostome Internal opening of a metanephridium. (See also *nephridiopore*.)

Neurohumoral system System of neurosecretory cells and hormone-secreting glands that regulates development in insects (also known as the **neuro-endocrine system**.)

Neuropodium Ventral process of a polychate parapodium. (See also *notopodium*.)

Nidamental glands Glands in the mantle cavity of cephalopod molluscs that secrete gelatinous coatings covering the eggs during oviposition (nidamental = nest-making.)

Notopodium Dorsal process of a polychaete parapodium. (See also *neuropodium*.)

Nucleated Containing one or more nuclei.

Nurse cell Cell associated with a developing oocyte and providing nutrients for oocyte growth.

Obligate Necessary for an event to proceed.

Ocellus Simple eye in arthropods.

Ommatidium Optical unit of the compound eye in arthropods; also in some molluscs and polychaetes.

Ontogeny The process of development.

Oostegites Large, plate-like processes on the coxae of certain thoracic legs in female peracaridan crustaceans, together forming a ventral brood pouch.

Oozooid Zooid in urochordates, formed by development from a fertilised egg.

Opisthosoma Posterior section (tagma) of the body in chelicerate arthropods. (See also *mesosoma, metasoma, prosoma*.)

Oral surface Body surface, in radially symmetrical animals, on which the mouth is located. (See also *aboral surface*.)

Osmoconformer Aquatic animal in which the internal osmotic concentration matches that of the surrounding water. (See also *osmoregulator*.)

Osmoregulator Aquatic animal which regulates and controls its internal osmotic concentration in relation to changes in the surrounding water. (See also *osmoconformer*.)

Osphradium Sense organ in the mantle cavity of aquatic gastropod and bivalve molluscs, detecting chemical stimuli and sediment levels in the inhalant water flow.

Ostium 1. Inhalant opening in the wall of an asconoid sponge. 2. Pore in the lamellae of eulamellibranch gills in bivalve molluscs. 3. Lateral opening in the wall of the heart in arthropods.

Ovariole Egg-producing unit of the ovary in insects.

Ovigerous Carrying eggs externally on the body.

Ovipositor Egg-laying apparatus, forming part of the external genitalia of a female insect.

Pallio-visceral Describing the lateral nerve cords of chitons, which innervate the mantle (pallium) and viscera (internal organs.)

Papulae Small evaginations of the soft body wall, scattered over the surface of echinoderms.

Paraphyletic Arising from a single ancestor, but not including all of the descendant groups with this origin. (See also *monophyletic, polyphyletic*.)

Parapodium Segmental appendage of polychaete annelids, containing an extension of the coelom. (See also *lobopodium*.)

Paratenic Host of a parasite which is not the normal host.

Pectinate In the form of a comb. (See also *bipectinate, monopectinate*.)

Pectines Comb-shaped sense organs on the mesosoma of scorpions.

Pedicellaria Small, specialised, jaw-like appendages on the body surface of some asteroid and echinoid echinoderms.

Pedipalp Second prosomal limb in arachnid chelicerates.

Peltate Shield-like.

Pen Internal, reduced shell of true squids (Teuthoida.)

Pereiopods Thoracic legs in malacostracan crustaceans that lie posterior to any thoracic legs that have been incorporated into the head as maxillipeds (e.g. in decapods, with three anterior pairs of maxillipeds, there are five pairs of pereiopods; in isopods, with one pair of maxillipeds, there are seven pairs of pereiopods). (See also *pleopod*.)

Perignathic Surrounding the jaws.

Perineurium Sheath of connective tissue covering nervous tissue.

Periostracum External organic layer of a calcareous shell.

Periproct Region around the anus.

Peristaltic Contractions passing along a tubular structure as alternating waves of narrowings and and widenings.

Peritrophic membrane Thin chitinous membrane secreted by the midgut epithelium in insects and other arthropods, surrounding the food mass in the midgut.

Pectinase Enzyme that digests pectin, a glue-like polysaccharide between adjacent cell walls in plant tissues.

Phagocytosis Active uptake of particles, taken into vesicles through the plasma membrane of a cell. (See also *endocytosis, pinocytosis*.)

Phenotypic Referring to the external form, or phenotype, of an organism.

Phyllobranch Leaf-like crustacean limb. Also used to describe crustaceans with limbs of this type.

Pilidium The ciliated larva of some nemertean worms.

Pinocytosis Active uptake by a cell of a small droplet of extracellular fluid. (See also *endocytosis*.)

Plankter Planktonic animal.

Planktonic Floating or swimming in the water column, but unable to resist positional displacement by water currents.

Planktotrophic Feeding on plankton. (See also *lecithotrophic*.)

Planula Free-swimming ciliated larva of cnidarians.

Plasmid Small circle of bacterial DNA that is separate from the single, bacterial chromosome and can replicate independently.

Plastron Permanent film of air held by hydrofuge hairs over part of the body in some aquatic insects, serving as a physical gill.

Pleopod Abdominal swimming limb in malacostracan crustaceans. (See also *pereiopod*.)

Pleural sclerites Hardened plates in the cuticle of the pleura of hexapodous and myriapodous arthropods. (See also *pleuron*.)

Pleuron Lateral cuticular plate of an arthropod segment. (See also *tergum, sternum*.)

Pluripotent Able to perform many functions. (See also *totipotent*.)

Pluteus Ciliated larva of ophiuroid and echinoid echinoderms.

Pneumostome Opening leading into the lung in pulmonate gastropod molluscs.

Podomere Section of an arthropod limb between successive joints.

Podium Tube foot of echinoderms.

Polyp Attached, tentaculate, cylindrical body form in cnidarians.

Polyphyletic Arising from a number of unrelated ancestors. (See also *monophyletic, paraphyletic*.)

Polypide Combined internal structures of an ectoproct zooid (lophophore, gut, muscles, etc.). (See also *cystid*.)

Previtellogenesis Early growth phase of an oocyte, before the onset of yolk storage. (See also *vitellogenesis*.)

Proctodaeum Hindgut of arthopods, lined by cuticle; also used for the embryonic hindgut of many animal groups. (See also *stomodaeum*.)

Procuticle Inner layer of the cuticle of arthropods, under-lying the thin epicuticle; often divided into an outer exocuticle and inner endocuticle.

Proglottid Section of the body of a eucestode (tapeworm), containing a set of reproductive organs.

Proline An amino acid, prominent in collagen fibres.

Promotor movement Forward swing of a limb on the body. (See also *remotor movement*.)

Prosoma Anterior section (tagma) of the body in chelicerate arthropods. (See also *opisthosoma*.)

Prostatic gland Gland in male reproductive system contributing secretions to seminal fluid.

Protandry Maturation of the male reproductive system before the female system in hermaphrodite animals. (See also *protogyny*.)

Proteoglycan Component of the ground substance of connective tissue. Proteoglycan molecules have a protein core with sulfate side chains.

Protobranch Generalised form of bipectinate ctenidium (gill) found in some bivalve molluscs. (See also *lamellibranch*.)

Protobranchiate Bivalve mollusc with protobranch gills.

Protogyny Maturation of the female reproductive system before the male in hermaphrodite animals. (See also *protandry*.)

Protonephridium Nephridium in which the inner end terminates in flame cells or solenocytes and there is no nephrostome.

Protonymphon Larval stage in pycnogonid chelicerates.

Protoscolex Developing anterior end (scolex) in a larval eucestode (tapeworm).

Protosome Anterior section of the body in a trimeric animal. (See also *protosome, mesosome*.)

Protostome Animal in which the mouth is formed from the blastopore during embryonic development and the anus develops as a secondary invagination. (See also *deuterostome*.)

Protozoea Larval stage in decapod crustaceans, intervening between the nauplius and zoea stages.

Pseudolamellibranch Type of lamellibranch gill in bivalve molluscs in which the gill filaments are joined by ciliary junctions and some tissue junctions. (See also *filibranch, eulamellibranch*.)

Pseudotracheae Air-filled respiratory tubules in isopod crustaceans, resembling insect tracheae.

Pterygote Winged (especially in relation to insects).

Ptilinum Bladder-like structure that is inflated at the front of the head in some flies (Diptera) to break open the puparium, facilitating emergence.

Pulvillus Pad or lobe beneath each pretarsal claw on the thoracic leg in some insects. (See also *arolium*.)

Puparium Hardened, barrel-shaped last larval cuticle within which some flies (Diptera) pass the pupal stage.

Quinones Compounds formed by replacement of two hydrogen atoms in an aromatic ring by two oxygen atoms.

Remotor movement Backward swing of a limb on the body. (See also *promotor movement*.)

Renette Excretory cell type in some nematode worms.

Respiratory tree Branching respiratory organ arising from the hindgut in holothuroid echinoderms.

Retinaculum Structure holding the furcula (spring) of Collembola in place; formed by the fused limbs of the third abdominal segment. (See also *furcula*.)

Rhabdiferous Rod-producing.

Rhabdomere Component of the rhabdom (optic rod) in an ommatidium of an arthropod compound eye.

Rheophilic Stream-dwelling.

Rheotactic Responding to current flow.

Rotulae Five radial plates forming part of the Aristotle's lantern of echinoid echinoderms.

Sarcomere Functional contractile unit in a striated muscle fibre.

Schizocoel Coelomic cavity formed by splitting within mesoderm. (See also *enterocoel*.)

Sclerosponges Demosponges which contain both siliceous and calcareous skeletal secretions.

Scyphistoma Polypoid stage in the life cycle of a scyphozoan cnidarian. (See also *ephyra*.)

Sedentary Sluggish lifestyle, involving little movement from place to place. (See also *motile, sessile*.)

Septibranchiate Condition in some bivalve molluscs in which the ctenidia (gills) are modified as muscular septa.

Septum Transverse partition in annelids, separating the coelomic cavities of successive segments.

Serine An amino acid.

Sessile Attached to a substratum, with no movement from place to place. (See also *motile, sedentary*.)

Setae Bristles, especially in crustaceans and other arthropods but also in polychaetes. (See *chaetae*.)

Setose Bearing setae.

Siphonoglyph Ciliated groove in the wall of the pharynx of anthozoan cnidarians.

Solenocyte Type of flame cell containing a flagellum instead of cilia.

Spadix Spade-shaped copulatory organ of a male nautilus, form by the fusion of four arms.

Spermatid Late stage in spermatogenesis, differentiating into a spermatozoon.

Sphaeridium Stalked ovoid body on the ambulacrum in echinoid echinoderms, containing a statocyst.

Spherulous Containing numerous small spheres.

Sporosac Medusa bud of hydrozoan cnidarian, remaining attached and reduced to a simple, gamete-producing sac. (See also *gonophore*.)

Stadium Period between moults in a developing insect.

Statocyst Gravity-sensing organ.

Stemma Simple eye of a holometobolous insect larva; a lateral ocellus.

Stereoblastula Blastula in which the wall of cells fills the interior, leaving no fluid-filled blastocoel. (See also *coeloblastula*.)

Sternum Ventral cuticular plate of an arthropod segment. (See also *tergum, pleuron*.)

Stolon Root-like extension in hydrozoan and anthozoan cnidarians, ectoprocts and ascidians, from which new zooids arise by budding.

Stomodaeum Foregut of arthropods, lined by cuticle; also used for the embryonic foregut of many animal groups. (See also *proctodaeum*.)

Strobilation Proliferation of a sequential chain of similar body units, as in a scyphozoan scyphistoma or a eucestode tapeworm.

Style 1. In arthropods, a vestigial abdominal limb. 2. In molluscs, a rotating rod associated with the stomach.

Subumbrella Under-surface (oral surface) of a cnidarian medusa. (See also *exumbrella*.)

Symplesiomorphy The sharing among taxa of structures of basic, unmodified form (plesiomorphic structures); not indicative of monophyly. (See also *synapomorphy*.)

Synapomorphy The sharing among taxa of structures of derived, modified form (apomorphic structures); indicative of monophyly. (See also *symplesiomorphy*.)

Syncytium Cells fused together as a multinucleate entity, without internal cell boundaries.

Tagma Compound body section of an arthropod, comprising several segments which together form a functional unit (e.g. head, thorax or abdomen).

Tarsomere Subunit of the tarsus in the legs of insects.

Tarsus Terminal region of the leg in insects, distal to the tibia.

Tegmen Hardened leathery forewing of insects of the orders Dictyoptera, Orthoptera and Cheleutoptera. (See also *elytron*.)

Telson Posterior, postsegmental part of an arthropod, behind the last definable segment. (See also *acron*.)

Tentilla Side branch of a ctenophore tentacle.

Tentorium Endoskeleton in the head capsule of an insect.

Tergum Dorsal cuticular plate of an arthropod segment. (See also *pleuron, sternum*.)

Trabecular Forming a supporting process or strut.

Totipotent Able to perform all functions. (See also *pluripotent*.)

Trimeric Having a body composed of three parts: protosome, mesosome and metasome (also referred to as **trisomic** or **oligomeric**).

Triploblastic Describing an animal in which the body

is developed from three cell layers: ectoderm, mesoderm and endoderm. (See also *diploblastic*.)

Trochus Preoral ciliary band at the apex of a rotifer. (See also *cingulum*.)

Trombiculids A family of mites.

Tryptophan An amino acid.

Tunicin Cellulose-like molecule forming a constituent of the external tunic of urochordates.

Ultrastructural Referring to structures revealed by electron microscopy.

Umbo Remnant of the initial larval shell at the apex of a shell valve in bivalve molluscs.

Uncini Short, hook-like chaetae (setae) of some polychaete worms.

Uniramous (Of arthropod limbs) having a single branch or ramus. (See also *biramous*.)

Uropods Posterior, terminal pair of limbs of the abdomen of malacostracan crustaceans. (See also *pleopod*.)

Vacuolated Containing one or more internal spaces filled with fluid (vacuoles.)

Velarium Median shelf of tissue around the bell margin in cubozoan cnidarians. (See also *velum*.)

Velum 1. Median shelf of tissue around the margin of the bell in the medusa of hydrozoan cnidarians. (See also *velarium*.) 2. Bilobed, ciliated swimming organ of molluscan veliger larvae.

Vitellaria larva Barrel-shaped, yolky, non-feeding larva of crinoid and holothuroid echinoderms.

Vitelline Referring to yolk (vitellus) in eggs.

Vitellogenesis Deposition of yolk in the cytoplasm of an oocyte.

Zoea Larval stage in decapod crustaceans, in which the abdomen is developed but lacks pleopods, and the posterior part of the thorax is still rudimentary.

Zooid Basic body unit in a colonial animal.

Zooxanthellae Symbiotic, unicellular, dinoflagellate algae occurring in the tissues of corals, soft corals and many sea anemones, and in tridacnid clams.

Index

Page numbers in **bold** refer to figures

Aaptus (Porifera)   22
Abarenicola (Annelida)   **191**
Acanthaster (Asteroidea)   369, 382
Acanthocephala   5, 88, 111–14, **113**
  development   114
  life cycles   114, **113**
  reproduction   112
acarines   324, 342, **325**, **331**
  feeding   331
  parasitism   331
  reproduction   338
Achatina (Gastropoda)   **132**, 133
Acholades (Platyhelminthes)   61, 63
Acoela   79
acron   226, 232, 283, 320
Acteon (Gastropoda)   **131**
actinotroch larva   348, **352**
actinula larva   52
Aculifera   173
Adenophorea   88, 92, 94, 98, 114
Allanaspides (Crustacea)   **289**, 290
Alloeocarpa (Urochordata)   402
Ammonoidea   141, 142
amphiblastula larva   **21**, 22
Amphipholis (Ophiuroidea)   **367**
Amphipoda   292, 293, **294**, 312, **317**
Amphiporus (Nemertea)   **84**
Anabrus (Insecta)   **239**
Anadara (Bivalvia)   155
Anaspides (Crustacea)   290
Ancylostoma (Nematoda)   102
Androctonus (Chelicerata)   329
Anisops (Insecta)   250
Annelida   6, 7, 174–203
  asexual reproduction   195
  chaetae   **176**, 177, **178**
  circulation   188, **189**
  Clitellata   175

cuticle   183
development   198, **199**
digestive system   189
evolution and phylogeny   201, 431
excretory system   194, **195**
eyes   **186**, 187, 188, **188**
feeding   191–3
Hirudinea   179
locomotion   183, **185**
nephridia   194, 195
nervous system   **186**
osmoregulation   194
parasitism   193
Pogonophora   181, 182
Polychaeta   175, 178, **180**
reproduction   195
respiration   188
sense organs   186
Anomalodesmata   152, **153**
Anoplotaenia (Platyhelminthes)   63, 70, **76**
Anostraca   297, 298, 317
Anthozoa   33, 54
  life cycles   48
Aphelenchoides (Nematoda)   **100**
Aplacophora   **123**, 167–70, 173
  development   171
  digestive system   170
  evolution and phylogeny   168
  excretion   170
  feeding   170
  nervous system   170
  reproduction   170
  respiration   168
  sense organs   170
  spermatozoa   170
Aplidium (Urochordata)   401, 402
Aplysia (Gastropoda)   131
Arachnida   319–42, **325**
  book lungs   323
  cuticle   323
  development   337, 338
  feeding   328, 329, **329**

locomotion   326, **327**
Malpighian tubules   334
reproduction   336
silk   330
tracheae   323, 324
venom   328, 330
Araneae   324, 341
  book lungs   324
  distribution   324
  reproduction   336
Araneus (Chelicerata)   331
Arca (Bivalvia)   153
Archaeognatha   265
Archiannelida   175
Architeuthis (Cephalopoda)   147
Arcitalitrus (Crustacea)   294
Arenicola (Annelida)   **184**, 192
Argonauta (Cephalopoda)   **145**
Argulus (Crustacea)   298, **299**
Arion (Gastropoda)   **132**
Aristotle's lantern   372, **372**, 383
Armandia (Annelida)   **188**
Artemia (Crustacea)   297, **298**
Arthropoda   7, 8, 222–342, **224**
  acron   226
  Chelicerata   *See* Chelicerata
  Crustacea   *See* Crustacea
  evolution and phylogeny   432
  general description   222–7
  See also Ecdysozoa
  Hexapoda   *See* Hexapoda
  Insecta   *See* Insecta
  jaws   225, **225**
  limbs   224, **225**
  molecular phylogeny   434, **435**
  monophyly   433
  Myriapoda   *See* Myriapoda
  polyphyly   433
  segmental organs   **224**, 226
  telson   226
Artioposthia (Platyhelminthes)   77
Aschelminthes   86–115
  classification   114

453

evolution and
phylogeny 430
See also Ecdysozoa
Ascidia (Urochordata) 401
Ascidiacea 399, 401–2, **403**
ascidian tadpole larva 405
asexual reproduction 406–9
circulation 405
development 405, **408**
digestion 404
feeding 404
fission and fusion 406
metamorphosis 405, **407**
respiration 405
reproduction 405
asexual reproduction
annelids 195
ascidians 406–9
cnidarians 46
echinoderms 379, 393
enteropneusts 413
lophophorates 361
platyhelminths 69, 72–5
poriferans 22, **22**, **23**
turbellarians 69
urochordates 406–409
Aspidobothrea 60, 61
Asymmetron
(Cephalochordata) 409
Asterias (Asteroidea) 369
Asteroidea 367, **367**, **369**,
369–370, **381**, **382**, 394
development 390, **391**, **392**
digestive system 381
feeding 382
larvae 390, **391**
locomotion 380
pedicellariae 370
reproduction 389, **390**
respiration 386
tube feet 369, **369**
water vascular system 375,
377
Astropecten (Asteroidea) 369
Atubaria (Hemichordata) 412
Atya (Crustacea) 295
Aurelia (Cnidaria) 33
life cycle 48, **49**
auricularia larva 392
Australonereis (Annelida) 186,
192
Australonuphis (Annelida) 192
Austramphilina
(Platyhelminthes) **60**, 63, 74,
76
Austrobilharzia
(Platyhelminthes) 70, 74, 77
autapomorphy 424

Autolytus (Annelida) 198

Balanoglossus
(Hemichordata) 412
Balanus (Crustacea) 300, **300**
Bathynella (Crustacea) 291
Bathynomus (Crustacea) 292
Bathypolypus (Cephalopoda) **145**
Bdelloura (Platyhelminthes) 61
Bennettiella (Nemertea) **81**
bioluminescence
Cephalopoda 147
Ctenophora 56
Echinodermata 388
Thaliacea 398
bipinnaria larva 390, **391**
Birgus (Crustacea) 303, 307, 313
Bivalvia 123, 148–56, **149**, **153**
adductor muscles 154
Anomalodesmata 152, **153**
byssus 154, 172
Cryptodonta 150, **153**
ctenidia 155
development 156
evolution and phylogeny 154
feeding 155
Heterodonta 151, **153**
kidneys 156
locomotion 154
nervous system **149**, 156
Palaeotaxodonta 150, **153**
Paleoheterodonta 151, **153**
Pteriomorphia 151
reproduction 156
respiration 155
sense organs 156
Septibranchia 152, 155
shell 148, **149**, 152
spermatozoa 156
zooxanthellae 156
blastula larva **20**
Blattodea 266
body cavities 4
coelom 5, 80, 117, 118, 419
See also coelom
gastrovascular cavity 31, 33
haemocoel 8
See also haemocoel
pseudocoel 4, 87, 120, 419, 430
Bonellia (Echiura) **181**
Boophilus (Chelicerata) 332
Botrylloides (Urochordata) 401,
402
brachiolaria larva 390, **391**, **392**
Brachiopoda 344–5, 349–355,
351, **352**, **353**, 361–4
circulation 355
defence 364

development **351**, 355
digestive system 353, **354**
evolution and phylogeny 345,
437
feeding 350, 352, **352**, 353
larvae **351**, 356
nephridia 355
nervous system 355
reproduction 355
shell 349, **351**
Branchinella (Crustacea) 297
Branchiostoma
(Cephalochordata) **408**, 409,
410
Branchiura 298, **299**, 317
Brissus (Echinoidea) 372
Brugia (Nematoda) 104
Buthus (Chelicerata) 328

Caenogastropoda 129, **161**
Caenorhabditis (Nematoda) **89**,
91, 96, 430
development 96, **97**
Calanus (Crustacea) 299
Calcarea 13, 26
development 21
Callapa (Crustacea) 295
Callianassa (Crustacea) **296**
Callistochiton
(Polyplacophora) 167
Calloria (Brachiopoda) **351**, 352,
353, 364
Calmanostraca 297, 317
Caobangia (Annelida) 197
Carcinonemertes (Nemertea) 80
Caris (Crustacea) 295
Cassidula (Gastropoda) **132**
Catostylus (Cnidaria) 33
Centrostephanus (Echinoidea) 383
Centruroides (Chelicerata) 329
Cephalocarida 297, **298**, 317
Cephalochordata 409, **410**, 415
development **408**, 411
digestion 411
feeding 411
reproduction 411
Cephalodiscus
(Hemichordata) 412
Cephalopoda 123, 139–47, **140**,
143, 172
chromatophores 146
circulation 143, **145**
ctenidia 140
development 148
digestive system 144, **145**
evolution and phylogeny 142,
143
excretory system 146

eyes **140**, **145**, 146
feeding **140**, 144, **145**
ink sac 142
kidneys 146
locomotion 141, **145**
nervous system **145**, 146
reproduction 140, **145**, 147
respiration 143
sense organs 146
shell 139, **140**, 141
Ceramonema (Nematoda) **91**
cercaria larva 72
Cestoda 62, **75**
Cestodaria **60**, 62
Chaetognatha **6**, **410**, 414–15
   development 415
   evolution and phylogeny 415, 436
   nervous system 415
   reproduction 415
Cheilostomata **357**, 358, 365
Chelicerata 227, 319–42, **320**, **322**, **333**, 341
   Arachnida 321
   book lungs 332
   circulation 334
   development **337**, 338
   digestive system 332, **333**
   evolution and phylogeny 321, 341
   excretion 334
   eyes 335
   feeding 322, 328, 340
   locomotion 321, **322**, 326, 340
   Merostomata 321
      See also Limulus
   nervous system **333**, 335
   parasitism 331
   Pycnogonida 321, 339, 342
   reproduction 335, **337**
   respiration 332, **337**
   segmental organs 226
   sense organs 335
   silk 330
   tracheae 334
Chicoreus (Gastropoda) **135**
Chilognatha 273, 285
Chilopoda 270-1, **272**, 284
   See also Myriapoda
Chiltonella (Crustacea) 298
Chirocephalus (Crustacea) 297
Chironex (Cnidaria) 33, 36, 37
Chiton (Polyplacophora) **164**
chlorocruorin 189
Chondrilla (Porifera) 27
Chordata 5, 397, 398–411, 415
   Cephalochordata
      See Cephalochordata

evolution and phylogeny 436
   Urochordata *See* Urochordata
Chortoicetes (Insecta) **233**
Cinachyra (Porifera) **22**, 23
circulation
   Annelida 188, **189**
   Brachiopoda 355
   Chelicerata 334
   Cnidaria 43
   Crustacea **304**, 305
   Ctenophora 57
   Echinodermata 386
   Ectoprocta 360
   Hemichordata 413
   Hexapoda 247
   Mollusca
      Aplacophora 168
      Bivalvia 155
      Cephalopoda 143
      Gastropoda 133
      Monoplacophora 162
      Polyplacophora 165
      Scaphopoda 159
   Myriapoda 278
   Onychophora 206
   Phoronida 347
   Urochordata 405
Cirripedia 300, **300**, 318
cladistics 424, **425**
Cladocera 297, **298**, 317
Cladorhiza (Porifera) 15, **16**
classification
   Acanthocephala 115
   Annelida 180, 201, **203**
   Aplacophora 173
   Arthropoda 227
   Aschelminthes 114
   Bivalvia 172
   Brachiopoda 365
   Cephalopoda 140, 172
   Chaetognatha 415
   Chelicerata 321, 341
   Chordata 415
   Clitellata 180, 202
   Cnidaria 53
   Crustacea 289–300, **315**, 316–18
   Ctenophora 57
   Echinodermata 394
   Ectoprocta 365
   Gastropoda 127, 171
   Hemichordata 415
   Hexapoda 265
   Hirudinea 180, 202
   Insecta 265
   Kinorhyncha 115
   Loricifera 115
   Mollusca **123**, 171

Monoplacophora 172
Myriapoda 284
Nematomorpha 115
Nematoda 114
Nemertea 83
Onychophora 215
Phoronida 364
Platyhelminthes 79
Porifera 26
Polychaeta 180, 202
Polyplacophora 173
Rotifera 115
Scaphopoda 172
Sipuncula 119
Tardigrada 221
Clathrina (Porifera) 26
Clavelina (Urochordata) 402
Clio (Gastropoda) **131**
Cliona (Porifera) 17, 24, 27
Clitellata 175, 179, 180, 202
   development 201
   reproduction 200
Clypeaster (Echinoidea) **367**, 371
Cnemidocarpa (Urochordata) 402
cnidae 35–37, **39**, **51**
Cnidaria 29–30, 31–52
   Anthozoa 33
   asexual reproduction 46
   circulation 43
   coral reefs 52
   Cubozoa 33
   development 50
   digestion 42
   evolution and phylogeny 30, 41, 428
   excretion 43
   feeding 39
   Hydrozoa 32
   larval development 50
   life cycles 32, 46, **47**, 49
   locomotion 37
   medusa 31, 32, **32**, 38, **49**
   mesogloea 29, 33
   musculature 38, **39**
   nervous system **39**, 43
   planula larva **47**
   polyp 31, 32, **32**, **49**, **51**
   ocelli 41
   osmoregulation 43
   regeneration **39**, 46
   reproduction 46
   Scyphozoa 33
   sense organs 26, 41
   skeletons 33, **36**
   statocysts 38, 41
   symbiotic algae 41, 43, **51**
   toxins 37
   zooxanthellae 43, **51**

455

coelom 5, 8, 9, 80, 117, 118, 419
   Annelida 177, 183
   Athropoda 223
   Cephalochordata 411
   Echinodermata 386, 390
   Ectoprocta 358
   Hemichordata **408**
   lophophorates 344
   Mollusca 124, 429
   Nemertea 429
   Onychophora 207, 214
   origin in evolution 431
   Sipuncula 118
   Tardigrada 220
coelomoducts
   Annelida 194
   Arthropoda 223
   Onychophora 147
Coenoplana (Platyhelminthes) 63, 69, 76
Coleoptera 237, 267
Collembola 236, **237**, 265
Comanthus (Crinoidea) 374
compound eyes
   Chilopoda 270, 279
   Chelicerata 335
   Crustacea 308
   Insecta 253, 254
   Polychaeta 187, **188**
Concentricycloidea 367, 374, 394
Conchifera **169**, 171
Conchostraca 297, **298**, 317
Conus (Gastropoda) **130**, **135**
Convoluta (Platyhelminthes) 63, 65
Copepoda 299, **299**, 317
Cornirostra (Gastropoda) **131**
coracidium larva 74
coral reefs 50, 52
Corella (Urochordata) 401, **403**
Corophium (Crustacea) 292
Crassostrea (Bivalvia) **153**
Craterostigmorpha 271, **272**, 284
Craterostigmus (Chilopoda) 271
Crinoidea 367, **368**, 374, 395
   digestive system 385
   evolution 375
   feeding 378, 385
   larvae **392**
   locomotion 392
   reproduction 389
   tube feet 374, 378
   water vascular system 376
Crustacea 7, 8, 227, 286–318
   Amphipoda 293, 294
   Anostraca 297, **298**
   antennal organ 304, 313
   Branchiura **298**, 299

Cephalocarida 297, **298**
circulation 305, 315
Cirripedia 300, **300**
Cladocera 297, **298**
Conchostraca 297, **298**
Copepoda 299, **299**
Decapoda 295, **295**, **296**
digestive system 288, 303
Eumalacostraca 290
Euphausiacea 294, **294**
evolution and phylogeny **315**, 316
excretion **304**, 306, 313, **314**
eyes 308, 313
feeding 301, **302**
fresh-water crustaceans 311
gills 289, 304, **304**, 313
Hoplocarida 290
Isopoda 292, **293**
larval development **311**
locomotion 301
lungs 313
Malacostraca 290
Maxillopoda 298, **299**
moulting 309
Mysida **291**
nervous system 307, **308**
Notostraca 297, **298**
osmoregulation 306
Ostracoda 298, **299**
Phyllocarida 297
Phyllopoda 297, **298**
Remipedia 289
reproduction 310, 313
respiration 304, **314**
Rhizocephala 300
segmental organs 226
sense organs 308, 313
Stomatopoda **289**, **302**
Syncarida **289**
Tanaidacea **291**, 292, 317
terrestrial crustaceans 311
Thecostraca 300
cryptobiosis 111, 219
Cryptochiton (Polyplacophora) 164
Cryptodonta 150, **153**
ctenidia
   Aplacophora 170
   Bivalvia 55
   Cephalopoda 140
   Gastropoda 125, 133, 136
   Monoplacophora 162
   Polyplacophora 165
Ctenodiscus (Asteroidea) 382
Ctenophora 2, 29–30, 54–7
   bioluminescence 56
   circulation 57

colloblasts 55
development 57
digestion 57
evolution and phylogeny 30, 428
excretion 57
feeding 55
locomotion 55
nervous system 57
reproduction 57
respiration 57
sense organs 57
Ctenostoma **357**, 358, 365
Cubozoa 32, 41, 48, 54
Cucumaria (Holothuroidea) 373
Cucumerio (Bivalvia) **153**
Cumacea **291**, 292, 317
cuticle
   Annelida 183
   Crustacea 288
   Gastrotricha 106
   Insecta 230, **231**
   Kinorhyncha 106
   Myriapoda 275
   Nematomorpha 105
   Nematoda 90
   Onychophora 206
   Priapula 120
   Sipuncula 117
   Tardigrada 217
Cyclistus (Crustacea) **314**
Cyclops (Crustacea) 299, **299**
cyphonautes larva **357**, 361
Cypraea (Gastropoda) **124**
Cypris (Crustacea) **298**
Cystodytes (Urochordata) **401**

Daphnia (Crustacea) 297, **298**
Decapoda 295, **295**, 317
Demodex (Chelicerata) **331**, 332
Demospongiae 20, 27
Dermaptera 237, 266
Dermatophagoides (Chelicerata) 331
Dermoergasilus (Crustacea) **299**
deuterostomes 5, 9, 390, 397, 435
development
   Acanthocephala 114
   Annelida 198, 199
   Ascidiacea 405
   Brachiopoda 355
   Cephalochordata 411
   Cestoda 74
   Chaetognatha 415
   Chelicerata 338, 341
   Cnidaria 50
   Crustacea 310

Ctenophora 57
Digenea 72
Echinodermata 390, 391, 392
Ectoprocta 361
Entoprocta 85
Gastrotricha 106
Hemichordata 413
Insecta 230, 259, 261, 262
Kinorhyncha 108
Mollusca
   Aplacophora 171
   Bivalvia 157
   Cephalopoda 148
   Gastropoda 139
   Monoplacophora 163
   Polyplacophora 167
   Scaphopoda 160
Monogenea 74
Myriapoda **280**, 283
Nematomorpha 105
Nemertea 83
Onychophora 213
Phoronida 348
Platyhelminthes 72
Priapula 121
Rotifera 110
Sipuncula 118
Tardigrada 219
Trematoda 72
Turbellaria 72
Urochordata 405
See also embryonic development
Diadema (Echinoidea) 388
Didemnum (Urochordata) 401, 406
Digenea 61
digestive system
   Brachiopoda 352
   Cephalochordata 411
   Chaetognatha 414
   Chelicerata 332, 340
   Cnidaria 42
   Crustacea 303
   Ctenophora 57
   Echinodermata 381, 385
   Ectoprocta 359
   Entoprocta 84
   Gastrotricha 106
   Hemichordata 413
   Insecta 242, **243**
   Mollusca
      Aplacophora 170
      Bivalvia 155
      Cephalopoda 144
      Gastropoda 134
      Monoplacophora 162
      Polyplacophora 165
      Scaphopoda 159

Myriapoda **274**, 276
Nematoda 92
Onychophora 208
Phoronida 346
Porifera 14
Priapula 120
Rotifera 109
Sipuncula 118
Tardigrada **217**
Urochordata 404
Diopsiulus (Diplopoda) 275
diploblastic structure 29, 55, 418
Diplopoda 270, 272, 285
   See also Myriapoda
Diplozoon (Platyhelminthes) 63
Diplura 236, **237**, 265
Diptera 237, 241, 267
Dirofilaria (Nematoda) 104
Dolichoglossus
   (Hemichordata) 412
doliolaria larva 392
Doliolum (Urochordata) **399**, 400
Dorylaimus (Nematoda) **93**
Dracunculus (Nematoda) 102
Drosophila (Insecta) 230, 259, 260, 261, 262, 434
Dugesia (Platyhelminthes) 63

Ecdysozoa 434, **435**
Echinocardium (Echinoidea) 371
Echinococcus (Platyhelminthes) 63, 71, 75, **75**, 77
Echinodera *See* Kinorhyncha
Echinoderes (Kinorhyncha) **103**, 106
Echinodermata 8, 366–95
   asexual reproduction 393
   Asteroidea *See* Asteroidea
   autotomy 378
   connective tissue 378
   circulation 386
   Concentricycloidea
      See Concentricycloidea
   Crinoidea *See* Crinoidea
   development 390
   digestive system 381, 385
   Echinoidea *See* Echinoidea
   evolution and phylogeny 375, 394, 436
   excretion 386
   haemal system 387
   Holothuroidea
      See Holothuroidea
   larval development 390
   life cycles 390–392
   locomotion 380
   metamorphosis 392
   nervous system 387

Ophiuroidea *See*
   Ophiuroidea
   regeneration 379
   reproduction 388
   sense organs 388
   skeleton 377
   water vascular system 375, **377**
Echinoidea 367, **367**, 371, 372, **383**, 394
   Aristotle's lantern 372, 383
   digestive system 383
   evolution 375
   feeding 383
   irregular echinoids **367**, 372, **385**
   larvae 391, **393**
   locomotion 380
   pedicellariae 371, **373**
   reproduction 389
   respiration 386
   tube feet 372, **373**
   water vascular system 376
echinopluteus larva **393**
Echiura 8, 178, 180, **181**, 182
   evolution and phylogeny 431
Ecteinascidia
   (Urochordata) 401, 402
Ectoprocta 6, **344**, 344–5, 356, 356–60, **357**, 361–4, 365
   asexual reproduction 361
   circulation 360
   defence **362**, 363
   development 361
   digestive system 35
   evolution and phylogeny 345, 437
   excretion 360
   feeding 359
   growth **362**, **362**
   larvae **357**, 361
   polymorphism 362, **362**
   regeneration 362
   reproduction 360
   zooid diversity 356, 358, 362, **362**
Edriophthalma 292, 317
Electra (Entoprocta) **84**
Eledone (Cephalopoda) **145**, 149
Ellipura 236, 265
Elminius (Crustacea) 300
Embioptera 266
embryonic development
   Ascidiacea 405, **408**
   Cephalochordata 411, **408**
   Chelicerata **337**, 338

Cnidaria 50
Echinodermata 390, **391**
Insecta 259, **261**
Myriapoda **280**, 283
Nematoda 96
Onychophora **212**, 213, **213**
Polychaeta 198
Porifera 21, **21**
Tardigrada 220
Endopterygota 238, 267
Enterobius (Nematoda) 102
Enteropneusta **6**, **410**, 412
  circulation 413
  development 413
  digestion 413
  feeding 413
  nervous system 413
  reproduction 413
  respiration 273
  tornaria larva 413
Enteroxenos (Gastropoda) 135
Entoprocta 4, 83, 83–4, **84**
  development 84, **84**
  digestion 84
  excretion 84
  feeding 83
  nervous system 84
  reproduction 84, **84**
Ephemeroptera 237, 238, 266
Epiperipatus (Onychophora) **211**
Eucestoda 62
Eucypris (Crustacea) **299**
Eudistoma (Urochordata) **403**
Eupentacta (Holothuroidea) **368**, **379**, **385**
Eukyphida 295, **299**, 317
Eumalacostraca 290, 316
Eumetazoan characteristics 2
Eunice (Annelida) 186, **186**
Euperipatoides (Onychophora) **205**
Euphausiacea 294, **294**, 317
Euplectella (Porifera) 26
Eurypterida **320**, 321, 341
Euscorpius (Chelicerata) 328
Euzygida 295, 317
evolution
  cladistic methods in 424, **425**
  gene sequences and 425, **426**
  molecular data and 426, **427**
  of animal groups:
    Annelida 201, 419
    Arthropoda 226, 432, 434
    Aschelminthes 418, 430
    Ascidiacea 405
    Brachiopoda 345, 437
    Chaetognatha 415, 436
    Chelicerata 321, 341, 433, 434

Chordata 436
Cnidaria 41, 418, 427
Crustacea 316, 434
Ctenophora 30, 419, 427
deuterostomes 436
Ecdysozoa 434
Echinodermata 375, 392, 436
Entoprocta 429
Ectoprocta 345, 437
Gnathostomulida 85
Hemichordata 397, 436
Hexapoda 241
lophophorates 345
Metazoa 427
Mollusca 125, 421, 429
  Aplacophora 168
  Bivalvia 154
  Cephalopoda 142
  Gastropoda 133
  Monoplacophora 162
  Polyplacophora 165
  Scaphopoda 159
Myriapoda 284, 434
Nematoda 430
Nemertea 82, 429
Onychophora 215, 432, 434
Phoronida 345, 437
Platyhelminthes 62, 428
Priapula 120
Porifera 418, 428
protostomes 431
Tardigrada 220, 434
excretion
  Annelida 194, **195**
  Brachiopoda 355
  Chelicerata 334
  Cnidaria 43
  Crustacea 306, **314**
  Ctenophora 57
  Echinodermata 386
  Ectoprocta 360
  Entoprocta 84
  Insecta 246, **253**
  Mollusca
    Aplacophora 170
    Bivalvia 156
    Cephalopoda 146
    Gastropoda 136
    Monoplacophora 162
    Polyplacophora 166
    Scaphopoda 159
  Myriapoda 277
  Nematoda 92
  Nemertea 81
  Onychophora 209
  Phoronida 347
  Platyhelminthes 67

Priapula 121
Rotifera 110
Sipuncula 118
Tardigrada 218
eyes
  Annelida 187, **188**
  Cephalopoda **140**, 146
  Chelicerata 329, 335
  Chilopoda 270, 279
  Crustacea 308
  Diplopoda 272, 279
  Echinodermata **369**, 388
  Gastropoda **126**, 137
  Insecta **253**, 254
  Myriapoda 279
  Onychophora 210
  Polyplacophora 166
  See also compound eyes

Fabricinuda (Annelida) **178**
Fasciola 63, 71, 76, 77
Filarioidea 104
Filograna (Annelida) **199**
Fissurella (Gastropoda) **128**
flame cells 67, **68**, 81, 110, 112, 194
flight in insects 239, 239, 255
Florometra (Crinoidea) **368**, **378**
Fritillaria (Urochordata) 400

Galeodes (Chelicerata) 328
Gastropoda 123, **124**,125-38, **126**, **128**, **130**, **131**, **132**, 171
  circulation 133
  ctenidia 125, 133, 136
  development 127, 139
  digestive system 134, **135**
  evolution and phylogeny 133
  excretion 136
  eyes 126
  feeding 134
  Heterobranchia 129, **131**
  kidneys 136
  nervous system **126**, **131**, 136
  Opisthobranchia 131, **131**
  Prosobranchia 128, **128**, **130**
  Pulmonata 132, **132**
  reproduction **131**, **135**, 137
  respiration 133
  sense organs 137
  shell 125, **126**
  spermatozoa **135**, 138
  torsion **126**, 127
Gastrotricha 5, **103**, 105, 115
  development 106
  digestive system 106
  nervous system 106
  protonephridia 106

reproduction 106
sense organs 106
Gemmula (Gastropoda) **130**
Geonemertes (Nemertea) 83
Geophilomorpha 271, **272**, 284
giant fibres
    Cephalopoda 146
    Cnidaria 44
    Echinodermata 389
    Insecta 251
    Onychophora 210
Gigantocypris (Crustacea) 298
Gladioferens (Crustacea) 299
Globodera (Nematoda) 99
glochidium larva 157
Glossobalanus
    (Hemichordata) 412
Glycymeris (Bivalvia) **153**
Gnathia (Crustacea) **293**
Gnathophausia (Crustacea) **291**
Gnathostomulida 4, 85–6
Goniopora (Cnidaria) **39**
Gordius (Nematomorpha) **107**
Gorgonocephalus
    (Ophiuroidea) **367**, 370
Grylloblattodea 237, 266
Gymnolaemata 358
Hadronyche (Chelicerata) 330
Haemadipsa (Annelida) **181**
haemerythrin 118, 120, 355
haemocoel 8
    Arthropoda 223
        Chelicerata 334
        Crustacea 305
        Hexapoda 247
        Myriapoda 278
    Mollusca 124
    Onychophora 206
    Tardigrada 217
haemocyanin 124, 133, 144, 278, 334
haemoglobin 124, 155, 189, 347, 386
Haliclona (Porifera) 27
Halicryptus (Priapula) 120
Hapalochlaena (Cephalopoda) 144
Haploscoloplos (Annelida) **184**
Harpiosquilla (Crustacea) **289**
heart
    Arthropoda 223
        Chelicerata 334
        Crustacea 306
        Hexapoda 247
        Myriapoda 278
    Mollusca
        Aplacophora 170
        Bivalvia 155
        Cephalopoda 143

        Gastropoda 133
        Monoplacophora 162
        Polyplacophora 165
        Scaphopoda 159
    Onychophora 206
Hedleyella (Gastropoda) 133
Heliocidaris (Echinoidea) **372**, 383, **383**, 393
Helix (Gastropoda) 132,133
Hemicaridea 292, 316
Hemichordata 6, 411, 412–14, 415
    development **408**
    Enteropneusta
        See Enteropneusta
    evolution and phylogeny 414, 436
    Pterobranchia 412
Hemilepistus (Crustacea) 313
Hemiptera 237, **241**, 241, 267
Henricia (Asteroidea) **377**, 382
Heterobranchia 129, **131**, 136, 139, 171
Heterodera (Nematoda) 99
Heterodonta 151, **153**, 172
Heterorhabditis (Nematoda) 101
Heteroteuthis (Cephalopoda) **143**
Heteroxenia 39
hexacanth larva 74
Hexactinellida 16, 26
Hexapoda 8, 228–68, 265
    abdomen 233
    ametabolous hexapods 229
    brain 232
    circulation 247
    Collembola 236, **237**, 265
    cuticle 230, **248**
    Diplura 236, **237**
    Ellipura 236
    evolution and phylogeny 241
    fat body 245, 246
    head 153, 232
    hemimetabolous hexapods 229
    holometabolous hexapods 229
    metamorphosis 229, 262
    moulting 229
    mouthparts 233, 240
    Protura 236, 265
    reproduction 255
    segmental organs 226
    See also Insecta
Hickmania (Chelicerata) 330
Hippolyte (Crustacea) 295
Hirudinea 179, 180, 202
Holothuria (Holothuroidea) 373, 394
Holothuroidea 367, **367**, 368, 373, **385**, 395

    digestive system 384
    evisceration 379, **379**
    feeding 384
    larvae 392
    locomotion 380
    reproduction 389
    respiration 386
    tube feet 374
    water vascular system 376
homology 423–4
    autapomorphy 424
    symplesiomorphy 424
    synapomorphy 424
homoplasy 423
Hoplocarida 290, 316
hormones
    Cnidaria 44
    Crustacea 309
    Insecta 251
    Myriapoda 279
Hutchinsoniella (Crustacea) 297
Hydra 32, 36–8, **39**, 44, 48
Hydrozoa 32, 47, 48, 53
Hymenolepis 63, 71
Hymenoptera 265, 268
hypodermic impregnation 106, 110, 197, 201, 212
Hypodistoma (Urochordata) 403
Hypselodoris (Gastropoda) 131

Iasis (Urochordata) 400
Illies (Insecta) **256**
Insecta **232**, **234**, 236, 265
    aquatic insects 250
    cuticle 230, **231**, 235
    development 230
    digestion 242–245, **243**
    eggs and oogenesis 258, **261**
    epidermis 231, **231**
    evolution and phylogeny 234, **234**, 241
    excretion 246, **247**
    eyes **253**, 254
    fat body 245, 246
    feeding 42–243
    flight 239, **239**
        control mechanisms 255
    hormonal control 251
    larval development 260
    locomotion 238
    Malpighian tubules 246, **247**
    mating 257, 258
    metamorphosis 262, **262**
    mouthparts **233**, **237**, 240, **241**
    nervous system 251, **252**
    orders of insects 237, 266
    osmoregulation 246
    oviposition 259

Pterygota 236, 237, 266
reproductive system 255, **256**
respiration 248
sense organs 242, 252, **253**
    See also eyes; flight
social insects 264
spermatozoa 258
symbiosis 245
thermoregulation 263
Thysanura 236, 266
tracheal system 248, **249**
water conservation 235
wings 233, 234, **234**, 238
Ircinia 27
Isopoda 292, **293**, 312, 317
Isoptera 237, 266
Ixeuticus (Chelicerata) 330
Ixodes (Chelicerata) 328, **331**, 332

Keratella (Rotifera) **107**
Kinorhyncha 5, 87, 106, **107**, 115
    cuticle 106
    development 108
    digestive system 108
    locomotion 108
    moulting 108
    nervous system 108
    protonephridia 108
    reproduction 108

Lactrodectus (Chelicerata) 330, 331, 338
Laevipilina (Monoplacophora) 161, 163
Lamellisabella (Annelida) 24
Lampona (Chelicerata) 330
Larvacea **399**, 400
Ledella (Bivalvia) 153
leeches 179, 180, **181**, 185, 200, 202
Lepas (Crustacea) 300, **300**
Lepidochitona (Polyplacophora) **164**, 166, 167
Lepidopleura (Polyplacophora) 166
Lepidoptera 268
Leptogorgia (Cnidaria) 39
Leptosynapta (Holothuroidea) 373
Leptotrombidium (Chelicerata) 331
Lernaeocera (Crustacea) **299**
Ligia (Crustacea) **293**
Lima (Bivalvia) **153**
Limnadia (Crustacea) 297, **298**, 312
Limulus (Chelicerata) **320**, 321, **322**, 333
    feeding 322

larva **337**
life cycle 338
mating 336
reproduction 323
respiration 323
Lingula (Brachiopoda) 350, **351**
Linkia (Asteroidea) 393
Lithobiomorpha 271, **272**, 284
Littorina (Gastropoda) **126, 135**
Lobostomata (Platyhelminthes) 60, 63, 76
locomotion
    Annelida 183, **186**
    Chelicerata 321, **322**, 326, **327**, **339**, 340
    Cnidaria 37
    Crustacea 301
    Ctenophora 55
    Echinodermata 380
    Insecta 238
    Kinorhyncha 108
    Mollusca
        Bivalvia 154
        Cephalopoda 141
        Gastropoda 125
        Scaphopoda 158
    Myriapoda 274, **274**
    Nematoda 88
    Nemertea 80
    Onychophora 208
    Rotifera 109
Loligo (Cephalopoda) **145**
Lophogastrida 291, 316
lophophorates 344
    evolution and phylogeny 345, 437
lophophore 6, **344**, 346, **347**
Loricifera 5, 87, 108
Loxostomella (Entoprocta) 84
Lumbricus (Annelida) **179, 185,** 189
Lygumis (Annelida) **179**

Macranthorhynchus (Acanthocephala) 111, **126**
Macrodasyida 105
Malacobdella (Nemertea) 83
Malacostraca **288**, 290-7, 316
Malpighian tubules 246, **247**, 277, 335
Manocuma (Crustacea) **291**
Mantodea 237, 266
Marphysa (Annelida) **179, 191,** 199
Maxillopoda **298**-301, **298**, 317
Mecoptera 237, 267
medusa 31, 34, 39
Megaloptera 237, 267

Megascolides (Annelida) 179
Meloidogyne (Nematoda) 99, 100
Membranipora (Ectoprocta) 363
Merostomata 320, 321, 341
Mesanthura (Crustacea) 293
mesoderm
    4d mesoderm **59**, 60, 72, 85, 119, 198
    Annelida 177, 198
    Brachiopoda 344
    Chelicerata 338
    Echinodermata 390
    Entoprocta 85
    Hemichordata 412
    Insecta 260
    Nemertea 82
    Onychophora **214**
    Platyhelminthes 60, 72
    Sipuncula 118
    Tardigrada 220
    Urochordata 405
Mesorhabditis (Nematoda) **89**
metacercaria larva **73**
Metacrinus (Echinodermata) 374
Metazoa
    ancestors of 428
    characteristics 2
    genes and phylogeny in **427**, 427
    phylogeny of 417, **422**, 427
Micropilina (Monoplacophora) 160, 161, 163
Millepora (Cnidaria) 32
miracidium larva 72
Missulena (Chelicerata) 330
Molgula (Urochordata) 402
Mollusca 122-73
    Aplacophora
        See Aplacophora
    Bivalvia *See* Bivalvia
    Cephalopoda
        See Cephalopoda
    evolution and phylogeny **124**, 125, **169**, 429, 431
    Gastropoda *See* Gastropoda
    Monoplacophora
        See Monoplacophora
    Polyplacophora
        See Polyplacophora
    Scaphopoda *See* Scaphopoda
Moniliformis (Acanthocephala) 111, **113**
Monobryozoon (Ectoprocta) 356
Monogenea **60**, 62
Monogononta 109, 110, 115
Monoplacophora 123, 160-3, **161**, 172

circulation 162
development 163
digestive system 162
evolution and phylogeny 162
excretory system 162
feeding 161
nervous system 162
reproduction 163
respiration 162
sense organs 162
Monoposthonia (Nematoda) **100**
moulting
  Crustacea 309
  Hexapoda 231
  Insecta 251
  Kinorhyncha 106
  Myriapoda 284
  Nematoda 90
  Onychophora 206
  See also Ecdysozoa
Mya (Bivalvia) **153**
Myriapoda 8, 227, 269–85
  aquatic myriapods 278
  Chilopoda 270, **272**
    feeding 276
    locomotion 271, **274**
    mouthparts 270, **271**
  circulation 278
  cuticle 275
  development **280**, 283
  digestive system **274**, 276
  Diplopoda 270, 272, **272**, **273**
    defences **274**, 280
    eyes 272
    feeding 276
    genitalia 273
    legs 273
    mouthparts **271**, 272
  evolution and phylogeny 284
  excretion 277
  eyes 272, 279
  feeding 276
  hormones 279
  locomotion 274
  mating 282
  moulting 284
  nervous system **274**, 279
  Pauropoda 274, **274**
    feeding 276
  reproductive system **277**, 281
  respiration 278
  segmental organs 226
  Symphyla 273, **274**
    feeding 276
    mouthparts 274
  tracheal system 278
  water balance 277
Myrmecia (Insecta) **230**

Mysida 291, **291**, 316
Mysis (Crustacea) **291**
Mystacocarida 317
Mytilus (Bivalvia) **153**
Myzostomida **181**, 182, 194
Myzus (Insecta) **230**

Nardoa (Asteroidea) **367**
nauplius larva 310
Nautiloida 140, 172
Nautilus (Cephalopoda) 140, **140**, 141, 143, 144, 146, 147, 148
Nebalia (Crustacea) **297**
Necator (Nematoda) 102
Nectonema (Nematomorpha) 107
Nematoda 2, 5, 88–104, **89**, 114
  Adenophorea 88
  cuticle 88
  digestion 90
  embryonic development 96, 97
  epidermis 88
  evolution and phylogeny 430
  excretory system 92
  feeding **93**, 99
  locomotion 88
  moulting 90
  musculature 90
  nervous system 94
  ocelli 94
  osmoregulation 92
  parasitic life cycles 100–104
  parthenogenesis 95
  reproduction 94
  Secernentea 88
  sense organs **93**, 94
Nematomenia (Aplacophora) 171
Nematomorpha 5, 88, **103**, 105, 115
Nemertea 4, 79–83, **81**, 83
  circulation 82, 429
  development 82
  digestion 81
  evolution and phylogeny 82, 429
  excretion 81
  feeding 80
  locomotion 80
  nervous system **81**, 82
  proboscis 80
  reproduction 82
  sense organs 82
Neodermata 63
Neogastropoda 129
Neomeniomorpha 167, 173
Neopilina (Monoplacophora) 160, **161**, 161, 163
Neoptera 238, 266

Neotrigonia (Bivalvia) 151, **153**
Nephila (Chelicerata) 331, 338
nephridia
  Annelida 194, **195**
  Brachiopoda 355
  Phoronida 348
  Sipuncula 118
Nereis (Annelida) 184, **184**
Nerita (Gastropoda) **128**
nervous system
  Acanthocephala 112
  Annelida 186, **186**
  Ascidiacea 404
  Bivalvia **149**
  Brachiopoda 355
  Cestoda 69
  Cephalopoda **145**
  Chaetognatha 415
  Chelicerata 335
  Cnidaria 43
  Crustacea 307, **308**
  Ctenophora 57
  Echinodermata 387
  Ectoprocta 360
  Entoprocta 84
  Gastrotricha 106
  Hemichordata 413
  Insecta 251
  Kinorhyncha 108
  Mollusca
    Aplacophora 170
    Bivalvia 156
    Cephalopoda 146
    Gastropoda 136
    Monoplacophora 162
    Polyplacophora **164**, 166
    Scaphopoda 160
  Myriapoda **274**, 279
  Nematoda 94
  Nematomorpha 105
  Nemertea 82
  Onychophora 210
  Platyhelminthes 68
  Priapula 120
  Rotifera 110
  Sipuncula 118
  Tardigrada 219
  Trematoda 69
  Turbellaria 68
Neuroptera 267
Nicolea (Annelida) 198
Nippostrongylus (Nematoda) **100**
Notochiton (Polyplacophora) 166
Notoplana (Platyhelminthes) 63
Notostraca 297, **298**, 317
Nuttallochiton (Polyplacophora) 166
Obelia (Cnidaria) 32, **47**, 48

461

ocelli 41, 69, 82, 110, 156, 187, 254, 279, 308, 335
Octopus (Cephalopoda) **145**, 148
Ocypode (Crustacea) 313, **314**
Odonata 237, 238, 266
Oikopleura (Urochordata) **399**, 400
Olios (Chelicerata) 330
Onchidium (Gastropoda) 132
Onchocerca (Nematomorpha) 104
oncomiracidium larva 74
Oniscus (Crustacea) **293**
Onychophora 8, 205–15, **205**, **206**, **207**
  body wall 206, **207**
  circulation 206
  coxal organs 206
  cuticle 206
  development 212–13, **213**
  digestive system 208
  evolution and phylogeny 215, 432, 434
    See also Ecdysozoa
  excretory system 209
  eyes 210
  feeding 208
  locomotion 208
  mouthparts 206
  musculature 207
  nervous system 210
  osmoregulation 206, 209
  reproduction 211
  respiration 210
  segmental organs 209
  sense organs 210
    slime glands 208
  tracheae 210
Ophiocoma (Ophiuroidea) 388
Ophiomyxa (Ophiuroidea) **390**
Ophionereis (Ophiuroidea) 370, **379**, 382, **389**
ophiopluteus larva 391, **393**
Ophiothrix (Ophiuroidea) **370**, **382**, **389**, 393
Ophiura (Ophiuroidea) 388
Ophiuroidea 367, **367**, 370, **370**, **379**, 382, 394
  digestive system 382
  feeding 382
  larvae 391, **393**
  locomotion 380
  reproduction 389, **389**, 390
  respiration 386
  tube feet **370**, 371
  water vascular system 376
Ophryotrocha (Annelida) 198
Opiliones 324, **325**, 342
Opisthobranchia 131

Opisthorchis (Platyhelminthes) 76
Opisthoteuthis (Cephalopoda) 147
Ordgarius (Chelicerata) 330
Orthoptera 237, 266
Ostracoda 298, **299**, 317

Pagurus (Crustacea) **296**
Palaemon (Crustacea) **294**, 295
Palaeotaxodonta 150, **153**, 172
Paleoheterodonta 151, **153**, 172
Palinurus (Crustacea) **296**
Palola (Annelida) 196
Pandalus (Crustacea) 295
Paragordius (Nematomorpha) **103**
Paralongidorus (Nematoda) 93
Paraneoptera 238, 267
Parartemia (Crustacea) 297
parasitism
  Acanthocephala 111
  acarines 331
  Annelida 193
  Cestoda 62
    effects on host 78
  energy metabolism in helminths 66
  Gastropoda 135
  Nematoda 100
  Rhizocephala 300
  Rotifera 111
  Trematoda 61
    host reactions 74
    effects on hosts 77
  vaccination against helminths 78
Paratanais (Crustacea) **291**
Parazoa 21
parenchymella larva 20, 21
Pareurythoe (Annelida) **184**
parthenogenesis
  Cladocera 297
  Insecta 257
  Nematoda 95, 101
  Onychophora 211
  Rotifera 110
Patella (Gastropoda) **126**, **135**
Patiriella (Asteroidea) 367, 369, **369**, 381, 382, **382**, **390**, **391**, 393
Pauropoda 270, 274, **274**, 285
  See also Myriapoda
Pecten (Bivalvia) **149**
Pedicellina (Entoprocta) **84**
Pedinogyra (Gastropoda) 133
Pelagia (Cnidaria) 48
pelagosphaera larva 119, **119**
Penaeus (Crustacea) 295
Penicillata 273, 285

Pentastomida 434
Peripatus (Onychophora) 215
Periplaneta (Insecta) 252
Perophora (Urochordata) **399**, 401
Petrobius (Insecta) 236
Petrolisthes (Crustacea) **296**
Phascolosoma (Sipuncula) 119, **119**
Phascolosomida 119
Phasmatodea 237, 266
Pheretima (Annelida) **176**
Philophthalmus (Platyhelminthes) 70
Phlyctenactis (Cnidaria) 38
Pholcus (Chelicerata) 331
Phoronida 6, 344–5, 346–8, **347**, 361–4
  actinotroch larva 348, **352**
  circulation 347, **347**
  development 348
  digestive system 346
  evolution and phylogeny 345, 437
  excretion 347
  feeding 346
  regeneration 361
  reproduction 348
Phoronis (Phoronida) **347**, 348
Phreatoicus (Crustacea) **293**
Phthiraptera 237, 267
Phyllidea (Gastropoda) 131
Phyllocarida 297, 317
Phyllopoda 297, **298**, 317
phylogenetic methods 422
  cladistics **419**, 424
  gene sequences 425
  molecular analysis 426
  phylogeny 416–38
Physa (Gastropoda) **132**
Physalia (Cnidaria) 32, 35, 37, 38, 40, **47**, 131
pilidium larva **81**, 82
Pita (Bivalvia) **153**
Placiphorella (Polyplacophora) 165
Planipapillus (Onychophora) **209**
planula larva 32, **45**, 50
Platyhelminthes 4, 59–78, **64**
  asexual reproduction 69
  Cestoda 62, 75
    life cycles 74, 78
  development 72
  Digenea
    life cycles 72, 77
  digestion 65
  egg production 71
  evolution and phylogeny 62, 428

excretion 67, **68**
feeding 65
  in agriculture 77
  in medicine 77
  life cycles 72, 74, 76
  Monogenea
    life cycles 64
  nervous system 68
  osmoregulation 67
  protonephridia 67, **68**
  reproduction 69
  sense organs 69
  Trematoda 61, **65**
    host reactions 74
    life cycles 72, 77
  Turbellaria 61
Plecoptera 237, 266
Plesiastrea **39**, 46
Pliciloricus (Loricifera) **103**
Plotohelmis (Annelida) **188**
pluteus larva 391
Podoclavella (Urochordata) 401
Pogonophora 8, 180, 182, 193, **203**
Polychaeta 175, 178, **179**, 180, **180**, **184**, 201, **203**
  buccal organs **180**, 189, **191**
  larval development 195, **199**
  life cycles 200
  locomotion 183, **185**
  reproduction 196, **197**
Polycladida 61
Polyclinum (Urochordata) **407**
Polydora (Annelida) **184**
Polymastia 20, 23, **23**, 27
Polyplacophora 123, 163, **164**, 173
  circulation **164**, 165
  development 167
  digestive system **164**, 165
  evolution and phylogeny 165
  excretion 166
  feeding 165
  nervous system **164**, 166
  reproduction 166
  respiration 165
  sense organs 166
  shell 164
  spermatozoa 166
  trochophore larva **164**
Pomatoceros (Annelida) **189**, **191**, **195**
Porifera 2, 10–27, **13**
  amphiblastula larva 22
  asexual reproduction 22
  Calcarea 21, 26
  cell aggregation 18
  cell functions 14, **15**

choanocytes 12, **13**, 14, **18**
Demospongiae 20, 27
development 21
digestion 15
ecology 24
embryonic development 20
evolution and phylogeny 27, 427
feeding 14, **16**
gemmules 17
grafting 18
Hexactinellida 16, 26
histocompatibility system 18
metabolites 25
reproduction 19, **20**, **21**, **22**, **23**
secretions 16
settlement and metamorphosis 21
spicules 17, **19**
spongin 17, **17**, **18**
symbiotic associations 24
toxicity 25
viviparity 20
Priapula 3, 5, 117, 120–1
  development 121
  digestive system 120
  evolution and phylogeny 120
  nervous system 120
  pseudocoel 120
  reproduction 121
Priapulus (Priapula) 120
proglottid 62
Prosobranchia 128, **128**, 129, **130**
  reproduction 137
  spermatozoa 138
Prostoma (Nemertea) 83
protonephridia 55
  Acanthocephala 112
  Annelida 194, **195**
  Entoprocta 84
  Gastrotricha 105
  Kinorhyncha 108
  Nemertea 81
  Platyhelminthes 67, 68
  Priapula 121
  Rotifera 110
protostomes 9, 431
Protura 236, 265
Pseudoceros (Platyhelminthes) **60**, 63, **76**
pseudocoel 5, 87, 120, 419, 430
Pseudoscorpiones 323, 325, 341
Pseudothoracocotyla
  (Platyhelminthes) **60**, 63, **76**
Psocoptera 237, 267
Pteria (Bivalvia) **153**
Pteriomorphia 151, **153**, 172
Pterobranchia 412

Pterygota 236, 237, 266
Ptychodera (Hemichordata) 412
Pulmonata 132, **132**
Pycnoclavella (Urochordata) **403**
Pycnogonida 321, 339, **339**, 342
  circulation 340
  development 341
  digestive system 340
  evolution and phylogeny 341
  excretion 340
  feeding 340
  locomotion **339**, 340
  nervous system 340
  reproduction 340
  respiration 340
Pycnogonum (Pycnogonida) **339**
Pyrosoma (Urochordata) 398, **399**
Pyura (Urochordata) **399**, 401, 402

radial symmetry 3, 29, 368
radula 124
  Aplacophora **168**, 170
  Cephalopoda 140, 144, **145**
  Gastropoda 134, **135**
  Polyplacophora 163, **164**
  Scaphopoda **158**, 159
Ranina (Crustacea) **295**
Raphidioptera 237, 267
redia larva 72
regeneration
  Cnidaria 46
  corals 39
  Echinodermata 379
  lophophorates 361
  Porifera 22
Regimitra (Onychophora) **209**
Remipedia 289, **289**, 316
reproduction
  Acanthocephala 112
  Annelida 195
  Brachiopoda 355
  Chaetognatha 415
  Chelicerata 335, **337**, 340
  Chordata
    Cephalochordata 411
    Urochordata 400, 405
  Cnidaria 46
  Crustacea 310, 313
  Ctenophora 57
  Echinodermata 388
  Ectoprocta 360
  Entoprocta 84
  Gastrotricha 106
  Gnathostomulida **85**
  Hemichordata 413
  Hexapoda 255
  Insecta 255, 256, **256**
  Kinorhyncha 108

Loricifera 108
Mollusca
　Aplacophora 170
　Bivalvia 156
　Cephalopoda 147
　Gastropoda 137
　Monoplacophora 163
　Polyplacophora 166
　Scaphopoda 160
Myriapoda 277, 280
Nematoda 94
Nematomorpha 105
Nemertea 82
Onychophora 211
Phoronida 348
Priapula 121
Platyhelminthes 69
Porifera 19
Rotifera 110
Sipuncula 118
Tardigrada 220
Reptantia 296, 317
respiration
　Annelida 188
　Chelicerata 323, 332, 340
　Cnidaria 40
　Crustacea **294**, 303
　Ctenophora 57
　Echinodermata 386
　Hemichordata 413
　Insecta 248, **249**
　Mollusca 133, 143, 155, 159, 162, 165, 168
　Myriapoda 278
　Onychophora 210
　Tardigrada 219
　Urochordata 405
Rhabdias (Nematoda) 95
Rhabditis (Nematoda) 92
Rhabdocoela 60, 79
Rhabdopleura
　(Hemichordata) **410**, 412
Rhizocephala 300, 318
Riftia (Annelida) 193
Rotatoria (Rotifera) **107**
Rotifera 5, 87, **107**, 109–10, 115
　development 110
　digestive system 109
　Digononta 109, 115
　excretion 110
　feeding 110
　locomotion 109
　Monogononta 109, 115
　nervous system 110
　osmoregulation 110
　parasitism 111
　protonephridia 110
　reproduction 110

sense organs 110
Seisonacea 109, 115
Ruhbergia (Onychophora) **209**

Sabella (Annelida) **188**
Saccocirrus (Annelida) 198
Saccoglossus (Hemichordata) **408**, 412
Salinator (Gastropoda) **132**
Salpa (Urochordata) **399**, 400
Sarcoptes (Chelicerata) **331**, 332
Sarsostraca 297, 317
Scaphopoda 123, 157–60, **158**, 172
　circulation 159
　development 160
　evolution and phylogeny 159
　excretion 159
　locomotion 158, **158**
　nervous system 160
　reproduction 160
　respiration 159
　sense organs 160
　shell 158
　spermatozoa 160
　trochophore larva **158**
Schistosoma 60, 63, 74, 76, 77
Schizocardium
　(Hemichordata) 412
Scolopendromorpha 271, **272**, 284
Scoloplos (Annelida) **184**
Scorpiones 323, **325**, 336, 341
Scutigeromorpha 271, **271**, 284
Scutus (Gastropoda) **128**
Scylla (Crustacea) **302**
scyphistoma 48
Scyphozoa 32, 41, 48, 54
Secernentea 88, **89**, 92, 94, 99, 114
segmental organs 209, 223, 226, 334
Seisonacea 109, 115
Selenocosmia (Chelicerata) 330
sense organs
　Annelida 186
　Chelicerata 335
　Cnidaria 41
　Crustacea 308
　Ctenophora 57
　Echinodermata 388
　Gastrotricha 106
　Insecta 252, **253**
　Mollusca
　　Aplacophora 170
　　Bivalvia 156
　　Cephalopoda 146
　　Gastropoda 137
　　Monoplacophora 162

　　Polyplacophora 166
　　Scaphopoda 160
　Myriapoda 279
　Nematoda 94
　Nemertea 82
　Onychophora 210
　Platyhelminthes 68
　Rotifera 110
　Tardigrada 219
Sergestes (Crustacea) 295
Sesarma (Crustacea) 315
Siboglinoides (Annelida) **181**
Siboglinum (Annelida) 193
Sigilina (Urochordata) **403**
Siphonaptera 237, 267
Siphonaria (Gastropoda) **132**
Sipuncula 5, 117–19, **119**
　development 118, **119**
　digestive system 118
　evolution and phylogeny 431
　excretory system 118
　nervous system 118
　reproduction 118, **119**
　trochophore larva **119**
Sipunculida 119
Sipunculus (Sipuncula) 119
Solemya (Bivalvia) 150, **153**
Solen (Bivalvia) 153
Solifuges 324
Speleonectes (Crustacea) **302**
spermatophores
　Arachnida 336
　Cephalopoda **145**, 147
　Crustacea 310
　Gastropoda 139
　Hexapoda 258, **259**
　Hirudinea 201
　Myriapoda 282
　Onychophora 212
　Polychaeta 197
spicules
　Cnidaria 34
　Porifera 11, 17, **19**, 25
spiral cleavage 8, 59, **59**, 72, 82, 85, 110, 114, 118, 125, 160, 167, 198, 310
Spirobranchus (Annelida) **199**
Spirorbis (Annelida) 176
Spirula (Cephalopoda) 141, **143**, 144
Spondylus (Bivalvia) **153**
statocysts
　Cnidaria 41
　Crustacea **308**, 309
　Echinodermata 388
　Mollusca
　　Cephalopoda 146
　　Gastropoda 137

Monoplacophora 163
Scaphopoda 160
Polychaeta 187
Turbellaria 69
Steinernema (Nematoda) 100
Stenopus (Crustacea) 295
Stenostomum
(Platyhelminthes) 63
Stichopus (Holothuroidea) 368
Stictococcus (Insecta) 245
Stictodora (Platyhelminthes) 63, 72, 73
Stomatopoda 316
Stratiodrilus (Annelida) 193
Strepsiptera 237, 267
Strongylocentrotus
(Echinoidea) 367, 372, 373
Strongyloides (Nematoda) 100, 101
Styela (Urochordata) 402
Sycon 21, 21, 26
Sycozoa (Urochordata) 401
symbiosis 24, 43, 65, 156, 193, 245
See also zooxanthellae
symmetry
bilateral 30, 59
biradial 29, 30
radial 29, 30
Symphyla 270, 273, **274**, 284
See also Myriapoda
symplesiomorphy 424
synapomorphy 424
Syncarida 290, 316
Synoicum (Urochordata) **399**
Syrinx (Gastropoda) 129, **130**

Taenia (Platyhelminthes) 63, **75**, 78
Tanaidacea **291**, 292, 317
Tardigrada 8, 216–21, **217**, **218**, 221
development 220
digestive system 217
evolution and phylogeny 220, 434
See also Ecdysozoa
excretion 218
feeding 217

nervous system 219
reproduction 219
respiration 219
sense organs 219
Teleogryllus (Insecta) 237
telson 226, 271, 273, 283, 320
Temnocephala
(Platyhelminthes) 63
Temnocephalida 79
Terebra (Gastropoda) **130**
Teredo (Bivalvia) **153**
Tethya (Porifera) 21, 27
Thaliacea 398, **399**
Thaumastocheles (Crustacea) **302**
Theba (Gastropoda) 133
Thecostraca 300, 318
Theodorella (Urochordata) 402
thermoregulation in insects 263
Thysanoptera 267
Thysanura 236, 266
tornaria larva **408**, 413
Toxocara (Nematoda) **100**, 101
Toxopneustes (Echinoidea) 372
tracheal system
Arachnida 334
Hexapoda 248
Insecta 248, **249**
Myriapoda 278
Onychophora 210
Trematoda 61, 63, **73**
Trichinella (Nematoda) 104
Trichoptera 237, 268
Tricladida 63, 79
Tridacna (Bivalvia) **153**
Triops (Crustacea) 297, **298**, 312
triploblastic structure 59, 87, 418
trochophore larva 431
Annelida 199, **199**
Mollusca 125
Aplacophora 171
Bivalvia 157
Gastropoda **135**, 139
Polyplacophora **164**, 167
Scaphopoda **158**, 160
Sipuncula 119, **119**
Trochosa (Chelicerata) 328
Trochus (Gastropoda) **126**
Trypanosyllis (Annelida) **179**
Tularia (Gastropoda) **131**

Turbellaria 60, **60**
Turbonilla (Gastropoda) **131**

Uca (Crustacea) 295
Urnatella 85
Urochordata 398, 415
Ascidiacea. See Ascidiacea
development 400, 405, **407**, **408**
evolution and phylogeny 398
Larvacea 400
Thaliacea 398
Uropygi **325**, 326

Vampyroteuthis
(Cephalopoda) 141, **145**
Vanadis (Annelida) **188**
Velella (Cnidaria) 38, 131
veliger larva 125, **135**, 139, 157
venom
Cephalopoda 144
Chelicerata 330
Chilopoda 270, 276
Cnidaria 37
Gastropoda 134
Porifera 20
Vestimentifera 8, 182
vitellaria larva 391, 392
viviparity
Annelida 197
Asteroidea 393
Chelicerata 336
Echinodermata 393
Insecta 258
Mollusca 167
Onychophora 213
Porifera 20

water vascular system 375, **377**
wings **231**, 234, **234**, 238
Wuchereria (Nematoda) 104

Xestospongia (Porifera) 20
Xiphosura **320**, 321, **322**, 341
Xyloplax (Concentricycloidea) 374

Yoldia (Bivalvia) **153**

zooxanthellae 43, **153**, 156
Zoraptera 266

About the Authors

Emeritus Professor Donald T. Anderson, AO, FRS, retired from the Challis Chair of Biology at the University of Sydney in 1991. A graduate of the Universities of London (PhD, DSc) and Sydney (DSc), Professor Anderson joined the staff of the University of Sydney in 1958. His major research interests have included annelid and arthropod embryology, animal phylogeny, and the biology of barnacles. Professor Anderson has published three books and almost 100 papers on aspects of invertebrate biology.

Professor Dame Patricia R. Bergquist is Professor of Marine Zoology at the University of Auckland and Honorary Research Professor at Macquarie University, Sydney. Professor Bergquist is a world authority on sponges and has published a textbook and over 100 papers on aspects of sponge biology. She has pioneered the development of chemical and molecular taxonomy in the largest sponge class, Demospongiae.

Dr Maria Byrne has been involved in echinoderm research since her undergraduate studies at the University of Galway. She completed her PhD at the Victoria University of Canada and followed this with postdoctoral research in the United States before joining the staff of the University of Sydney in 1988. Dr Byrne has published 50 papers on aspects of echinoderm biology, with an emphasis on functional morphology and evolution. Her current interest is in the evolution of development in Australian echinoderms.

Dr Peter Doherty graduated from the University of Auckland and completed a PhD at the University of Sydney in 1981. Following postdoctoral research in marine science, he was a lecturer in environmental studies at Griffith University for several years before joining the research staff of the Australian Institute of Marine Science. Dr Doherty is now a Principal Research Scientist at AIMS. His research interests include population and community ecology, reproduction and genetic evolution in the sea, and marine environmental management, all with an emphasis on the sustainability of coral reef fish stocks. Dr Doherty has published 60 papers on aspects of this work.

Associate Professor Peter Greenaway, a PhD graduate of the University of Newcastle upon Tyne, UK, joined the staff of the University of New South Wales in 1974. His research has been predominantly on the biology of crustaceans, and he is best known for his studies on the physiological adaptations of crustaceans to terrestrial life. Professor Greenaway has published 70 papers on invertebrate physiology.

Associate Professor Dinah Hales received her PhD degree from the University of Sydney in 1969, and joined the staff of Macquarie University, Sydney, in 1972. Professor Hales is a leading expert on aphid biology and has published more than 90 papers on insect structure, physiology, genetics and ecology. She is a former President of the Australian Entomological Society.

Dr John M. Healy is an Australian Research Council Senior Research Fellow at the University of Queensland. Dr Healy, who is a graduate of the University of Queensland (PhD 1984, DSc 1996), has held research appointments at the universities of Sydney and Queensland, and at the Queensland Museum. A world authority on molluscan biology, Dr Healy has revolutionised the use of sperm ultrastructure in molluscan systematics, a field in which he has published over 70 papers. He has also worked extensively on the taxonomy and phylogeny of molluscs and other animal groups.

Associate Professor Rosalind T. Hinde completed her undergraduate and doctoral studies at the University of Sydney and did postdoctoral research at Oxford University on animal–plant interactions in sea slugs. Following her return to a lectureship in biology at the University of Sydney in 1972, Professor Hinde has focused her research on the physiology and biochemistry of symbioses between algae and marine invertebrates, especially corals and sponges, and has 35 publications in this field. She is also investigating regeneration in corals and the culture of sponge cells.

Emeritus Professor Warwick Nicholas graduated from the University of Liverpool, UK, in 1951 and gained his PhD at the Liverpool School of Tropical Medicine.

Following a Rockefeller postdoctoral fellowship at the University of California, Berkeley, Professor Nicholas taught zoology and parasitology at the University of Liverpool, the University of Illinois and the Australian National University, Canberra, before his retirement in 1992. He has published numerous papers and a book on aspects of aschelminth biology.

Professor Rudolf A. Raff graduated in biochemistry from Pennsylvania State University in 1963 and completed his PhD at Duke University in 1967. Following a postdoctoral fellowship at the Massachusetts Institute of Technology, Professor Raff joined the staff of the Biology Department of Indiana University in 1971, and was appointed Professor of Biology at that University in 1979. Since 1983 he has also been the Director of the Institute for Molecular and Cellular Biology at Indiana University. Professor Raff is a world authority on genes, development and evolution. He has published more than 130 papers and written four influential books in this field, and is at the forefront of the new molecular and develop-mental studies of the phylogenetic evolution of animals.

Dr Greg Rouse, a graduate of the Universities of Queensland (MSc) and Sydney (PhD) is an Australian Research Council Postdoctoral Fellow at the University of Sydney and has held research posts at the University of Queensland and the Smithsonian Institute, Washington DC. His major research interest is in the evolution of life histories in marine invertebrates, particularly the polychaetes, a topic on which he has published numerous papers.

Dr Laura Stocker graduated from the University of Auckland in 1984 and completed her PhD studies at the University of Sydney in 1989. Her research has focused on the ecoological implications of fission and fusion in compound ascidians, on which she has published several papers. Dr Stocker has also carried out taxonomic studies on New Zealand ascidians and is currently writing a chapter on these animals for a text on New Zealand invertebrates. Since 1989, Dr Stocker has been on the staff of Murdoch University, as Lecturer in Ecologically Sustainable Development, and has published extensively on conservation matters.

Dr Noel Tait graduated from the University of Sydney in 1960 and gained his PhD at the Australian National University, Canberra. Dr Tait is a senior member of the Biology staff of Macquarie University, Sydney. His research has ranged widely across the biology of many invertebrate groups, from cnidarians to tunicates. Dr Tait's major focus is now on the biology of onychophorans, in which he and his collaborators are using genetic techniques to analyse a remarkable level of speciation among Australian onychophoran populations.

Dr John Walker, a PhD graduate of the University of Sydney, was for 20 years a medical parasitologist at the School of Public Health and Tropical Medicine of that University. He has carried out important research on the trematodes which cause cercarial dermatitis, and on the relationship between Australian freshwater snails and human schistosomes. Dr Walker is currently Head of the Parasitology Department in the Centre for Infectious Diseases at Westmead Hospital, Sydney.